An Introduction to

Fungi

4th Revised Edition

H C Dube

Former Professor and Head
Department of Life Sciences
Bhavnagar University
(Maharaja Krishnakumarsinhji Bhavnagar University)
Bhavnagar (Gujarat)
E mail hcdharish@yahoo.co.in

Published by:

New Delhi

4806/24, Ansari Road, Daryaganj, New Delhi - 110 002 (India)
E-mail: info@scientificpub.com Website: www.scientificpubonline.com

Jodhpur

5 A, New Pali Road, P.O. Box 91
Jodhpur - 342 001 Rajasthan (India)

An Introduction to Fungi 4th Revised Edition

HC Dube

Fourth Edition

Print - 2016
Print - 2018
Print - 2019
Print - 2020
Print - 2022
Print - 2023
Print - 2025
Print - 2026

ISBN: 978-81-72333-743-8 (HB)
eISBN: 978-93-86347-39-8 **(E/B)**

Printed in India

Dedicated with great respect

To

Prof. C.V. Subramanian

And with prayers for his good health.

President, International Mycological Association (1977-1983), ***Member,*** Executive Committee, International Union of Biological Sciences (1979-1982) and ***Former Director,*** Center for Advanced Study in Mycology and Plant Pathology, University of Madras.

Prof. B.P.R. Vittal

Foreword

I am happy Dr. Dube is bringing out the 4th edition of his now monumental book *An Introduction to Fungi*, first published in 1983. There was no comprehensive book on taxonomy and biology of fungi published by an Indian author until then. The book was written in most authentic and lucid manner. The 2nd and 3rd editions of the book were published in 1990 and 2005, respectively. The book has remained indispensable for students of mycology during the last three decades. The 4th edition has rightly raised high hopes and I am confident it will fully meet the expectations. Dr. Dube's contributions to mycology will be long remembered. He doesn't need a eulogy; nevertheless I can't resist expressing my deep respect and admiration for his enthusiasm and incessantly writing on varied areas of mycology. Knowing Dr. Dube personally and most closely since our University of Madras days, more than 40 years now, I am sure the new book will be tremendously popular.

B.P.R.Vitttal
Professor, CAS in Mycology and Plant Pathology
University of Madras

Preface to the Fourth Edition

I feel privileged to introduce in this (4th) edition the impact of the recent developments in mycology that have led to much better understanding of the organisms belonging to the three kingdoms of fungi and their evolution. The multigene sequence analysis by AFTOL (Arranging the Fungal Tree of Life) and "Deep hypha" projects sponsored by National Science Foundation of USA, as well as the Myconet, has revolutionized the taxonomy of fungi. For example, the phylum *Ascomycota*, which could not be classified above the rank of 'order' since 1983, has now (2008) 17 classes and 3 subphyla. No group has escaped such changes and this has resulted in a more natural classification of these diverse "organisms studied by mycologists".

The publication of Ainsworth and Sussman's 5 volumes entitled *The Fungi: an Advanced Treatise* (1966-1973) was a great event, a milestone in mycology. The first edition of *An Introduction to Fungi*, published in 1983, was largely based on information gleaned from these volumes. Ainsworth' classification was introduced in this edition which became most popular. Later, the molecular data, especially obtained from 18S rDNA sequence analysis, gave rise to a new phylogenetic classification, which was adapted in *The Dictionary of Fungi* (8th ed, 1995). Though emotionally difficult to break away from Ainsworth's familiar classification, the 3rd edition (2005) of this book adapted the DNA- based classification, popularly known as Hawksworth's classification. Now, we have a much clearer picture of the fungi provided by the mega projects (AFTOL and Deep Hypha) and their genetic-relatedness as never before. This information has been the basis of the new classification (Kirk *et al.*, 2008, *Dictionary of the Fungi*), which we have followed in the 4th edition. All that is worth-knowing is included here. The chapters on the biology of fungi have been thoroughly updated.

Needless to say, I have immensely enjoyed writing, and before that, reading the vast literature now available on one 'click'. To decipher, glean and put in order the scattered information was a challenge but it became easy due to help from friends. I profusely thank Elizabeth Frieders (University of

Wisconsin, Platteville, USA) and Anna Rosling (Univ. of Indiana, USA) for their help in solving knotty problems.

I remember (always!) with gratitude my teacher late Prof. KS Bilgrami, who in 1963 initiated me into research, and later, in 1976 made me an author in *A Text Book of Modern Plant Pathology*. One shock! My senior colleague at University of Madras, Prof. R N. Swamy passed away. He was a source of inspiration to me for over 40 years and his demise has been a great personal loss to me. Life Goes on!

I hope students would be benefited by this book and there in lays my happiness. You are welcome to discuss any mycological problem through my e mail hcdharish@yahoo.co.in. Thank you.

H C Dube

July 2012

Preface to the First Edition

Books, eversince Newton's *Principia* (which indeed reads like a textbook) and Darwin's *Origin of Species,* have had a respectable place in scientific literature. Their role and esteem have, however, been eroded in recent years with greater emphasis on journal articles. Even at the undergraduate level, students are provided with list of references to articles in journals with the argument that in rapidly advancing scientific fields good and up-to-date textbooks simply do not exist. This, to quote *Nature*[*] "Is a thin excuse for many teachers' laziness. The remedy is in their own hands, or pens." Journals and review articles, not withstanding their own merit, lack the historical perspective and the judicious evaluation of scientific publications. Also, they are less accessible and unportable. There is an urgent need for more books. Working scientists must withdraw for a time from the bench or "competition for space in the ordinary journals, and discharge their professional responsibility."

I have felt lonely on occasions. The pains of authorship became endurable only by remembering the fate of students in India, who have neither journals nor books available to them. I am offering them *An Introduction to Fungi* with great satisfaction. Without being unnecessarily humble, I feel happy that I am giving something which was long due.

I hope students will enjoy studying fungi which, since the beginning of man's quest for knowledge, have commanded great attention and have given rise to several additional disciplines like biochemistry and plant pathology. The enzymes, with which biochemistry started, were first studied in the yeasts (etymologically the word `enzyme' means *in yeasts*). Plant Pathology, to use Horsfall's phrase, was spawned in the backyard of mycology. Tatum an Beadle ushered in the biochemical genetics, which later developed in much wider realm of molecular biology, with *Neurospora.*

Mycology started with the study of mushrooms (mykes = mushrooms) which were highly delicious as food. Soon, some of them proved to be poisonous and got the name `toadestuhl' (death chair), which later became

---

* "Who will write more books"? 5 November 1981, pp. 17

`toadstool'. This necessitated the identification of edible mushrooms from poisonous toadstools. The speculations about their origin from a liquid accompanying thunder and storm or from the venom of snakes, may look odd today, but they did contribute by drawing attention to their study. The use of some fungi as source of hallucinations, which in no time could lift the soul and enable the patron to enjoy the sight and music of a different and bizarre world, has now been harnessed in the preparation of drugs for curing schizophrenia, and for `taking a trip'. More sobering and important contributions of fungi are in the realm of biodegradation, elucidation of complex life processes, their increasing role in contaminating the already threatened habitat by their toxins, and their prospects in biotechnology and industry, in which they have contributed more than any other group of organisms. Besides, the study of their own biology - the unending variability, their association in dual organisms, the novel alterantive of sex, the role of sex hormones in their reproduction, are all areas of interesting and challenging studies. Their study is never tiring; rather it is refreshing! Hope you will enjoy.

I remember with gratitude my teachers of mycology. Dr. G.C. Srivastava, Dr. P.C. Misra, Dr. R. Sahai, and Prof. K.S. Bhargava, are unforgettable. Prof. C.V. Subramanian, during my two years of stay at Madras, made an everlasting impression on me. Prof. Bilgrami's contributions are beyond repayment. But the students have not to be unduly careful to repay as their relation transcends acknowledgements. I am thankful to my students whom I taught for 12 years at Udaipur University. They, in their own way, also taught me mycology. Thanks to them. Most pleasant was my experience with my research colleagues, S.K. Doshi, D. Daniel, Ashok Shrivastava, T. Silas Paul and K. Prabhakaran, who always came forward to help me and never grudged the help. I thank them for the "Yes." Then, nothing can be done in a non-congenial atmosphere. The Biosciences Department of Sardar Patel University is an ideal place for research and serious work, and I congratulate my colleagues for achieving this.

Thanks are due to Prof Lichtwardt of University of Kansas, Lawrence, USA, Prof. U.K. Sinha of Patna University and the Academic Press, New York, for allowing me to use some of the figures. A word of thanks is most certainy due to my wife (better-half, to be more appropriate) who always cooperated and also typed the manuscript, not once but three times. Thanks to my children who never allowed peace to prevail in the house and made me immune to noise. Thank you all.

H C DUBE

Contents

Part I: Introduction

1. History of Mycology 1
2. Kingdoms of Fungi and their Classification 16
3. Fungi: General Characteristics 32

Part II: Kingdom *Fungi*

4. Phylum *Chytridiomycota* 57
5. Phylum *Neocallimastigomycota* 77
6. Phylum *Blastocladiomycota* 80
7. Phylum *Zygomycota* 90
8. Phylum *Glomeromycota* 123
9. Phylum *Ascomycota* 134
10. Subphylum *Taphrinomycotina* 153
11. Subphylum *Saccharomycotina* 166
12. Subphylum *Pezizomycotina* 177
13 Anamorphic Fungi 248
14. Phylum *Basidiomycota* 281
15 Subphylum *Agaricomycotina* 299
16 Subphylum *Ustilaginomycotina* 348
17 Subphylum *Pucciniomycotina* 371

Part III: Kingdom *Chromista* (=*Straminopila*)

18 Phylum *Hyphochytriomycota* 396

19 Phylum *Labyrinthulomycota* 402
20 Phylum *Oomycota* 406

Part IV:
Kingdom *Protozoa*

21. Phylum *Mycetozoa* 430
22 Order *Acrasida* 445
23 Class *Plasmodiophorea* 448

Part V:
Biology of Fungi

24. Symbiotic Associations of Fungi: I Mycorrhiza 455
25 Symbiotic Associations of Fungi: II Lichens 471
26. Nutrition of Fungi 487
27. Secondary Metabolites 508
28. Mycotoxins and Mycotoxicoses 523
29. Homothallism and Hetrerothallism 546
30. Heterokaryosis and Parasexual Cycle 555
31. Sex Hormones and Sex Morphogens 569
Author Index 582
Organism Index 586
Subject Index 595

1

History of Mycology

Man's interest in fungi started with the observation of the beautiful, umbrella-shaped mushrooms and toadstools growing on soils forming 'fairy rings.' Since these grew attached to the soil like plants, they were regarded as plants. Etymologically, mycology means the study of mushrooms (Gr. *Mykes* = mushroom, *logos* = discourse). Ever since, fungi have been objects of wonder and speculation. Man, "the hungry creature in search of food", soon discovered that these mushrooms were edible. The mycophagists started collecting and eating them.

The ancient Greeks and Romans, and surely their less civilized contemporaries, were fond of truffles, mushrooms and puff-balls. Mushrooms became such delicacies that they were the only food which wealthy people insisted on cooking themselves. Cases of accidental mushroom poisoning were also known to ancient Greeks and Romans as early as 500 B.C. The name 'evil ferments of the earth' was given to some of the mushrooms. The edible members were called **mushrooms**, while the poisonous varieties were termed "toadstools". The word **toadstool** is a distortion of the German name **todestuhl** which literally means "death chair." The deadly effects of these toadstools made them sure poisons. The Roman emperor Cladius Caesar (A.D. 54) was murdered by his wife who mixed his food with *Amanita phalloides* (see Wasson and Wasson, 1957). It necessitated the clear identification of the edible and poisonous varieties of mushrooms. **Dioocorides,** a 1st Century Greek physician listed the differences between the edible and poisonous mushrooms which were later followed by herbalists.

Mushrooms were popular among Roman potters and sculptors. A series of ancient stone sculptures (1000 B.C.–200 A.D.) bearing designs of mushrooms have been collected by **Wasson** and **Wasson** (1957) from Guatemala and other Central American countries. Lowe (1971) made new records of mushroom stones from Guatemala (Fig. 1.1).

Fig. 1.1 Ancient mushroom stones from Guatemala (courtesy B. Lowe)

There are records of plant diseases in the Vedas, the Hindu scripture (1200 B.C.) and the Bible. Nevertheless, their association with fungi was not suspected until the invention of the microscope by **Leeuwenhoek** in the late 17th Century. The microscope gave vision to the hitherto unknown world of microorganisms. With the invention of printing in Europe in 15th Century, started the period of renaissance (rebirth of learning). **Herbals** incorporating descriptions and illustrations of fungi started being published. **Clusius** in 1601 described many mushrooms with the help of watercolor drawings in his *Rariorium Plantarum Historia* which contained a section on fungi. The climax of mycological advance in herbals was the publication of *Theatrum Fungorum* in 1675 by **van Sterbeeck** (1630-1693) in Dutch. This was the first book devoted entirely to fungi. Sterbeeck's objective in writing this book was to aid correct identification of edible fungi from poisonous ones.

Curiosity demanded as to what these fungi were and where did they come from. **Pliny**, a Roman naturalist (about 3000 B.C.) could not decide whether fungi were living or non-living. The common belief was that they originated from thunder of clouds and a fluid accompanying the thunders penetrated the earth and gave rise to fungi (see **Buller** 1915). The physician poet **Nicander** (about 185 B.C.) called mushrooms, the evil ferments of the "earth" which had their origin from the breath of snakes. So dreadful an origin?

Porta (G. della Porta), as early as 1588, stated that the spores of mushrooms were their seeds, but he could not prove his point. He was followed by Malpighi (1679), Tournefort (1707) and Jursieu (1728).

Tournefort (1656–1708), who gave the concept of genus as a taxonomic category above the rank of species and is called the precursor of Modern Botany, however, maintained the cumbersome method of naming fungi by using a long descriptive phrase after the names of genera: e.g. *Fungus ramosus niger compressus parvus apicibus* and *albidis*.

Leeuwenhoek (1632–1723), the discoverer of bacteria, also saw the yeasts under his `simple' microscope. But the first illustrations of micro-fungi were first made by **Hooke** (1635–1703) in his book *Micrographia* (1665). Hooke drew the sporangia of *Mucor* and the teliospores of the rose-rust *Phragmidium mucronatum*. He believed that these spores were the seed pods rather than seeds and that the fungi initially arose spontaneously, which later produced seeds for further propagation. He also believed that the rust of rose originated from leaf tissues. There was vigorous controversy regarding propagation of fungi through seeds. An Italian botanist, **Micheli** (1679–1737), made full use of the microscope and made an extensive study of the fungi and their reproductive structures. In 1729 Micheli published *Nova Plantarum Genera* in Latin, on which rests his reputation as a mycologist.

PA Micheli

Micheli got the honour of being called the **father and founder of mycology**. He experimentally proved that fungi originated from their spores. Colonies of the same molds developed on the inoculated areas. This should have put an end the theory of origin of fungi from decaying materials; but it did not. Micheli was ahead time.

Linnaeus (Carl Von Line, 1707–1778), a Swede, who dominated the 18th century taxonomy of plants and animals and gave the binomial system of nomenclature, unfortunately failed to appreciate the importance of fungi. He called the ordering of fungi "a scandal to art" The few fungi which he took note of in his *Species Plantarum* (1753) were included by him among Vermes (worms) in the genus *Chaos*. Linnaeus had doubts about the affinity of fungi with plants.

Carolus Linnaeus

O'van Muenchhausen in 1750's proposed a theory that the fungi were composed of aggregations of "zoophytes" which separated out when soaked in water. Linneaus was highly impressed by Muenchhausen and made friends with him. Nonetheless, mycology did benefit from his **binomial nomenclature**. For example the fungus earlier called *Fungus ramosus niger compressus parvus apicibus albidis* was renamed as *Xylaria hypoxylon*.

The study of fungi was diligently pursued by several able systematists like Bulliard, Batsch, Persoon, Link, Schweinitz and Fries. The number of described species increased rapidly. Two important works of this period are the synopsis *Methodica Fungorum* (1801) by **Persoon** and *Systema Mycologicum* (1821–1832) by **Fries**. Persoon held no official position and on account of poverty sold his herbarium in 1825 to the Dutch Government. His book provided the framework on which Fries and all the later fungal systematists based their classifications. **Fries** came to be recognized the as the **Linnaeus of Mycology** and incidentally occupied the chair of Professor of Botany in the University of Upasala, earlier held by Linnaeus. The *Systema*, though produced without the help of compound microscope, continues to be one of the most important books on systematic mycology. It is especially important for Hymenomycetes. For over hundred years the treatment of *Basidiomycota* has been dominated by what is called the **Friesian approach**, which had and still has, the attraction of differentiating these fungi on the basis of field characters. Persoon and Fries, both students of Linnaeus, used the binomial system of nomenclature in their descriptions of fungi and put the taxonomy of fungi on firm footings. But it is most discomforting to record here what Fries thought of fungi and their origin! "These fungi" said Fries "depend on a diseased condition of plant rather than *vice versa*. From the study of many examples we learnt that they are hereditary and that they depend on the composition of the atmosphere. "*Every fear of their propagation by sporidia is superfluous*". Fries recognized fungi but didn't know them!

In the first 75 years of the 19th Century, the microscopic study of all sorts of fungi, e.g. rusts, smuts, 'imperfect fungi' helped in their proper identification. **Corda** (1809–1849) in his 6-volume work, *Icones Fungorum*, published between 1837–1854 showed the detailed structure of several large and microscopic fungi. The **Tulasne brothers** (L.R. Tulasne 1815–1885 and C. Tulasne 1816–1885) of Paris, in the three volumes of their monumental work, *Selecta Fungorum Carpologia* (1861–1865, in Latin) gave the basic, but revolutionary idea of **polymorphism** in fungi. **L.R Tulasne** is called the **reconstructor of mycology**. Tulasne brothers showed relation between the conidial stages and their perfect (sexual) fruit bodies. For the large number of fungi, which lacked the perfect stage, the group "Fungi Imperfecti" was created by **Fuckel** in 1869, four years after the appearance of the last volume of *Carpologia*. After this, much knowledge was gained on the structure of fungi. The emphasis shifted in the middle of the 19th Century to the study of life histories of fungi. Mycologists had earlier failed to recognize that they were studying the different stages of the same fungus. **De Bary** initiated the study of life history and soon it was discovered for several fungi. He also studied the life history of Myxomycetes, which he always considered as animals (Mycetozoa).

Anton De Bary (1831 88)

De Bary is called the **founder of modern mycology**. His mycological interests were more biological than taxonomic. His investigations ranged from study of life histories to elucidation of the subtle mechanisms of parasitism and saprobism (saprophytism) to the nature of lichens etc. He discovered **heteroecism** in rusts especially in *Puccinia graminis triticina*. He gave accounts of the development and sex in several fungi.

De Bary made a great impact on the later development of mycology. He recast the broad classification of fungi and wrote the well-known book *Morphologe and Biologie der pilze Flechten, and Myxomyceten* in 1866. Its English translation appeared in 1887 entitled *Comparative Morphology and Biology of the Fungi, Mycetozoa and Bacteria*. De Bary is also called the **father of Plant Pathology**, as it was he who proved that the fungus *Phytophthora infestans* was the cause of the potato blight. De Bary built a strong group of mycologists and 'plant doctors' around him. Chief among the mycologists were Brefeld and Woronin.

Brefeld (1839–1925) studied fungi by growing them in pure culture under varying conditions. He published his researches in 15 parts under the name "Under-suchungen aus dem Gesammt-gebeit der mykologie from 1872–1912. **Woronin,** a Russian, after several years of association with de Bary, returned to Russia and extended the life history studies with the help of his several able associates.

The Italian mycologist **P.A. Saccardo** (1845-1920) collected and brought together the scattered knowledge on systematic mycology under his massive 25 volume work, the *Sylloge Fungorum Omnium Hucusque Cognitorum* (1882–1925), whose first volume appeared in 1882 and the 25th in 1931. It was a Herculean job! With the publication of the **Sylloge**, as the work is usually called, systematic mycology took a big leap forward.

PA Saccardo (1840-1920)

Sylloge was not a mere compilation; it involved many difficult taxonomic decisions, which only a mycologist of Saccardo's stature could take! His classification of Fungi Imperfecti based on spore groups has been in use ever since it was published in 1880.

The last quarter of the 19th Century was most eventful. The **theory of spontaneous generation** breathed its last! The **germ theory of disease** was established by **Robert Koch**, **Louis Pasteur** and **De Bary**. The life histories of fungi were studied with the help of cytological and biochemical tools. The physiology, ecology and genetics of fungi became areas of active research. The process of fermentation (conversion of sugar solution into alcohol) became a subject of great controversy and even public discussion. Leeuwenhoek had observed yeast cells in fermenting solutions but he was mystified by the process of fermentation. Pasteur proved experimentally that yeasts and fermentation were inseparable. This was the **vital theory** which was initially propounded by the French Physicist **Cagniard de la Tour** in 1836 and supported by two German scientists **Theodor Schwann**, a physiologist, and **F.T. Kutzing**, an algologist. "No fermentation without life" declared Pasteur. **J. Von Liebig** (1803 –1873) a German chemist opposed and claimed that fermentation was brought about by a Labile organic substance, formed by interaction of air with plant juices. In 1897, **E. Buchner** found that the cell free extracts from yeast cells could bring about the fermentation. The chemical which caused fermentation was later named **zymase.** It was an enzyme, which etymologically means "in yeast". The discovery of enzyme gave birth to a new science, **Biochemistry.**

20TH CENTURY: MODERNIZATION OF MYCOLOGY

Physiology of Fungi

Though physiological studies with fungi had started by the middle of the 19th Century, it was restricted to few special aspects like the alcoholic fermentation and physiology of parasitism. That was the first intense research activity centered around microorganisms. **Koch and Pasteur** had projected bacteria and yeasts causing important human diseases and fermentation, respectively. De Bary, ushered in the era of **Physiological Plant Pathology** by suggesting the involvement of toxins and enzymes in plant disease.

Before the First World War, mycology was mostly devoted to taxonomy and cytology with the elucidation of the phylogenetic relationships as the main objective. In the second quarter of the present century greater attention was paid to other aspects like ecology, artificial cultivation, nutrition, physiology, and biochemistry of fungi. The emphasis in mycology, as was true for biology in general, shifted drastically from form to function.

Practically all corner-stones of fungal physiology can be traced to last Century, but if these researches could not develop much it was because the academic environment was not mature. The tools, the mechanisms and concepts of biochemistry, were not ripe. Otto Warburg, the famed German biochemist, had to abandon his researches on tumor physiology only for non-availability of the right type of tools. Some of those whose studies have a historical significance are: the studies of **Van Tiegham** (1867, conversion of tannin to gallic acid by fungi); **Raulin** (1867, Mineral nutrition of fungi); and **Wehmer** (1891 organic acid production from sugar). They were ahead of their times.

It is difficult to apportion credit for the modern application of molds to industry. Pasteur was "... convinced that a day will come when molds will be utilized in certain industrial operations on account of their power of destroying organic matter." **Industrial exploitation of fungi** stimulated investigations in several directions, as described below.

1. **Commercial Production of Enzymes**, especially proteolytic, pectolytic and diastase enzymes.
2. **Production of chemicals (citric acid)**. The best example of the industrial application of fungi as foreseen by Pasteur, was the manufacture of citric acid from *Aspergillus niger*. Many molds produce this acid, but *Aspergillus niger* has proved to be the most suitable. In 1992 about 90% of the world supply of calcium citrate was produced in Italy from citrus fruit juice, but within 8 years the export had practically stopped because of its commercial production by molds everywhere.
3. **Development of fungicidal research**. The arsenal of the plant pathologist, which had some important inorganic fungicides like the Bordeaux mixture, discovered by **Millardet** in 1882, was further enriched by the discovery of dithiocarbamtes in the 1930's by the Du pont company of U.S.A. This ushered in an era of organic fungicides. Later, antibiotics also joined the fungicide group of chemicals and made important impact as systemic fungicides.
4. **Agricultural – especially fungus transformations in soils**. Until 1915, it was not known that fungi inhabit the soil. **Waksman, Reinking, Manns** and **Garrett** discovered many important aspects of the role of fungi in soil. Now we know that the soil fungi are the primary, rather most important, degraders of cellulose, which forms the bulk of decomposing plant remains. These fungi are indispensable for the maintenance of the carbon and mineral cycles of nature. Ecology of soil fungi became a subject of great importance.

5. **Food manufacture, notably cheese**. The manufacture of mold-ripened cheese is an industry of unknown origin. Until the present century, the cheeses were named after the place of production, like the Roquefort, Camembert, Gorgonzola and Silton cheeses. The cause of variations in taste of these cheeses lay in the quality of the curd used and, more importantly, in the mold strain used, which are mostly species of *Penicillium*. Studies on the ripening process were first made in 1906 by **C. Thom.**

6. **Food spoilage**. Fungi growing on foodstuffs render them useless and some times even toxic. Before 1960 studies on fungal contamination of food centered mainly on the economic aspects of spoilage, rather than on the potential hazard to animal and human health. In this regard food microbiologists have given more attention to bacteria than to fungi, possibly because bacteria are more commonly associated with food poisoning and spread of diseases. The toxicity of moldy food has to be examined with greater emphasis.

7. **Rotting of textile fibers**. Cotton in its natural state is very resistant to enzymic degradation; but there are fungi which attack and degrade them. Fungi frequently spoil clothes in humid warm weather.

8. **Antibiotics**. The first antibiotic, reported by **Alexander Fleming,** was **penicillin** produced by a fungus, *Penicillium notatum*. Since then, many antibiotics have been discovered and, although most are derived from *Actinomycetes*, there are many reports of their production by fungi. Brian (1951) and Broadbent (1966) summarized the information of antibiotic production by fungi.

Alexander Fleming
(Nobel Prize 1945)

S.A. Waksman
(Nobel Prize 1952)

Presently, the studies on respiration, growth factors, nutrition, enzymes, intermediary products of metabolism constitute the intensely investigated areas in fungal physiology.

Fungi and Genetics (Heterothallism to Yeast Genome Project)

The 20th century ushered in the era of genetics, and shortly after the discovery of Mendel's laws of inheritance, **AF Blakeslee,** in 1904, discovered **heterothallism** in mucors. Now, the power of Bipolar and Tetra polar heterothallism in ensuring out-breeding is well known. A major factor in the rapid development of fungal genetics was the discovery of *Neurospora* by **Dodge**. Lindegren (1932-1934) focused attention on tetrad analysis that was possible due to the linear arrangement of the ascospores. **Beadle** and **Tatum** used *N. crassa* as the model organism to elucidate the physiological effects of the hereditary factors and gave the 'one-gene, one enzyme hypothesis'. This led to a new **Era of biochemical and Molecular Genetics.** In 1952, Dodge said: "The results, as reported, are such as would have been unbelievable ten years ago". The same statement can be made today, after the emergence of **a new biology** in 1953, with the discovery of double-helical structure of DNA. Now genetics has become transparent and power of genetics and recombinant DNA technology has enabled man to play GOD! For harnessing the power, a knowledge where the gene sits on the chromosome was a prerequisite. The choice fell on *Saccharomyces cerevisiae*. It became the first eukaryotic organism whose genome was sequenced. More than 600 scientists from Europe, North America and Japan participated in the **Yeast Genome Project** initiated in 1989 and completed in 1995 end.

Fungi in Soils

When in 1901 **Oedummans** and **Koning** described fungi isolated from soils, it was received with much skepticism. Whether fungi were capable of growing and producing mycelium in soil was a matter of controversy. Fungi were believed to be present in the soil merely as spores drifted from air or as colonizers of surface debris. **Waksman** played a major role in proving that fungi were a natural flora of soil and played important role in the ecology of soil as decomposers of dead organic matter. **Reinking** and **Manns** (1933) demonstrated the difference between **soil inhabitants** and **soil invaders. Garrett** gave the concept of **competitive saprobic ability** as the characteristic for differentiating **root-inhabiting fungi** (which were **soil invaders**) from the **soil-inhabiting saprobic fungi.** Another important contribution of soil mycology was the concept of **inoculum potential**, by Garrett.

S.D. Garrett

Fungi as Biodegraders Much work was done on cellulose biodegradation by *Trichoderma* spp be **Reese** and **Mandles**. Lignin biodegradation by *Phanerochaete chrysosporium* (The *E. coli* of lignin biodegradation) was elucidated by **Kirk**. The structure of plant cell wall elucidated by **Albersheim** and associates enabled the elucidation of enzymic degradation of cellulose, hemi-celluloses and rhamnogalacturonans (pectic materials).

Parasex and Sex Hormones

Pontecorvo and **Roper** discovered **parasexuality** in *Aspergillus nidulans* in 1952. It was a novel method of genetic recombination and was basically **somatic recombination,** not associated with sex organs or gametes. **Sex hormones** (pheromones), proposed by **de Bary** (1881) in *Achlya*, is now a well-researched area in mycology. Sirenin, antheridiol, trisporic acid and yeast alfa-factor are well understood sex hormones. Sex morphogens are also well established chemicals that trigger fruit body formation.

Mycotoxins

Mycotoxicoses, the "neglected diseases", got a shot-in-arm after discovery of **aflatoxins** the causal agent of the mysterious turkey "X disease" in England in 1963. The term **mycotoxin** was coined in the aftermath of the turkey X disease. The decade following the aflatoxin discovery was an intense period of research on mycotoxins, which appeared as a new threat to health of animals and humans.

Fungi in Biotechnology

Fungi are important in biotechnological production of food, extra-cellular enzymes, secondary metabolites and organic acids. In each case, fungi compete well with bacteria or cell cultures or chemical syntheses. Filamentous fungi are now used in the production of various organic acids, edible oils, antibiotics (beta lactam and polyketides). Many fungal enzymes, cell-associated or cell-free, are used in biotechnology. Cell-associated enzymes are used in the production of mushrooms and fermented foods and beverages. Amylases and glucoamylses, several plant cell wall polysacc-haridases, lipases, nucleases and proteases are produced biotechnologically using fungi.

Lichens get a Home and Mycorrhizae are Omnipresent

As the nature of lichens was not known until the end of the last century, lichenology was a rather independent study. Its integration with fungi (and algae and cyanobacteria) started in 1970s. Now the lichen-forming fungi are bonafide organisms having their rightful place in the fungal classification.

REFERENCES

Broadbent, D., 1966, Antibiotics produced by Fungi, *Bot. Rev.,* 2. 219-242.

Buller, A.H.R., 1915, The Fungus lore of the Greeks and Romans, *Trans. Brit. Mycol. Soc.,* 5: 21–66.

De Bary, A, 1887, Comparative Morphology and Biology of the Fungi, Mycetozoa and Bacteria, English Translation, Clarendon Press, Oxford.

Lowe, B., 1971, New records of mushroom stones from Guatemala *Mycologia* 63: 982–993.

Saccardo, P.A. 1882-1931, *Sylloge Fungorum Omnium Hucusque cognitorum*, 25 vols. Published by the author, Pavia.

Wasson, R.G. 1971, *Soma, Divine Mushroom of Immortality*. Harcoust, Brace. N.Y.

Wasson V.P. and Wasson, R.G. 1957, *Mushrooms, Russia and History*. Pantheon Books N.Y. Vols. 1 and 2.

SUMMARY (Points to Remember)

1. Stone sculptures bearing designs of mushrooms, and 'mushroom stones' belonging to 1000 BC have been found in Guatemala and other Central American countries.
2. The first book on mycology, *Theatrum Fungorum*, was written by van Sterbeeck in 1675. It enabled correct identification of edible and poisonous mushrooms.

3. Leeuwenhoek first observed yeasts in his 'simple' microscope, but first microscopic illustrations of micro-fungi, like yeasts, mucors and rust teliospores, were made by Robert Hooke, in his book *Micrographia* (1665) with the help of his 'compound' microscope,

4. Micheli experimentally proved that fungi originated from their spores and with this ended the various fanciful beliefs of their origin from thunder, venom of snakes, dead organic matter in the soil or from the dead plant tissues. His book *Nova Plantarum Genera* (1729) in Latin is another great contribution as a mycologist. Micheli is rightly called the "father and founder of mycology".

5. Linnaeus, who gave the binomial system of nomenclature (1753) for plants and animals, considered fungi as worms and placed all of them in the genus *Chaos*. He ridiculed their classification as "a scandal to art", But, later the binomial system was used in naming of fungi and, interestingly, by his own students, Persoon and Fries.

6. Fries is called "Linnaeus of Mycology". His book *Systema Mycologicum* (1821) is a milestone in the taxonomy of fungi. But his understanding of the origin of fungi "from diseased plat tissues and not *vice versa*", despite Micheli's clear demonstration of their origin from their spores, was unbelievably unscientific.

7. Tulasne brothers (LR Tulasne and C Tulasne) described "polymorphism" in fungi and showed relation between conidial and sexual (perfect) states of the same organism. Fuckel (1869) created the group "fungi imperfecti" for the large number of conidial fungi that lacked the sexual state.

8. Anton de Bary, known as "founder of modern mycology" emphasized study of the biology of fungi. His investigations included study of life cycles of fungi, mechanisms of parasitism and saprobism, nature of lichens, heteroecism in rust and sex in several fungi. He proved that late blight disease of potato was caused by *Phytophthora infestans*, which led to the development of germ theory of disease. Thus, de Bary came to be also called "father of plant pathology".

9. PA Saccardo (1882) brought together the scattered information on fungal taxonomy under his monumental 25 volume set of books, the *Sylloge Fungorum*. Also, his classification of *Fungi imperfecti* published in 1880 was highly popular.

10. Alexander Fleming in 1928 discovered the antibiotic now known as penicillin in the culture filtrate of *Penicillium notatum*.

11. AF Blakeslee in 1904 discovered heterothallism in *Rhizopus nigricans*.

12. BO Dodge (1932) discovered the fungus *Neurospora*, which dethroned *Drosophila* as the tool in the study of genetics.
13. Beadle and Tatum used *Neurospora crassa* to elucidate the physiological effects of hereditary factors, and with this started the era of "biochemical or molecular genetics"
14. The Yeast Genome Project, initiated in 1989 to sequence the entire genome of *Saccharomyces cerevisiae*, was completed in 1995.
15. SA Waksman established the role of fungi in the soil biology (biodegradation) and with this ended the belief that fungi were not integral part of the soil biota.
16. SD Garrett gave the concept of Competitive Saprobic Ability in soil fungi which differentiated the root-infecting fungi from soil-inhabiting fungi.
17. Pontecorvo and Roper discovered parasex in fungi (*Aspergillus nidulans*).
18. Aflatoxins were discovered in 1963 as the cause of the death of thousands of turkey poults in England due to consumption of moldy peanuts infected with *Aspergillus flavus*. After this several mycotoxins and mycotoxicoses were reported.

QUESTIONS

Q.1. Match the book with its correct author:

(a) Theatrum Fungorum	(i) Fries
(b) Micrographia	(ii) Linnaeus
(c) Nova Plantarum Genera	(iii) Hooke
(d) Sylloge Fungorum	(iv) Van Sterbeeck
(e) Systema Mycologicum	(v) Saccardo
(f) Species Plantarum	(vi) Micheli

Q.2. Fill in the blank:

a. That the fungi originated from their spores was experimentally proved by ----------------------.

b. The use of binomial nomenclature of fungi was introduced by -------- and -------.

c. The fungal group, *Fungi imperfecti* was created by --------------.

d. ---------- shifted the emphasis of study from taxonomy to biology of fungi.

e. *Neurospora* was discovered by----------.

f. ---------- and ----------- called penicillin the "miracle drug".

g. That fungi are the natural components of the soil biota was demonstrated by ------.

h. The concept of Competitive Saprobic Ability was given by -------.

i. Stone sculptures (1000 BC) bearing images of mushrooms and 'mushroom stones' were discovered from ------------- and other ----------------------- countries,

j. Toadstool is the distortion of the German word ------------ ,

k. Brefeld and Woronin were students of --------------.

Q.3. Write briefly on the important contributions of:

(i) Robert Hooke, (ii) PA Micheli, (iii) EM Fries, (iv) Tulasne brothers, (v) Anton de Bary (vi) PA Saccardo (vii) SA Waksman (viii) SD Garrett.

Q.4. Distinguish between:

(i) Root-infecting and soil-inhabiting fungi

(ii) Soil inhabitants and soil invaders.

Q.5. Answer in brief:

a. Which was the fungus whose genome was sequenced first?

b. When was the term mycotoxin coined?

c. Who in 1601 described many mushrooms with water colour drawings?

d. Who called classifying fungi a "scandal to art"?

e. Who, on account of poverty, sold his herbarium to Danish Government?

f. Who is called the "Linnaeus of mycology"?

g. Who gave the revolutionary idea of polymorphism in fungi?

h. Who created the group "fungi Imperfecti"?.

2

Kingdoms of Fungi and their Classification

(Fungi, Chromista and *Protozoa)*

When the organisms were divided into plant and animal kingdoms (Linnaeus, 1735), fungi were included among the plants. Dissatisfaction with the two kingdom system was felt especially while dealing with unicellular organisms. Haeckel, a German biologist, proposed a third kingdom ***Protista*** (Protisto meaning very first) (1866) to include organi-sms which lacked tissues namely algae, fungi, bacteria, protozoa etc. Protists themselves were later divided into lower and higher protista. The lower protista, which included bacteria and blue-green algae, was named the fourth kingdom, **Monera**. Thus, the living world came to have four kingdoms – Plantae, Animalia, Protista and Monera (Copeland, 1938, 1956; Barkley, 1968).

Whittaker (1969) proposed a five-kingdom classification (Fig. 2.1) and placed fungi in a separate kingdom. The five kingdoms were: **Monera** (for prokaryotic organisms viz. bacteria and cyanobacteia), **Protista** (for unicellular eukaryotes), **Plantae**, **Fungal kingdom** and **Animalia**. According to Whittaker, three nutritionally-distinct lines have developed from the protista, which characterized the three kingdoms (Earlier, only autotrophic and heterotrophic (ingestive) modes of nutrition were considered for recognizing separate kingdoms. The absorptive mode was ignored).

1. The plant kingdom, characterized by photosynthesis.
2. The animal kingdom, characterized by ingestive nutrition, and
3. The fungi, characterized by absorptive nutrition.

Three Domains and More Kingdoms

Woese (1990), at the University of Illinois, USA, erected a new taxon – the **domain** (meaning empire) and divided all organisms into three domains: ***Archaea***, ***Bacteria*** and ***Eukarya***. The archaea, like bacteria, are prokaryotes, but differ in base sequences of their 16S ribosomal RNA and in the extreme temperature, salinity and anaerobic conditions in which they live. Thus, they are more representative of the ancient forms of life that emerged three billion years ago, about one billion years after the earth cooled down to physiological temperatures. The archaea also differ from bacteria in the chemical composition of their cell wall and metabolic pathways. Both archaea and bacteria have independent origin from a common ancestor, the **Progenote**. The *Eukarya* (eukaryotes) appeared much later (1–2 billion years later) independently as *Urkaryote* (nucleated organisms, now extinct) from the same Progenote and later received the chloroplast and mitochondria by symbiotic union with certain bacteria (Fig. 2.2). Thus, the chloroplast and the mitochondria had an origin different from that of the cell in which they exist. The independent origin of the three domains is supported by molecular evidences. Thus, the "tree of life" has three stems and not one.

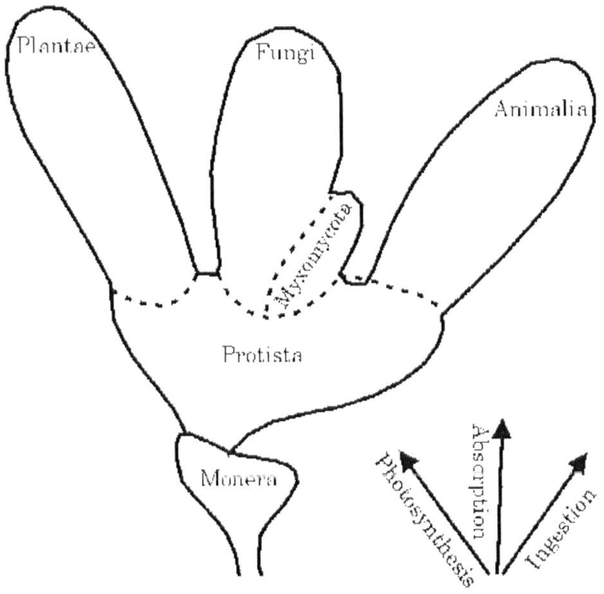

Fig. 2.1. Five kingdoms of the living world (after Whittaker, 1969).

The domains are further divided into kingdoms. The domain ***Bacteria*** is not spilt into kingdoms, while ***Archaea*** has three kingdoms: ***Crenarchaeota, Euryarchaeota***, and ***Korarchaeota***. The ***Eukarya*** has six kingdoms; ***Animalia, Chromista, Fungi, Plantae, Protozoa*** and ***Straminopila***. There are eight "candidate kingdoms" which are yet to be recognized as kingdoms. Besides, there are many protozoans that are not yet placed in any candidate kingdom. The *Chromista, Protozoa* and *Straminopila* comprise of unicellular (or simple multicellular) eukaryotic organisms which were earlier placed in kingdom *Protista* (or *Protoctista*), and which is not a recognized kingdom now. Likewise, *Monera* is no more a recognized kingdom, and is replaced by two domains *Archaea* and *Bacteria*. Kingdom *Protista* was an assemblage of diverse types of organisms, a "catch all" kingdom for all eukaryotes which did not fit in other 3 kingdoms. It was **polyphyletic**, unlike the kingdoms *Animalia, Fungi* and *Plantae,* which were **monophyletic**, i.e., *groups of organisms that have evolved from a single common ancestor*, eukaryotes. This necessitated their replacement by the 3 new unicellular, monophyletic kingdoms – *Chromista, Protozoa*, and *Straminopila*, which along with the 3 multicellular kingdoms (*Animalia, Fungi* and *Plantae*), comprise the eukaryotic crown of the 'tree of life'.

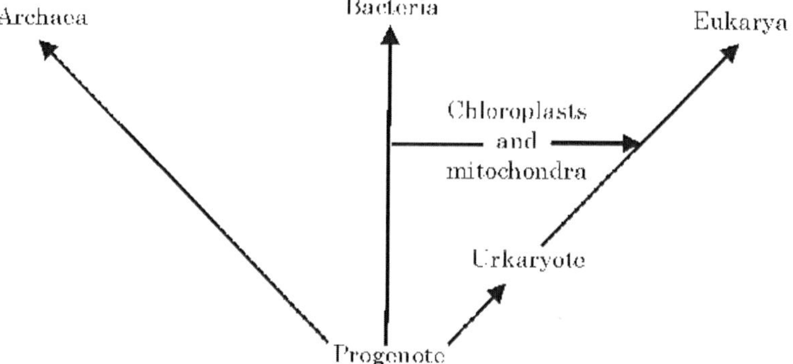

Fig. 2.2. Major evolutionary lines of descent proposed after the discovery of *Archaea.*

Classification and Phylogenetic Relationships of Fungi

The "organisms, studied by mycologists", and called fungi, are now separated and classified under three kingdoms: *Fungi*, *Protozoa* and Chromista/*Straminopila**. This is done on phylogenetic considerations to establish monophyletic groups. The organisms collectively called as fungi, though similar in morphology, ecology and mode of nutrition, are highly

---

*Dick (2001) proposed kingdom *Straminopila* to accommodate most organisms previously generally referred to the kingdom *Chromista* and primarily characterized by the presence of tripartite tubular hairs on flagella or cysts. If this classification becomes generally accepted the *Hyphochytriomycota* and *Oomycota* will have to be referred here (Kirk *et al.*, 2008)

dissimilar in the base sequences of their 18S rRNA, (or more precisely in the DNA coding for it), which is now considered as most valued parameter (chronometer) of determining genetic relatedness. Thus, *the fungi are a polyphyletic assemblage of unrelated organisms.*

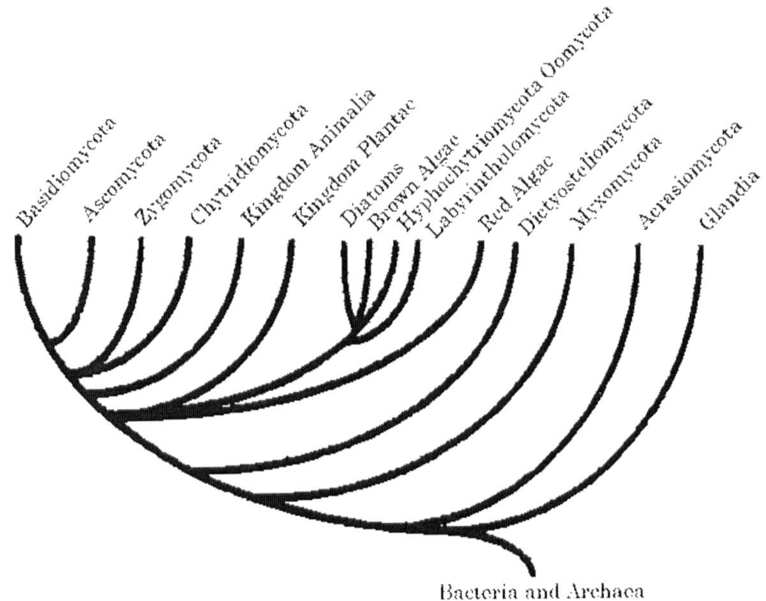

Fig. 2.3. Evolutionary tree of organism studied by mycologists. Updated and redrawn from Bruns *et al.,* 1991.

Fig. 2.3 shows the polyphyly of the fungi based on the sequence analysis of their 18S rDNA. The present kingdom *Fungi*, comprising of 8 phyla is monophyletic i.e. it has a common ancestor and all are its descendants. *Fungi* and *Animalia* also jointly form a monophyletic group, but there is no common name for them, and are called as **sister groups** (two **clades** resulting from a common lineage). This is suggested by comparison of several factors like 18S rDNA, elongation factor, cytoskeleton proteins, shared chemical features (storage of glycogen, chitin, mitochondrial codon UGA coding for tryptophan instead of termination), and phenotype of the motile stage (found in fungal zoospores, animal male gametes). Thus, fungi are nearer to animals than to plants. The various slime molds (acrasid, plasmodial, dictyostelid) and straminopilar fungi (*Oomycetes, Hyphochytriomycetes, Labyrinthulomycetes*) are now clearly distant groups and polyphyletic i.e., they do not show a close common ancestor. Thus, the earlier clubbing them together under one lot, fungi, was an artificial and not a natural classification. Within kingdom *Fungi* also, all organisms are not equally close: *Basidiomycota* and *Ascomycota* are closer among themselves than they are to *Zygomycota*, and all these are closer to each other, than they are to *Chytridiomycota*, *Neocallimastigomycota* and *Blastocladiomycota*. Also, *Blastoladiomycota* appears to be a sister group of some orders of *Zygomycota* (Lutzoni *et al.*, 2004).

How can fungi fit into 3 kingdoms? The answer lies in the way we define the term fungus. **Fungi are eukaryotic, heterotrophic, absorptive organisms that develop a rather diffuse, branched, tubular body (or are more rarely single-celled) and reproduce by means of spores.** This describes not a single phylogenetic line, but rather a way of life shared by organisms with different evolutionary backgrounds. Thus, we can call them as **eumycotan** fungi, **protozoan** fungi and **chromistan fungi**. They are, however, united by their morphology, ecology and mode of nutrition.

Natural and Artificial Classification

A natural classification attempts to place organisms in an orderly arrangement on the basis of overall resemblances, preferably on genetic basis. Classifications based on one or few convenient characters to serve special purposes, for example ease of identification, are called artificial classification. It may have no genetic significance. The natural classifications reflect degrees of evolutionary relationships and, therefore, may be phylogenetic.

The Taxa

Taxon is the general term for any taxonomic group, whatever its rank. The various taxa are:

Domain
  Kingdom
    Phylum
      Class
        Order
          Family
            Genus
              Species

The domain is the largest taxon. It comprises of kingdoms which include many phyla and this goes on up to the species, which is the unit of classification. Each of these categories may have a sub-taxon e.g. a phylum may have sub-phylum. Similarly, we have sub-classes and sub-orders. Species are also divisible into varieties, biological strains and physiological races and *formae speciales* (f. spp.). Sub-ranks sub-phylum, and sub-class are being used for kingdom Fungi. However, the rank of special form (*forma specialis*) (f.sp.) can be used to separate morphologically 'identical' fungi by host reaction.

The various taxa have standard 'endings' as shown below.

Taxa		***Standard endings* (Example)**
Phylum	-	mycota (*Basidio**mycota***)
Sub-phylum	-	mycotina (*Agarico**mycotina***)
Class	-	mycetes (*Agarico**mycetes***)
Sub-class	-	mycetidae (*Agarico**mycetidae***)
Order	-	ales (*Agaric**ales***)
Family	-	aceae (*Agaric**aceae***)

Genera and species have no standard endings. The technical name of an organism consists of two parts. It is called the *Binomial Nomenclature*. The first or the generic name is a noun and the second, denoting the species, is an adjective, describing the genus. The names of genera begin with a capital letter and that of species with a small letter. The genus and species are printed in italic type and when written are always underlined.

The technical name is followed by the name (full or abbreviated) of the discoverer of the species. The full names and abbreviations of most of these authors can be found in *Ainsworth and Bisby's Dictionary of the Fungi*, published by Commonwealth Mycological Institute. When the name of an organism is changed, the name of the original author is written in parenthesis, followed by the name of scientist responsible for the present name, e.g. *Mucor mucedo* (Linne) Brefeld. The species *mucedo* is the new name given by Brefeld to the species earlier named differently by Linne.

SOME OLD CLASSIFICATIONS OF FUNGI (Saccardo, Alexopolous, Ainsworth)

The true fungi were divided by **Saccardo** (1886) into four classes – Phycomycetes, Ascomycetes, Basidiomycetes and the 'form class' Deutero-mycetes, on the basis of the presence or absence of septa and characteristic sexually-produced spores (viz. oospores, zygospores, ascospores and basidiospores). The Phycomycetes (algal fungi) are also sometimes loosely called as 'lower fungi' and the fungi belonging to Ascomycetes, Basidiomycetes and Deuteromycetes as the 'higher fungi'.

Alexopoulos (1962) used "nine-class" classification in his book, the immense popularity of which made this classification widely accepted. Alexopoulos put fungi in division Mycota under two sub-divisions, the Myxomycotina (Myxomycetes) and Eumycotina (true fungi). The nine classes of Eumycotina were: Chytridiomycetes, Hyphochytriomycetes, Plasmodiophoromycetes, Oomycetes (all four earlier belonging to class Phycomycetes), Zygomycetes, Trichomycetes, Ascomycetes, Basidiomycetes and Deuteromycetes (fungi imperfecti).

The next important change came in 1966, when **Ainsworth** proposed a classification of fungi, which was used in the *Dictionary of the Fungi* (6th edition 1971) and also followed in *The Fungi, An Advanced Treatise* Vol. IV (1973). The kingdom fungi, was divided into two divisions: the Myxomycota, for plasmodial forms, and Eumycota, for non-plasmodial forms, which are usually mycelial (true fungi). The true fungi were divided into five sub-divisions, *viz.* Mastigomycotina, Zygomycotina, Ascomycotina, Basidiomycotina and Deuteromycotina. The outline of this classification is given below.

Kingdom – Fungi

Division	**Division**
Myxomycota	Eumycota
(4 classes)	(5 sub-divisions)
Acrasiomycetes	1. Mastigomycotina
Hydromyxomycetes	(3 classes)
Myxomycetes	Chytridiomycetes
Plasmodiophoromycetes	Hyphochytriomycetes
	Oomycetes
	2. Zygomycotina (2 classes)
	Zygomycetes
	Trichomycetes
	3. Ascomycotina
	(6 classes)
	Hemiascomycetes
	Loculoascomycetes
	Plectomycetes
	Laboulbeniomycetes

Pyrenomycetes
Discomycetes
4. Basidiomycotina
(3 classes)
Teliomycetes
Hymenomycetes
Gasteromycetes
5. Deuteromycotina
(3 classes)
Blastomycetes
Hyphomycetes
Coelomycetes

Divisions end in – mycota; sub-divisions in – mycotina; classes in – mycetes; sub-classes in – mycetidae; orders in-ales and families in-aceae.

CURRENT CLASSIFICATION (Kirk *et al*., 2008)

In view of the information on phylogenetic relationships of fungi (suggested by 18S (small subunit) ribosomal DNA), the recognition of domain as the taxon above kingdom, and creation of more kingdoms in *Eukarya,* Hawksworth *et al.* (1995), recast the classification of fungi by placing them under three kingdoms: *Protozoa, Chromista* and *Fungi*, under the domain *Eukarya*. The taxon **division** was replaced by **phylum** (1993), a term earlier used only by zoologists. In plant classification, the taxon corresponding to phylum is division. This arrangement is now universally accepted and followed in the subsequent 9th and 10th editions of the *Dictionary of the Fungi* (2001 and 2008). The standard endings of the taxa remain unchanged. However, now names of all taxa, domain to species, are italicized in print and underlined when written by hand (Earlier only genus and species were italicized).

The shift from a homogeneous kingdom *Fungi* to three kingdoms is a drastic change indeed. But, it is neither surprising, nor shocking. From the very beginning; plasmodial and amoeboid (cellular) slime molds, having phagotrophic nutrition, were put separately from "true fungi" under Myxomycota. These were also studied by zoologists who placed them under Mycetozoa (Protozoa). Also, *Hyphochytriomycetes* and *Oomycetes*, because of their cellulose cell wall, were put, along with algae in former kingdom Protista.

We shall mostly follow the classification of **Kirk** *et al.* as given in the 10th edition of the *Dictionary of the Fungi* (2008). It is important to mention here that the 2008 edition of *the Dictionary* has been divided into three parts; *Dictionary of the fungi*, *Dictionary of the chromistan/straminopile fungi-like organisms*, and a *Dictionary of the protozoan fungi-like organisms*. The classification (Kirk *et al.*, 2008) is largely based on the results from the **AFTOL** (Assembling the Fungal Tree of Life) project. The multi-gene sequence analysis undertaken in the AFTOL project has revolutionized our understanding of how the kingdoms should be classified. The AFTOL project, started in 2002, is funded by National Science Foundation of USA. It involved more than 100 collaborators from 20 countries. The project aims to enhance understanding of evolution of organisms in the kingdoms of fungi. There are eight recognized phyla of the kingdom *Fungi*. An outline of the classification (Kirk *et al.*, 2008) of the three groups of fungi up to phyla is given below.

Domain ***Eukarya***

Kingdom ***Fungi*** (8 phyla)	Kingdom ***Chromista*** (3 phyla)	Kingdom ***Protozoa*** (3 clases)
1. *Chytridiomycota* (2 classes)	1. *Hyphochytriomycota* (1 class)	1. *Protostelea* 2. *Myxogastrea*
2. *Neoallimastigomycota* (1 class)	2. *Labyrinthulomycota* (=*Labyrinthulea*) (1 class)	3. *Dictyostelia*
3. *Blastocladiomycota* (1 class)	3. *Oomycota* (1 class)	
4. *Zygomycota* (4 subphyla)		
5. *Ascomycota* (3 subphyla:15 classes)		
6. *Glomeromycota* (1 class)		
7. *Microsporidia* (taxa above genera not recognized)		
8. *Basidiomycota* (3 subphyla; 16 classes)		

Kingdom *Fungi*

These are also referred as "true fungi", and some authors (Barr, 1992) use *Eumycota* as kingdom name. But the kingdom name *Fungi* is preferred as most fungi belong here and the name fungi is immediately familiar. Members of the kingdom are characterized by (1) Absorptive mode of nutrition, (2) unicellular or filamentous somatic phase (3) cell wall containing chitin and beta-glucans, (4) presence of only whiplash flagella (5) mitochondria with flattened cristae, and (6) presence of peoxisomes and Golgi bodies. There are 8 phyla: *Chytidiomycota*, *Neocallimastigomycota**, *Blastocladiomycota**, *Zygomycota, Ascomycota, Glomeromycota, Microsporidia* and *Basidiomycota.*

The 'imperfect fungi', which reproduce only asexually, are not assigned an independent taxon. They are not a monophyletic unit, but are fungi which have lost their sexual (perfect) phase, or which are anamorphs of 'filamentous' *Asco-* and *Basidiomycota.* Many have been shifted to above groups after discovery of the teleomorph (sexual phase). But still two thirds of the 3000-odd remain "orphan taxa", unassigned to the parent organism. Such asexually-reproducing fungi are now referred as **anamorphic fungi** (mitosporic fungi). DNA sequencing and ultrastructural techniques now make it possible to place the anamorphic fungi with their parent organisms.

The ***Microsporidia***, earlier placed in *Protozoa* are now placed in *fungi.* This is supported by studies of α and β tubulins and other genes. These are obligate parasites of all groups of animals; most commonly insects, crustaceans and fish. Its classification does not recognize taxa above genera. *Nosema locustae* is commercially used for biological control of locusts and grasshoppers, while *N. apis* is cosmopolitan parasite of silk worms. About eight species cause diseases in immuno-compromised humans.

Kingdom *Chromista* or the Chromistan fungi are characterized by (1) unicellular or filamentous somatic phase, (2) absorptive mode of nutrition, (3) cellulose in cell wall, (4) presence of tinsel type of flagella, (5) tubular mitochondrial cristae, (6) presence of peoxisomes and Golgi bodies. The characters 1, 2, and 6 are same as in "true fungi". The difference lies in the nature of the cell wall and the flagella. In 18S rDNA phylogenetic studies these fungi show closer genetic relatedness to brown algae and diatoms than to true fungi. There are three phyla: *Hyphochytriomycota*, *Labyrinthulomycota* (=*Labyrinthulea*), and *Oomycota.*

---

* These phyla were segregated from *Chytridiomycota* because of phylogenetic distinctness (Hibbett *et al.*, 2007)

Kingdom *Protozoa* * or "Protozoan fungi" differ drastically from other fungi in form, function and life cycle. The somatic phase is plasmodial or amoeboid and there is no cell wall in the somatic phase. Nutrition is by ingestion (engulfment or phagocytic) except in plant parasites where the plasmodium obtains nutrition by absorption. The mitochondrial cristae are tubular. Spore walls are cellulosic or chitinous. Spores germinate to produce 1-2 flagellated cells. Earlier these organisms were placed under four phyla (Hawksworth, 1995) and later under three phyla (Kirk *et al.,* 2001). But recent data from molecular and ultrastructural studies suggest that these organisms are polyphyletic and appear to belong to five separate lineages (Kirk *et al.,* 2008), which are represented in 3 classes: *Protostelea, Myxogastrea* and *Dictyostelea.*

Nomenclatural Types

The specimen or specimens, on the basis of which the name of a taxonomic group is based, is called "nomenclatural type(s)". The original specimen is preserved and called **type specimen** or type collection. The important nomenclatural types are the following:

1. *Holotype* – is the specimen (or a part of it) on which the author based the new taxon.
2. *Syntype* – any specimen (when more than one) of author's original type material when no holotype was named.
3. *Isotype* – one of the specimens of the type collection of a species or variety other than the single specimen designated as the holotype.
4. *Lectotype* – a type taken later from the original material for a group for which the author named no type.
5. *Neotype* – specimen selected to replace the holotype when all original material is lost or destroyed.
6. *Paratype* – any specimen other than the holotype on which the first account of a taxon is based.

Species Concept

The species represent the lowest principal rank in the classification. There are several concepts of species, based on morphology, inter-breeding,

---

* Kingdom *Protozoa* has 18 phyla; the protozoan fungi fall under *Phylum Mycetozoa*

phylogeny (evolution), ecology and a combination of characters. The **morphological species** (or morpho species) are based on morphological characters, and most of the species in fungi are morphological species.

The inter-breeding populations are called **biological species**. These are restricted to sexually-reproducing organisms. Cultures established from single species, that are morphologically alike, are combined in cultures for testing inter-breeding. *Neurospora sitophila* was defined in this way. **Phylogenetic** (evolutionary) **species** are based on differences in biochemical, molecular or any other characters assessed by phylogenetic methods. The DNA – complementarity test is one such method, which was first used in speciation of yeasts. This is best method for asexual fungi. The **ecological species** are based on adaptation to particular niches. The species that are based on a combination of several characters are called **polythetic** species.

Forma specialis (f. sp.) is an informal rank (not regulated by code of nomenclature) used for distinguishing morphologically similar parasitic fungi by their ability to cause disease in a particular host. *Fusarium oxysporum* f.sp. *lycopersici* and *F. oxysporum* f.sp. *cubense* are identified by their ability to cause wilt disease in tomato and banana, respectively. These are comparable to 'pathovar' of plant pathogenic bacteria.

REFERENCES

Ainsworth, G.C. 1966. A general purpose classification of fungi. *Bibl. Syst. Mycol.* No. 1: 1–4.

Ainsworth, G.C. 1973. Introduction and keys to higher taxa. In: *The Fungi. An Advanced Treatise* (G.C. Ainsworth, F.K. Sparrow and A.S. Sussman eds.). Vol. IV A. 1–7. Academic Press, New York, London.

Barkley, F.A. 1968. *Outline Classification of Organisms*. 2nd ed. Hopkins Press. Providence, Massachusetts.

Burns, T.D. *et al.* 1991. Fungal molecular systematics. *Annu. Rev. Ecol. Syst.* **22:** 524–564.

Copeland, H.F. 1938 The kingdoms of organisms. *Quarterly Review of Biology*. **13**:383-420

Copeland, H.F. 1956. *The Classification of Lower Organisms*. Pacific Books. Palo Alto, California.

Doolittle, W.F. 1999. Phylogenetic classification and the universal tree. *Science*. 284: 2124–2128.

Haeckel, E.H. 1866. *Generelle Morphologie*. 2 vols. Berlin.

Hawksworth, D.L. *et al.*, 1995. *Ainsworth and Bisby's Dictionary of the Fungi* (8th ed.) C.A.B. International, Oxon, U.K.

Hibbett, D.S. *et al.*, 2007. A higher level phylogenetic classification of the Fungi. Mycol. Res. 111: 509-547

James, T.Y. *et al.,* 2006. A molecular phylogeny of flagellated fungi (*Chytridiomycota*) and description of a new phylum (*Blastocladiomycota*). *Mycologia*. 98 (6): 860-871

Kirk, P.M. *et al.,* 2001. *Ainsworth and Bisby's Dictionary of the Fungi* (9th ed.) C.A.B. International, Oxon, U.K.

Kirk, P.M. *et al.,* 2008. *Ainsworth and Bisby's Dictionary of the Fungi* (10th ed.) C.A.B. International, Oxon, U.K.

Lutzoni, F.F. *et al.,* 2004. Assembling the fungal tree of life: progress, classification and evolution of subcellular traits. *Amer. Jr. Botany* (91): 1446-1480

SUMMARY

Fungi were classified as plants when there were only two kingdoms: *Plantae* and *Animalia*. When the kingdom *Protista*, was created (Haeckel, 1886) for unicellular, tissue-lacking organisms, fungi as well as protozoa, bacteria and blue green algae were placed in this kingdom. Later, *Protista* was divided into 'higher' and 'lower' protista in the 4-kingdom classification (Copeland, 1956) and fungi were placed in the higher protista (*Protista)* and bacteria and blue-green algae in lower protisa (*Monera*). Whittaker (1969) in his five-kingdom classification (*Plantae*, *Animalia*, *Fungal kingdom*, *Protista* and *Monera*) made a new kingdom for fungi (including myxomycetes) based on their absorptive mode of nutrition.

Woese (1990) erected a new taxon **domain**, above the rank of kingdoms, and recognized three domains, *Archaea*, *Bacteria* and *Eukarya*. *Archaea*, like bacteria, are prokaryotes but much different in their 16S rRNA sequences and in the extreme temperature, salinity and anaerobic conditions in which they live. They, in fact, represent the harsh, archaic mode of life that existed at the beginning of the living world 3 billion years ago. Both bacteria and the archaea (archaeobacteria) originated independently form a common ancestor, the **Poregenote**. The eukarya originated independently from the progenote later (1-2 billion years ago) as **Urkaryote** (nucleated organisms, now extinct), which during symbiotic association with the bacteria developed chloroplast and the mitochondria.

The *Eukarya,* based on phylogenetic data, are divided into six monophyletic kingdoms: *Plantae, Animalia, Fungi, Chromista*, *Protozoa* and *Straminopila* (The organisms earlier placed under Protista are now placed under *Chromista*, *Protozoa and Straminopila,* while those belonging to Monera are placed among *Archaea* and *Bacteria*). Fungi, deftly defined as "organisms studied by mycologists" because of their heterogeneity and polyphyletic nature, are now placed in three monophyletic kingdoms: *Fungi* (true fungi or eumycotan fungi), *Chromista* (chromistan fungi) and *Protozoa* (protozoan fungi). In "phylogenetic tree", the fungi are closer to animals, and

Fungi and *Animalia* jointly form a monophyletic group, but there is no common name for them, and are called as **sister groups**. The chromistan fungi are closer to brown algae and diatoms, while the protozoan fungi are closer to *Protozoa*. Thus, the fungi represent a way of life shared by organisms with different evolutionary backgrounds. They are united by their morphology, ecology and the mode of nutrition. All the three groups of fungi share the characteristics of fungi, which are defined as *eukaryotic, heterotrophic, absorptive organisms that develop a rather diffuse, branched, tubular body* (*or are more rarely single-celled*) *and reproduce by means of spores*.

The older classifications of fungi (by Saccardo, Alexopoulos and Ainsworth) were artificial classifications, based on similarity of morphological characters and which were practically convenient for identification. The recent classifications adopted by the *Dictionary of the Fungi* (Hawksworth, 1995; Kirk *et al.*, 2001; Kirk *et al.,* 2008), are based on phylogenetic relationships as evidenced by 18S rRNA sequences. The classification (Kirk *et. al.,* 2008) is largely based on the results from the AFTOL (Assembling the Fungal Tree of Life) project. The multi-gene sequence analysis undertaken in the AFTOL project has revolutionized our understanding of how the kingdoms should be classified. In this classification the kingdoms are divided into phyla and classes. The Kingdom *Fungi* has eight phyla, while *Chromista* has three. The Protozoan fungi could no be put under monophyletic phyla; and have been classified under five separate lineages, which are represented in 3 classes.

There is much debate on how to define a species. Depending on the concept of delimiting species, there are several types of species. There are morphological species, biological species, phylogenetic species, ecological species and polythetic species. Forma specialis is an informal rank used to distinguish same species differing by their host specialization. It is comparable to pathovars of bacterial plant pathogens.

Test Questions

1. Give an outline classification of fungi up to phyla. No description is required.
2. Describe the characters that distinguish the fungi belonging to three different kingdoms?
3. Answer in brief:
   i. In Whittaker's classification which three nutritionally-distinct lines have developed from the protista?

ii. Which kingdoms have replaced the former Protista?

iii. Which two domains have replaced the former kingdom Monera?

iv. Which two kingdoms in Whittaker's classification were polyphyletic?

v. In which three kingdoms are the fungi placed?

vi. Who erected the new taxon 'domain'? How many domains are there?

vii. Which is the common ancestor for the three domains of life? Which domain represents the ancient mode of life and why?

viii. Which two kingdoms form a monophyletic group but in absence of a common name are called sister groups?

ix. What are 'Urkaryotes'? Which domain arose from them and how?

x. Which three characters are shared by *fungi* and *Chromistan fungi*?

xi. Why are *Microsporidia* now placed in kingdom *Fungi*?

4. Write in brief on:

(i) AFTOL,

(ii) Progenote,

(iii) Types of fungal species

(iv) Origin of *Eukarya* from the Progenote.

5. Distinguish between

(i) Polyphyletic and Monophyletic

(ii) Artificial and Natural classification.

(iii) Holotype and Neotype.

(iv) Former kingdoms Protista and Monera.

6. Fill in the blanks.

(i) The fungi belonging to three kingdoms are united by their --------------, ------------, and -------------.

(ii) The former 'imperfect fungi' are now called------------------

(iii) The 10th edition of *Dictionary of the Fungi* (2008) is divided in three parts. These are *Dictionary of the Fungi*, *Dictionary of the* -----------, and *Dictionary of the* ---------------.

(iv) The protozoan fungi-like organisms are not divided into phyla but assigned to five ------------, which are represented in 3 clases

(v) Saccardo divided fungi under --------- classes, while Alexopoulos put them under ------- classes.

(vi) The full name of the *Dictionary of the Fungi* is ------------- and --- ----------'s *Dictionary of the Fungi.*

(vii) The taxon --------- has been replaced by phylum in the classification of fungi.

(viii) Most of the species of fungi are ----------------- species.

(ix) Forma specialis is an informal rank similar to ------------ of plant pathogenic bacteria.

(x) The *Archaea* represent life that existed ------- billion years ago. Fungi appeared ------- billion years later.

3

Fungi – General Characteristics

Fungi are a big heterogeneous group of organisms spread over three kingdoms: *Fungi, Chromista* and *Protozoa.* The members of first two kingdoms are much similar in their thallus construction and absorptive nutrition, in which the protozoan fungi are much different. Nevertheless, the unity in diversity is remarkable and they are inseparable by their morphology, ecology and nutrition. The main characteristics of fungi as follows.

Nutrition heterotrophic and absorptive; ingestion rare (phagotrophic in cellular slime molds).

Thallus on or in the substratum, and plasmodial, amoeboid or pseudoplasmodial; or in the substratum, and unicellular or filamentous (mycelial), septate or non-septate, typically non-motile (with protoplasmic flow through the mycelium) but motile states (e.g. zoospores) may occur.

Cell wall well-defined, typically chitinized in eumycotan fungi; cellulosic in chromistan fungi.

Nuclear status eukaryotic, multinucleate, the mycelium being homo-or heterokaryotic, haploid, dikaryotic or diploid, the last being usually of limited duration.

Life cycles simple to complex.

Sexuality asexual or sexual and homo- or heterothallic.

Sporocarps microscopic or macroscopic and showing limited tissue differentiation.

Habitat ubiquitous as saprobes, symbionts, parasites, or hyperparasites.

Distribution cosmopolitan.

NUTRITION

Fungi, are heterotrophic, dependent on external supply of food, like animals and some bacteria. These are *Chemoorganotrophs* i.e. derive their energy from oxidation of organic substances. Ingestion of food is rare and found only in slime molds, which in the somatic phase represented by myxamoebae and plasmodia (multinucleate amoeboid mass of cytoplasm with pseudopodia) feed on bacteria, yeasts and spores etc. by phagotrophy. The plasmodia can also absorb external nutrients. The absorptive mode of nutrition of fungi, (except the slime molds) resembles that part of animal nutrition, which involves the absorption of soluble substances. The extracellular enzymes convert the insoluble food into soluble form, which are then absorbed (Fig. 3.1).

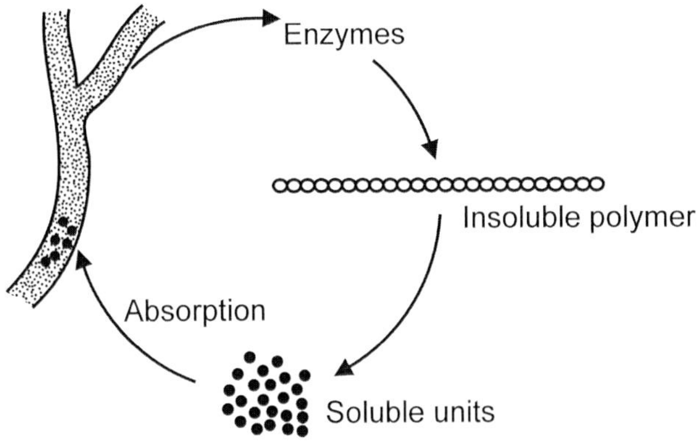

Fig. 3.1. Showing degradation of an insoluble polymer into its soluble units. (Holophytic nutrition)

Fungi are hetetotrophic for carbon compounds, i.e. they are unable to use carbon dioxide as their carbon source. If supplied with carbohydrates, they can synthesize their own proteins. Fungi have two major ways of obtaining their food–saprobism and parasitism. Saprobic fungi derive their food from dead organic matter, while parasitic fungi get their nutrition by infecting living organisms. The distinction between saprobism and parasitism is subtle and cannot always be demarcated. The parasitic fungi are able to invade living tissues which saprobic fungi are unable to do. Some parasitic fungi can be grown in artificial culture. These are termed facultative parasites.

The obligate parasites, which have no saprobic phase, on the other hand, cannot be cultured on artificial media, for reasons not yet known. The facultative parasites, though ecological parasites, are nutritional saprobes.

Fungal parasites cause diseases in plants, animals and human beings. Some fungi live in plants as endophytes (symptomless parasites) or as symbiotic partners with algae (lichens), enabling them to grow under harsh conditions which they could not do otherwise. The large majority of fungi decompose dead organic material. They are particularly important in the degradation of cellulose, which forms the bulk of decomposing plant materials.

THALLUS

The thallus of fungi usually consists of microscopic tubular filaments, called **hyphae** (sing. hypha); mass of **hyphae** constituting the thallus is called **mycelium** (pl. mycelia).

The hyphae are branched (rarely unbranched), and the branches ramify on or inside the substratum to form a three dimensional network. Cytoplasmic streaming in fungal hyphae is unidirectional, towards the tip, where growth takes place. Hyphae grow entirely at their tips. The key to the fungal growth lies in the apex. There is built-in mechanism at the hyphal apex for varying its width and shape. It is the site of active differentiation and nuclear divisions. The thin and plastic hyphal tip, 50-100 μm in the apical region, is the zone of elongation and is filled with protoplasm (Fig. 3.2). The portion behind this is vacuolated and incapable of elongation. But it helps in growth by synthesizing the cytoplasm, which is transported to the tip. Lateral branches arise behind the apex in acropetal order. Each of these daughter branches behave like the parent hypha and forms secondary and tertiary branches behind their growing tips.

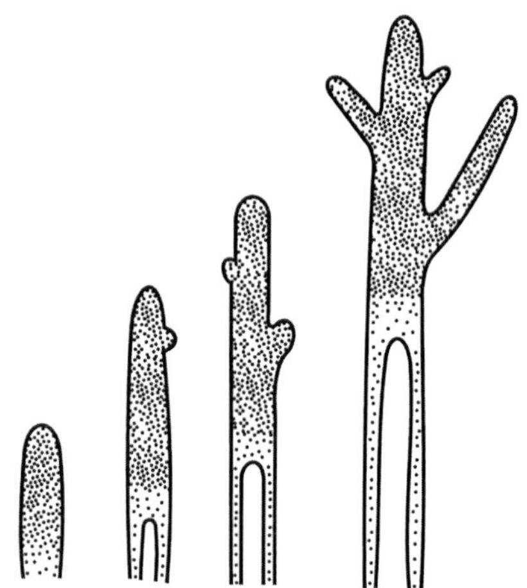

Fig. 3.2. Showing apical growth of a hypha.

The hyphae may be septate or unseptate (Fig. 3.3). The unseptate hyphae have nuclei scattered in the cytoplasm. This condition is known as **coenocytic**. Even in the aseptate hyphae, septa are formed to cut off old, empty portions of hyphae in the hinder region, and to delimit the sex organs. Such septa are formed as a result of changes in the local concentration of cytoplasm and are called **adventitious septa**. The other type of septa, the **primary septa**, are formed in association with nuclear divisions, appearing regularly and producing septate hyphae, having cells of roughly comparable size. The adventitious septa are complete, while the primary septa have a pore in the center, allowing cytoplasmic flow between the cells of the hyphae. A septum is formed by an annulus of wall material, which invaginates the plasmamembrane and grows centripetally inwards. The plasma membrane is continuous around the pore (Fig. 3.4). A 'dolipore septum' having an annular thickening around the pore, has been reported in some basidiomycetes.

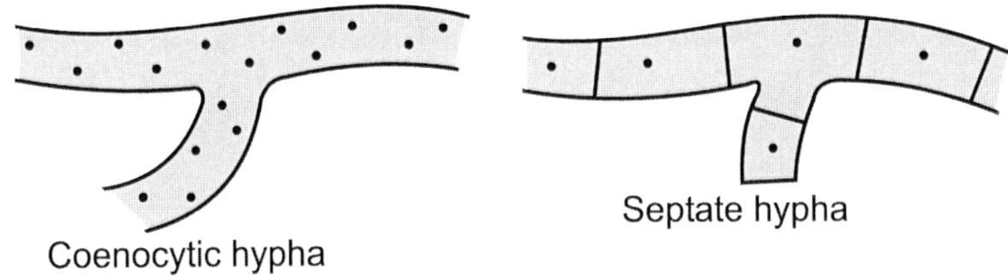

Fig. 3.3. Portions of coenocytic and septate hyphae

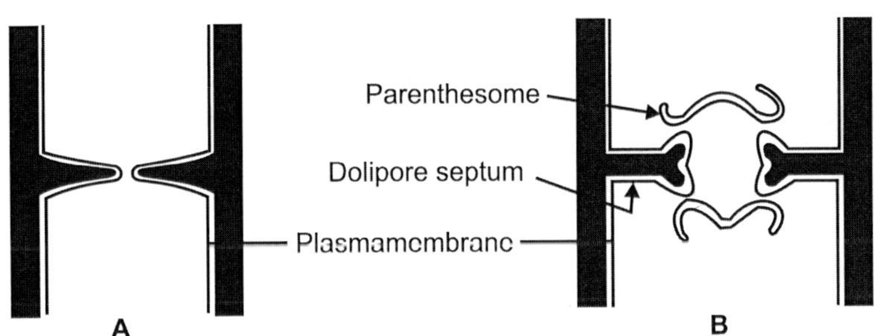

Fig. 3.4. Primary septa. **A.** Perforated septum; **B.** doliphore septum.

Fungi have eukaryotic cell structure. They have double membrane-bound cell organelles like nucleus and mitochondria, tubular endoplasmic reticulum (ER), the golgi bodies and the ribosomes. However, there are some differences from typical eukaryotic cells, like the ribosomes, lying free in the cytoplasm and not attached to the ER. The nuclei of some show unusual features. Lomasomes (plasmalemasomes), appear as particles or vesicles in pockets, formed between the cell wall and the plasmamembrane. Their function is not yet known.

Though the hypha is the characteristic unit of the structure in fungi, there are fungi whose thallus may consist of a single cell. Some fungi show **dimorphism** and exist in both mycelial and unicellular forms in different environments.

Fungal tissues are called **plectenchyma**. These are formed by branching, interweaving, anastomosis and coiling of hyphae. During these events, septation occurs and the hyphae divide and become short-celled and compressed. Fungal tissues are, thus, relatively simple and limited in variety. They differ mainly in the compactness of organization of the hyphae and degree of division into short cells. These are of two types (Fig. 3.5).

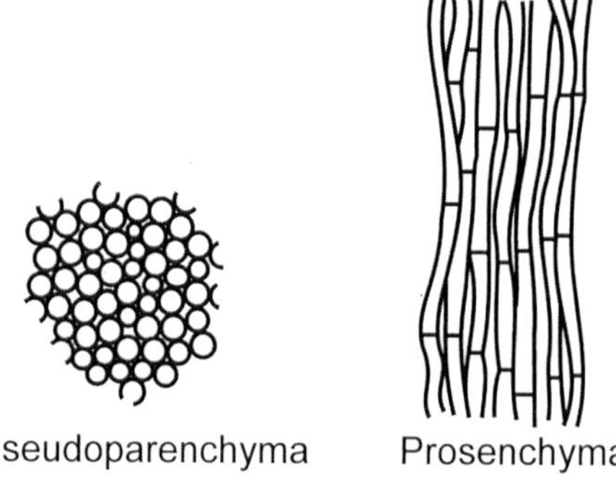

Fig. 3.5. Plectenchyma (= fungal tissues).

1. **Prosenchyma**, in which component hyphae are loosely woven and can be recognized as such.
2. **Pseudoparenchyma**. It consists of closely packed hyphae, which under sections resemble the parenchyma of higher plants. The hyphae divide into short cells, which by mutual pressure, become isodiametric.

Plectenchymatous tissues in higher fungi are formed in somatic structures, such as mycelial strands, rhizomorphs, pseudorhizas, sclerotia, pseudosclerotia and stromata. In general, these structures are divisible into two anatomical regions, an outer hard rind, called **cortex**, which is made of pigmented pseudoparenchyma, and the inner soft, hyaline, prosenchymatous core, called **medulla**.

The **plasmodium**, which is a multinucleate, motile mass of protoplasm without a cell wall, represents the thallus of slime-molds. These are of various types viz. **protoplasmodium,** an undifferentiated microscopic plasmodium, which gives rise to a single sporangium; **aphanoplasmodium**, a plasmodium composed of a network of undifferentiated strands of non-granular protoplasm; **phaneroplasmodium**, a plasmodium composed of a well-differentiated advancing fan and thick strands of granular protoplasm exhibiting ecto- and endoplasmic regions; and **aggregate plasmodium** or **pseudoplasmodium,** a structure formed in *Acrasiales* by the massing of separate myxamoebae or cells before reproduction (a grex). A plasmodium, called **fibroplasmodium** is a net-like pseudoplasmodium is found in *Labyrinthulales* belonging to *Chromista*.

CELL WALL

Fungal cell walls contain 80-90% carbohydrates; the remainder being proteins and lipids. A typical feature of fungal cell wall is the presence of chitin (a prominent constituent of the exo-skeleton of insects) and not cellulose, the universal cell wall material of plants. Cellulose does occur in the cell walls of *Oomycetes* and *Hyphochytriomycetes*. The widespread occurrence of chitin and cellulose in the cell walls of fungi, and its relation to fungal taxonomy, has intrigued mycologists for about a century. Chitin in the wall is accompanied by glucose polymers, other than cellulose, and mannans (polymer made of mannose sugar). The composition of the cell wall is variable and is modified by age and environmental factors.

Chitin is a linear polmer of β N-acetyl glucosmine units, linked by β 1-4 glycosidic bonds (Fig. 3.6). The chitin may be overlain or intermeshed with other glucans, like callose, which consists of β 1-3 linked glucose residues.

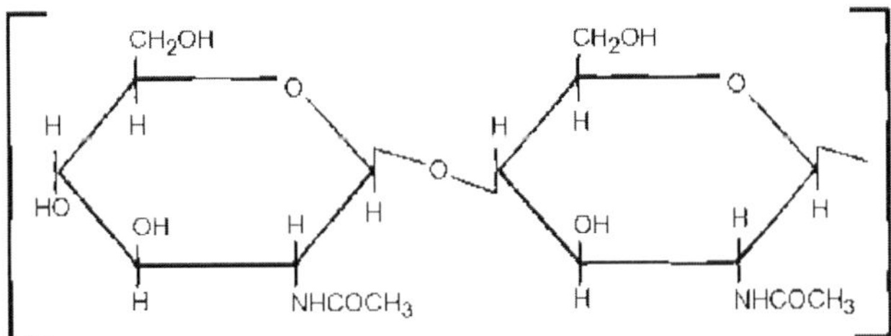

Fig. 3.6. Chitin, polymer of β N-acetyl glucosamine units.

The fungal membrane, like those of other eukaryotic membranes, is made up of a phospholipid bilayer with into which are anchored proteins. Sterols are also present which control membrane fluidity, membrane-associated enzymes and transport mechanisms. In mammalian cells, the chief sterol is cholesterol; in fungi it is ergosterol (but in *Chytridiomycota* and *Oomycota*, it is cholesterol). This difference in the sterol component of fungal and mammalian cell membranes has allowed the development of two classes of antifungal antibiotics-the **polyenes** (e.g. nystatin and amphotericin) and **azole** compounds (triazoles and imidazoles). The polyene antibiotics bind to the ergosterol and form a pore causing damage to membrane permeability, while the azoles inhibit ergosterol biosynthesis which results in membrane dysfunction and cell death.

NUCLEAR STATUS

The nuclei of fungi show important characteristics. They are extremely small-near the limit of resolution of the light microscope. This makes the karyology of fungi a difficult subject to pursue. The nuclear divisions show marked differences from normal mitotic divisions. Electron microphotographs have revealed that the nuclear membrane does not disorganise during division and the nucleus divides by constriction in the center (Fig. 3.7). The nuclei become dumb-bell-shaped when the nuclear membrane is drawn out between the separating daughter nuclei. This type of non-mitotic nuclear division has been termed **karyochorisis** (Moore, 1964). However, typical mitosis does occur in yeasts. Meiotic divisions of fungi are similar to that of higher plants and animals.

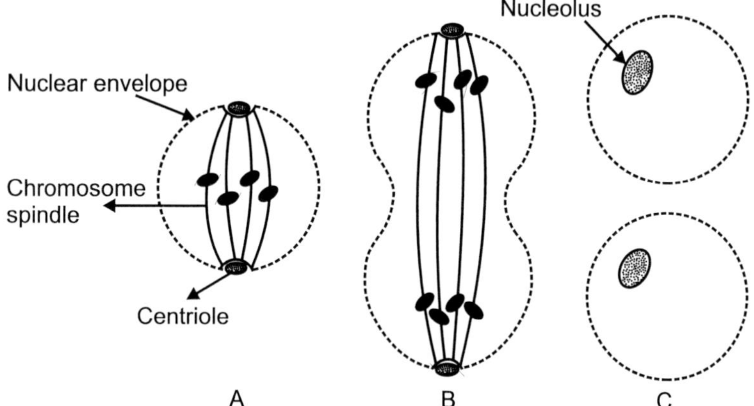

Fig. 3.7. Showing some stages of karyochorisis.
A. Mataphase; B. telphase; C. interphase.

In aseptate hyphae, nuclei lie scattered throughout the cytoplasm. In septate hyphae, the cells may have one, two or more nuclei. In most fungi, the nuclei are haploid throughout the life cycle, and the diploid stage is represented only by the zygote nucleus which does not multiply. Prolonged diploid phase, similar to that occurring in higher plants and animals, is shown by some fungi e.g. *Allomyces*, *Oomycetes* and yeasts. In the higher fungi, like the *Asco-* and *Basidiomycota*, the nuclei of opposite mating types form dikaryons in each cell of the hyphae.

Heterokaryosis

The nuclear cycle is not always as clear-cut as described above. All the nuclei are not of the same kind. Mutation freely converts a homokaryotic mycelium (containing similar nuclei, derived from a single parent nucleus) into heterokaryotic. Occurrence of genetically dissimilar nuclei in the same cell is called **Heterokaryosis**. It is brought about by: (1) mutation (2) anastomosis and (3) karyogamy between dissimilar nuclei. We shall study this in details in Chapter 30.

LIFE-CYCLES

Many fungi, like other sexually-reproducing organisms, show an alternation between haploid and diploid nuclear phases in their life cycles. The haploid phase begins with the completion of meiosis, while the diploid phase starts with the fusion of two haploid nuclei during sexual reproduction. In many higher *Asco-*, and *Basidiomycota* a third phase - the **dikaryophase**, intervenes the haploid and diploid phases. The nuclei associate in pairs to form dikaryons, which multiply simultaneously (by conjugate divisions) and the daughter pairs of nuclei form new cells. The multiplication of the gametic nuclei increases the number of sexual fusions and this, after karyogamy, results in the formation of numerous diploid nuclei. A life cycle, involving dikaryons, is found only in fungi and nowhere else. Fungi display different types of life cycles, centered around the sexual reproduction. Raper (1954) has described seven basic types of life cycles in fungi (Fig. 3.8), which are as follows:

1. *Asexual cycle* The entire group of anamorphic fungi (fungi imperfecti is now obsolete term) shows this type of life cycle. There is no alternation of haploid and diploid nuclear phases. The diploid (2n) phase is lacking.
2. *Haploid cycle* The life cycle is completely haploid (n); the diploid phase is restricted only to the zygote nucleus. This is the most common type of life cycle found in majority of fungi and ascomycetes.
3. *Haploid cycle with restricted dikaryon* This type of life cycle is found in higher ascomycetes, forming dikaryotic (n+n) ascogenous hyphae, which are completely dependent on the

haploid mycelium. This is predominantly a haploid life cycle, but it differs in the separation of plasmogamy and karyogamy in space and time. The gametic nuclei pair to form dikaryons, which multiply by conjugate mitotic divisions in the ascogenous hyphae. The dikaryons finally undergo karyogamy and meiosis in the ascal primordia.

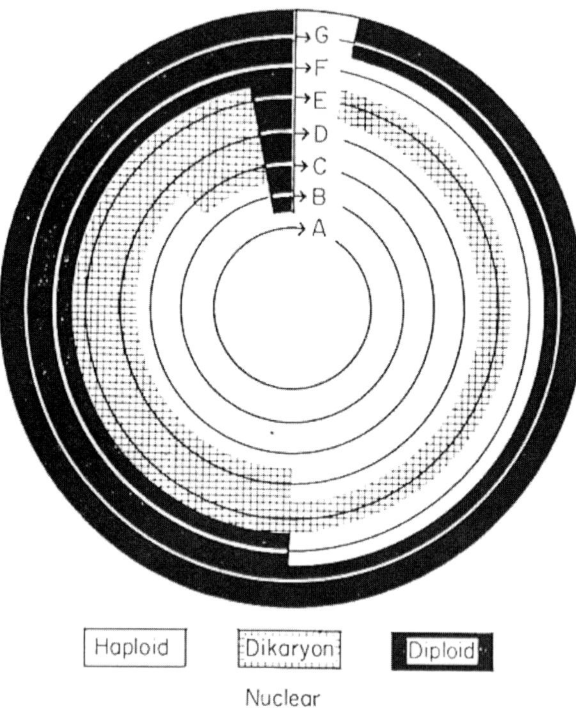

Fig. 3.8. The seven types of life cycles in fungi.

4. *Haploid-dikaryotic cycle* Most of the basidiomycetes, excluding the smuts, exhibit this type of life cycle in which the dikaryophase (n + n) is more extensive and also independent of the haploid phase. The cycle comprises two roughly equivalent phases and terminates in a single diploid nuclear generation, represented by the diploid zygote nucleus.

5. *Dikaryotic cycle* The complete life cycle is passed in dikaryotic phase, and both the haploid and diploid generations are represented by a single nuclear generation. The immediate products of meiosis viz. the ascospores or the basidiospores, fuse

to initiate the dikaryophase, which persists until karyogamy occurs. Examples are the rusts and several smuts.

6. *Haploid-diploid cycle* This type of life cycle, which is characteristic of algae, occurs only in two groups of fungi viz. the *Blastocladiales* and *Saccharomyces cerevisiae*.
7. *Diploid cycle* This type of life cycle, that is completely diploid, except for the gametes, is reported in a number of diplobiontic yeasts, *Mycetozoa*, and the *Oomycota*.

SEXUALITY

Fungi reproduce both asexually and sexually. In some fungi, e.g. fungi imperfecti, (anamorphic fungi), reproduction is by asexual methods only. In some others, it is parthenogenetic. But in majority of fungi, reproduction occurs in two well marked phases - the asexual and sexual. Differing from the sexual reproducion, is **parasexuality**, demonstrated by Pontecorvo and Roper (1952) in *Aspergillus* and *Penicillium*. It has since been reported in other groups of fungi as well. But its extent of prevalence and role in development of new species in nature is not assessed.

Asexual reproduction is being used here in a broader sense, including somatic (=vegetative also) reproduction. Usually, the term somatic reproduction is used for the development of new individuals from unspecialized detached portions of the thallus. In 'asexual' reproduction, the new individuals arise from specialized reproductive bodies, produced by the thallus, without involving any nuclear fusions. However, as biologically there is no difference in the somatic and asexual reproductions (both being non-sexual), they are put together under asexual reproduction.

Asexual Reproduction

Fungi exhibit the following methods of asexual reproduction (1) fragmentation (2) budding, (3) fission, and (4) spores.

Fragmentation A detached fragment of the hypha, in suitable conditions, gives rise to a new individual. This happens frequently in nature and is the method used for cultivation of fungi in the laboratory.

Budding Budding of somatic cells is common in 'yeasts'. A soft zone appears on the cell wall which bulges out, constricts and finally pinches off, to form a daughter cell. Budding may occur on spores also.

Fission This is characteristic of bacteria, and occurs only in 'fission yeasts' e.g. *Schizosaccharomyces*. In fission, the cell divides in transverse plane into two cells (by constriction) and later formation of a cell wall in-between.

Spores Spores are the most common method of asexual reproduction in fungi. The term spore is used for any small propagative, reproductive or survival unit, which separates from a hypha or a sporogenous cell and gives rise to a new individual. In contrast to the somatic mycelium, the spore is characterized by:

- ceasation of cytoplasmic movement,
- small water content,
- slow metabolism, and
- lack of vacuoles.

In mycology, spore is a general term, capable of wide interpretation. It is qualified by prefixes to indicate the particular type. These may be produced asexually or sexually. Thick-walled zygospores (or oospores), ascospores and basidiospores are spores formed as a result of sexual reproduction. These are called **meiospores**. Asexual spores are called **mitospores**.

Asexual spores are of two main types: sporangiospores and conidia. **Sporangiospores** are formed within a sac-like structure, called sporangium (pl. sporangia), which is borne on an undifferentiated or specialized hyphal structure, called sporangiophore. **Conidia** are formed externally on hyphae or more commonly on morphologically differentiated conidiophores. Sporangiophores and conidiophores are sometimes characteristically branched and serve as a useful taxonomic criterion.

Sporangiospores (Fig. 3.9) may be motile or non-motile. The motile sporangiospores are called **zoospores** and are provided with one or two flagella. The non-motile sporangiospores are called **aplanospores**; they lack flagella. Zoospores are naked spores lacking cell wall which after a swarming period, encyst i.e. secrete a wall, and germinate to form a germ tube that develops into the thallus. The flagella are of two types: (1) **tinsel type**, and (2) **whiplash type**. The tinsel type flagella have a large number of hair like outgrowths (tinsels) on their surface. The whiplash flagellum resembles a whip, having a broader and rigid basal portion and thinner, small, flexible distal region. The flagella originate from a kernel-like granule called **blepharoplast** (also called centriole, kinetosome or basal body) lying in the cytoplasm close to the plasmamembrane. The blepharoplast is connected to the nucleus by a strand, called **rhizoplast**.

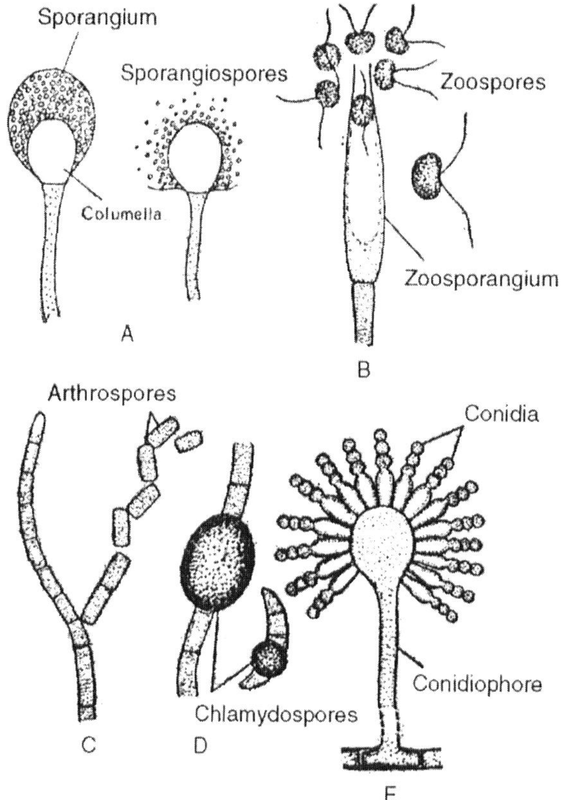

Fig. 3.9. Asexual reproduction through spores. A. sporangiospores; B. zoospores; C. arthrospores; D. chlamydospores; E. conidia.

A cross section of the shaft of the flagellum shows in electron - microphotographs a 9+2 structure (Fig. 3.10), which is a character- istic feature of flagella of all eukaryotic organisms. The flagellum is composed of eleven fibers, 2 **central fibers** covered by a **central sheath**, and 9 separate **outer fibers** in the peripheral region. A membrane surrounds the 11 fibers. Between the outer and central fibers, there are nine thinner **secondary fibers**. Each of the 9 outer fibers is composed of **sub-fibers**. One of the sub fibers of each outer fiber bears two short projections, designated as **arms** which point in the clockwise direction. The two central fibers are connected to the blepharoplast. The membrane surrounding the 9 + 2 fibers is continuous with the plasmamembrane of the spore. In the whiplash flagellum, the two central fibers extend beyond the nine peripheral fibers and give rise to the thin apical piece.

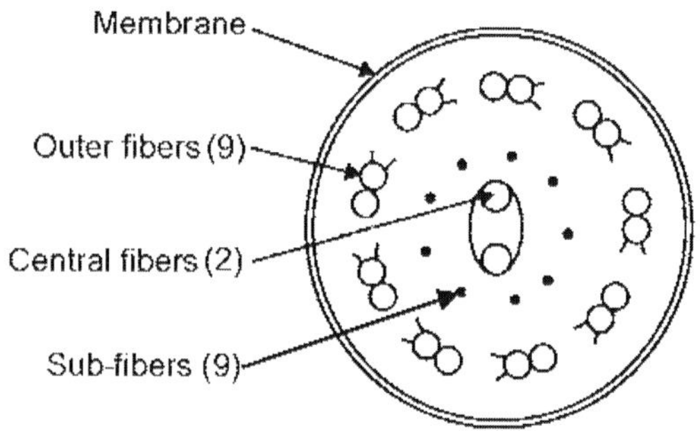

Fig. 3.10. The 9 + 2 structure of eukaryotic flagellum.

The flagella show characteristic insertion on the zoospores and form the basis on which four types of zoospores (Fig. 3.11) are differentiated viz:

1. uniflagellate zoospore with a posterior whiplash flagellum
2. uniflagellate zoospore with an anterior tinsel-type flagellum
3. biflagellate zoospore with two anterior whip-lash flagella.
4. biflagellate zoospore with one flagellum of each type, attached apically or laterally

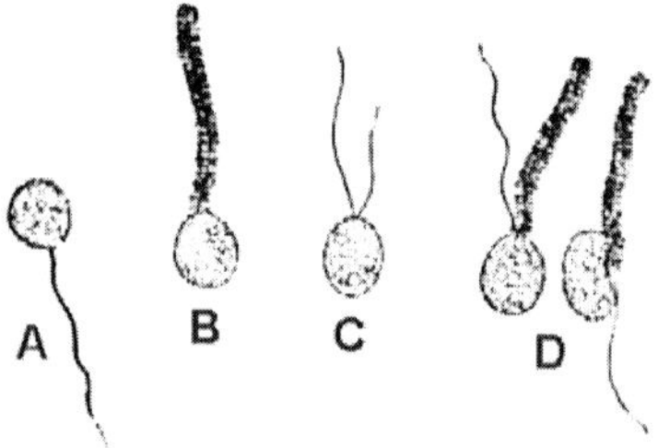

Fig. 3. 11 The four types of zoospores, based on nature and position of the flagella.

Conidia (see Fig. 3.9) Conidia are of two main types:

1. **Thallospores** (e.g. arthrospores, chlamydospores), and
2. **Conidiospores**, or true conidia.

Thallospores are formed by transformation of pre-existing cells of the thallus and are detached by decay of the hypha, or disarticulation of the cells. The **conidiospores**, on the other hand, are formed as new structures on the thallus and are easily detachable.

Arthrospores (syn=oidia) are formed by close septation of the distal end of the hypha in a basipetalous succession (proceeding from apex to the base of the hypha). Both exogenous and endogenous arthrospores are known to exist as shown in Fig. 3.12.

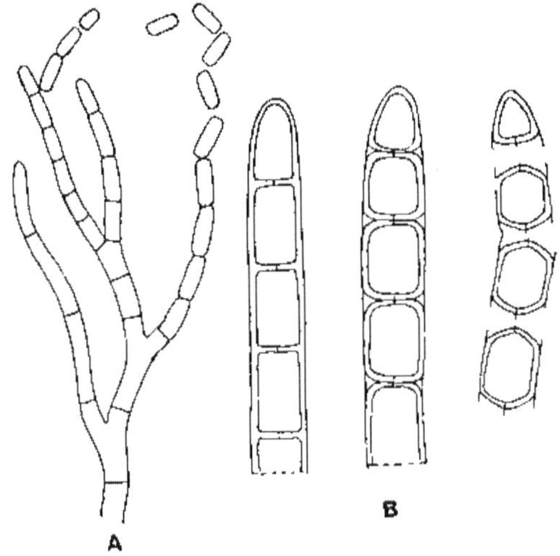

Fig. 3.12. Arthrospores; exo-, and endogenous.

The exogenous arthrospores separate by splitting of the transverse septa. In endogenous formation of arthrospores transverse septa are laid down, as in the exogenous arthrospores, the neighboring or alternate cells develop a thin wall. The intermediate cells lose their contents and the thin lateral walls break to free the arthrospores. Small pieces of hyphal remnants remain attached at both ends of the arthrospores.

Chlamydospores These are thick-walled, resistant spores formed by terminal or intercalary cells of the hypha and are released after its death. Occasionally, chlamydospores may be formed from cells of other kinds of spores e.g. in *Fusarium*, chlamydospores are formed on the macroconidia (see Fig. 3.9). In the formation of chlamydospores, the cells round off, usually enlarge and develop a thick, often coloured wall. The cytoplasm becomes dense with enough reserve food for consumption during unfavourable environment. Chlamydospores are resistant to lysis chemicals produced by other soil organisms. It has been reported that chlamydospores, formed on the macroconidia of *Fusarium*, survive under conditions in which conidia and the hyphae are quickly lysed.

Conidiospores (true conidia) These are most common in *Ascomycota* and anamorphic fungi. In the 'lower' fungi some genera form true conidiospores, while in some others e.g. *Albugo* and downy mildews, the sporangia, in dry weather, behave like conidia and germinate directly through a germ tube, rather than by forming zoospores. Such fungi are considered transitional forms between aquatic and terrestrial fungi, showing the possible evolution of conidia from the sporangia. Conidia, which may be uni- or multicelluar, display variety of shapes and are really thrilling to look at by virtue of their beauty and colour. The conidiophores may be free or aggregated to form compound sporophores viz. **synnemata**. (sing. **synnema**) or **sporodochia** (sing. **sporodochium**). The conidiophores of a large number of fungi are borne inside structures called conidiomata (sing. **conidioma**) which may be saucer-shaped **acervuli** (sing. **acervulus**) or flask-shaped or globuler **pycnidia** (sing. **pycnidium**) opening by an ostiole. Details about these structures will be studied in chapters on the *Ascomycota* and anamorphic fungi (Chapter 9 and 13), where the conidia have reached their zenith.

Sexual Reproduction

Sexual reproduction involves three principal events: **plasmogamy**, **karyogamy** and **meiosis**, occurring in a cyclic manner. Plasmogamy is the fusion of two sexual cells – a process, which results in coming together of compatible, sexually differentiated nuclei in a single protoplast, in preparation for nuclear fusion. Karyogamy (fusion of nuclei) results in the production of a diploid zygote nucleus. Meiosis promotes genetic recombination through random assortment of chromosomes or crossing over, resulting in the production of genetically different individuals.

The sexual cycle of fungi may vary due to variation in timings of the above events. In many ascomycetes and most basidiomycetes, the plasmogamy and karyogamy are separated in space and time. The compatible nuclei do not undergo karyogamy immediately after plasmogamy, but remain paired, and called a **dikaryon**. The dikaryon may persist for a varying period by conjugate divisions of the two nuclei. This nuclear phase in the life cycle is called the **dikaryophae**.

Some fungi produce morphologically distinct male and female sex organs, while in others e.g. *Mucor*, these are indistinguishable. Most of the basidiomycetes do not produce any morphological sex organs and the sexual function is taken up by cells of the somatic hyphae. Sex organs of fungi are called gametangia. These may form sex cells, the male and female gametes. The male gametangium is called the antheridium and female, the oogonium. The motile male gamete is called **antherozoid** (sperm) and the female gamete, contained inside the oogonium, the egg or oosphere. Morphologically similar gametangia and gametes are called isogametangia and isogametes, respectively.

Plasmogamy There are five basic modes which bring the compatible nuclei in juxtaposition (Fig. 3.13). These modes of sexual fusions are often referred to as methods of sexual reproduction.

1. **Planogametic copulation**. This involves the fusion of two naked, free gametes, one or both of which may be motile. The motile gametes are called planogametes. Depending on size and motility of the fusing gametes there are three types of planogametic copulation - **isogamy, anisogamy** and **heterogamy (=oogamy)**. When the two fusing gametes are of the same shape and size, they are called isogametes and their fusion isogamy. In anisogamy, which occurs only in one genus-*Allomyces,* a chytrid, the two planogametes are morphologically similar but different in size. Heterogamy involves the fusion of a motile male gamete with a non-motile female gamete (oosphere, contained in an oogonium). The motile gamete, the antherozoid, enters the oogonium and fertilizes the egg or oosphere. Heterogamy (=oogamy) occurs only in one genus, *Monoblepharidomycetes*.

2. **Gametangial contact**. In this type of plasmogamy, the male gamete is usually represented by the nucleus contained inside the antheridium, while the female gamete is represented by the egg contained in the ooginium. When the two gametangia come in contact, the male gametic nucleus (or nuclei) migrates into the oogonium either through a pore dissolved at the point of contact or through a fertilization tube especially developed for the purpose by the male gametangium. The two gametangia do not fuse but retain their identity. The oogonium undergoes post-copulation changes while the antherifium usually disintegrates. For terms, ooplast, oosphere and oospore see *Oomycota* (Chapter 20).

3. **Gametangial copulation**. The entire contents of the two gametangia fuse and become one. It occurs in two ways:

   (a) **Direct fusion of gametangia**. As exemplified by *Mucor*, *Dipodascus*, *Eremascus* etc, the two gametangia fuse and become one cell.

   (b) **Migration of entire protoplast of one gametangium into the other through a pore**. The entire gametangia act as gametes. The recipient gametangium is called the female, while the gametangium that empties its contents is the antheridium. Example - *Rhizophydium* and some holocarpic chytrids.

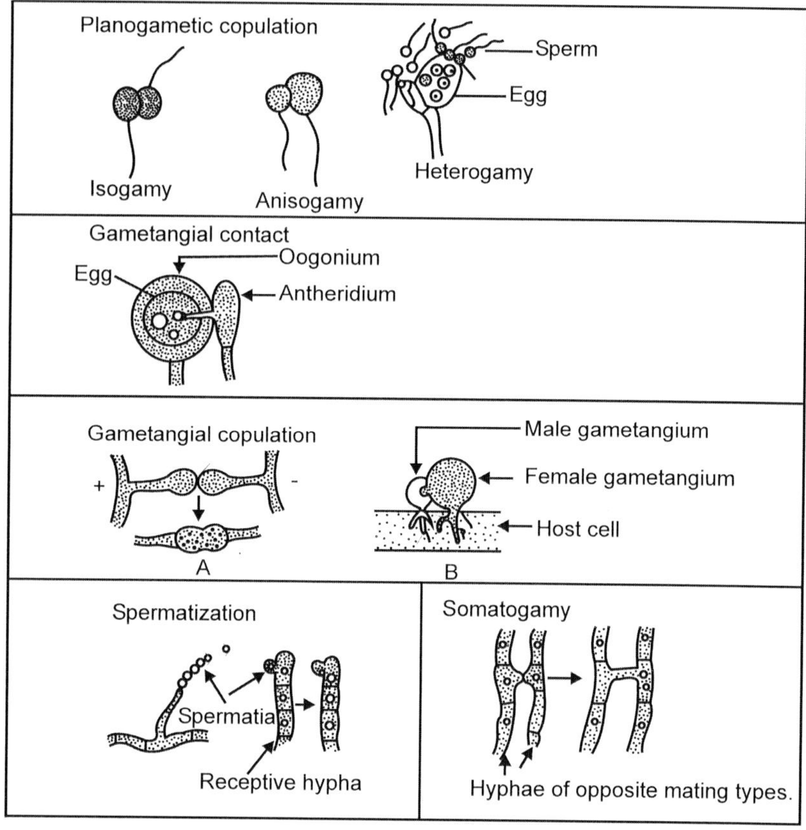

Fig. 3.13. Sexual reproduction: various types of plasmogamy.

4. **Spermatization**. This occurs in some ascomycetes (e.g. *Neurospora*) and basidiomyctetes (e.g. *Puccinia*). Minute conidia-like, male gametes, called **spermatia** (sing. *spermatium*) are produced on **spermatiophores**, borne externally on hyphae (spermatiophore) or inside cavities, called **spermogonia** (sing. spermogonium). On contact with the female organ, the spermatia empty their contents through a pore. The female organ may be a female gametangium, a specialized receptive hypha or even a somatic hypha. The spermatia are carried to the female organ through wind, water, insects etc.. The spermatia differ from conidia in being smaller and their incapability to germinate and give rise to a new thallus.

5. **Somatogamy**. Fusion between undifferentiated somatic cells or spores is called **somatic copulation** or **somatogamy**. In mycelial forms, **anastomosis** (fusion between somatic hyphae) a phenomenon common in higher fungi but absent in lower fungi, is followed by reciprocal nuclear migration; each mate fertilizing the other. The compatible nuclei usually start a dikaryotic association for an indefinite period prior to karyogamy.

Karyogamy After coming together, the two compatible nuclei may undergo fusion (karyogamy) immediately to give rise to a diploid (2n) nucleus or may start a dikaryotic (n+n) association (dikaryophase), extending for varying periods in different fungi. The two nuclei, eventually, fuse in asci or basidia. In *Basidiomycota*, the dikaryophase is extended through dikaryotic spores, like the aeciospores, urediniospores, and teliospores.

Meiosis The diploid nucleus in the zygote may undergo meiosis immediately or after an interval. Rarely, the diploid nucleus in some fungi e.g. *Allomyces* and members of *Plasmodiophorales*, undergoes mitosis forming a diploid phase in the life cycle.

In many lower fungi, the diploid zygote undergoes a resting period for varying duration therefore called the 'resting spore'. In such fungi sexual reproduction takes place when conditions become unfavourable for growth. The resting spores serve for survival. On the return of favourable conditions, the resting spores germinate indirectly, forming zoospores or directly through a germ tube. Meiosis occurs during the germination.

The meiosis in *Asco-*, and *Basidiomycota* occurs in characteristic structures, the asci and the basidia. The products of meiosis are the meiospores viz. ascospores and basidiospores, which on germination, give rise to the haploid mycelium.

SPOROCARPS

Any spore-bearing organ, which produces spores, is called a sporocarp. The majority of asco- and basidiomycetes produce sporocarps of various sizes and shapes, called **ascoma** and **basidioma** respectively. Sporocarps are not formed by aquatic fungi. The water provides a supporting medium and the spores are well adapted to dispersal by water. The thallus is usually small and composed of a loose mycelium, which tends to break easily in the turbulant, unstable water medium. The asco-and basidiomycetes, on the other hand, grow in situations where spore-bearing structures have to be raised above the substratum for efficient liberation and dissemination of spores. Also, the spores have to be protected from desiccation. Nevertheless, many higher fungi produce conidia on simple sporophores, borne directly on the hyphae and are efficiently disseminated. Although the morphological varieties of the sporocarps are staggering, they show striking similarities in microscopic features and tissue differentiation.

HABITAT

Fungi are ubiquitous as saprobes, symbionts, parasites or hyper-parasites and are found in any conceivable habitat. They are versatile by virtue of their high degree of adaptability. Anything that can be decomposed to yield energy will find some fungi to colonize it and, therefore, it is difficult to enumerate their numerous habitats. It has to be remembered that habitat is not only substratum but includes other features also. For example, it develops its characteristic community through the process of selection, involving factors originating from fungi and other colonizing organisms, e.g. antagonism.

Aquatic fungi

Fungi are found in all kinds of aquatic habitats (freshwater and marine), growing both in open waters (pelagic) and at the bottom (benthic zones). They live as saprobes or parasites of plants and animals. Saprobic fungi grow on decaying plant and animal remains, submerged branches, twigs, fruits, leaves etc. The parasitic fungi greatly influence the ecology of aquatic habitats by causing sudden epidemics of phytoplanktonic algae and other populations, like fishes. Unicellular algae are most susceptible to these attacks.

Terrestrial fungi

The terrestrial fungi can be conveniently classified into several ecological groups based on their habitats, e.g. soil fungi, symbionts and parasites.

Soil fungi. In 1917 Waksman asked "Is there any fungus flora of the soil"? It was not known then whether the fungi, isolated from the soil, were the bonafide residents or temporary intruders. Today, we know that the soil is not an inert medium but a site of great microbial activity, and that the soil fungi outnumber the fungi that grow elsewhere. Fungi of all groups, from chytrids to agarics, and of all kinds – saprobes, parasites, symbionts, predators etc., occur in the soil. Soil fungi are classified into two categories 1. **soil inhabitants**, and 2. **root inhabitatnts**, depending on their **competitive saprobic ability**, a concept given by Garrett. The former are the free-living soil saprobes, while the latter have limited saprobic ability.

The soil-inhabiting fungi, on the basis of the substrate utilized, are classified into several substrate groups e.g. sugar fungi, cellulose decomposing fungi, lignin-decomposing fungi and some special substrate groups, like coprophilous fungi (growing on dung), predaceous fungi, entomogenous fungi etc.

Symbionts

Fungi form two important symbiotic associations, called **lichens** and **mycorrhiza**.

Lichens Some fungi (mostly ascomycetes) and some algae or cyanobacteria, which fail to grow independently, enter into a symbiotic relation and form a thallus, which is different from either of the components. It is not a casual association between two hapless organisms, but a product of evolution, which is prevented from being lost through propagules containing both fungal and algal elements like **soredia** and **isidia**. Lichens are first colonizers of rocks, and survive extremes of climatic conditions. Because of their sensitivity to air pollution, their study may be useful as "an early warning system".

Mycorrhiza (fungus root) is a symbiotic association between fungi and roots of higher plants. All saprobic higher plants, a number of mono and dicotyledons, conifers, some pteridophytes and bryophytes are mycorrhizal.

Fungal Parasites

Fungi include well-known parasites of plants, animals and human beings. Rust and smut diseases of plants have been known since the recorded history. Ainsworth (1968) remarked that "the fungal diseases of plants are of greater economic and social importance than fungal diseases of man and animals. No parallel is found among mycotic diseases of man and animals to black rust of wheat and potato blight, which are the equivalents of such major human diseases, as tuberculosis and Malaria".

Fungal Hyperparasites

These are parasites of other fungal parasites and are found in all classes of fungi. *Ampelomyces quisqualis*, an anamorphic fungus, is parasitic on the powdery mildew fungi, and *Endarlucka caricis*, an ascomycete, is parasite of many rust fungi.

Fungi in Atmosphere (Air spora)

Air does not serve as a habitat. Only, fragments and spores of terrestrial fungi adapted for aerial dispersal, constitute the "air spora". The composition of the air spora is governed only by physical factors of the moving air and not by any nutritional factors. Fungi in the atmosphere are mostly present in the **troposphere**, up to a height of 10 km and are rarely found in the **stratosphere**.

DISTRIBUTION AND NUMBERS OF FUNGI

Fungi are cosmopolitan in distribution. The geographical distribution of fungi has not been worked out well. Distribution of the known fungi is related to the distribution of the mycologists, who explore them. The total number of known phanerogams is about twice the number of the reported fungi, though in areas, where fungi have been explored, the situation is the other way round. The number of fungal species invariably exceeds the phanerogams.

The problem of the number of fungi has many semantic problems, regarding the individual fungi, kinds of fungi and names of fungi. Then, there are doubts about the bonafides of the genera and the species. According to Kirk *et al.* (2008) there are 97,330 species of fungi. But there has been a continuous upward trend during the last 50 years in the number of described species. Bisby and Ainsworth (1943) estimated that there are at least 100,000 species, which was considered an "exceptionally conservative estimate" by Martin (1951). Martin suggested that the number is at least as great as the number of phanerogams, then believed to be not less than 2,50,000. Based on extrapolations of three independent data sets (ratio of number of fungi in all habitats to plants, numbers restricted to particular habitats, and a community studied in depth), Hawksworth (1991) suggested that a conservative working figure of 1.5 million species to be adopted. This 1.5 million estimate has been supported by subsequent data. Currently, this figure remains the most widely accepted (Heywood, *Global Biodiversity Assessment,* 1995). The gap between described and estimated number is big. But, new species are regularly been found in all countries of the world.

FUNGAL BIODIVERSITY

Biological diversity is defined as the *variety and value of life on earth at genetic, organism and ecological level.* From these view points fungal biodiversity is no less than of animal and plants. But, keeping in view the enormous gap in the number of discovered and estimated species (only 97,330/1.5 million), the fungal diversity is highly under estimated. There are no "hot spots' known for fungi. Many ecosystems remain unexplored from fungus point of view. The **UN Convention on Biological Diversity** (1992) though includes all biodiversity, proper attention has not been given to fungi in preparation of inventories for biodiversity prepared by different countries. Studies on fungal diversity, its conservation and sustainable use need more attention.

A GENERALISED LIFE CYCLE OF FUNGI

The fungi are mostly haploid organisms, having haploid nuclei in their cells. The thallus reproduces itself by asexual methods (Fig. 3.14). Eventually, through sexual reproduction, a diploid zygote is formed. The zygote represents the diploid phase. This phase is terminated by meiosis forming haploid spores. These spores, on germination, form the haploid somatic phase.

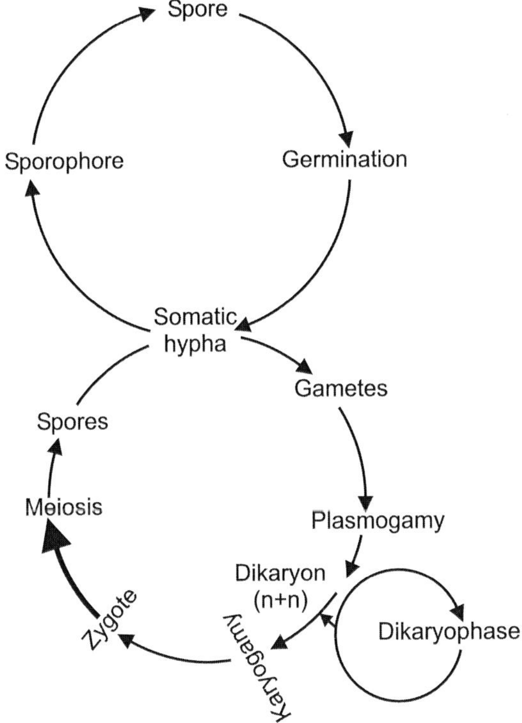

Fig. 3.14. A generalized life cycle of haploid fungi. The relative lengths of these stages may vary from fungus to fungus.

In *Ascomycota* and *Basidiomycota*, plasmogamy is not followed by karyogamy. The two nuclei of opposite mating types remain as paired nuclei called **dikaryons**, which increase in number by simultaneous or conjugate divisions and give rise to dikaryotic hyphae. They are of limited duration in *Ascomycota*. Only a small portion of the mycelium viz., the **ascogenous hyphae**, are dikaryotic, but in *Basidiomycota*, these extend through the major portion in the life cycle. The monokaryotic mycelium is of very short duration and its function is only to establish the dikaryotic mycelium, as soon as possible. Thus, the dikaryophase, which is nonexistent in coenocytic fungi, and present for short duration in *Ascomycota*, are extensive in the *Basidiomycota*.

REFERENCES

Ainsworth, G.C. 1968. The number of fungi. In: *The Fungi. An Advanced Treatise*, (G.C. Ainsworth and A.S. Sussman, eds.). Vol. III. pp. 505-514. Academic Press, New York and London.

Ainsworth, G.C. 1973. Introduction and keys to higher taxa. In: *The Fungi. An Advanced Treatise*. (G.C. Ainsworth, F.K. Sparrow and A.S. Sussman eds.). Vol. IVA. 1–7. Academic Press, New York, London.

Bisby, G.R. and Ainsworth, G.C. 1943. The numbers of fungi. *Trans. Brit. Mycol. Soc.*, 26: 16–19.

Moore, R.T. 1964. Fine structure of mycota. 12. Karyochorisis—somatic nuclear division in *Cordyceps militaris Z. Zellforsch*. 63: 221–237

Raper, J.R. 1954. Life cycles, sexuality and sexual mechanisms in the fungi. In: *Sex in Microorganisms* (D.H. Wenrich, I.F. Lewis and J.R. Raper eds.). *Amer. Assoc. Adv. Sci.* Washington. pp. 42–81.

Pontecorvo, G. and Roper, J.A. 1952. Genetic analysis without sexual reproduction by means of polyploidy in *Aspergillus nidulans*. *J. Gen. Microbiol*, 6: VII (Abstr.).

Waksman, S.A. 1917. Is there any fungus flora of the soil? *Soil. Sci.*, 3: 565–589.

SUMMARY

Fungi are a heterogeneous group of organisms placed in 3 kingdoms: *Fungi*, *Chromista* and *Protozoa*. These are characterized by their nutrition, thallus, cell wall, nuclei, life cycle, sexuality and habitat. Nutrition is absorptive and ingestion is rare (in slime molds by phagotrophy. Thallus is cellular, mycelial or plasomodial. The hyphae form tissues and well expressed in somatic structures (rhizomorphs, sclerotia and stroma) and the fruit bodies of *Asco-*, and *Basidiomycota*. Nuclei are very small, lying near the power of resolution of the light microscope. The karyology of fungi is, therefore, poorly understood. There are seven types of life cycles. Reproduction is asexual (fragmentation, budding, fission, spores and conidia) and sexual (by gametes or gametangia). Sex organs are gradually replaced by anastomosis between heterothallic compatible hyphae in asco-, and basidiomycetes. Parasexual phenomenon is well known in several organisms, but its extent and role in development of new species in nature is not fully assessed.

The biodiversity of fungi is not fully explored, as only 97,330 species of 1.5 million estimated species are presently discovered. Many ecosystems are not yet explored from fungus point of view.

TEST QUESTIONS

1. What are the main characteristics of fungi? Describe the nutrition, thallus and cell wall of fungi
2. Describe the various types of asexual spores in fungi.
3. Answer briefly:
   a. What is 9+2 structure of flagellum?
   b. Name the four types of zoospores, based on location of flagella.
   c. Name the seven types of life cycles in fungi.
   d. What are the two types of conidia?
   e. Name the three types of conidiomata.
   f. What are the two categories of soil fungi?
   g. Why much information is not available on fungal biodiversity?
   h. Name the three types of planogametic copulaltion.
   i. What are the two types of arthrospores?
4. Distinguish between:
   a. Adventitious and primary septa.
   b. Prosenchyma and pseudoparenchyma.
   c. Mitospores and meiospores.
   d. Sporangiospores and conidia.
   e. Tinsel and whiplash flagellum.
   f. Exogenous and endogenous arthrospore
   g. Gametangial contact and gametangial copulation.
5. Fill in the blanks:
   a. Fungal nutrition is -------------.
   b. Fungal tissues are called ----------------.
   c. Diploid life cycle occurs in ------- -------- ------- .
   d. Haploid life cycle is found in ------------

e. Flagella originate from ------------ lying in the cytoplasm.

f. Rhizoplast connects ---------- to the -----------.

g. The number of described species in fungi is ----------, while the estimated number is -------.

6. Write notes on :

(i) Ultrastructure of flagellum, (ii) Somatogamy,

(iii) Number of fungi, and (iv) Fungal biodiversity

4

Kingdom *Fungi*

Phylum *Chytridiomycota*

Flagellated, Zoosporic *Fungi* I
(Chytrids, Core Cytrids or True Chytrids)

Chytridiomycota are the smallest, simplest and most primitive fungi. They emerged soon after Precambian period, and are ancestors to all fungi. The name chytrid is derived from 'chytridion' meaning 'little pot'. The distinguishing feature of this phylum* is the production of motile zoospores having a single posterior whiplash flagellum, with unique flagellar 'root' system. The other characters, though not unique to chytrids are: (1) *The coenocytic thallus*, which may be an oval multinucleate cell or a small elongated hypha, and (2) *the zygote,* which is converted into a resting spore or a resting sporangium. There are two classes, *Chytridiomycetes* and *Monoblepharidomycetes*.

The thallus may consist of a simple spherical structure (often without cell wall in endoparasites), which is entirely converted into a reproductive organ. Such a thallus is called **holocarpic**. When, on the other hand, only a portion of the thallus is used in the formation of reproductive structure, and the remaining portion performs vegetative functions, the thallus is called **eucarpic**. From the vegetative part, fine thread-like, distally-tapering structures, called **rhizoids**, arise which serve to anchor the unicellular thallus to the substratum and also absorb food. Some species produce a network of rhizoids called **rhizomycelium** which is usually non-nucleated.

---

* The presence of flagellate zoospores suggests inclusion of these fungi under kingdom *Chromista*, but are retained in *Fungi* on (1) ultrastructural (flattened mitochondrial cristae), (2) chitinous cell wall, and (3) molecular phylogenetic considerations.

The rhizomycelium, in such fungi, functions as the somatic part which bears one or more reproductive structures. If the thallus consists of rhizomycelium and a single reproductive structure, the thallus is called **monocentric**; when the reproductive structures are many, it is **polycentric**. These reproductive structures are never produced in a tuft at one place but lie scattered and are connected by the rhizoidal hyphae.

The thallus may be **endobiotic** (growing inside a host cell), **epibiotic** (growing attached externally but sending rhizoids inside host cells), or **interbiotic** (attached to many hosts) through the rhizomycelium.

Reproduction

Both asexual and sexual reproductions are found. Asexual reproduction takes place by means of zoospores. The whole thallus, or a part of it, is converted into a zoosporangium. Zoospores, formed inside the sporangium, escape through one or more discharge papillae or orifices, formed on the sporangial wall or at the tip of discharge tubes arising from the sporangium (Fig. 4.1). In some species the discharge papillae form a distinct circular, lid-like structure, called the **operculum**, which is pushed out, or thrown off, during the release of the zoospores. Species forming operculum are called as **operculate**. In most of the species the discharge papilla dissolves forming a pore. Species which release the zoospores through such a pore are called **inoperculate**.

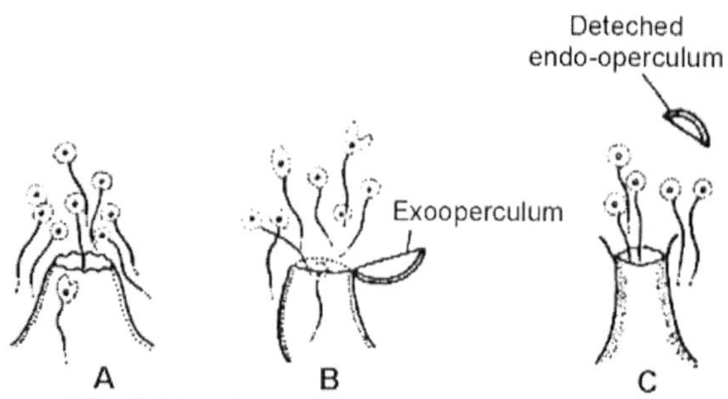

Fig. 4.1. Modes of dehiscence of zoosporangia.

The Zoospore

The flagellar apparatus of zoospores of *Chytridiomycota* shows a pair of kinetosomes, one associated with the flagellum while the other is a vestigial remnant called dormant or non-functional kinetosome. (Also called flagellar-, and non-flagellar kinetosomes). This suggests the possible origin of the chytrids from biflagellate ancestors by losing one flagellum. The flagellum, about 15 mm long, shows the typical eukaryotic 9 + 2 structure with its enclosing sheath that is continuous with the plasmamembrane of the zoospore body. The fibrils are connected to the kinetosome.

The ultrastructural characters of the zoospores, which have been studied extensively, provide a reliable criterion for the demarcation of the class into orders. The characters useful in defining the orders are: (1) the scattered or aggregated nature of the ribosomes, (2) the number and location of lipid bodies, (3) organization of Microbody-Lipid globule Complex (MLC) in association with mitochondria and membrane cisternae, (4) orientation of the dormant kinetosome in relation to the kinetosome, (5) organization of microtubules and in some cases fibrillar structures, called **rhizoplasts**, and (6) last, but most importantly, the spatial relationship of the nucleus and kinetosome.

Sexual Reproduction

Sexual reproduction may be brought about by any of the following methods.

1. *Planogametic copulation*-(isogamy, anisogamy or heterogamy). Heterogamy (= oogamy) involves the fusion of a motile male gamete with a non-motile female gamete (oosphere, contained in an oogonium). In *Monoblepharidomycetes*, heterogamy is the exclusive mode of plasmogamy.
2. *Gametangial copulation.*
3. *Somatogamy*, by fusion between rhizoidal filaments.

The diploid zygote, formed after plasmogamy and karyogamy, is converted into a resting spore or resting sporangium. Zygotic mitosis is characteristic of *Chytridiales*. Resting spores are also formed asexually.

Isolation and Culture

The saprobic chytrids of soil and water can be isolated easily by simple baiting techniques. Soil or water samples are put in Petri dishes and covered with sterile water. The "bait", such as snake-skin casts, decalcified shrimp skeleton, defatted hair, pollen grains, boiled cellophane, split boiled hemp or mustard seeds etc., is then added. Chytrid zoospores, if present in the water or soil sample, grow on these substrates and give rise to mature thalli, which can be taken in culture.

Classification (Kirk *et al.*, 2008)

There are two classes, *Chytridiomycetes* and *Monoblepahari-domycetes*. The thallus is simple, unicellular and spherical in *Chytridio-*

mycetes, while it is mycelial in the latter. The *Chytridiomycetes* has three orders, which are further divided in families based on thallus construction, zoospore ultrastructure, mode of plasmogamy and resting spores. The three orders are *Chytridiales* (4 families), *Rhizophydiaceae* (three families) and *Spizellomycetales* (three families). The class *Monoblepharidomycetes* has one order *Monoblephaidales* and four families. An outline of the classification up to families is given below:

Phylum *Chytridiomycota* (2 classes)

Class *Chytridiomycetes*
(3 orders)

Chytridiales (4 families)

- *Chytridiaceae*
- *Cladochytriaceae*
- *Endochytriaceae*
- *Synchytriaceae*

Rhizophydiales (3 families)

- *Kappamycetaceae*
- *Rhizophydiaceae*
- *Terramycetaceae*

Spizellomycetales (3 families)

- *Caulochytriaceae*
- *Olpidiaceae*
- *Spizellomycetaceae*

Class *Monoblepaharidomycetes*
(1 order)

Monoblephaidales (4 families)

- *Gonapodiaceae*
- *Harpochytriaceae*
- *Monoblepharidaceae*
- *Oedogoneomycetaceae*

Class *Chytridiomycetes*

Chytridiomycetes is the major class of the phylum. The fungi included in this class are usually parasites or saprobic generally in water and also in soil. As parasites, they grow internally in algae, microfauna and water molds. Frequent destruction of unicellular aquatic algae is caused by these chytrids. They parasitize the terrestrial higher plants (*Synchytrium endobioticum* on potato), but it is rare. As saprobes, they grow on pollen, plant debris, chitin and keratin. *Batrachochytrium dendrobatidis*

(*Spizellomycetaceae*) infects amphibians (frogs) causing "chytridiomycosis' which appears as thickened epidermis due to hyperkeratosis. The skin, which gets discoloured and rough, peels off. It is a fatal disease resulting in decline of the amphibian population.

Order *Chytridiales*

The thalus is endobiotic or epibiotic, monocentric or poly-centric, and the vegetative parts, when present, are rhizoidal; never mycelial. The zoosporangium may be operculate or inoperculate. Germination of the zoospore is unipolar. The resting spore fills the cavity of its container. On germination, it functions as a sporangium or a prosporangium.

The ultrastructural characters of the zoospore (Fig. 4.2) include aggregation of the ribosomes(R) in the center and loosely bound by endoplasmic reticulum. Closely associated with this are the nucleus(N), MLC—Microbody-Lipid (L) globule Complex, two mitochondria and rumposome (tooth-like in section and honey-comb like in surface view; an organelle required for flageller activity). Rumposome is also found in the zoospores of *Mooblepha-ridomycetes*.

It also shows the non-functional kinetosome connected to functional kinetosome and a microtubule extending from one side of the kinetosome into the cytoplasm.

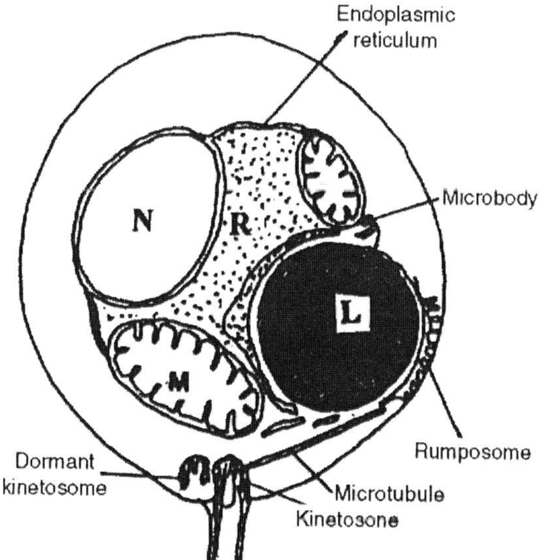

Fig. 4.2. Ultrastructure of zoospores in *Chytridiales*

Family *Chytridiaceae*

The thallus is monocentric. The sporangium develops inside the zoospore cyst; operculate or inoperculate; rhizoids endo- or exogenous. There are 33 genera. We shall study the genus *Polyphagus*.

Genus *Polyphagus*

The zoospores are released from the sporangium through an apical pore. Resting spores are formed sexually in a conjugation tube formed between two mating thalli. The important species are *P. euglenae* (parasite on encysted *Euglena*, a protozoan, and *Chlamydomonas*) and *P. parasiticus* on *Tribonema*.

P. euglenae is found abundantly in foul ditches and sewage, parasitizing simultaneously (interbiotic) on several encysted euglenae or other unicellular algae.

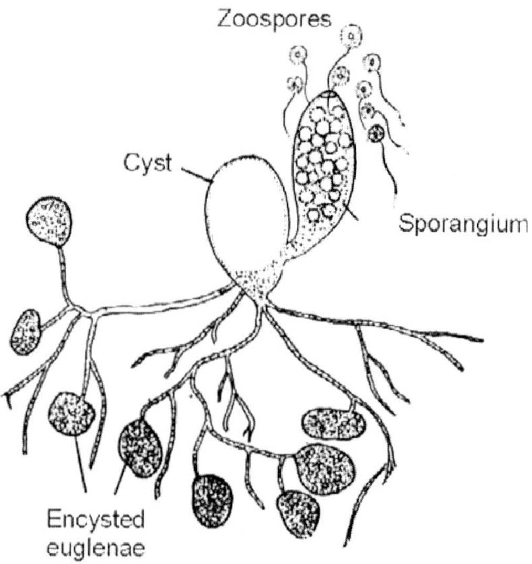

Fig. 4.3. *Polyphagus euglenae*: habit of mature vegetative thallus, attached to several euglenae and bearing a dehiscing sporangium.

The posteriorly uniflagellate zoospores settle down in still waters (not directly on any substratum), secrete a wall and form richly-branched rhizoids which extend in all directions (Fig. 4.3. On coming in contact with encysted euglenae, the rhizoid tips penetrate and parasitize on them. As many as 50 euglenae are parasitized simultaneously by the extensive branched rhizoids.

The enlarged body of the encysted zoospore (the cyst) is the core of the thallus. After a certain period of growth, asexual reproduction starts. The cyst functions as a prosporangium and gives rise to a lateral outgrowth which develops into the sporangium. The contents of the prosporangium are emptied into the outgrowth. The empty prosporangium is cut off from the sporangium by a septum. After many nuclear divisions, the cytoplasm is cleaved into hundreds of uninucleate portions which develop into posteriorly uniflagellate zoospores. A pore is dissolved at the apex of the sporangium, through which the zoospores escape and swim away.

Sexual reproduction, by gametangial copulation, occurs when conditions become unfavourable for growth of the fungus. Two thalli, consisting of the enlarged body and rhizoids, take part in the sexual reproduction (Fig. 4.4). The enlarged bodies (cysts) act as gametangia. The smaller cyst acts as the male gametangium and the bigger one as the female gametangium. The male gametangium puts out a slender lateral filament, the conjugation tube, which, on contact with the female gametangium, enlarges at the tip.

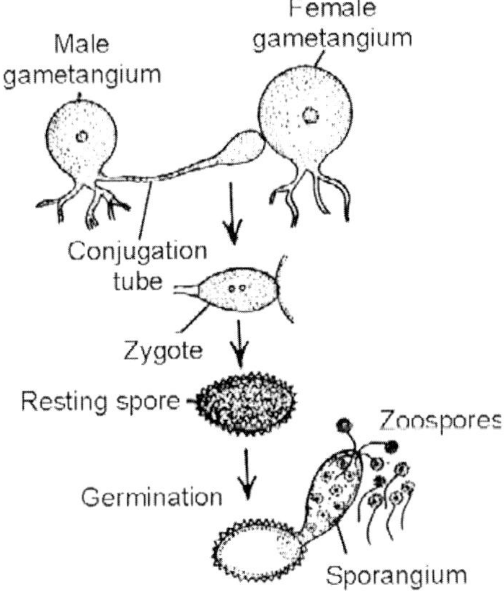

Fig. 4.4. *P. euglenae*: stages in sexual reproduction

The contents of the male gametangium migrate into the enlarged tip of the conjugation tube, wherein flows the protoplast of the female gametangium through a pore, dissolved at the point of contact. Connections with the conjugating thalli are severed. The wall becomes thick and the zygote transforms into a resting spore. Karyogamy does not occur at this stage. The resting spore lies dormant until the environment conducive for growth returns. During germination of the resting spore, a vesicle comes out of the resting spore through a pore carrying the protoplast. The two nuclei, which had remained apart until now, undergo fusion and a diploid nucleus is formed. Meiosis follows and the haploid nuclei, thus formed, are multiplied by repeated mitotic divisions. Finally, zoospores, each containing a single haploid nucleus, are formed. These are liberated through an apical pore on the sporangium.

Family *Synchytriaceae*

The five genera recognized in the family, are endobiotic, holocarpic parasites. The thallus is divided into several sporangia (or gametangia), which are covered by a common membrane and, therefore, called a sorus, e.g. *S. taraxaci*. In some species e.g. *S. endobioticum*, the thallus, instead of forming the sorus by itself, may pass its contents into a vesicle which later forms the sorus. Thus, the thallus behaves as a prosorus. In still other cases, the thallus instead of transforming into a sorus or a prosorus, may form a thick-walled resting spore. There are 5 genera. We shall study only one genus, *Synchytrium*.

Genus *Synchytrium*

Synchytrium is the largest genus, which induces galls on vascular plants. There are 120 species which are all plant parasites. *S. taraxaci* causes galls and hypertrophy in leaves of *Taraxacum offcinale* (dandelion).

S. endobioticum causes the black wart disease of potato, which is reported from all the potato-growing areas of the world, especially areas having cool, and moist climate. In India, the disease is restricted to Darjeeling district of West Bengal and adjoining areas of Nepal. Strict quarantine measures have prevented spread of the disease to plains where the disease can occur in the cooler parts. The fungus cannot survive in hot areas.

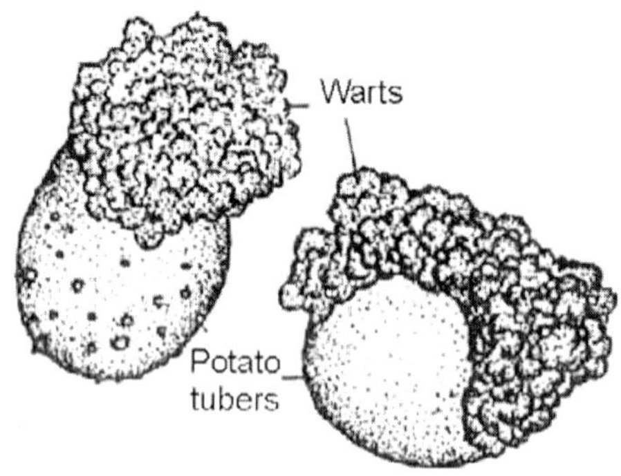

Fig. 4.5. Showing symptom of 'Wart disease' of potato

Dark brown or black cauliflower-like outgrowths (Fig. 4.5) appear on the tubers. On shoots, the outgrowths appear as green, convoluted leafy structures. This fungus serves as a vector of potato virus X (PVX) and potato mop top virus.

Order *Rhizophydiales*

The order was created in 2006 by Letscher *et al.* (2006) and James *et al.* (2006) based on molecular phylogeny. There are only 4 genera distributed among three families; *Kappamycetaceae* (1 genus, *Kappamyces* and 1 spp.), **Rhizophydiaceae** (1 genus, ***Rhizophydium*** with 100 spp), *Terramycetaceae* (2 genera, 2 spp, *Terramyces*).

Genus *Rhizophydium*

The thallus consists of an epibiotic zoosporangium with rhizoids developed to varying degree inside the substratum. The species (100) grow on a variety of living or dead plant and animal substrata. Some of the important species are: *R. sphaerotheca*, on pollen; *R. keratinophilum*, on keratin; *R. planktonicum*, on the diatom *Asterionella* and *R. couchii*, on *Spirogyra* (a green alga).

R. couchii parasitizes *Spirogyra* spp. The swarming zoospores settle on the *Spirogyra* filament (Fig. 4.6), lose the flagellum, encyst and develop a germ tube which enters the algal cell and forms a tuft of fine branched rhizoids. The rhizoids anchor the cyst to the host and also derive nutrition for the cyst. The body of the zoospore cyst enlarges and transforms into a zoosporangium. When the sporangium is mature, the rhizoids degenerate. Some weak spots appear on the sporangial wall, which bulge out as papillae and later dissolve to form pores through which the numerous zoospores are liberated. The zoospores remain motionless for sometime at the mouth of the exit papillae; later they dart away.

Sexual reproduction takes place by gametangial copulation. The gametangia also develop in the same way as do the zoosporangia. When two thalli grow together, one of them grows bigger than the other and acts as the female gametangium (oogonium); the smaller acts as the male gametangium (antheridium). The entire content (protoplast) of the antheridium migrates into the oogonium through a pore, formed at the point of contact between the two gametangia, or alternatively, through a fertilization tube originating from the antheridium. The oogonium develops a thick wall and becomes a resting spore (or resting sporangium). Karyogamy is delayed until the time of germination of the resting spore. A pore is formed on the thick wall of the resting spore through which the protoplast moves out, surrounded by a thin wall, to form a zoosporangium. The zoospores are liberated through an apical exit papilla.

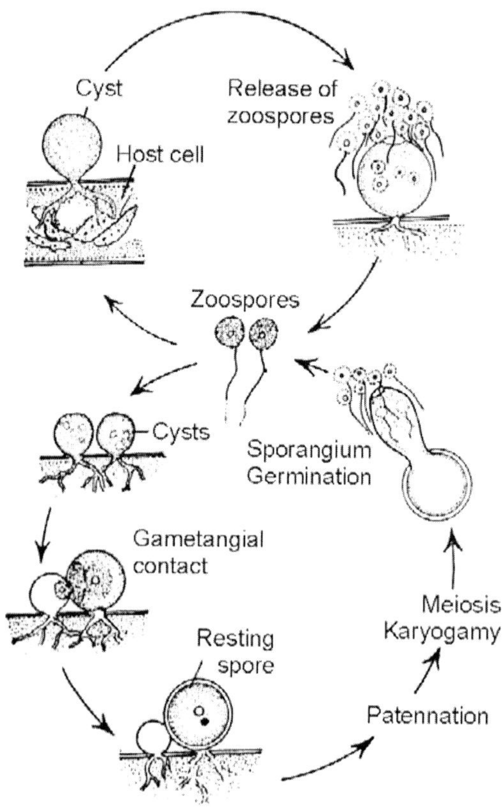

Fig. 4.6. Life cycle of *Rhizophydium couchii*

Order *Spizellomycetales*

The members of this order, like *Rhizophydiales,* earlier belonged to *Chytridiales*, but were segregated into a separate order due to differences in zoospore ulrastructure. The ribosomes are dispersed (not aggregated as in *Chytridiales*), there are several lipid globules and presence of microbodies in the anterior region of the zoospore (Fig. 4.7). The nucleus (N) lies loosely associated with the kinetosome and several microtubules radiate from the kinetosome into the cytoplasm. Another important feature is the amoeboid movement of the zoospores of some members (Barr, 1992). There are three families, *Caulochytriaceae*, *Olpidiaceae* and *Spizellomycetaceae*. The inclusion of family *Olpidiacae* in this order is contested (James *et al.*, 2006) and it is considered *incertae sedis* i.e. has uncertain taxonomic position.

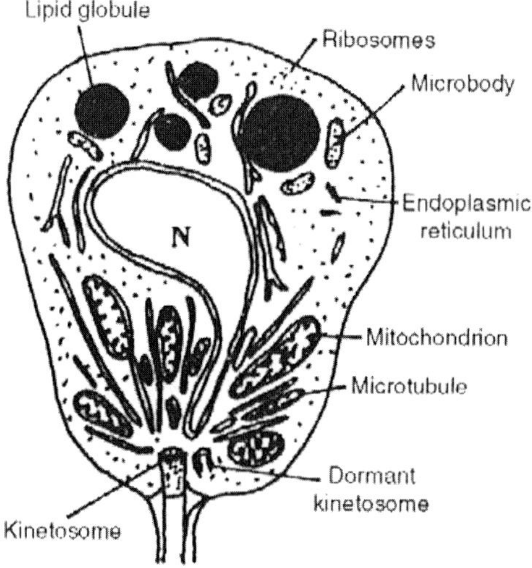

Fig. 4.7. Ultrastructure of zoospores in *Spizellomycetales*

Family *Olpidiaceae*

The thallus is endobiotic (remains inside host cells) and holocarpic i.e. the entire thallus is converted into a zoosporangium or a resting sporangium, which on germination gives rise to zoospores. The sporangium is inoperculate. sexual reproduction is by fusion of two isogametes. The biflagellate diploid zygote penetrates the host. There are 5 genera in the family.

Genus *Olpidium*

About 40 species of *Olpidium* are known which occur in water or wet soils and parasitize algae and roots of higher plants. *O. hyalothecae* is a parasite of the desmid *Hyalotheca*. The important root parasites of vascular plants are: *O. brassicae*, causing damping off of crucifer seedlings and other unrelated hosts, and *O. viciae*, which infects leaves and stems of *Vicia unijuga*, causing chlorosis.

O. brassicae serves as a vector of at least three viruses in soil viz. lettuce "big vein" virus (BVV), tobacco necrosis virus (TNV) and tobacco stunt virus (TSV). The viruses are disseminated by zoospores and also survive in the resting spores for several months. Whether the viruses multiply also in the zoospores is not known.

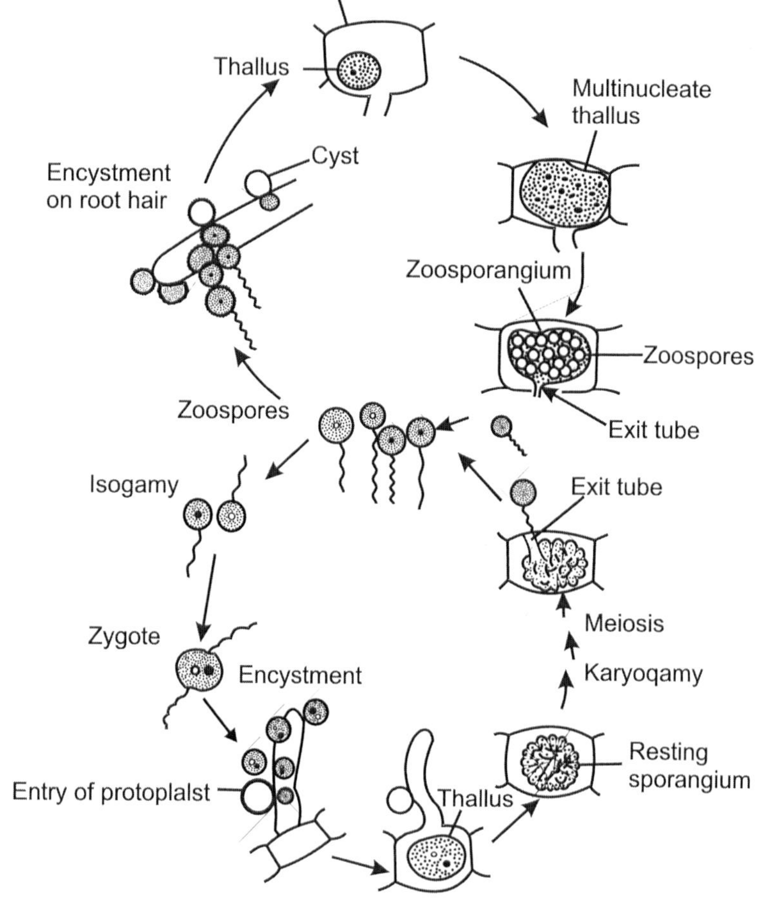

Fig. 4.8. Life cycle of *Olpidium brassicae*.

The posteriorly uniflagellate zoospores swim actively in the film of water. When they come in contact with the roots of the host plant, they settle down on the root hairs (Fig. 4.8), lose the flagellum and encyst i.e. develop a wall. A pore is dissolved on the root hair, the protoplast enters the cell while the cyst is left behind. Inside the cell, by growth and nuclear divisions, it develops into a multinucleate single-celled thallus. In 4 to 5 days the whole thallus is transformed into a zoosporangium. The contents cleave into zoospores. Exit tubes originate from the thallus which penetrate the host cell wall and open outside. The tip dissolves; the zoospores rush out and reinfect the root hairs.

Zoospores also behave as gametes. The fungus exists as two distinct strains, and zoospores produced by thalli of opposite strains only copulate. Zoospores, which are of the same size, copulate outside the host. It is called isogamy. A biflagellate zygote is formed which after swimming for sometime, encysts and infects root in the same way as described for the zoospores. Inside the host cell, the zygote forms a thick-walled resting sporangium, which remains dormant for many months. At the time of germination, the two nuclei fuse and undergo meiosis. Large number of zoospores are formed which are released by exit tubes.

Class *Monoblepharidomycetes*

The class, which includes the highest evolved members of phylum *Chytridiomycetes* are characterized by (1) a well-developed mycelial thallus, (2) a unique type of heterogamy (oogamy) in sexual reproduction, and (3) presence of **rumposome** in the zoospores. The thallus may be extensive or simple and unbranched. They frequently have a holdfast at the base. These are the only chytrids with non-motile female gametes and smaller, motile gametes (antherozoids). *This type of oogamy is not known elsewhere in fungi.*

Resting sporangia are not formed. Presence of rumposomes is a character that the group shares with *Chytridiales*, and based on this, Cavalier-Smith (1987) and James *et al.* (2006) suggested a new class *Rumpomycetes* for *Monoblepharidales* and *Chytridiales* (excluding *Spizellomycetales*)

The zoospores show the ribosomes aggregated around the nucleus, and surrounded by a loose endoplasmic reticulum. There are numerous lipid gloules in the anterior region and several mitochondria in the posterior region of the zoospore. There is a rumposome (tooth-like in section and honey-comb like in surface view) involved in the regulation of fllagellar activities.

The class has one order *Monolepharidales*, four families and five genera: *Monoblepharidaceae* (1 genus, *Monoblepharis*), *Gonapodyaceae* (2 genera, *Gonapodya* and *Monoblepharella*), *Harpochytriaceae* (1 genus, *Harpochytrium*) and *Oedogoniomycetaceae* (1 genus, *Oedogoniomyces*). We shall study two examples: *Monoblepharis* and *Monoblepharella* which represent the typical character of the two families, *Monoblepharidaceae* and *Gonapodyaceae*. In the former, the zygote remains inside the oogonium and encysts within or outside at the top of the oogonium. In *Gonapodyaceae* the zygote propels out and has a period of motility before encystment.

Family *Monoblepharidaceae*

Genus *Monoblepharis*

Monoblepharis (12 spp.) grows in still waters, attached to the substratum by means of rhizoids. It can be isolated by floating pieces of twigs etc., collected from a pond in a Petri dish containing distilled water. Low temperature (about 3°C) favours the development of *Monoblepharis*.

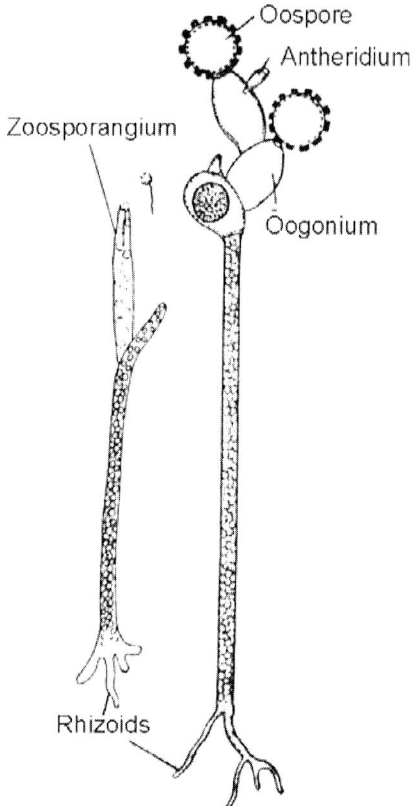

Fig. 4.9. *Monoblepharis polymorpha*

The thallus is hypha-like. The mycelium, arising from a tuft of rhizoids, is variously branched. The cytoplasm is highly vacuolated and has a foamy appearance (Fig. 4.9).

Elongated sporangia, which are of the same diameter as the hypha, are formed at the tips of hyphal branches (Fig. 4.10). The multinucleate cytoplasm is cleaved into uninucleate portions that develop into posteriorly uniflagellate, wall-less zoospores. They are released from the sporangium through an apical pore, and after completing the swimming period settle down on the substratum and germinate to give rise to the thallus. After the sporangium is emptied, another may be formed on a branch at the base of the old sporangium.

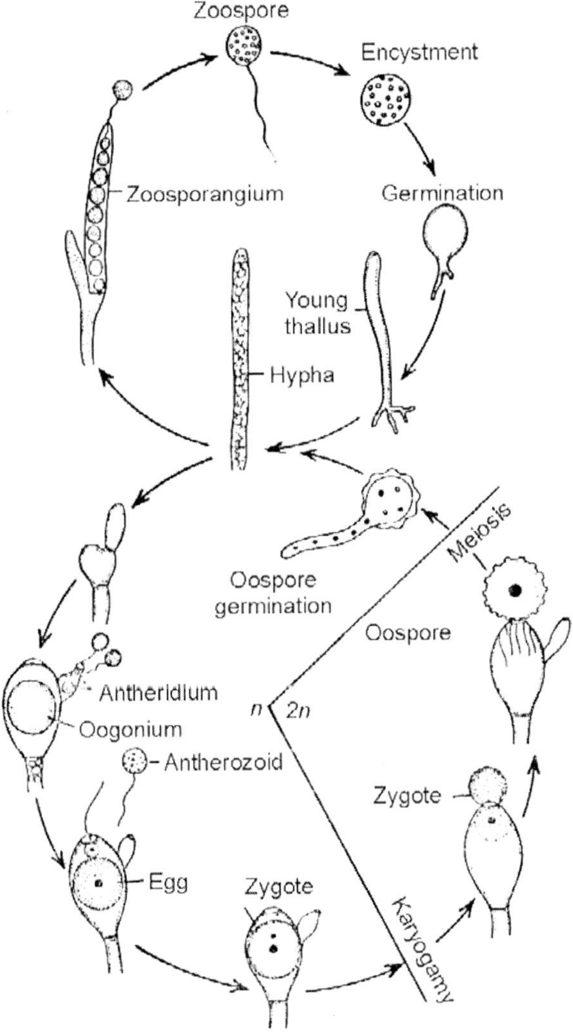

Fig. 4.10. Life cycle of *Monoblepharis polymorpha*

Higher temperature, which is unfavourable for growth and zoosporangia formation, induces formation of gametangia on the same thallus. The antheridium is initiated from a terminal cell, cut off by a septum. A second segment, oogonial initial, is formed immediately below the antheridial cell. It enlarges somewhat asymmetrically, as a result of which, the antheridium is displaced to take a lateral position on the oogonium. The antheridia are narrow and elongated while the oogonia are globular and much bigger. At maturity, the antheridia appear as if inserted on the oogonia, which appear in chains. Posteriorly uniflagellate antherozoids (4-8 in number), which are smaller than the zoospores, are released from the antheridium through an apical pore. The protoplast of the oogonium rounds up and forms a single, uninucleate, naked, non-motile female gamete, called the egg. The antherozoids are released much before the oospheres are ripe. At maturity, an apical receptive papilla is formed on the oogonium which dissolves to form a pore. An antherozoid enters the oogonium, retracts the flagellum and fuses with the oosphere to form the zygote.

Oospore formation is exogenous; the zygote oozes out and encysts at the oogonial orifice. The oospore remains attached to the receptive papilla or the oogonium. Karyogamy occurs during encystment. The oospore may be smooth or thick-walled with warts, and undergoes a resting period. At the return of cooler temperatures, it germinates and gives rise to the haploid thallus. Meiosis probably takes place at the time of germination.

Family *Gonapodyaceae*

Genus *Monoblepharella*

There are two species in this genus, *M. taylori* (Fig. 4.11) and *M. mexicana*, which occur in tropical soils as saprobes growing on organic debris. They can be isolated from soil using hemp seeds as bait.

The thallus, consisting of branched hyphae, lies attached to the substratum by means of rhizoids. The hyphae are non-septate and highly vacuolated giving a foamy appearance to the cytoplasm. The tips of the hyphae develop into zoosporangia or gametangia, depending on the favourable or unfavourable environment, respectively.

The zoosporangia, which are narrow and ovate, are borne on sympodially-branched hyphae. The posteriorly uniflagellate zoospores are released through an apical pore. After the end of a swimming period, they settle down and germinate to give rise to the thallus.

The sexual reproduction starts with the onset of unfavourable conditions. Gametangia (antheridia and oogonia) are also formed on sympodially-branched hyphae. The antheridia are cylindrical while the oogonia are globular or obovate. Antherozoids, which are smaller than the zoospores, are liberated through an apical pore. They swim and reach the adjacent oogonium. The mature oogonium usually contains a single egg. A receptive spot develops at the apex of the oogonium which dissolves to form a pore. The antherozoid enters the oogonium and fuses with the egg. The flagellum of the antherozoid remains functional and propels the zygote out of the oogonium and then farther away from it. Karyogamy does not follow immediately. Eventually, the swimming zygote settles down, secretes a wall (encysts) and becomes transformed into an oospore.

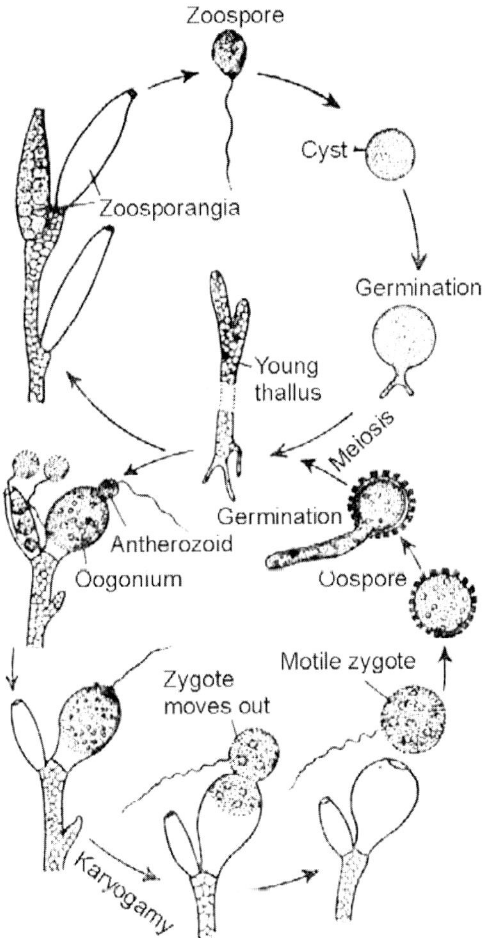

Fig. 4.11. Life cycle of *Monoblepharella taylori*

Karyogamy occurs at the time of encystment. The oospore is hard and withstands the unfavourable environment. At the return of favourable conditions, it germinates and gives rise to the thallus. Meiosis occurs during germination.

REFERENCES

Barr, D.J.S. 1992 Evolution and kingdoms of organisms from the perspective of a mycologist. Mycologia. 84: 1-11

Cavalier-Smith, T. 1987. In: *Evolutionary Biology of the Fungi*. Ed. Raymer *et al.* BMS Symposium. (April 1986)

Letscher, 2006. Mycol. Res. 110: 898

James, T.Y. *et al.* 2006. A molecular phylogeny of flagellated fungi (*Chytridiomycota*) and description of a new phylum (*Blastocladiomycota*). *Mycologia*. 98 (6): 860-871

James, T.Y. *et al.* 2006. Restructuring the early evolution of Fungi using a six gene phylogeny. *Nature* 443: 818-822

Kirk, P.M. *et al.* 2008. *Ainsworth and Bisby's Dictionary of the Fungi* (10th ed.) C.A.B. International, Oxon, U.K.

SUMMARY

Phylum *Chytridiomycota* contains the 'core' or 'true' chytrids left after removal of phylogenetically-dissimilar chytrids into two new phyla, *Neocallimastigomycota* and *Blastocladiomycota.* These true chytrids represent the smallest and most primitive fungi, which are ancestors to all fungi. The distinguishing feature of this phylum is the production of motile zoospores having a single posterior whiplash flagellum, with unique flagellar 'root' system. The other characters, though not unique to these chytrids are: *the coenocytic thallus*, which may be an oval multinucleate cell, or a small elongated hypha, and *the zygote,* which is converted into a resting spore or a resting sporangium.

The simple spherical thallus may be wholly or partly transformed into a reproductive structure (zoosporangium). When it is wholly converted into asexual reproductive body, the thallus is called holocarpic (e.g. *Synchytrium*). In eucarpic, only a portion of the thallus forms sporangium, rest, vegetative part, forms the rhizoids, which anchors the sporangium to the substrate or host (e.g. *Olpidium*). The rhizoids may grow and become extensive, then it is called rhizomycelium (*Polyphagus*). The rhizoids or rhizomycelium is the vegetative or somatic part of the thallus. If it bears one sporangium, the thallus is monocentric, and if there are many sporangia distributed over, the rhizomycelium, it is called polycentric. There are more terms also. The thallus may be **endobiotic** (growing inside a host cell), **epibiotic** (growing attached externally but sending rhizoids inside host cells), or **interbiotic** (attached to many hosts, as in *Polyphagus*) through the rhizomycelium. Zoospores, formed inside the sporangium, escape through one or more discharge papillae or orifices, formed on the sporangial wall or at the tip of discharge tubes arising from the sporangium. In some species the discharge papillae form a lid, called the **operculum**, which is pushed out, or thrown off, during the release of the zoospores. Species forming operculum are called as **operculate**. In most of the species the discharge papilla dissolves forming a pore. Species which release the zoospores through such a pore are called **inoperculate**.

The ultrastructural characters of the zoospores are used in formation of classes and orders. The characters useful in defining the orders are: (1) the scattered or aggregated nature of the ribosomes, (2) the number and location of lipid bodies, (3) organization of Microbody-Lipid globule Complex (MLC) in association with mitochondria and membrane cisternae, (4) orientation of the dormant kinetosome in relation to the kinetosome, (5) organization of microtubules and in some cases fibrillar structures called **rhizoplasts**, and (6) the spatial relationship of the nucleus and kinetosome.

Sexual reproduction may be brought about by (1) *Planogametic copulation*-(isogamy, anisogamy or heterogamy = oogamy). Heterogamy, which occurs in *Monoblepharidomycota* involves the fusion of a motile male gamete with a non-motile female gamete (oosphere, contained in an oogonium), (2) *Gametangial copulation* (3) *Somatogamy*, by fusion between rhizoidal filaments. The diploid zygote, formed after plasmogamy and karyogamy, is converted into a resting spore or resting sporangium. Zygotic mitosis is characteristic of *Chytridiales*. Resting spores are also formed asexually.

There are two classes, *Chytridiomycetes* and *Monoblepaharidomycetes*. The thallus is simple, unicellular and spherical in *Chytridiomycetes*, while it is mycelial in the latter. The *Chytridiomycetes* has three orders, which are further divided in families based on thallus construction, zoospore ultrastructure, mode of plasmogamy and resting spores. The three orders are *Chytridiales* (four families), *Rhizophydiaceae* (three families) and *Spizellomycetales* (three families). The class *Monoblepharidomycetes* has one order *Monoblephaidales* and four families.

Test Questions

1. Describe the classification of *Chytridiomycota* up to orders. Which ultrastructural characters of the zoospores are used in defining of the orders?

2. Describe the distinguishing features of phylum *Chytridiomycota*. Write on the differences between its two classes with reference to thallus, sexual reproduction, and rumposome.

3. Describe the life cycle of any **two** of the following:

   (a) *Polyphagus*, (b) *Rhizophydium*,

   (c) *Olpidium*, and (d) *Monoblepharis*.

4. Distinguish between the following:

   (a) holocarpic and eucarpic thalli, (b) endobioic, epibiotic and interbiotic thalli, (c) monocentric and polycentric thalli, (d) operculate and inoperculate release of zoospores, (e) zygote behaviour in the two families *Monoblepharidaceae* and *Gonapodyaceae* of *Monoblepharidales*.

5. Write notes on:

   (a) rhizomycelium, (b) ultrastructure of zoospores of *Chytridiales*, (c) rumposome, (d) reasons for keeping *Chytridiomycota* in kingdom *Fungi* and not *Chromista*

6. Fill in the blanks:

   (a) The phyla, *Chytridiomycota, Neocallimastigomycota and Blastocladiomycota* once belonged to phylum ----------------.

   (b) The zoospores of members of class *Monoblephridomycetes* and the order--------------- have rumposome.

   (c) *Batrachochytrium*, causing chytriomycosis of amphibians (frogs) belongs to family -----------------------.

   (d) The wart disease of potato is caused by -----------------------.

   (e) A new class ------------, was proposed based on the presence rumpsomes.

   (f) Rhizomycelium represents the -------------------- part of the thallus.

5

Phylum *Neocallimastigomycota*

Flagellated, Zoosporic *Fungi* II (Obligately Anaerobic Rumen Chytrids)

These fungi earlier belonged to phylum Chytridiomycota, but because of phylogenetic distinctness they were segregated under a new phylum

(Hibbett, 2007).

The members are anaerobic saprobes, inhabiting the gut (rumen and caecum) of herbivore animals, including ruminant and non-ruminant mammals and herbivore reptiles. The thallus is mono-or polycentric, and the zoospores are mono- or polyflagellate. The production of posterior (9-14) polyflagellate zoospores by some species is a very important feature of these fungi. The ultrastructure of the zoospore shows an aggregate of ribosomes and lack of MLC and mitochondria (The absence of mitochondria explains the anaerobic nature of these organisms). The highly specialized habitat, obligate anaerobic nature, differences in flagellation, ultrastructure of the zoospore, and differences in the base sequences of 18S rDNA, make them distinct from other chytrids.

These organisms were first described by Orpin in 1975 from the rumen of sheep. At that time the microbial population of the rumen was believed to consist only of bacteria and protozoa. Since then, these organisms have been isolated from the digestive tracts of 50 ruminant and non-ruminant herbivores. They reproduce in the stomach of ruminants through zoospores. They have a simple life cycle alternating between a motile multiflagellate zoospore and an extensive vegetative stage usually found associated with particulate plant material in the alimentary canal. In absence of mitochondria, they use 'hydrogenosomes' to oxidize NADH to $NADH^+$, releasing hydrogen as a by product (Yarlett *et al.*, 1986). Hydrogenosomes are vesicles 0.5-2 µm in diameter, sometimes elongated with a fine granular matrix and often containing a dense core. These are considered as reduced mitochondria which were adapted to an anoxic environment. They act as hydrogen evolving, ATP-generating and oxygen-scavenging organelle. Hydrogenosomes have been used as a model to understand the metabolic processes of obligately anaerobic eukaryote.

These organisms are highly fibrolytic and produce polysaccharide-degrading enzymes that degrade the most recalcitrant plant fiber polymers, including unlignified cell walls. The enzymes are organized into multi-protein complex, similar to bacterial 'cellulosome'.

The phylum (Kirk *et al.,* 2008) has one class, *Neocallimastigomycetes*, one order, *Neocallimastigales*, and one family *Neocallimastigaceae*, with 6 genera and 16 species. These are: *Aneromyces, Caecomyces, Cyallamyces, Neocallimastix, Orpinomyces* and *Piromyces*. *Neocallimastix* is the type genus and also the best studied one.

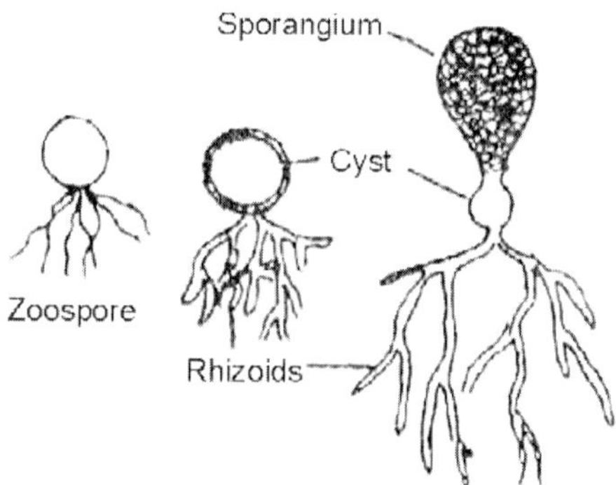

Fig. 5.1. *Neocallimastix*

Genus *Neocallimastix*

The genus has 5 species which can be obtained from fresh and dried faecal materials as well as caecum and saliva of herbivore animals, mostly sheep and cattle. The zoospores possess 9–14 posterior flagella (Fig. 5.1), which provide a spiral movement to the spore. The zoospores settle down (as seen in nutrient agar) and form a cyst around the cell. The rhizoids emerge and develop into an extensively branched system (vegetative phase) that draws nutrients, derived from fiber degradation. When mature, the cyst enlarges and transforms into a sporangium, which may be sessile or stalked. The sporangium forms haploid, polyflagellade zoospores that are released through a pore. The zoospores encyst immediately near the sporangium. Under unfavorable conditions, as in drying faeces, the encysted zoospores develop into resting sporangia, which are aerotolerant and have diploid nuclei. The resting spores survive for long periods under dried conditions and later they germinate and form zoosporangia as discussed above. Meiosis occurs during zoospore formation but the events of plasmogamy, karyogamy and meiosis have not been observed.

Sexual reproduction is not reported.

REFERENCES

Hibbett, D.S. *et al.* 2007. A higher level phylogenetic classification of the Fungi. Mycol. Res. 111: 509-547

Kirk, P.M. *et al.* 2008. *Ainsworth and Bisby's Dictionary of the Fungi* (10th ed.) C.A.B. International, Oxon, U.K.

Orpin, C.G. 1975. Studies on the rumen flagellate *Neocallimastix frontalis*. *J. Gen. Microbiol.* 91 (2); 249-262

Yarlett, N. *et al.* 1986. Hydrogenosomes in the rumen fungus *Neocallimastix patriciarum*. *Biochem. J.* 236; 729-739

SUMMARY

The phylum *Neocallimastigomycota* includes obligately anaerobic flagellated fungi that thrive saprobically in the rumen of herbivore mammals and some reptiles.

The thallus is mono-or polycentric, and the zoospores are mono-or polyflagellate. The ultrastructure of the zoospore shows an aggregate of ribosomes and lack of MLC and mitochondria. The absence of mitochondria explains the anaerobic nature of these organisms. The highly specialized habitat, obligate anaerobic nature, differences in flagellation, ultrastructure of the zoospore, and differences in the base sequences of 18S rDNA, make them distinct from other chytrids. In absence of mitochondria, they use 'hydrogenosomes' to oxidize NADH to $NADH^+$, releasing hydrogen as a by product. These are considered as reduced mitochondria adapted to an anoxic environment.

These organisms are highly fibrolytic and produce polysaccharide-degrading enzymes, which are organized into multi-protein complex, similar to bacterial 'cellulosome'. The phylum has one order, one family and six genera, *Neocallimastix* being the type genus.

TEST QUESTIONS

1. What are the characters that make *Neocallimastigomycota* distinct from other chytrids?
2. Write short notes on: (a) hydrogenosomes, (b) fibrolytic activity, (c) *Neocallimastix*.

6

Phylum *Blastocladiomycota* Flagellated

Zoosporic *Fungi* III
The Blastocladian Chytrids

The blastocladian chytrids were traditionally classified under Chytridiomycota, but they were transferred to new phylum because of phylogenetic distinctness James *et al.*, 2006.

The phylum is characterized by: (1) formation of true mycelium, (2) alternation of generations between haploid gametothallus and diploid sporothallus, (3) formation of resting sporangia (or resting spores) with typical pitted walls (minute, inwardly directed, cone-shaped pits or pores) (4) isogamous to anisogamous sexual reproduction with both gametes motile and (5) the posteriorly uniflagellate zoospores (or gametes) having two characteristic structures, a "nuclear cap" ("ribosomal cap") and a 'side-body' (mitochondrion) (Fig. 6.1). The nuclear cap contains all the cellular ribosomes and is surrounded by a double-layered membrane. The 'side body' is a giant-size mitochondrion. About a dozen lipid granules, arranged in a single layer, lie attached to the mitochondrion. These are surrounded on the outer side by a double membrane of their own. The cytoplasm contains several **gamma particles** which store protein. These are single-membrane-bound, cup-shaped, dense bodies found only in zoospore, which disappear after germination of the zoospore. The phylum has one class, *Blastocladiomycetes*, one order, *Blastocladiales*, and five families viz. *Blastocladiaceae* (5 genera), *Catenariaceae* (3 genera), *Coelomomycetaceae* (2 genera), *Physodermataceae* (1 genus), and *Sorochytriaceae* (1 genus).

Family *Blastocladiaceae*

There are five genera in this family which show great variation in structure. Usually, the thallus is divisible into hold fast, consisting of branched rhizoids, which anchors the fungus to the substratum, a stout or slender basal body, and dichotomous branches above the basal body which bear the reproductive organs. In some cases the reproductive organs are borne directly on the distal end of the body. The basal body may be much reduced or even absent. The five genera are *Blastocladiella*, *Blastocladia*, *Blastocladiopsis*, *Allomyces*, and *Microallomyces*. The simplest thallus is that of *Blastocladiella*.

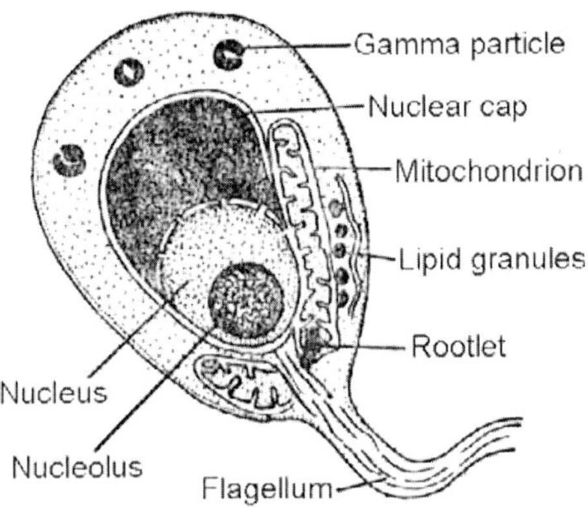

Fig. 6.1. Ultrastructure of zoospore of *Blastocladiella emersonii*

Genus *Blastocladiella*

The thallus is simple, consisting of a spherical sac-like structure, attached to the substratum by means of rhizoids. It resembles a monocentric chytrid. The spherical body either itself becomes converted into the reproductive organ, or bears one at its apex.

Species of *Blastocladiella* occur mostly in soil; *B. stubenii*, *B. simplex* and *B. emersonii* are saprobes growing in soil on plant and animal debris. A parasitic species growing on a cyanobacterium, *Anabena*, has been reported from Britain. *B. emersonii* possesses the characters useful in developmental (morphogenetic) studies and, therefore, has been studied extensively.

The zoospores give rise to four different kinds of thalli, whose proportions depend upon environmental conditions (Fig. 6.2). The thalli are distinguishable by the sporangia, which may be (1) orange thin-walled (O), (2) early colourless, thin-walled (EC), (3) resistant sporangial (RS), which are brown thick-walled, pitted and (4) late colourless, thin-walled (LC).

The early stages of growth are similar in all the four developmental lines. A cell developing along one line, can be induced to form a different sporangial type by changing the conditions before a particular period, called "point of no return".

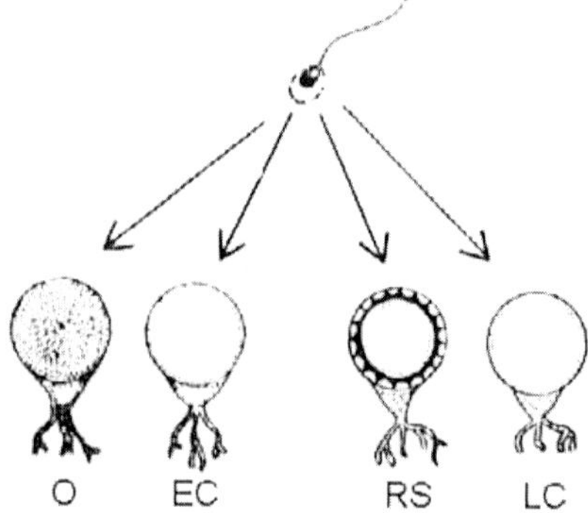

Fig. 6.2. Showing four developmental paths that may be taken by spores of *B. emersonii*

The life cycle begins and ends with a posteriorly uniflagellate zoospore (Fig. 6.3). After a variable period of swimming, the zoospores settle down, round up, retract the flagellum and develop into a cyst by secreting a wall. The cyst sends out a germ tube which forms non-septate rhizoids. A daughter nucleus passes into the rhizoids. The cyst enlarges enormously. Later, a cross wall is laid down delimiting the cyst into a basal cell bearing the rhizoids and an apical multinucleate fertile cell, which develops into a thin-walled zoosporangium. The basal cell is devoid of protoplasm at maturity. The zoosporangium gives rise to hundreds of uninucleate zoospores which are released through pores formed by dissolution of papillae developed on the wall.

Under unfavourable conditions (like crowding or high bicarbonate or cation concentration), the fertile cell instead of forming a thin-walled, colourless zoosporangium, develops into a thick-walled globular, dark brown, pitted resting sporangium. At the return of favourable conditions, the resting sporangium germinates and releases zoospores that are released through pores. In addition to the above two, other common types of thalli, are also encountered in nature. These form orange coloured, thin-walled zoosporangia or late colourless, thin-walled zoosporangia.

The orange colour of the sporangia is due to the presence of carotenes in the cytoplasm, while the brown colour of resistant sporangia is due to melanin pigment in the wall and gamma carotene in the cytoplasm. The development of colourless and orange zoosporangia is related to the amount of gamma particles present in the zoospores from which they develop. Zoospores having lesser number of gamma particles, develop into orange thalli, while those having high concentration, develop into white thalli.

Sexual Reproduction is not known.

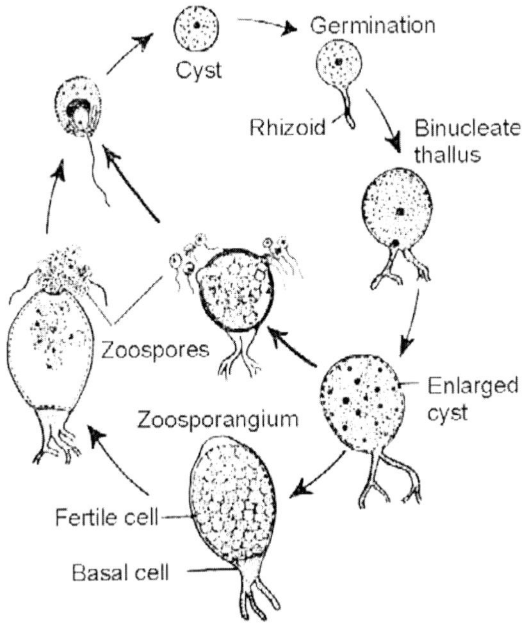

Fig. 6.3. Life cycle of *B. emersonii.*

Genus *Blastocladia*

Blastocladia (Fig. 6.4) shows greater development of the thallus than *Blastocladiella*. The basal part is well developed in this genus and bears a rhizoidal system at the lower end. Cylindrical zoosporangia and subspherical or ovoid resting sporangia, interspersed with setae, are borne at the distal end (*B. pringsheimii*). In some species e.g. *B. ramosa*, the basal part is dischotomously branched and the zoosporangia and resting bodies are borne on the branches.

Zoospores produced in the zoosporangia or the resting sporangia are liberated through a single, large, dish-shaped papilla. These give rise to new thalli. The resting sporangia are meant for survival under unfavourable conditions. Sexual reproduction is not known.

The important species are *B. pringsheimii* and *B. ramosa*, which grow on plant debris in water. *Blastocladia* can be obtained by sinking small tins with holes in the bottom, containing apples or tomatoes, in a pond for about 15 days during winter. Characteristic white pustules are formed on the fruits, which can be cut and grown in liquid media.

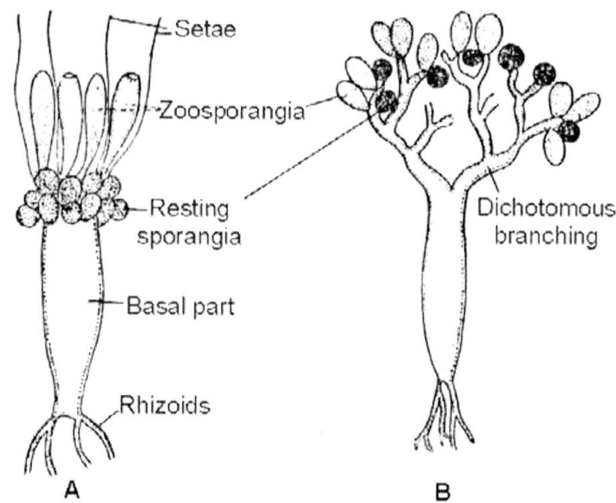

Fig. 6.4. Thalli of *B. pringsheimii* (A) and *B, ramosa* (B), showing the basal part with rhizoids at the lower end and reproductive structures at the distal end.

Genus *Allomyces*

Allomyces is another fungus, which has been extensively used in morphogenetic studies. It was discovered in India in 1911 by E.J. Butler. There are nine species, the better known are: *A. anomalus, A. arbuscula, A. javanicus, A. macrogynus, A. moniliformis, A. neomoniliformis, & A. reticulatus*, The species *A. javanicus* and *A. neomoniliformis* grow commonly on organic debris in soil and water. These can be obtained by placing dry soils in a Petri dish containing water and boiled hemp seeds as bait. After a few days, the seeds are colonized by zoospores, which can be pipetted out and grown on agar media.

There are two types of thalli (the diploid sporothallus and haploid gametothallus), which grow independently. Both the thalli are isomorphic, i.e. morphologically indistinguishable until the formation of the reproductive bodies, the gametangia or the sporangia. Both thalli have a rhizoidal system and a trunk-like region which bears pseudoseptate dischotomous branches. Alternation of generations is not obligatory and both generations can multiply immediately. The thallus has a distinct basal part, which has a rhizoidal 'hold fast' at the proximal end and dichotomously-branched, pseudoseptate hyphae at the distal end. The pseudosepta are incomplete and wheel-like with radiating strands. The zoosporangia and resting sporangia are borne on the dichotomous branches.

The sporothallus bears two kinds of sporangia (Fig. 6.5), the thin-walled zoosporangia, and oval, thick-walled, brown, and pitted resting sporangia. The brown colour is due to melanin pigment. The zoosporangia produce diploid zoospores, which are released through pores. On germination, the diploid zoospore gives rise to diploid sporothallus. The resting sporangia germinate only after a rest period of 2-8 weeks and produce haploid zoospores. The haploid zoospores, which are slightly smaller than the diploid zoospores, on germination give rise to gametothalli.

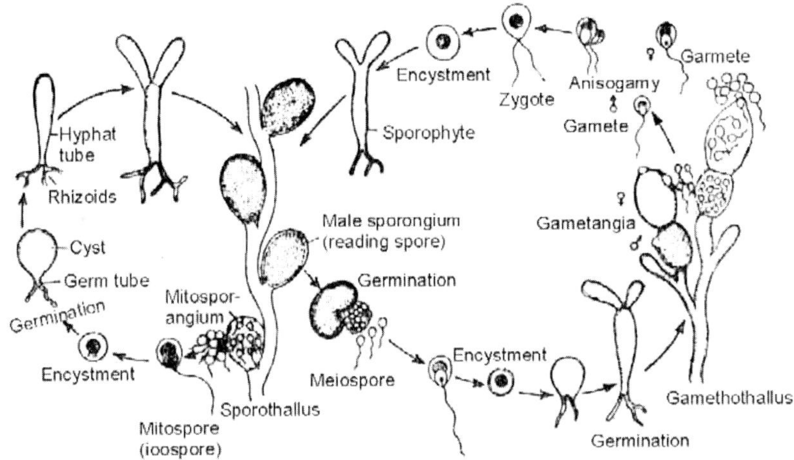

Fig. 6.5. Life cycle of *Allomyces arbuscula*.

The gametothalli on maturity form both male and female gametangia on the same thallus. Thus, the thallus is monoecious. The gametangia lie one above the other. In *A. arbuscula*, the female gametangium lies above the male gametangium (Fig. 6.6); the opposite occurs, in *A. macrogynus* i.e. male gametangium lies over the female. The male gametangia are smaller and orange-coloured, while the female gametangia are bigger and colourless. Both gametangia form posteriorly uniflagellate gametes, which are released through 1-3 discharge papillae. The male gametes are orange-coloured, smaller and twice or thrice as numerous as the female gametes, which are colourless and much bigger in size. A hormone, **sirenin,** (a sperm attractant) is produced by the female gametes (while still inside the gametangium), which diffuses in the surrounding water and exerts a chemotactic influence on the male gametes (Chapter 31). The female gametes are surrounded by several male gametes as soon as they are released from the female gametangium. The male and female gametes fuse in pairs; karyogamy follows and a biflagellate diploid zygote is produced. Female gametes, which fail to pair, can give rise to haploid gametothallus parthenogenetically. The diploid zygote comes to rest, sheds the flagella, becomes round and encysts.

It produces a germ tube, which develops into rhizoids. The cyst elongates, divides dichotomously and develops into a diploid sporothallus.

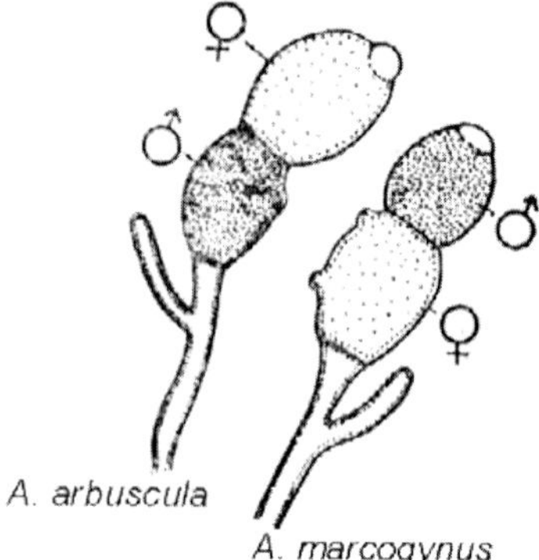

Fig 6.6. Showing position of male and female sex organs in *A. arbuscula* and *A. macrogynus.*

Family *Coelomomycetaceae*

Genus *Coelomomyces*

The family consists of two genera, *Coelomomyces* and *Zografia*. *Coelomomyces* is an obligate parasite inhabiting the coelome (body cavity) of mosquito larvae with great potential as biological control agent of mosquitoes. There are 70 species which are wide-spread. *C. anophelescia* and *C. indiana* attack Anopheles mosquitoes. The fungus shows alternation of generations and a heteroecious life cycle i.e., the life cycle is completed in two alternate hosts.

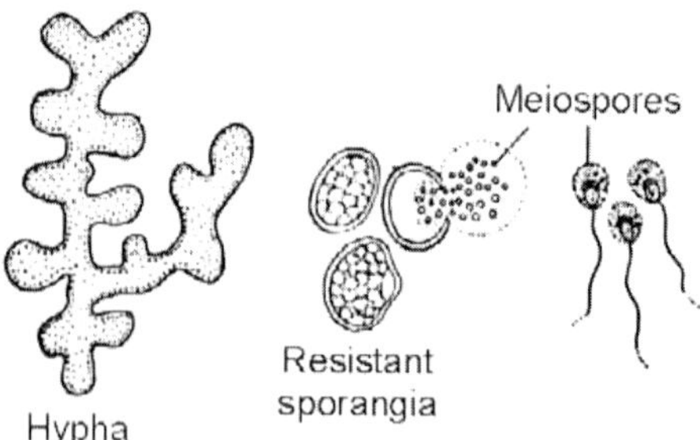

Fig. 6.7. *Coelomomyces psorophorae* thallus, resistant sporangia and meiospores.

The thalli, both sporo-, and gametothalli consist of non-septate, irregularly branched or lobed hyphae which lack a cell wall (Fig. 6.7), an unusual character not found elsewhere in kingdom *Fungi*. Fragmentation of the thallus results in the spread of the fungus within the host and formation of more thalli.

C. psorophorae completes its life cycle in two alternate hosts (Fig. 6.8). The diploid sporogenus stage occurs in mosquito larvae (*Culistea informata*) and the haploid gametogenous stage in the copepod - *Cyclops vernalis*.

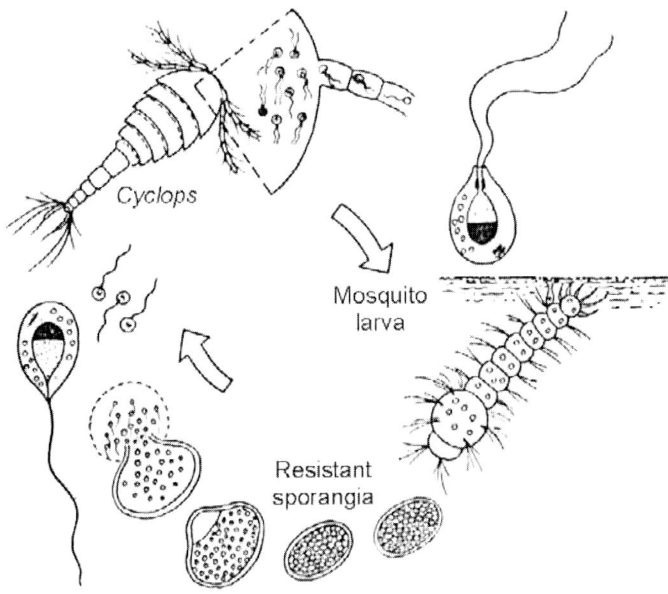

Fig. 6.8. Life cycle of *C. psorophorae*. Resistant sporangia from the body of the mosquito release meiospores, which infect *Cyclops*, and the biflagellate spores from the *Cyclops* / *Callimastix* phase infect mosquito larvae.

Mosquito Phase. Biflagellate motile zygotes released from *Cyclops* infect mosquito larvae and cause infection. The fungus develops in coelome and appears as small, unwalled hyphal bodies which later differentiate into thick-walled resting sporangia. The morphological features of the resting sporangia provide diagnostic features for the identification of the species. The resting sporangia increase in number, and, in advanced stages of infection, completely fill the body of the larvae. These are released on death of the larvae. On availability of appropriate conditions, the sporangia germinate and release haploid spores (suggesting occurrence of meiosis). The haploid meiospores are of two types, + and –. These spores are unable to infect mosquitoes and the search for the solution of this riddle led to the discovery of the alternate host, *Cyclops* (Whisler *et al.*, 1975).

***Cyclops* Phase.** The free, swimming + and – meiospores released by dead mosquito larvae infect *Cyclops* (a copepod). They encyst before penetration, and give rise to wall-less + and – gametothalli in the coelome of the copepod. The thalli grow and ultimately form uniflagellate + and – gametes that fill the coelome of the infected *Cyclops*. The gametes fuse and give rise to biflagellate zygotes which are released after the death of *Cyclops*. These zygotes infect the mosquito larvae and in this way life cycle goes on involving two alternate hosts.

REFERENCES

Butler, E.J. 1911. On *Allomyces*, a new aquatic fungus. *Ann. Bot.* (London), 25: 1023–1035.

Hibbett, D.S. *et al.* 2007. A higher level phylogenetic classification of the Fungi. Mycol. Res. 111: 509-547.

James, T.Y. *et al.* 2006. A molecular phylogeny of flagellated fungi (*Chytridiomycota*) and description of a new phylum (*Blastocladiomycota*). *Mycologia*. 98 (6): 860-871.

Kirk, P.M. *et al.* 2008. *Ainsworth and Bisby's Dictionary of the Fungi* (10th ed.) C.A.B. International, Oxon, U.K.

Lutzoni, *et al.* 2004. Assembling the fungal tree of life: progress, classification and evolution of subcellular traits. *Amer. Jr. Botany* (10): 1446-1480.

Whisler, H.C. *et al.* 1975. Life History of *Coelomomyces psorophorae*. *Proc. Natl. Acad. Sci*. 72: 69–696.

SUMMARY

The blastocladian chytrids were transferred from *Chytridiomycota,* to new phylum because of phylogenetic distinctness. The phylum is characterized by: (1) formation of true mycelium, (2) alternation of generations between haploid gametothallus and diploid sporothallus, (3) formation of resting sporangia (or resting spores) with typical pitted walls (minute, inwardly directed, cone-shaped pits or pores) (4) isogamous to anisogamous sexual reproduction with both gametes motile and (5) the posteriorly uniflagellate zoospores (or gametes). The ultrastructure of the zoospores and gametes shows a "nuclear cap" ("ribosomal cap"), a 'side-body' (mitochondrion), about a dozen linearly arranged lipid granules, surrounded on the outer side by a double membrane. The cytoplasm of the zoospore contains several gamma particles which are organelles for storage of proteins. These are single-membrane-bound, cup-shaped, dense bodies found only in zoospore, which disappear after germination of the zoospore. The phylum has one class, *Blastocladiomycetes*, one order, *Blastocladiales* and five families viz *Blastocladiaceae*, *Catenariaceae*, *Coelomomycetaceae*, *Physodermataceae*, and *Sorochytriaceaei.*

Test Questions

1 Give the diagnostic features of phylum *Blastocladiomycota*. Describe the ultrastructure of the zoospore.

2. Describe the life cycle of *Coelomomyces* **or** *Allomyces*.

3. Describe briefly: (a) Gamma particles and pigmentation of zoosporangia and resistant sporangia of *Blastocladiella*, (b) the four types of thalli of *Blastocladiella*, (c) two types of thalli of *Blastocladia*, (c) the thallus of *Coelomomyces.*

4. (a) What is the unusual feature in the thallus of *Coelomomyces* which is not found elsewhere in fungi?

   (b) What is the difference in the position and colour of the male and female gametangia in *Allomyces arbuscula* and *A. macrogynus*?

   (c) Which gametes produces sirenin in *Allomyces* and for what purpose?

   (d) What is the difference in the flagellated body that infects mosquito larvae and *Cyclops* in the life cycle of *Coelomomyces*?

7

Phylum *Zygomycota*

(Zygospore-forming *Fungi*)

Fungi, which lack motile cells and produce **zygospores** as the sexual spore, are placed under phylum *Zygomycota*. Zygospore defines and distinguishes the phylum from rest of the fungi. It is formed by fusion of two gametangia (gametangial copulation). These fungi reproduce asexually by means of **aplanospores** (also called sporangiospores, formed inside sporangia) or **condia**. A further point of unity of these fungi is the nature of the **cell wall** which is made of chitosan chitin, a partially deacetylated form of chitin. Members of *Mucoromycotina* and *Zoopagomycotina* form coenocytic hyphae. Septa are formed at regular intervals in rest of the subphyla to varying extent.

Classification Prior to the use of molecular phylogenetics in defining taxa, the phylum was divided into two classes: *Zygomycetes* and *Trichomycetes*. But now, there are no classes. The former class Trichomycetes is now non-existent. Among four of its orders two (*Amoebidiales* and *Eccrinales*) are now placed under kingdom *Protozoa* and the other two, (*Asellariales* and *Harpellales*), which are true fungi, are kept in *Zygomycota* (sub-phylum *Kickxellomycotina*). The order Glomales is removed from *Zygomycetes* and elevated to phylum *Glomeromycota*, due to lack of zygospore formation, mycorrhizal habitat, and lack of DNA sequence homology (Schuessler *et al.,* 2001). Molecular phylogenetics also reveals that the phylum *Zygomycota*, even in its present form, is polyphyletic and could see a split into several new phyla in coming years (Tanabe *et al.,* 2000, 2004).

Presently (Kirk *et al.,* 2008) the phylum *Zygomycota* has four subphyla, which are monophyletic, but the relationships among them are not resolved. The subphyla have 9 orders and 27 families (no class). These are:

Phylum *Zygomycota*

Subphyla (4, given below)

Mucoromycotina (3 orders) *Mucorales, Endogonales,* and *Mortierellales.*

Entomophthoromycotina (1 order) *Entomophthorales.*

Kickxellomycotina (4 orders) *Asellariales, Kickxellales, Dimargaritales* and *Harpellales.*

Zoopagomycotina (1 order) *Zoopagales.*

Subphylum *Mucoromycotina* (3 orders)

Order *Mucorales* (9 famiies)

It is a large order comprising 9 families and 51 genera, which are saprobes, growing on decaying organic matter. They are mycoparasites, or facultative parasites of plants or animals, including humans. Mucorales are associated with **mucoromycosis** (=phycomycosis) of domestic animals and humans.

The thallus consists of typical coenocytic mycelium. Asexual reproduction is varied and forms the basis of division into families and identification of the genera and species. The sporangia or conidia are borne apically on simple or characteristically-branched sporangiophores or conidiophores. A typical sporangium is a globular structure, divisible into a central sterile portion, columella, and an outer fertile sporogenous portion containing spores. The two portions are separated by a dome-shaped septum formed by the arching up of the septum that originally separated the sporangium from the sporangiophore. The sporangia form numerous, walled sporangiospores by internal cleavage of the sporangial cytoplasm.

Smaller sporangia with a few sporangiospores only, are termed **sporangioles** (sing. sporangiole). Both sporangia and sporangioles may occur on the branches of the same sporangiophore. There are sporangioles with a single spore. When the wall of such single-spored sporangiole is fused with that of the spore, it is termed a conidium. The two walls, however, can be distinctly stained and differentiated. Cylindrical sporangioles, containing spores in a row, are called **merosporangia** (sing. merosporangium). The merosporangia too, like the sporangioles, may have in some cases a single spore, free or with fused walls. The sporangiospores are released after dissolution of the sporangial wall and usually disseminated by wind. In *Pilobolus* (shot gun fungus) the sporangium is violently discharged.

Sexual reproduction involves fusion of two morphologically similar gametangia, that transform into a thick-walled 'resting spore' called **zygo-**

spore (=resting sporangium), whose wall is derived from the walls of the two gametangia. The hyphal branches (copulation branches) that bear the gametangia are called **progametangia**, while the special hypha that bears copulation branches (progametangia) is called **zygophore**. A terminal cell is cut off at the apex of each progametangium, which develops into a gameta-mgium. The lower portion of the progametangium is called **suspensor**. The zygospore is held between the two suspensors which are of great help in identification of the genera. They vary in shape, size and presence or absence of appendages on them. Zygospore represents the only diploid phase of the fungus. Meosis occurs during germination, but this has been demonstrated only in *Phycomyces blakesleeanus*. The germination of zygospore, which is infrequent, may be by formation of a germ tube, which gives rise to a hypha or alternatively, through a sporangium, called "germ sporangium", formed at the tip of the germ tube.

Economic Importance

Many saprobic *Mucorales* are important in industry. Various acids e.g. fumaric and lactic acids, are now manufactured from species of *Rhizopus* under anaerobic conditions. Species of *Rhizopus* and *Mucor* are used in industrial fermentations at the initial stages for converting starch into sugar (yeasts lack the enzymes required for breaking starch into glucose). Some members are capable of rearranging steroid molecules into more useful compounds. Strains of *Rhizopus* find use in the preparation of a delicious food 'Tempeh' from soyabeans in Indonesia. 'Sufu' or 'Chinese cheese' is prepared from soyabeans with the help of *Actinomucor*. *Blakeslea* may be used as a good source of carotenes.

Many diseases of fruits that occur during storage and transport are caused by the *Mucorales*. *R. stolonifer* (=*R. nigricans*) causes soft rot of sweet potatoes and 'leak' (due to soft rot) of strawberries, raspberries, peaches etc. Some species of *Rhizopus*, *Mucor* and *Absidia* attack stored grains. Spoilage of bread by *Rhizopus* is so common that *R. niger* is called the "bread mold".

Classification

The order is divided into 9 families (Kirk *et al.,* 2008), which are mainly based on nature of the sporangium and sporangiophores and to lesser extent on the habitat and sexual reproduction. Since sexual reproduction is infrequent, and also difficult to induce in laboratory, it does not find much importance in the identification of these fungi. The 9 families and the genera that we shall study are given below.

1. *Mucoraceae* (*Mucor, Rhizopus, Absidia, Zygorhyncus, Thamnidium*)
2. *Choanephoraceae* (*Choanephora, Blakeslea)*
3. *Cunninghamellaceae* (*Cunninghamella*)
4. *Mycotyphaceae,*
7. *Phycomycetaceae* (*Phycomyces*)
6. *Pilobolaceae* (*Pilobolus*)
7. *Radiomycetaceae* = *Saksenaeaceae* (*Saksenaea*)
8. *Syncephalastraceae,* and
9. *Umbelopsidaceae.*

Family *Mucoraceae*

Members of this family possess large, multi-spored, columullate sporangia; there are no sporangioles. There are 25 genera that grow on different habitats. Some are exclusively parasitic on mushrooms (*Syzygites*), while others e.g. *Zygorhyncus* and many species of *Mucor* and *Absidia* are strictly soil fungi. Some species of *Mucor* are coprophilous i.e. dung-inhabiting. Several species of *Rhizopus* and *Absidia* attack stored grains, decaying fruits and vegetables, and are also found in compost and air-spora. Various species of *Mucor*, *Rhizopus* and *Actinomucor* attack fruits in storage and transport. These are used for making fermented foods (tempeh, sufu etc.) and in steroid transformations. *M. ramosissimus* and species of *Rhizomucor* cause diseases in human beings.

Genus *Mucor*

This is a large genus of over 50 species. The hyphae are coarse, coenocytic and richly branched, usually tapering to fine points. Septa appear in older hyphae to cut off the dead parts or to delimit the sporangia, gmetangia etc. Chlamydospores are formed between such septa.

In anaerobic liquid cultures, especially in presence of carbon dioxide, the fungus exists in **Torula condition** in which the hyphae break to form yeast-like bodies. It reverts to the filamentous condition when aerobic condition is restored. Hyaline, variously-branched sporangiophores bear typical collumullate globular sporangium. On maturity, when the sporangial wall dissolves, the spores remain firmly adhered to the collumella in a drop of fluid and are not blown off even by strong winds. The spores are disseminated by insects and ants. Species of *Mucor*, therefore, do not cause contamination of air and laboratory cultures. The zygospores are always brown to black, with rough blunt projections or warts. The species may be homo- or heterothallic. The life cycle of *Mucor* is depicted in Fig. 7.1.

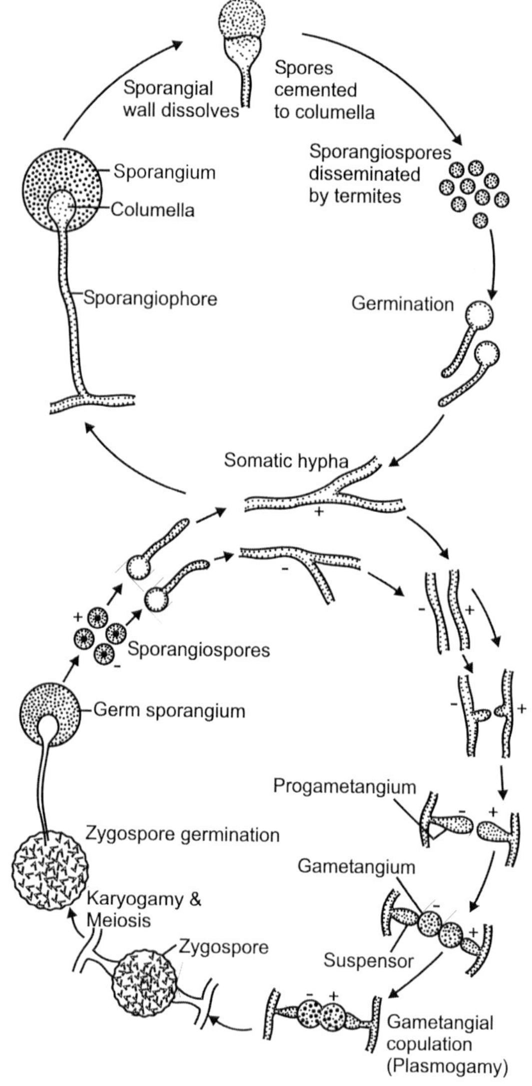

Fig. 7.1. Life cycle of *Mucor* sp.

Genus *Rhizopus*

Rhizopus (Fig. 7.2) resembles *Mucor*, but it differs in the following points: the mycelium is distinguishable into aerial hyphae called **stolons**, and rook-like **rhizoids**, which remain submerged in the substratum. The sporangiophores, which are shorter, stouter and stiffer than those of *Mucor*, arise from the junction of the stolons and the rhizoids. The sporangiospores are dry and easily blown by wind. The species of *Rhizopus* are, therefore, to be handled with care to avoid contamination in laboratory. The zygospore is of *Mucor*-type. There are 9 species that are commonly found in soil, decaying fruits and vegetation. *Rhizopus stolonifer* (= *R. nigricans*) is a common saprobe and facultative parasite of mature fruits and vegetables. Spoilage of bread by *Rhizopus* is caused by *R. niger,* hence called the "bread mold".

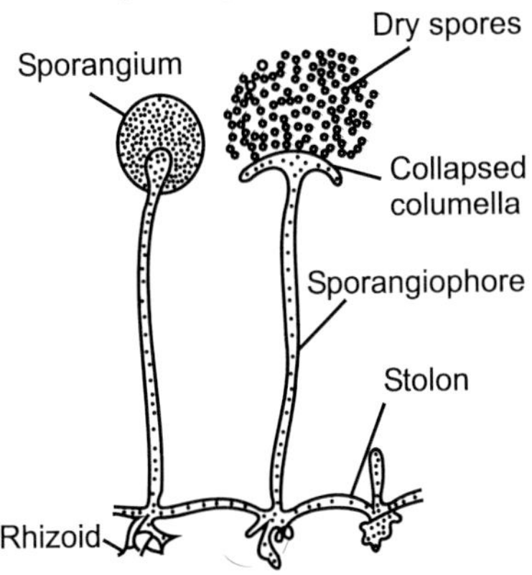

Fig. 7.2. *Rhizopus*, hyphae and sporangia

Genus *Absidia*

Absida (18 spp) is widespread in soils. Like *Rhizopus*, the mycelium comprises of stolons and rhizoids but with the difference that the sporangiophores arise in groups of two to five from internodes of the stolons, and not opposite the rhizoids (Fig. 7.3). The sporangia are pyriform and thin-walled. The columella occasionally has an apical projection. Circinate appendages arise from one or both the suspensors enveloping the zygospores which are borne on the stolons. The species may be homothallic (*A. spinosa*) or heterothallic (*A. glauca*).

Genus *Zygorhyncus*

The sporangia are of *Mucor*-type (Fig. 7.4). The conjugating gametangia and the suspensors are unequal. One suspensor is small and straight, while the other is curved and thicker (bulbous). Eight species that are reported are homothallic and soil-inhabiting.

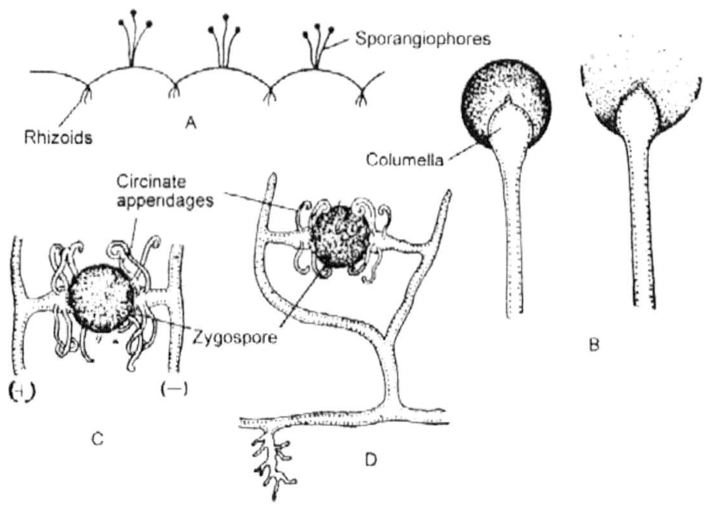

Fig. 7.3. *Absidia glauca* (A, B, C) and *A. spinosa* (D, homothallic). A, sporangiophores arising in groups from internodes; B sporangia with columella that has an apical projection; C, zygospore and circinate appendages.

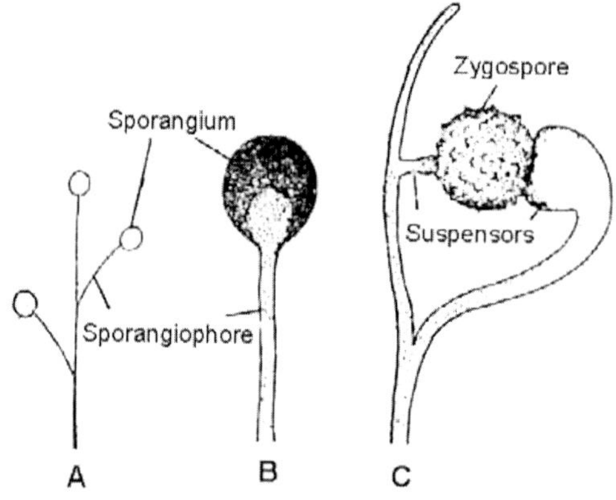

Fig. 7.4. *Zygorhyncus*. A, showing habit; B, sporangiophores with columullate sporangium and C, zygospore.

Genus *Thamnidium*

The erect sporangiophore bears a terminal *Mucor*-type sporangium (Fig. 7.5). Whorls of dichotomous branches arise from the lower portion of the sporangiophore, which terminate in few-spored sporangioles. The zygospore is of *Mucor*-type. There is one species, which is heterothallic, grows saprobically in soil on leaf-litter, dung and meat-products stored at low temperatures.

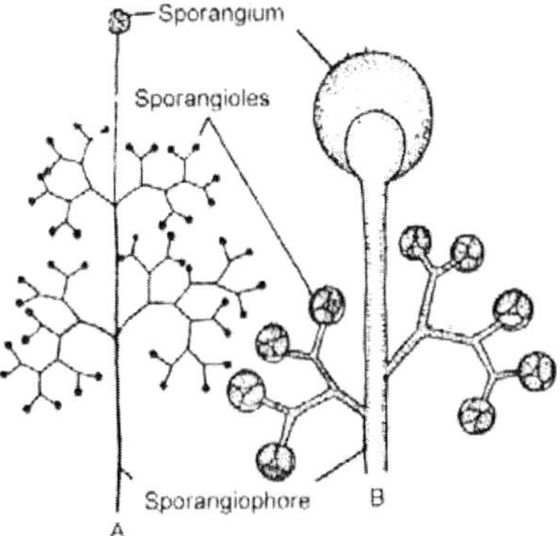

Fig. 7.5. *Thamnidium elegans.* A, sporangiophore with terminal sporangium and whorls of lateral dichotomies bearing sporangioles; B, magnified view of a portion of sporangiophore.

Family *Choanephoraceae*

There are three genera; *Choanephora*, *Blakeslea*, and *Poitrasia* which grow on fruits and flowers of higher plants. These possess large, columullate sporangia borne in circinate vernation when young. *Choanephora* and *Blakeslea*, in addition to the sporangia, form 2-5 spored sporangioles, usually 3-spored in *Blakeslea* (Fig. 7.6) and 1-spored (called conidia) in *Choanephora*. *C. cucurbitarum* (Fig. 7.7) causes fruit rot of cucurbits and other fruits.

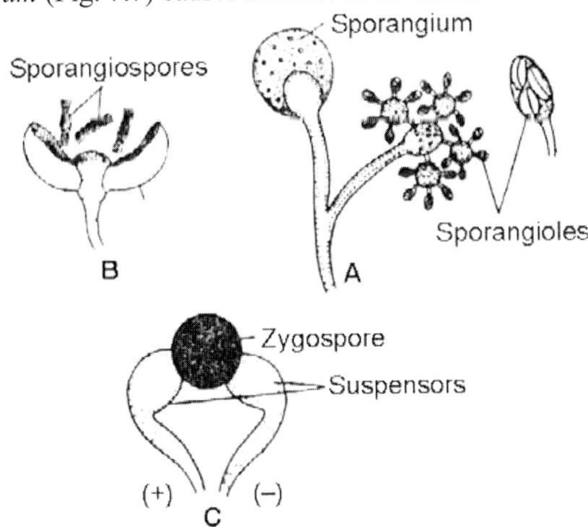

Fig. 7.6. *Blakeslea trispora.* A, columullate sporangium and 3-spored sporangioles; B, dehisced sporangiole; C, zygospore

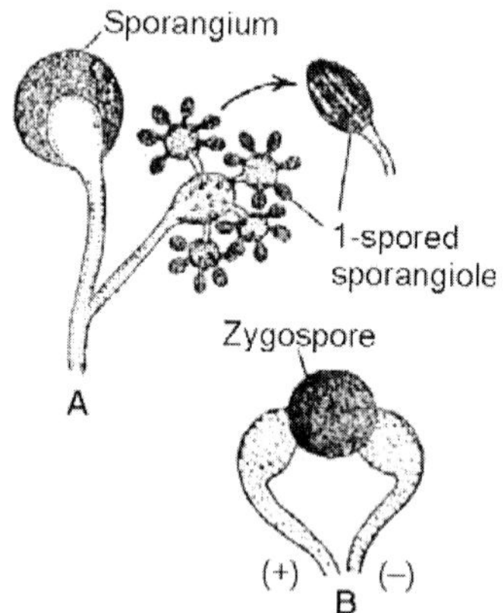

Fig. 7.7. *Choanephora cucurbitarum*. A. columullate sporangium and 1- spored sporangioles B. zygospore.

Family *Cunninghamellaceae*

There are five genera. *Cunninghamella*, has 7 species, which are found frequently in soils. These produce (Fig. 7.8) 1-spored sporangioles (conidia). Zygospores, where known, are of *Mucor*-type i.e. borne on opposed suspensors. The conidia are hyaline and borne singly on globbular vesicles, formed apically on simple or branched conidiophores of indefinite length.

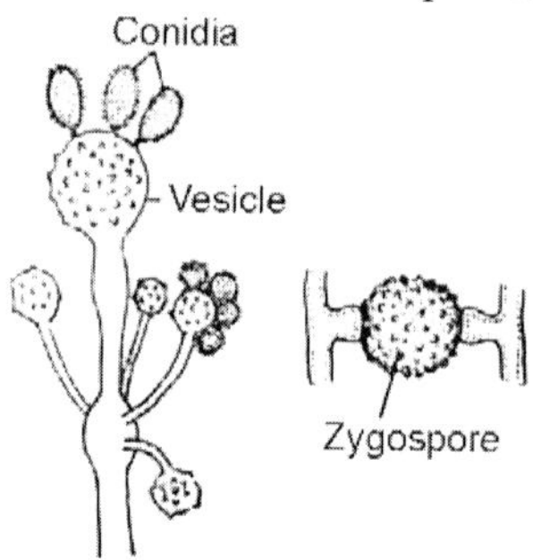

Fig. 7.8. *Cunninghamella echinulata.*

Family *Phycomycetaceae*

Genus *Phycomyces*

There are 2 genera. *Phycomyces* (3 species), grows on dung, wood and substances rich in fatty materials. *P. blakesleeanus* (Fig. 7.9) is heterothallic and shows highly sensitive phototropic behaviour. Its sporangial development has been used as a model system to study the genetic basis of various tropisms, including the strong phototropic responses to blue light.

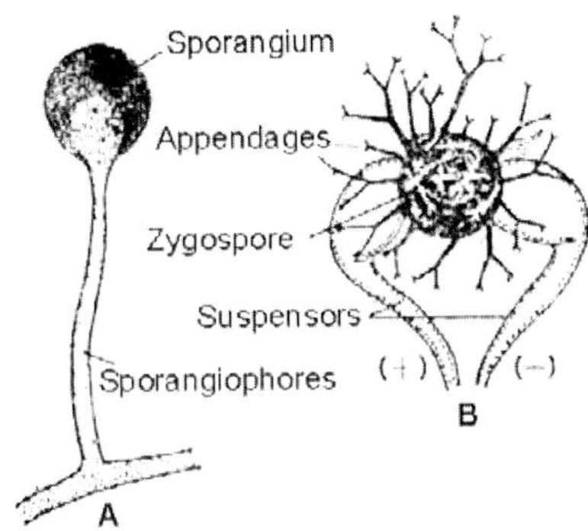

Fig. 7.9. *Phycomyces blakesleeanus*. A, sporangiophore with columullate sporangium; B, zygospore.

The sporangiophores are of two types, called **microphores** and **macrophores** depending on their length. The microphores are about 1mm long and bear small sporangia containing about 100 spores. The macrophores, which have a metallic lusture, are 100-500 mm long and bear big sporangia containing 100,000 spores.

The sporangia are columullate with thick, persistent walls. Zygospores are borne between tong-shaped suspensors, which bear thorn-like, dichotomously branched appendages near the junction with zygospore. Meiosis occurs during germination. This has been demonstrated in *Phycomyces blakesleeanus*.

Family *Pilobolaceae*

The two genera included in this family viz. *Pilobolus* and *Utharomyces* are coprophilous (i.e. grow on dung). Sporangia are dark with a flat columella which is not very conspicuous. The sporangiophores are long and phototropic. Suspensors are tong-shaped, holding the zygospore between them. All are heterothallic.

Genus *Pilobolus*
(Hat thrower or Shot gun fungus)

The genus, comprising of 5 species, is found on dung of herbivore animals. It is the only genus of *Mucorales* that shows violent discharge of spore. The other points of interest are: (1) the specialized nutrition, (2) photropic nature of the sporangiophore, and (3) the periodicity of sporangial discharge.

Pilobolus can be isolated by incubating fresh horse or cow dung in a glass-covered humid dish and placed near a window. After one to three days a large number of sporangiophores, bent towards the incident light, can be seen through a hand lens. These may persist up to 15 days. *P. kleinii* is the most common species inhabiting horse dung, but for purposes of study *P. sphaerosus* is preferred, as it can be more easily grown in the laboratory.

The coenocytic, much-branched hyphae ramify and feed on the dung. Erect aerial sporangiophores (Fig. 7.10) arise from a carotene-rich **trophocyst**, delimited from the feeding mycelium by one or two septa depending on its terminal or intercalary position. The sporangiophore, which is 0.5 to 1cm long, bears apically a black, button-shaped, columullate sporangium subtended by much larger, crystal-clear **sub-sporangial vesicle**. The sporangium, which is estimated to contain 15 to 30 thousand spores, is violently shot off up to a distance of 2–2.5 meters. A remarkable feat indeed for a tiny mold! *Pilobolus* is aptly called the "Hat thrower" or the "Shot-gun" fungus.

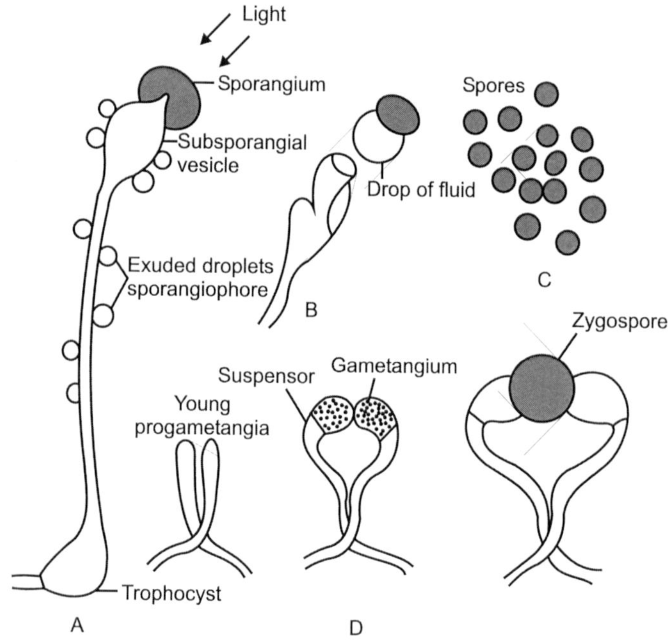

Fig. 7.10. *Pilobolus kleinii*

The mature sporangiophore, including the tryphocyst, is a single turgid cell. The stretched cell wall has a thin peripheral layer of cytoplasm, which is considerably thicker at the base of the sub-sporangial vesicle. This region is rich in oil droplets, containing carotene that gives it a conspicuous orange colour. The large vacuole in the vesicle is occupied by cell sap, which has an osmotic pressure of as high as 7 atmospheres. The sub-sporangial vesicle bursts below the junction with the columella. The stretched wall contracts with lightning speed and squirts out a jet of cell sap which propels the sporangium to a distance of 2–2.5 meters.

The darting sporangium gets glued to the object it strikes, like grass leaves. The sporangia reach the alimentary canal of herbivore animal along with the grass it eats where the spores are released. The spores receive the stimulation required for germination in the alimentary canal, which later occurs in the freshly-deposited dung. Hesseltine *et. al.* (1952) reported the presence of a growth factor, **coprogen,** in the dung which was necessary for the growth of *Pilobolus*. Now coprogen is known to be a **siderophore** (iron-chelating compound, produced by microorganisms under iron–deficient environment).

Sporangial production is a rhythmic phenomenon requiring alternate light and dark periods. The sprangiophore formation is initiated during night and is completed by the morning. By mid-day the sporangia are discharged. Successive crops of sporangia appear each morning at a definite interval of 24 hrs. It is an example of **Circadian Rhythm** or 24-hour **Biological Clock**. By adjusting the light and dark periods the 'clock' can be adjusted to get crops of sporangia at the desired time during day or night.

The genus has both heterothallic and homothallic species. Fusion of two multinucleate gametangia of opposite strains leads to the formation of a thick, smooth-walled zygospore. The suspensors are twisted around each other and then separated like jaws of a pair of tongs. The zygospore is held by their tips. After a period of dormancy, the zygospore germinates and forms a sporangiophore and the "germ sporangium". Meiosis occurs and haploid spores are produced in the germ sporangium. The germination of the zygospore, however, has been infrequently seen. It can also germinate directly producing a mycelium.

The other genus, *Utharomyces*, lacks the violent discharge of sporangia.

Family *Radiomycetaceae (= Saksenaeaceae)*

Genus *Saksenaea*

The genus *Saksenaea,* with one species, *S. vasiformis**, has gained much importance as a human pathogen. It forms a sporangium, which is

---

* *S. vasiformis* is a "complex" of species. Alvarez *et al.* (2010) have proposed 2 new pathogenic species of the genus, *S. oblongispora* (oblong spores and no growth at 42°C) and *S. erythrospora* (ellipsoidal spores).

unique in *Mucorales*. The genus (Fig. 7.11) was discovered from India in 1953 from the forest soils of Sagar (Madhya Pradesh) by Dr. S.B. Saksena who named it after his teacher Prof. R.K. Saksena. The sporangia are formed freely if a mycelial disc is cut from an agar culture and floated on water in a Petri dish. Sporangiophores arise from the hyphae above short rhizoid-like hyphae. The sporangium, which is formed singly or in pairs, is flask-shaped, with a distinct dome-shaped columella in the venter. The neck opens at the apex through a pore, which remains plugged by a mucilagenous substance. The spores are released through this pore after dissolution of the plug.

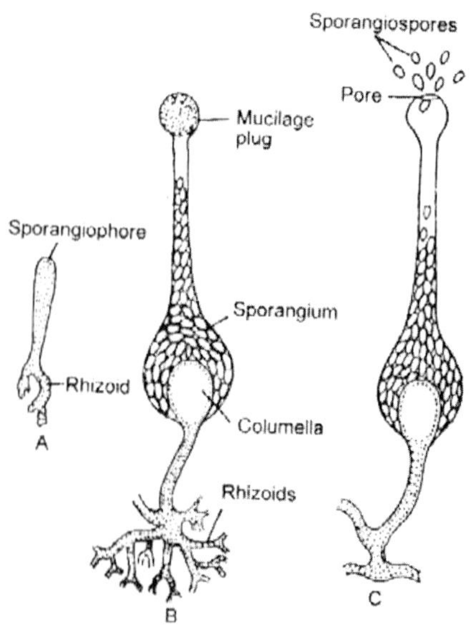

Fig. 7.11. *Saksenaea vasiformis*. A, young thallus; B, mature thallus with flask-shaped columullate sporangium; C, release of sporangiospores through an apical pore.

S. vasiformis has attained great medical importance. In 1971 a patient in USA, whose hand was amputated died due to infection of his arm stump by this fungus. Even additional amputations failed to arrest the spread of fungus. In another case (Ajello *et al.*, 1976), a patient, who had undergone plastic surgery for facial wounds, developed high fever and didn't respond to any antibacterial drug. Examination of the wound drainage materials indicated infection by this fungus, which was then promptly cured by treatment with amphotericin B. Several such cases of human infections by *S. vasiformis* have been reported since then from all over the world. Infection of burn wounds, development of soft necrotizing tissue infection (cellulolysis), with fat necrosis and vasculitis after minor trauma, are most common infections caused by *S. vasiformis* (Wilson, 2008).

Order *Endogonales* (1 family)

The order has one family, *Endoganaceae* with three genera : *Endogone*, *Sclerogone*, and *Youngiomyces*, which are saprobes or form **ectomycorrhizae**. Characteristic sclerotium-like structures, 2-3 cm in diameter, called sporocarps are formed underground which contain three types of reproductive bodies: chlamydospores, zygospores or non-columullate sporangia, but all never occur together in the same sporocarp. The sporocarps are mostly hypogeneous (underground) and interestingly, emit odour of onions, burnt sugar or fish.

These fungi have been taken in culture but do not show sporulation. Some do sporulate if grown with mycorrhizal host seedlings. The sporocarps have so far not been grown in culture.

Order *Mortierellales* (1 family)

Family *Mortierellaceae*

The order has one family *Mortierellaceae,* and seven genera, which produce one to many-spored sporangia with a vestigial columella. In some genera this may also be lacking. Chlamydospores are present. Zygospores, which are found only in a few species of *Mortierella*, are held between tong-shaped suspensors and surrounded by a weft of mycelium. *Mortierella*, (Fig. 7.12), the type genus, is common in soils.

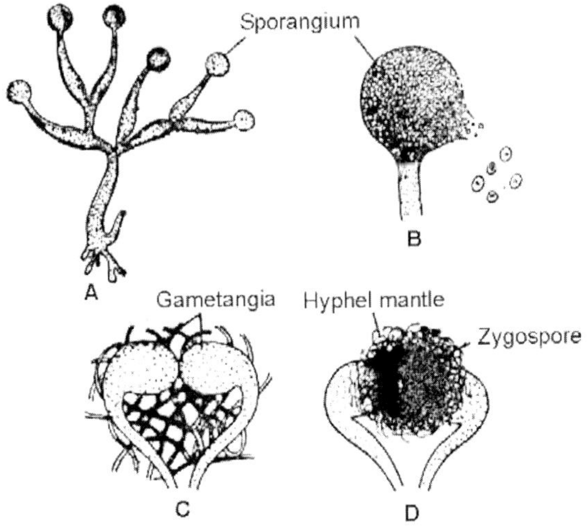

Fig. 7.12. *Mortierella*. A, habit; B, sporangium and sporangiospores; C, gametangial copulation; D, zygospore with a hyphal mantle.

Subphylum *Entomophthoromycotina* (1 order)

Insect destroyer (Greek; Entomo = insect, phthor = destroyer)

These fungi, all placed in one order *Entomophthorales*, differ morphologically from other members of *Zygomycota* in (1) formation of conidia (sporangia in some cases), and (2) in the frequent production of secondary conidia by budding.

Order *Entomophthorales* (6 families)

This order comprises a small group of fungi which are chiefly parasites of insects. A few attack nematodes and mites while some are saprobes in soil or dung. The mycelium is not extensive and consists of coenocytic coarse hyphae which in some genera e.g. *Entomophthora*, become septate and break into multinucleate cells called **hyphal bodies**. The hyphal bodies multiply by division or budding and develop conidiophore bearing a conidium. Alternatively, these might act as gametangia. In *Basidiobolus* and some other genera, the hyphae do become septate but never break up. Sexual reproduction involves fusion between two hyphal bodies, undifferentiated hyphae or enlarged adjacent hyphal segments. The zygospore develops as a lateral bud from either one of the two fused cells, or midway between them. Azygospores may be formed parthenogenetically from one of the gametangia.

Table 7.1. Genera of *Entomophthorales*: their habitats and chara-cteristics

Genus / Family	Habitats and characteristics
Ancylistes/ Anacylistaceae	Parasitic in desmids. (e.g. *A. closterii* in *Closterium*); mycelium limited, septate, hyphal bodies not formed.
Conidiobolus/ Anacylistaceae	Saprobic, common in soil, (*C. coronatus*); mycelium septate, but not breaking up into hyphal bodies.
Basidiobolus/ Basidiobolaceae	A mammalian pathogen and commensel in gut of frog and lizard. *B. ranarum causes* 'basidiobolomycosis' in man. Nuclei unusually large and visible under microscope; isolated in pure culture. Mycelium septate but does not break into hyphal bodies. Conidia violently discharged.
Completoria/ Completoriaceae	Parasitic on fern prothalli; mycelium septate, not breaking into hyphal bodies.
Entomophthora / Entomophthoraceae	Parasitic in arthropods (insects); mycelium breaks into multinucleate hyphal bodies; primary conidiophores basidium-like; conidia smooth, produced outside the host and violently discharged.
Massospora/ Entomophthoraceae	Parasitic in vertebrates; hyphal bodies formed; conidia produced inside host; passively discharged.

There are 6 families: *Entomophthoraceae, Ancylistaceae, Basidiobolaceae, Completoriaceae, Meristacraceae, and Neozygitaceae*. The organisms show varying modes of life. A few species of *Basidiobolus* and *Conidiobolus* have been successfully grown on synthetic media. The habitats and important characteristics of some important genera are given in Table 7.1.

Family *Entomophthoraceae*

Genus *Entomophthora*

The family, with 12 genera and 202 species, is the biggest family of the order. *Entomophthora* with 30 entomogenous species is the best known genus. *E. muscae* attacks houseflies and causes 'fly cholera', which is quite common during summers. Dead flies surrounded by conidia are frequently seen on window panes (Fig. 7.13). Healthy flies contact the sticky conidia and get infected. *E. maimaiga* is used as a biocontrol agent against gypsy moths.

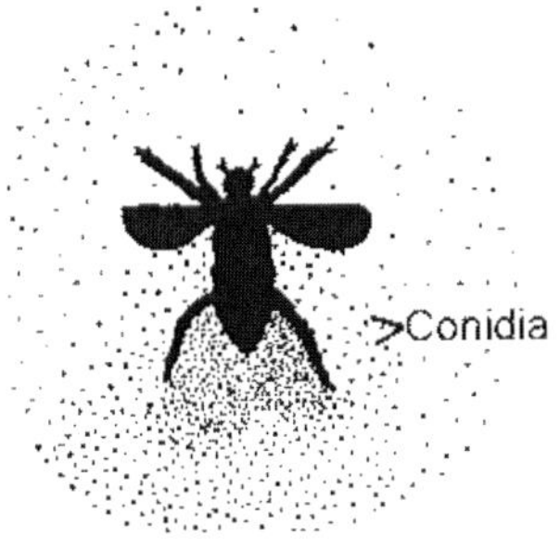

Fig. 7.13. *Entomophthora muscae*, dead fly on a glass pane with ejected conidia

Life Cycle of *E. Muscae*

The sticky conidia, when come in contact with a fly, adhere to its body with the help of the sticky muclilage that surround the conidia (Fig. 7.14). In absence of the host, the conidia produce secondary conidia by budding. The conidia germinate and the germ tube enters the body through thinner portions of the body, like the inter-segmental areas or through the joints of the appendages. Inside the body, the free end of the germ tube buds off cells which give rise to hyphae. These hyphae soon become septate and break up into hyphal bodies, which multiply by budding and soon fill the entire body of the fly. The insect becomes completely plugged with the fungus. Finally, all the internal structures are destroyed. Such fungus-stuffed flies cling to any substratum (glass panes, branches or leaves) in a firmly attached position. Flies die after 5–8 days of infection.

Shortly after death, the hyphal bodies form conidiophores, which emerge through the inter-segmental portions. The conidiophores are club-shaped and bear a single conidium at the tip. Conidia are discharged violently as in *Pilobolus*. A biconvex, water-filled cavity appears in the conidium above the septum, which generates a hydrostatic pressure that brings about the violent discharge. The conidiophore bursts shooting the conidium to a distance of 2-3 cm. The enormous number of the ejected conidia, form a white ring (halo) around the dead fly.

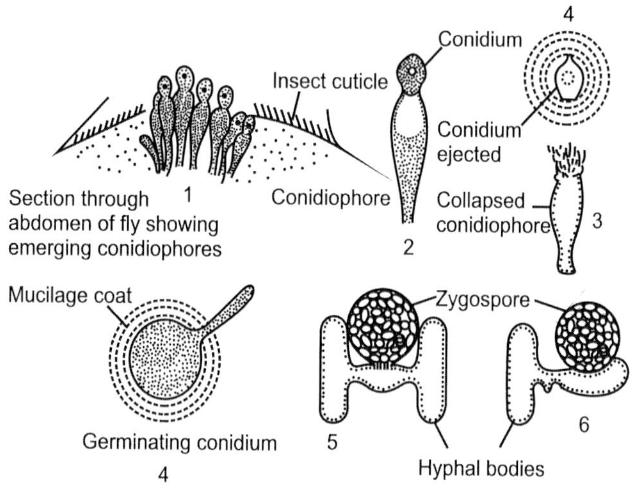

Fig. 7.14. *E. muscae*: 1-4, asexual reproduction; 5 and 6, sexual reproduction.

Conidium formation is followed by sexual reproduction that leads to the formation of zygospore. Hyphal bodies act as gametangia. The zygospore is formed by: (1) enlargement of one of the two gametangia, (2) an outgrowth arising between the two hyphal bodies, or (3) one of them after fusion.

The role of the zygospore is not known. It might germinate by a germ tube after a resting period. Parthenogenetic development of the zygospore (called azygospore) is also known.

Family *Basidiobolaceae*

The inclusion of this family under *Entomphthorales* is controversial. Molecular studies are inconclusive on their position. Some suggest that the *Basidiobalace* are closer to Blastocladian chytids and they should be shifted there, while some (Tanabe *et al.,* 2004, and James *et al.,* 2006) prefer to put them under *Zygomycota.*

Genus *Basidiobolus*

The family has one genus *Basidiobolus* with four species. It is commonly found in soil, decaying vegetables, and in the gut of amphibians, reptiles, fish and bat. *B. ranarum* causes a serious subcutaneous disease **basidiobolomycosis** of human beings. The disease occurs usually in children, less often in adolescents and rarely in adults. Singh *et al.* (2008) reported a case of basidioboloomycosis showing bilateral nasal block and nasal discharge. CT scan of sinuses revealed polypoidal mass in all the sinuses with extradural extension. The tissue biopsy examined histopathologically and microbiologically, revealed *B. ranarum*.

Basidiobolus can be easily isolated from the excreta of frogs. Place a frog in a jar containing water on a filter paper. Filter off the dung and place the moist filter paper on the under-surface of the lid of a Petri dish containing peptone agar in the lower lid. Within a few days coarse mycelium of *Basidiobolus* becomes visible on Petri plate.

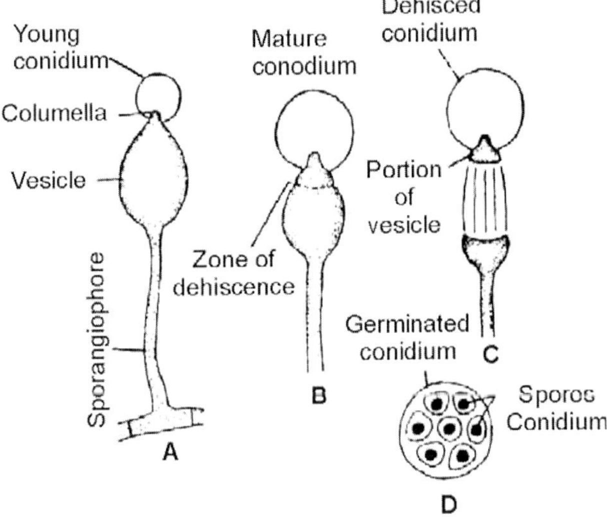

Fig. 7.15. *Basidiobolus ranarum*, showing A, conidiophore bearing conidium over a vesicle; B, mature conidium and zone of bursting of vesicle; C, explosive discharge of conidium; D, conidial germination and formation of 8 aplanospores.

Life Cycle of *B. ranarum*

The coarse hyphae become septate but do not break into 'hyphal bodies'. The vertical, light-sensitive conidiophores (Fig. 7.15) consist of a slender stalk and a characteristic inflated vesicle (not formed in

Entomophthoraceae) below the conidium. The tip of the vesicle protrudes into the conidium as a minute columella, as in *Pilobolus*. This inflated vesicle is turgid and bursts along a weak zone near the base of the conidium. As a result, the upper part of the vesicle, along with the conidium, is shot off as one unit; the part of the vesicle propels the conidium like a rocket. It is, however, detached during the flight in the air. The conidia, on germination, may form and explode a similar secondary conidium. Usually the primary conidia are eaten by beetles, and remain unchanged in the alimentary canal. But when the beetles themselves are eaten by frog, lizard, etc, the conidia get released into the alimentary canal. The 'conidium' acts as a sporangium and gives rise to usually eight, thin-walled, uninucleate aplanospores. The spores which germinate before or after excretion, give rise to the coarse mycelium.

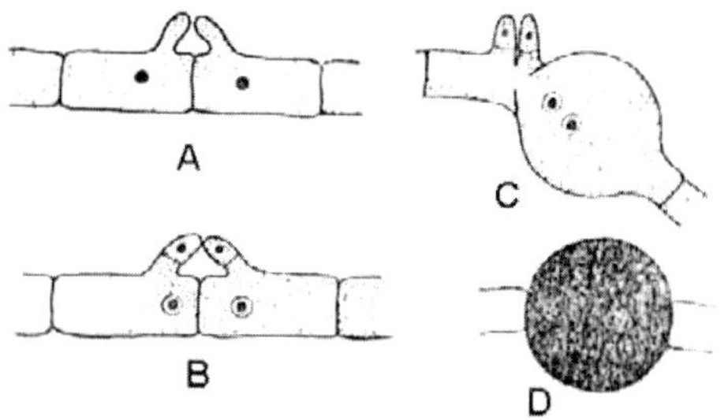

Fig. 7.16. Sexual reproduction in *Basidiobolus ranarum*.

Zygospores are formed by conjugation between two adjacent cells of a hypha (Fig. 7.16). Each of the conjugating cells puts out a beak-like outgrowth adjoining the septum. The single nucleus of each cell migrates into the beak and divides. One nucleus lies in the beak and is separated by a septum. One of the two original cells enlarges considerably; a pore is dissolved in the separating septum and the nucleus of the smaller cell migrates with some cytoplasm into the enlarged cell. Karyogamy takes place and the large cell develops into the zygospore. On germination, the zygospore gives rise to a germ tube, which develops into a conidiophore or a hypha.

Subphylum *Kickxellomycotina* (4 orders)

There are 4 orders: *Asellariales, Kickxellales, Dimargaritales* and *Harpellales*. The *Asellariales* and *Harpellales*, which earlier belonged to the erstwhile class *Trichomycetes*, are endocommensals of arthropods (*Insecta*), while members of *Kickxellales* are saprobes or rarely parasites. The members of *Dimargaritales* are mycoparasites. Analysis of RPB1 (RNA Polymerase B1) DNA sequences demonstrated that three orders *Kickxellales, Harpellales* and *Dimargaritales* form a monophyletic group "K-H-D clade"

(Tanabe *et al.,* 2004). The synapolymorphy (the shared derived morphological character that defines a clade) is the possession of **uniperforated septum** with a lenticular cavity (Fig. 7.17). A lens-shaped plug lies within the septal pore (lenticular cavity).

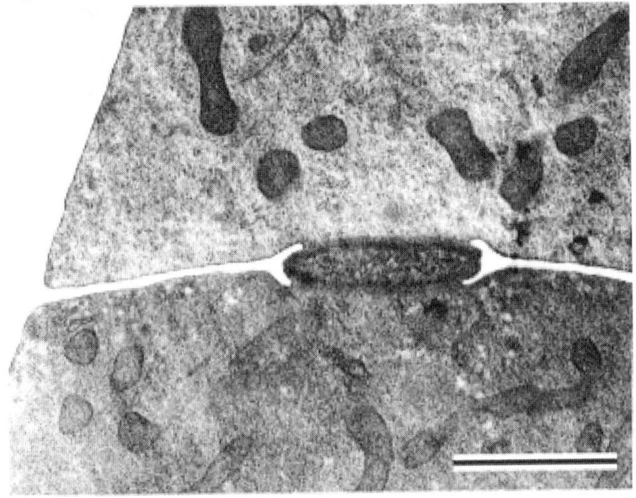

Fig. 7.17. Trannsmission electrom micrograph of vegetative hypha of *Kickxella* sp., A lens-shaped plug lies within the septal pore (=lenticular cavity). (Source: Tanabe *et al.,* 2004)

Order *Kickxellales* (1 family)

Family *Kickxellaceae*

The family has 12 genera, which are characterized by (1) presence of regularly septate hyphae, having uni-perforated septa, (2) formation of zygospores from progametangia that are not differentiated into suspensor and gametangium, (3) the formation of one-spored sporangia (sporangiole) laterally on one side of a septate or unseptate sporocladium, borne apically on the sporangiophore.

The genera are found in soil, on dung, dead insects, decaying debris etc. A few species are parasitic. These are differentiated on the basis of the nature of the sporangiophore and sporocladia. In *Kickxella* (Fig. 7.18), the sporocladia are sessile and formed simultaneously in a whorl at the apex of the sporangiophore. In *Coemansia* (Fig. 7.19) the sporocladia, which are stalked, though formed terminally are pushed to the lateral side, due to continued growth of the sporangiophore. The one-spored, elongate, elliptical sporangioles arise from pseudophialides arranged transversely on the lower surface of the boat-shaped sporocladia.

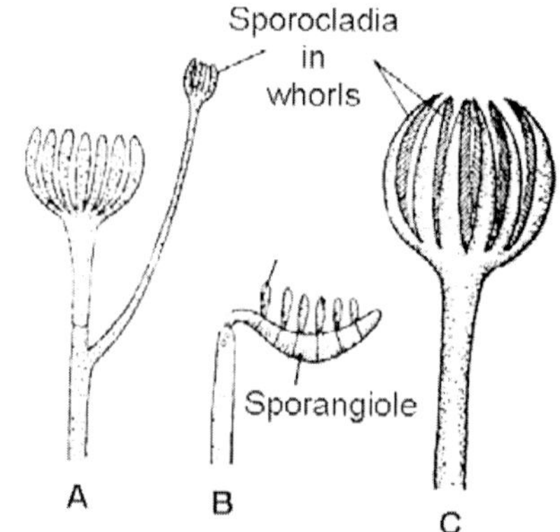

Fig. 7.18. *Kickxella* sp

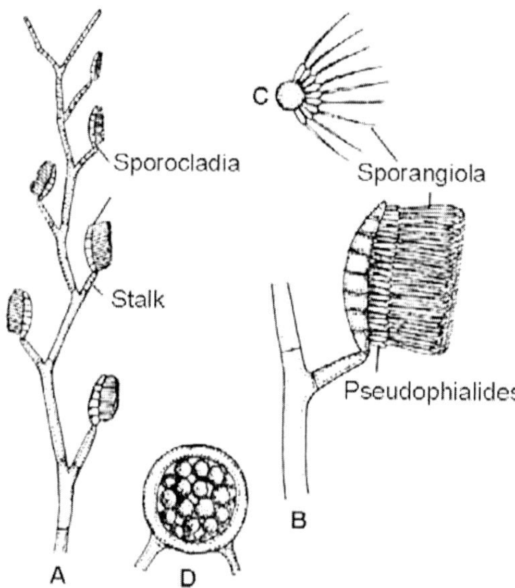

Fig. 7.19. *Coemansia pectinata* A. Sporangiophore, B. single sporocladium with pseudophialides and sporangioles, C. T.S. of sporocladium, D. zygspore

Order *Harpellales* (2 families)

There are two families: *Harpellaceae* and *Legeriomycetaceae* (=*Genistellaceae*), including 36 genera. The members are endo-commensals in the alimentary canal of fresh water arthropods (*Insecta*). The thallus is septaate, unbranched (*Harpellaceae*) or branched (*Legeriomycetaceae*) with uniperforated septa and a basal hold fast. Asexual reproduction is typically through trichospores. The zygospore development is by conjugation between two thalli. There is no special structure like progametangium.

The zygospores are typically biconical. The genera are identified by the shape of trichospores, the number of the appendages, zygospore morphology and hold-fast characteristics.

The order is characterized by the **trichospores**, which bear one or more hair-like appendages at the basal end. These appendages are not flagella, (lack the 9 + 2 structure); nevertheless, they help in the passive transmission of the spore. The trichospores are produced exogenously on **generative cells**, but the appendages arise from within the generative cells. The trichospore is actually an elongate monosporous sporangium. Once ingested by the host, it germinates, with the inner wall breaking out through the outer wall and attaching to the gut-wall. The zygospore too after a rest period germinates inside the host and gives rise to the thallus.

The two families differ in the nature of the thallus and part of the arthropod gut they inhabit. In *Harpellaceae* (5 genera) the thallus is unbranched and they inhabit the mid gut. Example: *Harpella* and *Stachylina* (Fig. 7.20). The trichospore of *Harpella* is cylindrical and coiled with four appendages, while in *Stachylina* (and *Smittium* of *Legeriomycetaceae*) it is ellipsoidal with a single appendage. In *Legeriomycetaceae* (31 genera) the thallus is branched and they inhabit the hind gut. *Smittium, Legeriomyces, Genistellospora* are some of the important genera.

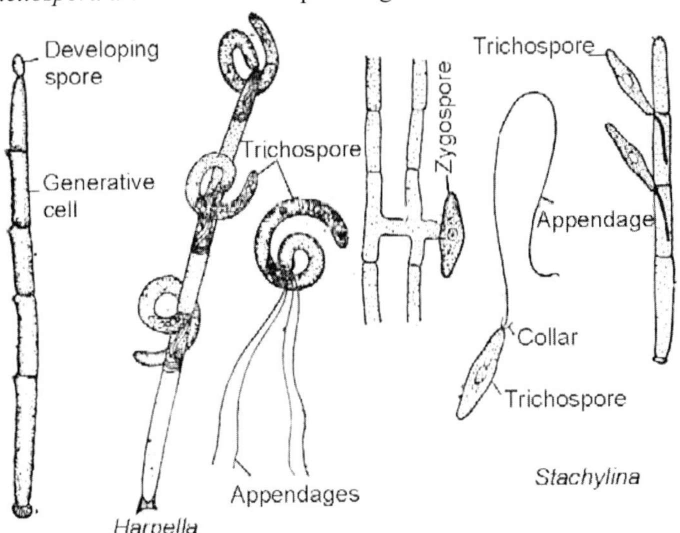

Fig. 7.20 *Harpella* and *Stachylina*

Order *Demargaritales* (1 family)

Family *Dimargaritaceae*

The members of this family, which are mycoparasites, are characterized by (1) presence of regularly septate hyphae having uni-perforated septa (2) obligately parasitic nature (on *Mucorales* and rarely on *Chaetomium*), and (3) presence of haustoria that penetrate into the host cells.

These fungi form merosporangia on simple or complex sporangiophores. (Cylindrical sporangioles, cotaining spores in a row, are called **merosporangia**). The merosporangia too, like the sporangioles, may have in some cases a single spore, free or with fused walls.

Zygospores are similar to those of *Kickxellales*, i.e. formed from progametangia that are not differentiated into suspensor and gametangium.

There are 3 three genera: *Dimargaris*, *Dispira*, and *Tieghemiomyces*.

These mycoparasites have special nutritional requirement, provided by the host mycelium. These can be grown in culture containing glycerol or other 3-carbon sugars, as these fungi do not utilize common sugars including glucose.

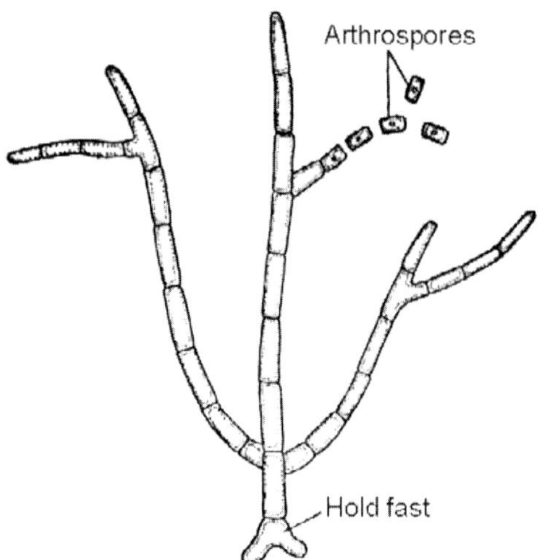

Fig. 7.21. *Asellaria*

Order *Asellariales* (1 family)

There is a single family *Asellariaceae*, comprising of two genera, which are endo-commensals in the alimentary canal of fresh water arthropods (*Insecta*). These possess branched and septate hyphae, with uniperforated septa. Arthrospores serve as the only means of reproduction, example: *Asellaria* (Fig. 7.21).

Subphylum *Zoopagomycotina* (1 order)

Order *Zoopagales* (5 families)

The single order includes fungi which are obligate predators or parasites of protozoans (mainly *Amoeba*), nematodes, and fungi (mostly *Mucorales*). The branched coenocytic hyphae are extensive and possess complex haustoria in the predacious genera. In the endoparasitic genera, the thallus is a coiled cylindrical hypha of usually one and a half turns. The thallus is further reduced in ectoparasites. *The formation of true conidia, which are passively discharged, is the characteristic feature of the order.* Fusion of two similar or dissimilar gametangia, results in the formation of a zygospore, which bears hemispherical warts. In the predacious genera, the gametangia are undifferentiated hyphal branches i.e. no suspensors and gametangia. In some species the zygospore is contained in a zygosporangium.

There are 22 genera placed in five families: *Zoopagaceae*, *Cochlonemataceae, Helicocephalidaceae, Piptocephalidaceae,* and *Sigmoideomycetaceae.* While *Helicocephalidaceae* includes obligate parasites of nematode eggs, *Sigmoideomycetaceae* too includes mycoparasitic fungi, but differ from *Piptocephalidaceae* in lacking haustoria. The *Piptocephalidaceae* are contact mycoparasites forming haustoria.

Since none of the zoopagales have been cultured on chemically-defined media, our knowledge about their physiology is almost negligible. Naturally, some chemicals present in the specific hosts are indispensable for their germination and growth. As a matter of fact, these fungi were discovered as late as 1935 by Drechsler. Duddington (1968) assigned two reasons for their late discovery: (1) their minute size, and (2) the reluctance on part of mycologists to examine rather 'dirty' plate-cultures of these fungi.

Family *Zoopagaceae* (Predacious *Zoopagles*)

The family includes 5 predacious genera (Fig. 7.22): *Acaulopage*, *Cystopage*, *Stylopage*, *Zoopage*, and *Zoophagus*, which capture and devour their prey, mostly amoebae and nematodes. (*Zoophagus* captures algae besides rotifers). The hyphae secrete a sticky substance that helps in holding the prey that comes in its contact. The fungus immediately pushes its haustoria into the body of the 'catch', and drains out all the available nutrients. The carcass of the dead amoebae remain attached to the hyphae.

Species that capture nematodes push trophic hyphae into the body of the prey.

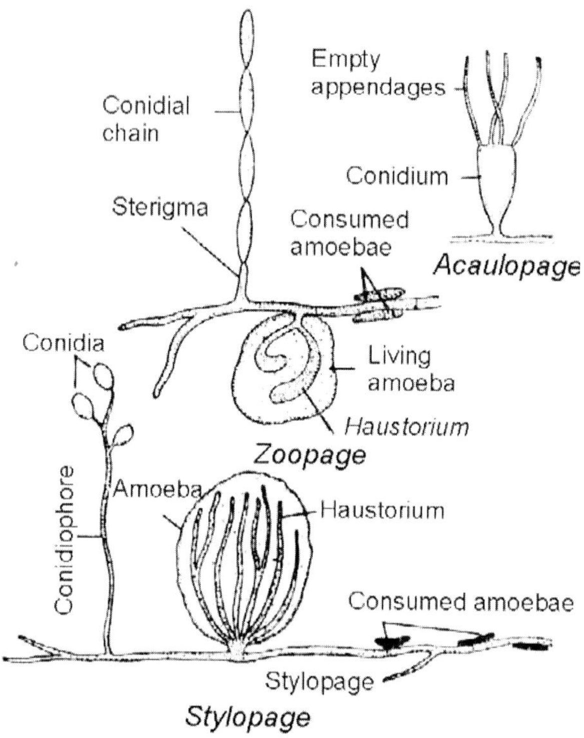

Fig. 7.22. Some predacious zoopagales, *Zoopage*, *Acaulopage* and *Stylopage*.

In *Zoopage*, the haustoria are inflated or lobed and the external unseptate mycelium bears chains of spindle-shaped conidia on short sterigmata. *Acaulopage* and *Stylopage* do not form the conidia in chains. *Acaulopage* forms conidia directly on the hypha or occasionally on short sterigmata. The conidia bear one or more empty, distal appendages. *Stylopage* produces conidia singly or in groups on erect and long conidiophores. It produces characteristic haustoria, which are bush-like in *S. rhicnacra* and pedicillate in *S. haploe*. *Cystopage* lacks conidia, but it froms chlamydospores. Sexual reproduction is unknown in many species. Zygospores, where formed, are warty and borne on spirally twisted suspensors.

Family *Cochlonemataceae* (Parasitic *Zoopagales*)

There are 6 genera: *Amoebophilus*, *Aplectosoma*, *Bdellospora*, *Cochlonema*, *Endocochlus* and *Euryancale*. Except *Amoebophilus*, which includes etcoparasitic species, the genera are endoparasites.

We shall study *Cochlonema* and *Endocochlus* both, endo-parasites of amoebae. The thallus is a broad, coiled filament of one and a half turns. These fungi feed on amoebae which eventually die. After the death of the amoebae, aerial unbranched and unseptate hyphae emerge from the coiled internal thallus, and bear conidia or gametangia. In *Cochlonema* (Fig. 7.23) the fertile hyphae transform into a chain of spindle-shaped conidia, from the apical end. In *Endocochlus* (Fig. 7.24), on the other hand, the terminal portions of the aerial hyphae become septate, and from some of the resulting cells, spindle-shaped, sessile conidia arise laterally. The detached conidia contact fresh amoebae and infect them. A germ tube, formed by the germinating conidium, enters the amoeba and develops into a globular swelling, the tip of which elongates and gives rise to the helical thallus. Infection may also occur by ingestion of the conidia by the amoebae. Occasionally, a single amoeba shows multiple infections, evidenced by the presence of several thalli.

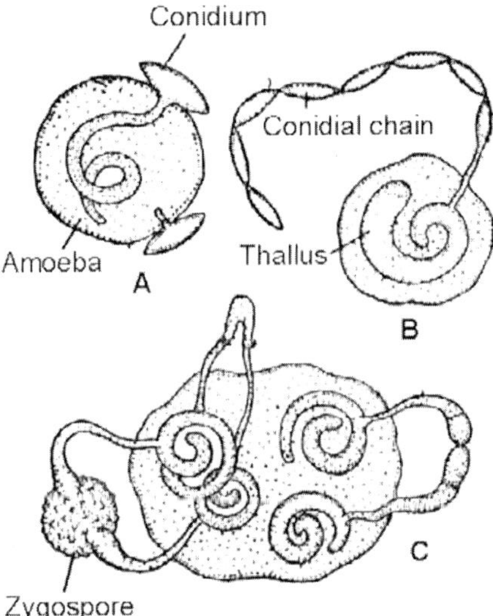

Fig. 7.23. *Cochlonema*, an endoparaste of *Amoeba*; A, infection through conidia; B, chain of external conidia and C. zygospore with a warty wall.

Sexual reproduction occurs by fusion of the apical portions of two aerial hyphae, originating from different individuals. This suggests that these fungi are heterothallic. The zygospore is thick-walled and enclosed in a zygosporangium, formed by extension of one of the copulating hyphae (progametangia). The wall of the zygospore is angularly lobed in *Endocochlus* and warty in *Cochlonema*. In some species the collapsed wall of the zygosporangium persists on the zygospore, which is helpful in identification of the species. The nuclear aspects remain unknown.

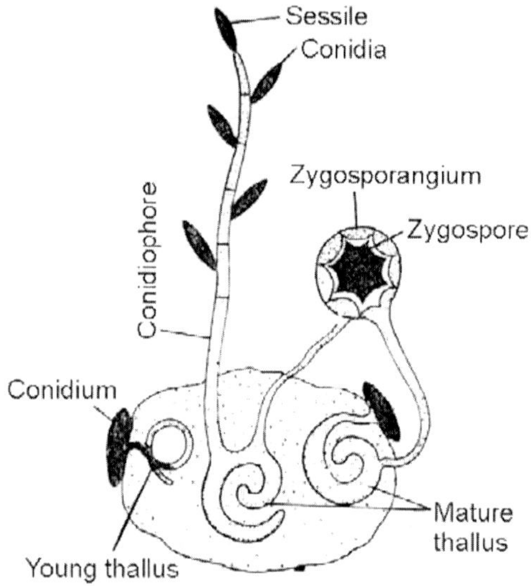

Fig. 7.24. *Endocochlus*, an endoparasite of *Amoeba*. Note the sessile conidia and zygospore.

Family *Piptocephalidaceae*

There are two genera, *Piptocephalis* and *Syncephalis*, which are parasites on other *Mucorales*. These are characterized by one to many-spored merosporangia, borne on vesicles at the tips of sporangiophores, which are branched (*Piptocephalis*) or unbranched (*Syncephalis*). The thin and highly-branched hyphae adhere to the host hyphae and form appresoria at certain places, from which haustoria originate and penetrate into the host to derive nutrition. The zygospres are held between suspensors that are tong-shaped and coiled.

In *Piptocephalis* (Fig. 7.25), the sporangiophore is dichotomously branched and bears merosporangia, which are deciduous. In *Syncephalis*, (Fig. 7.26) the sporangiophore is unbranched, thicker at the base and gradually tapering towards the apex, where the merosporangia (which are not deciduous) are borne on fertile vesicles. The erect sporangiophore is secured on the host hypha by means of suckers.

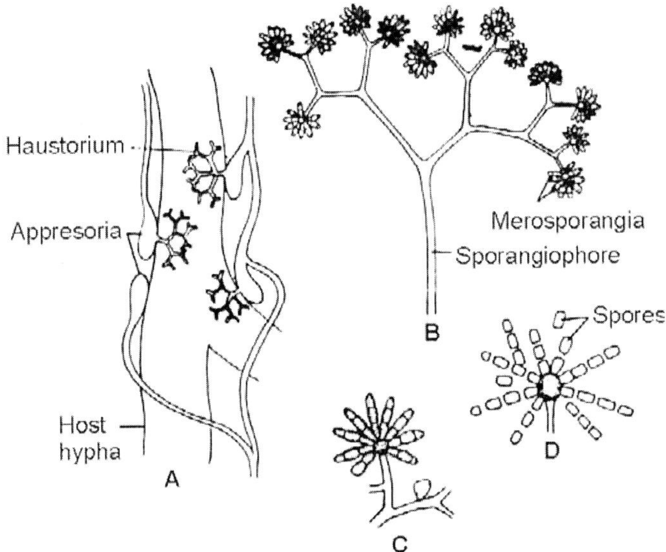

Fig. 7.25. *Piptocephalis*. A. habit; B. sporangiophore bearing merosporangia; C. magnified view of merosporangia; D, breakdown of merosporangia.

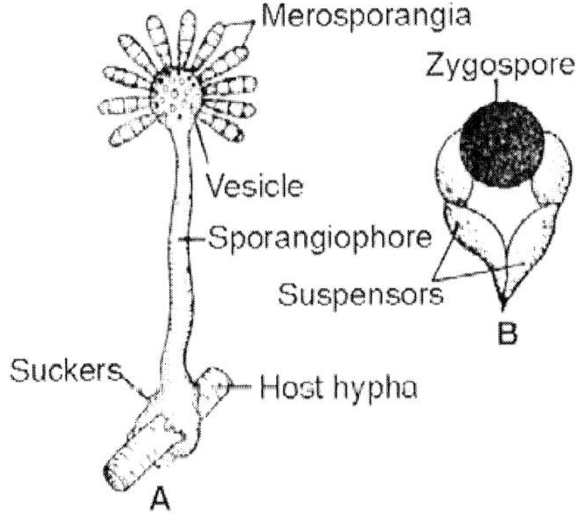

Fig. 7.26. *Syncephalis nana*. A. unbranched sporangiophore with merosporangia; B. zygospore with tong shaped suspensors.

REFERENCES

Ajello, L.D. *et al.* 1976. The zygomycete *Saksenaea vasiformis* as a pathogen of humans with a critical review of the etiology of zygomycosis. *Mycologia*, 68: 52–61.

Alvarez, E. *et al.* 2010 Molecular phylogeny and proposal of two new species of the emerging pathogenic fungus, *Saksenaea. Jr. Clinical Microbiol.* 48(12): 4410-16.

Drechsler, C. 1937. Some conidial phycomycetes destructive to terricolous amoebae. *Mycologia*, 27: 6–40.

Duddington, C.L. 1968. Predacious fungi. *In: The Fungi An Advanced Treatise*. Vol. III. (G.C. Ainsworth and A.S. Sussman eds.). pp. 239–251. Academic Press, N.Y. and London.

Duddington, C.L. 1973. Zoopagales. In: *The Fungi. An Advanced Treatise*. Vol. IV B. (G.C. Ainsworth, F.K. Sparrow and A.S. Sussman eds.). p. 231–234. Academic Press, N.Y. and London.

Hesseltine, C.W. *et al.,* 1952. Coprogen, a new growth factor for copro-philic fungi. *Jr. Am. Chem. Soc.* 14: 7.

James, T.Y. *et al.* 2006 Restructuring the early evolution of Fungi using a six gene phylogeny. *Nature* 443: 818-822

Kirk, P.M. *et al.* 2008 *Ainsworth and Bisby's Dictionary of the Fungi* (10th ed.) CAB International, Oxon, U.K.

Saksena, S.B. 1953. A new genus of the Mucorales. *Mycologia*. 45: 426–436.

Schuessler, A.D. 2001 A new phylum *Glomeromycota*: phylogeny and evolution. *Mycol. Res.* 105: 1413-1421

Singh, R. *et al.* 2008 Basidiobolomycosis; a rare case report. *Indian J Med Microbiol.* 26: 265-267

Tanabe, Y. *et al.* 2000 Molecular phylogeny of parasitic *Zygomycota* (*Dimargaritales, Zoopagales*) based on nuclear SSSU rDNA sequen-ces. *Mol. Phylogenet. Evol.* 16: 253-262

Tanabe, Y. *et al.* 2004 Molecular phylogeny of *Zygomycota* on EF-1 and RPB 1 sequences. *Mol. Phylogenet. Evol.* 30: 438-449

Wilson, 2008 Zygomycosis due to *Saksenaea vasiformis. Med. Jr. Australia*. 89: 521-522

SUMMARY

Phylum *Zygomycota* includes fungi that form zygospores after gametangial copulation. They reproduce asexually by non-motile aplanospores, the sporangispores or conidia. Their cell wall is made of chitosan chitin. The hyphae may be coenocytic (as in *Mucoromycotina* and *Zoopagomycotina*) or septate to varying degree (in other two subphyla). The classification of the phylum has undergone many big changes in recent years due to greater reliance on molecular phylogenetic data, indicating genetic relatedness, than on morphological resemblances. The former classes Zygomycetes and Trichomycetes are no more valid. The order Glomales of former class Zygomycetes, has been elevated to level of phylum (*Glomeromcota*), and two of the four orders of the former class Trichomycetes have been shifted to

Protozoan fungi and rest two retained under *Zygomycota*. Even now the phylum is polyphyletic and can see more changes in coming years.

Presently the phylum *Zygomycota* has four monophyletic subphyla, but the relationships among them are not resolved. The subphyla have 9 orders and 27 families (no class). These are:

Mucoromycotina (3 orders) *Mucorales, Endogonales,* and *Mortierellales.*

The thallus consists of typical coenocytic mycelium. Asexual reproduction is through sporangiospores formed inside a sporangium, sporangiole or merosporangium. It forms the basis of division into families and identification of the genera and species. The sporangia or conidia (one-spored sporangia with fused walls) are borne apically on simple or characteristically-branched sporangiophores or conidiophores. The sporangiospores are released passively after rupture of the sporangial wall and disseminated. The sporangia are discharged explosively in *Pilobolus*. Sexual reproduction involves fusion of two gametamgia subtended by suspensor, formed on conjugation hyphae (progameangium). The size, shape and appendages of suspensors are of great help in identification of the genera. The zygospore represents the only diploid phase in he life cycle. The order has 9 families. Some of the more important genera are; *Mucor, Rhizopus, Absidia, Zygorhyncus*, *Thamnidium*, *Choanephora, Blakeslea Cunninghamella, Phycomyces, Pilobolus and Saksenaea*.

Entomophthoromycotina (1 order) *Entomophthorales.*

These fungi, placed in one order *Entomophthorales*, differ morphologically from other members of *Zygomycota* in (1) formation of conidia (sporangia in some cases), and (2) in the frequent production of secondary conidia by budding. The order comprises a small group of fungi which are chiefly parasites of insects. The mycelium is not extensive and consists of coenocytic coarse hyphae which in some genera e.g. *Entomophthora*, become septate and break into multinucleate cells called **hyphal bodies**. The hyphal bodies multiply by division or budding and develop conidiophore bearing a conidium. Alternatively, these might act as gametangia. In *Basidiobolus* and some other genera, the hyphae do become septate but never break up. Sexual reproduction involves fusion between two hyphal bodies, undifferentiated hyphae or enlarged adjacent hyphal segments. The zygospore develops as a lateral bud from either one of the two fused cells, or midway between them. Azygospores may be formed parthenogenetically from one of the gametangia. Therw are 6 families. The inclusion of the family *Basidiobolaceae* under *Entomphthorales* is controversial. Molecular studies are inconclusive on their position. Some suggest that the *Basidiobalace* are closer to blastocladian chytids and they should be shifted there, while some (Tanabe *et al.*, 2004, and James *et al.*, 2006) prefer to put them under *Zygomycota*.

Kickxellomycotina (4 orders)

There are 4 orders : *Asellariales* (1 family), *Kickxellales* (1 family), *Dimargaritales* (1 family) and *Harpellales* (2 families). Analysis of RPB1 (RNA Polymerase B1) DNA sequences demonstrated that three orders *Kickxellales, Harpellales* and *Dimargaritales* form a monophyletic group "K-H-D clade". The synapolymorphy (the shared derived morphological character that defines a clade) is the possession of **uniperforated septum** with a lenticular cavity. A lens-shaped plug lies within the septal pore (lenticular cavity).

In ***Kickxellaceae***, whose members are saprobes (unlike other three orders), form zygospores by fusion of copulation hyphae (progametangium) and asexual reproduction is through sporangioles borne on sporocladia.

The ***Harpellales*** (commensals in arthropods) are characterized by formation of trichospores and conical zygospores (formed by fusion of undifferentiated thalli). The thallus is septate, unbranched in the family ***Harpellaceae***, and septate and branched in ***Legeriomycetaceae***.

Family *Dimargaritaceae* includes obligate mycoparasites on *Mucorales* (rarely on *Chaetomium*). These form haustoria that penetrate into the host cells. Merosporangia serve as asexual bodies. Zygospores, as in *Kickxellales*, are formed by fusion of progametangia.

Family ***Asellariaceae*** are endo-commensals in the alimentary canal of fresh water arthropods (*Insecta*). These possess branched and septate hyphae and arthrospores serve as the only means of reproduction.

Zoopagomycotina (1 order) *Zoopagales* (5 families)

The formation of true conidia, which are passively discharged, is the characteristic feature of this group. The single order *Zoopagales,* with 5 families, includes fungi which are obligate predators or parasites of protozoans (mainly *Amoeba*), nematodes, and fungi (mostly *Mucorales*). The branched coenocytic hyphae are extensive and form complex haustoria in the predacious genera. In the endoparasitic genera, the thallus is a coiled cylindrical hypha of usually one and a half turns. The thallus is further reduced in ectoparasites. Fusion of two similar or dissimilar gametangia, results in the formation of a zygospore, which bears hemispherical warts. In the predacious genera, progametangia fuse to from the zygospore (no suspensors and gametangia). In some species the zygospore is contained in a zygosporangium.

Zoopagaceae (includes predacious zoopagales forming haustoria), ***Cochlonemataceae*** (includes endoparasitic zoopagales), ***Helicocephalidaceae*** (includes obligate parasites of nematode eggs), ***Sigmoideomycetaceae*** (includes mycoparasitic fungi, lacking haustoria), ***Piptocephalidaceae*** (includes haustorial mycoparasites).

TEST QUESTIONS

1. What are the distinguishing features of phylum *Zygomycota*? Give its classification into subphyla and orders.
2. Give the classification of the order *Mucorales* into families. Mention the more important genera with their families.
3. Describe the characteristic features of subphylum *Entomophthoromycotina*. Describe the important features (not life cycle) and importance of *Entomophthora* or *Basidiobolus*.
4. What is K-H-D clade in *Kickxellomycotina*? What are the characteristic features of the members of the clade? How do the orders differ in their habitat?
5. Describe the important features of *Zoopagales* (*Zoopagomycotina*). In which two families are haustoria formed? Describe a few members of *Zoopagaceae* and *Cochlonemataceae*.
6. Answer **any four** in brief.
   (a) Why were Glomales (= Glomerales) removed from *Zygomycota* as a separate phylum?
   (b) Which subphyla in *Zygomycota* have coenocytic fungi?
   (c) Which members in *Zygomycota* discharge the sporangia explosively?
   (d) What is "Torula condition" in *Mucor*?
   (e) Members of which family in *Mucoromycotina* form ectomycorrhizae?
   (f) Which two characters distinguish *Entomopthoromycotina* from other *Zygomycota*?
   (g) Why is the family *Basidiobolaceae* considered a misfit under *Entomophthorales*?

(h) Members of which two families in *Zoopagomycotina* form haustoria?

7. Describe the genera *Saksenaea,* and *Basidiobolus* and give their importance in medicine.

8. Write on the following with reference to *Pilobolus*: biological clock, coprogen, and trophocyst.

9. Write short notes on **any three** of the following genera: *Phycomyces, Zygorhyncus, Thamnidium, Harpella, Zoopage* and *Cochlonema.*

10. Distinguish between **any four**:

    (a) Sporangia, sporangiole and merosporangia,

    (b) Zygophore and progametangia in *Mucorales*,

    (c) *Mucor, Rhizopus* and *Absidia,*

    (d) *Zoopage* and *Acaulopage*,

    (e) *Piptocephalis* and *Syncephalis*,

    (f) *Kickxella* and *Coemansia*,

    (g) *Choanephora* and *Blakeslea*,

    (h) Trichospores of *Harpella* and *Stachylina*,

    (i) *Harpellaceae* and *Legeriomycetaceae*,

    (j) Habitat of *Kickxellaceae* and *Dimargaritaceae.*

8

Phylum *Glomeromycota*

Coenocytic, Arbuscular Mycorrhizal (AM) Fungi

The phylum is characterized by: (1) obligate symbiosis, (2) formation of arbusules in plant root cells, (3) conocytic hyphae, and (4) large, multinucleate spores, with layered walls. Analysis of SSS r DNA data suggest that these fungi form a monophyletic group (Schuessler *et al.*, 2001) and are unrelated to *Zygomycota* with whom they were placed until 2001 under order Glomales (=Glomerales). The rDNA data also suggest that these fungi are closer to dikarya (*Asco*-, and *Basidiomycota*), sharing a common ancestor. However, RPB1 (RNA Polymerase B1) protein gene sequences, that have been available recently, don't support the above "sister group" relationship with Dikarya.

These fungi form obligate arbuscular mycorrhizal (endosymbiotic) association with about 80% land plants, except one organism, *Geosiphon pyriformis*, which forms a different type of obligate symbiotic association (endocyanosis) with *Nostoc*, a cyano-bacterium, as the endosymbiont. It forms single-celled, multinucleate cells (bladders or vesicles) that harbor the *Nostoc* filaments. Since these organisms are obligate symbionts, they have not been cultured in absence of the host plant. This prevents biochemical and genetic characterization of the organisms. However, root cultures have made it possible to obtain axenic fungal biomass. The first mycorrhizal gene obtained from the spores of several genera from root cultures, and sequenced, was the highly conserved SSU rRNA gene. The molecular clock approach, based on the substitution rates of SSU sequences, indicated that these fungi are over 350 million years old. This enforces the earlier belief, based on fossil records, that these fungi were instrumental in the colonization of land by plants (Fossils of plant roots 450 million years old clearly show hyphae and spores of glomeromycotan fungi, indicating this group to be oldest fungi).

Host specificity of AM fungi is very low, and many species colonize a wide range of host plants.

The AM fungi (as these fungi are called) grow in between the root cells and form finely divided, tree-like haustorial branches, called **arbuscules**, or hyphal coils (= pelotons) that fill the root cells. These serve as sites for nutrient exchange between the fungus and the host plant. Some AM fungi also form globular structures, called **vesicles**, intercellularly (rarely intracellularly) that serve as storage organs.

These fungi form coenocytic hyphae and for this reason, as well as, formation of big spores resembling zygospores, the AM fungi were placed under *Zygomycota* until 2001.

The AM fungi form asexual spores, called **glomerospores**, which serve as the only means of reproduction. These are large, multinucleate, with outer **spore wall**, and inner **germinal wall**. They are formed at the tip of the hyphae, inside or outside the root (The fungus forms a fine network of hyphae in the soil which is more efficient than the walled roots or the root hairs in mineral uptake). The spores are very large (80-400 μm) and can be seen under low power of a dissecting microscope. (They can be collected by shaking a small sample of soil in a container and then examining the floating particles collected from the surface). The spores are multinucleate (contain several hundreds to thousands), which is very unusual for a eukaryotic organism, as multicellular organisms are recreated from a uninucleate propagule. The spores may be formed (1) singly, (2) in clusters or (3) aggregated in so called "sporocarps" (consisting of spores and a peridium). The spores (their size, shape, number, and the mode of attachment to subtending hypha) and spore wall characteristics are used in the differentiation of genera and families. Recently SSS of rRNA gene have been used in establishing genera e.g. *Pacispora* and *Diversispora*.

Spore wall structure is used for identification of genera and species. The spore walls comprise of an outer **spore wall** and an inner **germinal wall**. The two walls differ in their origin and function. The outer spore wall is continuation of the hyphal wall from which the spore is formed. It is 2-layered, comprising of outer **surface layer** and inner **structural layer**. The surface layer, which may be mucilaginous, coloured, and ornamented, usually sloughs off and does not persist in mature spores. The inner **germinal walls** (1-3) are formed later and have no physical connection with the spore wall. Each of the 1-3 germinal walls too, in turn, is 1-3 layered. The germinal walls are colourless and flexible. These are associated with the germination process.

During germination the germ tube may emerge directly from the germinal wall, or from preformed, "pre-germination structures" **germination orb, germination shield,** and **germination structure**, formed on the innermost germinal layer. Germination orb, found in a few species of *Acaulospora* and *Entrophospora*, is a circular structure instrumental in penetrating the outer spore wall. Germination shield (found in *Scutellospora* and *Pacispora*) comprises thin lobes, which also helps in penetration of the spore wall by the germ tube. Germination structure, formed in *Archaeospora*, is irregular in shape.

There is **no evidence of sexual reproduction**, morphological (zygospore) or molecular. Studies using molecular marker genes indicate no genetic recombination.

LIFE CYCLE

The spores germinate under favourable conditions. The germ tube grows through the soil until it comes in contact with the root. Host specificity of AM fungi is very low, because many species colonize a wide range of host plants. It forms appressoria on the root surface, penetrates the root and grows between the root cells. It can also penetrate the root cell wall and enter the root cell. Inside the roots, the fungus forms intra-radical hyphae, arbuscules, and vesicles and later extraradical hyphae. Spores are formed on extra- and intraradical hyphae. There is no sexual reproduction and asexual spores are the only means of propagation.

CLASSIFICATION

The phylum *Glomeromycota* is divided into 4 orders and 9 families (Schuessler *et al.,* 2001; Kirk *et al.,* 2008). These are given below.

Glomerales: 1 family *Glomeraceae* (1 genus *Glomus*; Group A,B)

Diversisporales : 4 families

Acaulosporaceae (2 genera, *Acaulospora, Entrophospora*)

Diversisporaceae (1genus, *Diversispora*)

Gigasporaceae (2 genera, *Gigaspora, Scutellospora*)

Pacisporacese (1 genus, *Pacispora*)

Archaeosporales: 3 families

Archaeosporaceae (1 genus, *Archaeospora*)

Ambisporaceae (1 genus, *Ambispora*)

Geosiphonaceae (1 genus, *Geosiphon*)

Paraglomerales:1 family *Paraglomeraceae* (1 genus *Paraglomus*)

The Fig. 8.1 shows a phylogenetic tree based on analyses of small subunit sequences (SSS) of rRNA gene (From Schwarzott *et al.*, 2001)

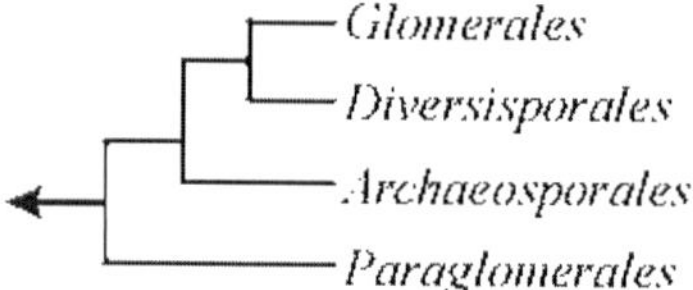

Fig. 8.1. Phylogenetic classification of *Glomeromycota*

Order *Glomerales* (1 Family, Glomeraceae, 1 genus *Glomus*)

Genus *Glomus*

Glomus, the lone genus in *Glomerales,* is the most common and largest genus (85 spp), forming about 53% of the AM fungi. It forms arbuscules and vesicles (but not commonly) and intra- and extra-radical hyphae. Spores are formed blasticaslly (i.e. by budding) at the tip of a sporogenous hypha singly, in clusters or aggregated in so called "sporocarps" (consisting of spores and a peridium). The sporogenous hypha may be straight, recurved or funnel-shaped (Fig. 8.2). The spores are formed inside the root as well as in the soil on extraradical hyphae. These are large (80-400 μm), multinucleate, with one outer, 2-layered spore wall, which is continuation of the wall of the sporogenous hypha. The surface of the spores may be smooth or differently ornamented. There are no germinal walls. The spores germinate by formation of a germ tube that emerges through the lumen of the subtending hypha or the spore wall or both (The former genus *Sclerocystis* is now merged in this genus). The *Glomus*-type spore, besides *Glomus*, is formed in several lineages namely *Paraglomus, Archaeospora, Pacispora, Diversispora* and *Geosiphon.*

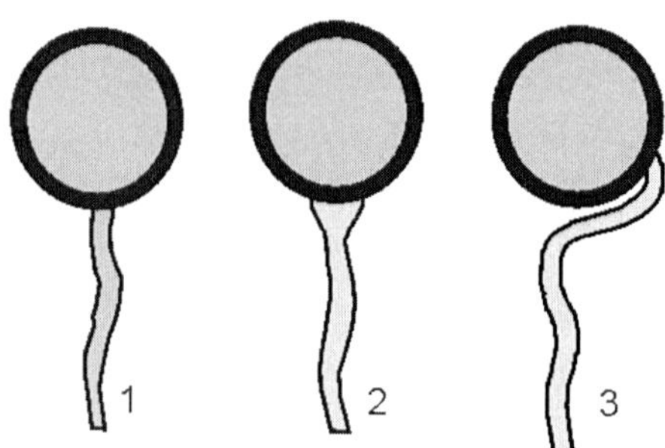

Fig. 8.2 *Glomus*: the sporogenous hyphae bearing Chlamydospores.

Glomus is also the most diverse genus and many of its species have been designated as new genera (*Pacispora, Diversispora*). SSS r RNA gene is now playing great role in defining species. The genus comprises 3 clades, called **phylogenetic groups, Glomus groups A, B** and **C**. (Schwarzott *et al.*, 2001). Group C was separated as a new genus *Diversispora* under a new family *Diversisporaceae*, under order *Diversisporales* (Walker and Schuessler 2004). Group A contains 2-3 subclades (a,b,c), while Group B is phylogenetically uniform. The two *Glomus* groups A & B are represented by *G. mosseae* and *G. claroideum*, respectively. The groups A and B are monophyletic but relatively distant, and may be separated as two families (Schuessler *et al.*, 2001).

Several species, including *G. aggregatum* are cultured on root and sold as mycorrhizal inoculants for agricultural soils. One species, *G. macrocarpum*, (and possibly also *G. microsporum*) causes tobacco stunt disease.

Order *Diversisporales*: 4 families

Many members of the order form vesicles, auxiliary cells and a wide range of spores and hence the name *Diversisporales*.

Family *Acaulosporaceae* (2 genera, *Acaulospora, Entrophospora*)

The two genera, *Acaulospora* and *Entrophospora*, form spores below a swollen "soporiferous saccule" borne apically on the sporogenous hypha (Fig. 8.3). They differ in the position of the spore. In *Acaulospora*, the spore is formed laterally from the saccule neck, while in *Entrophospora* it is formed inside the hypha below the saccule. As the spore matures, the saccule loses its contents (not emptied into the spore) and eventually sloughs off so that it is not attached to the mature spore. The spores can be formed singly or in clusters inside a sporocarp. The spores in both genera have germinal walls. A germination orb is formed on the inner germinal wall through which the germ tube emerges and penetrates the spore wall.

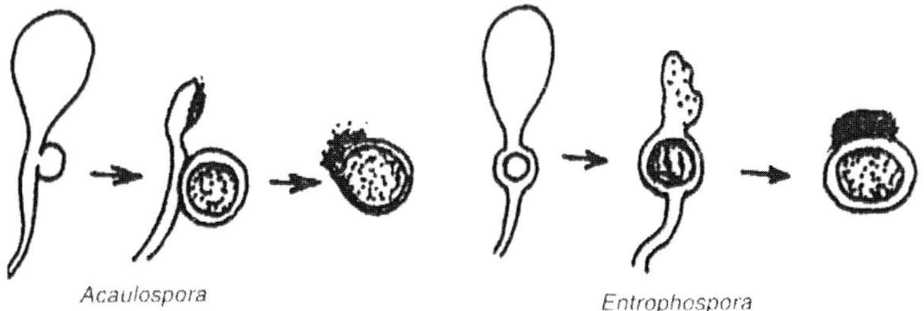

Fig. 8.3. Genus Acaulospora and Entrophospora.

Family *Diversisporaceae* (1 genus, *Diversispora*)

The family is monophyletic with other families of the order, *Acaulosporaceae, Gigasporaceae* and *Pacisporaceae* and has specific SSS rRNA signature. The family differs from other families of the order in germination of its spores. It is not accompanied by formation of a germination shield (as in *Scutellospora* of *Gigasporaceae*) or a germination orb (*Pacisporaceae* and *Acaulosporaceae*).

The genus ***Diversispora*** is monotypic i.e. has only one species, *D. spurca*, which forms typical *Glomus* type spore. Earlier it constituted *Glomus* group C, before its designation as a new genus. It differs from *Glomus* because of absence of vesicles and distinct molecular properties.

Family *Gigasporaceae* (2 genera, *Gigaspora, Scutellospora*)

The two genera *Gigaspora* and *Scutellospora* form spores above a bulbous sporogenous cell on the sporophore (Fig. 8.4). These genera also form auxiliary cells (= auxiliary bodies) of unknown function, but no vesicles. These are formed after spore germination, before the formation of mycorrhizae, and later on extra-radical hyphae in the soil.

The two genera differ in their germination process and the auxiliary cells. In *Scutellospora* the germ tube emerges through a germination shield formed on the innermost germinal wall. *Gigaspora* lacks the germinal walls and germinates directly through a newly-formed opening in the spore wall. The auxiliary cells of *Gigaspora* are formed in aggregates of 4-20, sub-globsoe or ovoid, with narrow spiny projections and borne on tightly-coiled hyphae. In *Scutellospora*, the auxilary cells are knobby and formed in clusters of 10-20 on coiled, thin-walled hyphae

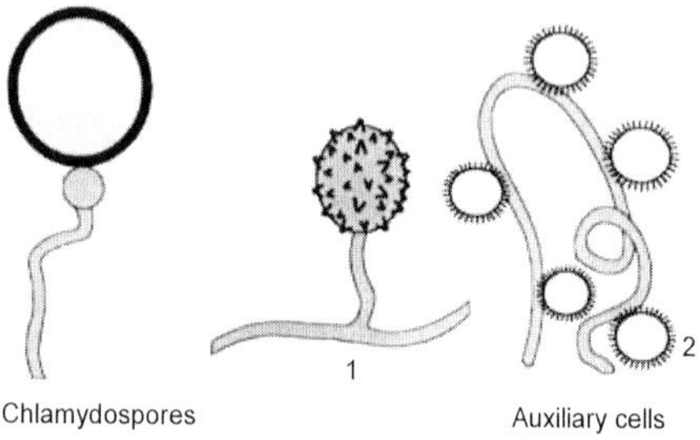

Fig. 8.4. Genus *Gigaspora* and *Scutellospora*

Family *Pacisporaceae* (1 genus, *Pacispora*)

The genus *Pacispora* is monotypic, i.e. has only one species, *P. scintillans*. It forms arbuscules, vesicles, intra-, and extra-radical hyphae,

Glomus-type spores and auxiliary cells, which are knobby resembling the auxiliary cells of *Scutellospora*. The spores have germinal walls from which they germinate through a germination shield (as in *Scutellospora*). The germ tube emerges through the spore wall. The genus resembles *Glomus* in forming spores in the same way, but differs in having germinal walls which are absent in *Glomus*. It resembles *Scutellospora* in forming similar auxiliary cells, and mode of germination, through a germination shield.

Order *Archaeosporales*: 3 families

Family *Archaeosporaceae* (1 genus, *Archaeospora*, 1 sp.)

Ar .trappei is the type species and also the only species in the genus. It is a dimorphic fungus, forming both acaulosporoid (*Acaulospora*-like) and glomoid (*Glomus*-like) spores. The former develops, as in *Acaulospora*, laterally from the neck of a "sporogenous saccule". When the spore detaches, there is no attached subtending hypha and, therefore, the spores are sessile. *Glomus*-like spores are formed at the tip or intercalary in the fertile hyphae. Spore wall comprises of 2 layers, while the germinal wall has 1 semi-flexible layer. The spore germinates by a germ tube developing from an irregular "germination structure", and emerges out from the spore wall.

Family *Geosiphonaceae* (1 genus, *Geosiphon*, 1 sp.)

Geosiphon pyriformis forms the only known fungal **endocyanosis** (= inclusion of cyanobacteria inside the cells of another organism). The coenocytic fungus forms unicellular, multinucleated cells, called **bladders** or **vesicles** (swollen hypha) of up to 2 mm in length at maturity (Fig. 8.5 and 8.6) harboring endosymbiotic filaments of *Nostoc punctiforme. Geosiphon* bladders are photosynthetically active and also capable of fixing nitrogen from the atmosphere. The wall of the bladder has a small pore for limited exchange with the environment. The fungus forms *Glomus*-type spores, which have a diameter of about 250 µm. This *Geosiphon-Nostoc* symbiosis is not a lichen symbiosis as it is an intracellular association. From functional and evolutionary points of view, it is nearer to AM symbioses than to lichens.

It appears, in the long run, the definition of *Glomeromycota* will have to be modified to accommodate non-mycorrhizal fungi. It is interesting to mention that it was the analyses of the phylogeny of *Geosiphon* that led to the erection of the phylum *Glomeromycota* and eventually, to the phylogenetic classification of *Glomeromycota*.

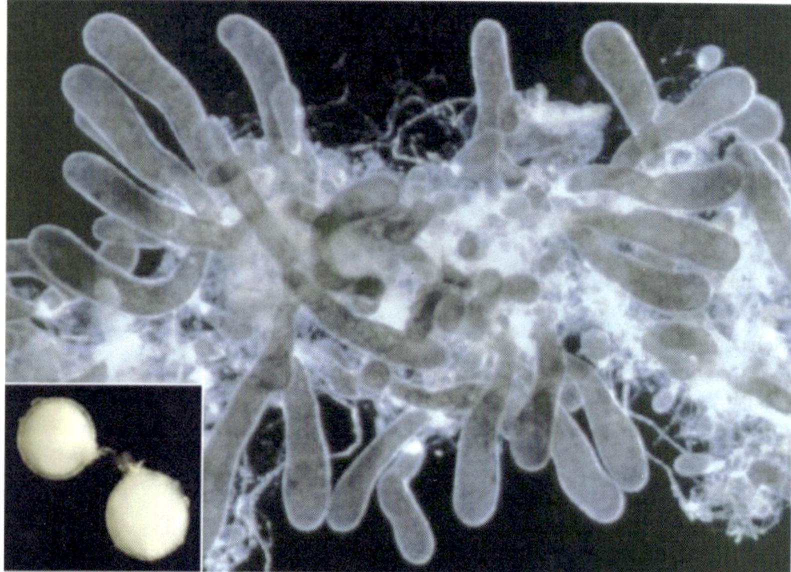

Fig. 8.5. *Geosiphon-Nostac* symbiosis: Mature bladders; the insert shows *Geosiphon pyriformis* spores (*Glomus*-type). (source : A. Schuessler)

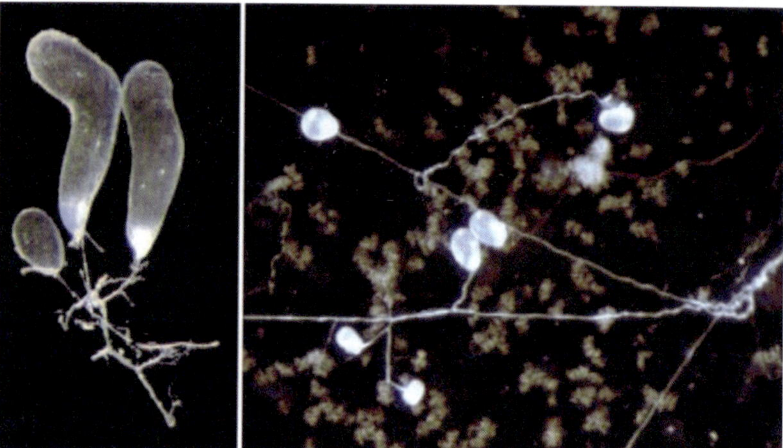

Fig. 8.6. Showing young and mature *Geosiphon* bladders formed at the fungal mycelium 7-10 days after uptake of the cyanobacterium. The picture on left is a magnified view. (source : A. Schuessler)

Order *Paraglomerales*: 1 family *Paraglomeraceae* (1 genus *Paraglomus*)

The genus has 3 species, which form *Glomus*-type spores. Like *Archaeospora, Paraglomus* is distinguished from *Glomus* by rDNA sequences, unique fatty acids and mycorrhizal morphology (differences in arbuscule morphology and absence of vesicles. It has no germinal walls. Based on 18S rDNA sequences, *Paraglomus* seems to be the most ancestral lineage within *Glomus*.

REFERENCES

Kirk, P.M. *et al.*, 2008 *Ainsworth and Bisby's Dictionary of the Fungi* (10th ed.) C.A.B. International, Oxon, U.K.

Schuessler, A.D. 2001 A new phylum *Glomeromycota*: phylogeny and evolution. Mycol.Res. 105: 1413-1421

Schwarzott, A. *et al.*, 2001 Glomus, the largest genus of the arbuscular mycorhizal fungi (Glomales) is non-monophyletic. Mol. Phylogen. and Evol. 21: 190-197

Walker, C. and Schuessler, A.D. 2004. Nomenclatural clarifications and new taxa in *Glomeromycota*. Mycol. Res. 108

SUMMARY

The phylum is characterized by: (1) obligate symbiosis, (2) formation of arbusules in plant root cells, (3) conocytic hyphae, and (4) large, multinucleate spores, with layered walls. The rDNA data suggest that these fungi form a monophyletic group, unrelated to *Zygomycota* and closer to *Asco*-, and *Basidiomycota*, sharing a common ancestor. However, RPB1 (RNA Polymerase B1) protein gene sequences, do not confirm any "sister group" relationship with Dikarya.

The glomeromycotan fungi form obligate arbuscular mycorrhizal (AM) association with plants, except one organism, *Geosiphon pyriformis*, which forms endocyanosis with *Nostoc*. The SSU rRNA gene sequences of these fungi confirm that these are oldest fungi, over 350 million years old, which were instrumental in the colonization of land by plants. Host specificity of AM fungi is very low, and many species colonize a wide range of host plants.

The AM fungi form only asexual spores, called glomerospores, at the tip of hyphae by budding, These are multinucleate and very large. The spores may be formed (1) singly, (2) in clusters or (3) aggregated in so called "sporocarps" (consisting of spores and a peridium). The spore characteristics are used in the differentiation of genera and families. Recently SSS of rRNA gene have been used in establishing molecular genera like *Pacispora* and *Diversispora*. The spore walls comprise of an outer spore wall and an inner germinal wall. The germ tube may emerge directly from the germinal wall, or from preformed, "pre-germination structures" **germination orb**, **germination shield** and **germination structure**, formed on the innermost germinal layer.

The phylum *Glomeromycota* is divided into 4 orders and 9 families. *Glomus* is the most diverse genus and many of its species have been designated as new genera (*Pacispora*, *Diversispora* .....). The *Glomus*-type spore, besides *Glomus*, is formed in several genera e.g. *Paraglomus, Archaeospora, Pacispora, Diversispora* and *Geosiphon.*

Some characteristic properties of the various genera are given in Table 8.1.

Table 8.1. Shows the various genera and some characteristic structures (+ present, - absent)

Genus	Vesicles	Auxiliary cells	*Glomus*- type spore	Germinal wall	Germination structures
Glomus	+	-	+	-	-
Acaulospora	+	-	-	+	Ger. Orb
Entrophospora	+	-	-	+	Ger. Orb
Diversispora	-	-	+	-	-
Gigaspora	-	+	-	-	-
Scutellospora	-	+	-	+	Ger. Shield
Pacispora	+	+	+	+	Ger. Shield
Archaeospora	-	-	+	+	Ger. Structure
Paraglomus	-	-	+	-	-
Geosiphon	-	-	+	-	-

TEST QUESTIONS

1. Write on the characteristic features of *Glomeromycota*. Give its classification up to orders.
2. Distinguish between any **five**: (1) *Acaulospora* and *Entrophospora* (2) *Gigaspora* and *Scutellospora* (3) *Glomus* and *Pacispora* (4) Intra-, and extra-radical hyphae (5) Symbiosis of *Geosiphion* and other AM fungi (6) *Glomus* and *Diversispora*, (7) *Glomus* and *Paraglomus* (8) Outer and inner spore walls of glomeromycotan fungi.
3. Write notes on any **three**: (1) Endocyanosis (2) Glomerospores (3) Spore walls in *Glomeromycota*, (4) Phylogenetic relationship of *Glomeromycota* with other phyla.
4. Fill in the blanks:
   a. Arbuscules provide ---------- while the vesicles serve as -----------.
   b. *Archaeospoa* differs from *Glomus* in forming ----------.

c. *Glomeromycota* is phylogenetically closer to ---------- and -------- than to---------------.

d. Host specificity in AM fungi is .... (low / high).

e. "Germination shield" is formed by ------------- and ----------.

f. "Germination structure" is formed by ---------------.

g. Inocula of *Glomus* are obtained from ----------- culture.

h. *Glomus* Group C was separated as genus ....

i. "Germination orb" is formed by ----------- and ------------.

5. Answer in brief any **five** the following:

a. Which 4 genera form vesicles?

b. Spores of which genera do not have germinal wall?

c. What are the 2 types of spore walls in glomeromycotan fungi?

d. Name the three preformed "pre-germination structures" in glomeromycotan spores.

e. Which genus in *Glomeromycota* does not form AM symbiosis?

f. What is endocyanosis ?

g. Which molecular data suggest AM fungi to be 350 million years old?

h. Which AM genus is dimorphic, forming 2 types of spores?

i. Which genera, besides *Glomus*, form *Glomus*-type spore?

j. Which 3 glomeromycotan genera form auxiliary cells?

9

Phylum *Ascomycota* : Characteristics and Classification

(Ascus-forming fungi)

The ascus and the two-layered cell wall are the diagnostic features of *Ascomycota*. The wall has a thin, electron dense outer layer and a, relatively thick, electron transparent inner layer. This enables the recognition of ascomycetous anamorphic fungi in absence of asci. The other characters, though not exclusive to these fungi are: the septate mycelium, absence of flagellated structures, formation of conidia as the main asexual spore, and the presence of a dikaryophase in the life cycle. The term ascomycetes has now no taxonomic status, but it has not lost its existence or use. We still say, for example, "It is an ascomycete, not a basidiomycete". The term is further abbreviated as "ascos"(and basidios).

The *Ascomycota* is the largest phylum of fungi having about 75% of all known fungi, including the conidial ascomycetes (former Deuteomycotina or fungi imperfecti, now obsolete) and the lichen-forming fungi. Comparative studies of 18S rDNA sequences and the two-layered cell wall structure have made the Deuteromycetes redundant as a separate taxonomic group. The enormous size makes it difficult to generalize the wide range of structures met within this phylum.

Somatic Structure

But for the yeasts (which comprise unicellular somatic cells or a pseudomycelium), these fungi possess well-developed septate and profusely-branched mycelium. The cell wall, which is two layered, is made of chitin. In the yeasts, mannans and $\beta 1-3$ glucans are the principal cell wall polysaccharides and chitin is present in very small amounts, and that too restricted to the bud scars. However, chitin has been detected in *Schizosaccharomyces*. The cells of the hyphae may contain one to several nuclei. The septa have a central pore through which the cell contents, including the organelles, migrate freely from one cell to another. If one cell anywhere in the mycelium becomes heterokaryotic, the rest of the cells of the hyphae are also rendered heterokaryotic by migration of the new daughter nuclei through septal pores. The septal pore may be blocked, when required, by various types of membrane-bound structures, like **Woronin bodies** (named after Woronin), and more complex **septal pore organelles**. The Woronin bodies are spherical or rectangular proteinaceous structures that plug the pore to ward off dead or ageing hyphae, while the septal pore organelles are pulley-wheel like in shape, and are used to shut off sexually-reproducing hyphae from somatic hyphae. The somatic hyphae are often organized into somatic tissues like **sclerotia**, **stroma** and **mycelial strands**.

Sclerotia (sing. sclerotium) are firm aggregations of modified somatic hyphae, which serve as resting bodies to overcome adverse environment. On germination, these form stroma bearing reproductive bodies (as in *Claviceps*) or directly give rise to the mycelium. Anatomically, the sclerotia show an outer **rind** made of thick-walled, closely-septate, fused hyphae and an inner **medulla** consisting of hyphae with comparatively thinner walls, but filled with dense contents which are mostly oil globules.

Stromata (sing. stroma) are compact somatic structures, which bear spores or the fructifications. These may be formed with or without the association of host tissues or substratum. A large number of ascomycetes, possess stromata. Anatomically, it may be prosenchymatous or pseudoparenchymatous and shows two distinct regions, the outer narrow **ectostroma** and an inner massive **endostroma**. Ectostroma serves mainly mechanical function, while the endostroma gives rise to the sporophores or ascomata.

Mycelial strands e.g. rhizomorphs, are linear hyphal aggregates capable of unlimited growth in one direction.

Reproduction

These fungi show extremes of behaviour in reproduction. While in the majority of genera, asexual spores, especially the conidia, play most important role, there are genera which multiply exclusively by sexual reproduction, i.e. through ascospores. The loss or degeneration of the sex organs, especially of the antheridium, is a conspicuous feature. Gametangial contact has been steadily taken over by spermatization and somatogamy, as the mode of plasmogamy. Thus, the sexual reproduction becomes subtle, but is certainly not lost.

Asexual Reproduction

The common means of asexual reproduction are **budding**, **fission**, **fragmentation**, **chlamydospores** and **conidia**. Of these, conidia are the most important. Budding and fission are characteristic of the yeasts.

Budding in ascomycetous yeasts is different from that in basidiomycetous yeasts, like *Rhodotorula* and it is helpful in identifying asco-, or basidiomycetous nature of the yeasts. It is also used to search phylogenetic relationship between the two groups. Budding is **hologenous** in ascomycetes i.e., the new wall layers of the bud cell is continuous with that of the parent cell wall. In basidiomycetous yeasts, the bud wall is newly synthesized, and is called **enterogenous**. Spores produced by budding, are termed **blastospores**, which etymologically means 'bud spores'. Fission occurs only in the yeast genus *Schizosaccharomyces*.

Conidia have reached their zenith in the *Ascomycota*. They are of various shapes and sizes and more often beautifully coloured. Conidia are borne on conidiophores which may be scattered on the mycelium or formed in special structures, the **conidiomata** (sing. conidioma), viz. acervuli (sing. acervulus), pycnidia (sing. pycnidium), and sporodochia (sing. sporodochium). The terms acervuli, pycnidia, sporodochia are now obsolete as nouns and are used as adjectives, e.g. acervular conidioma, pycnidial conidioma etc. The conidiophores may be simple, short, undifferentiated hyphae or specialized structures with characteristic shapes and branching, which in several cases are diagnostic of the genera and species. The conidiophores, occasionally, aggregate and fuse up to varying lengths to form structures called **synnemata** (sing. **synnema**) and **coremia** (sing. **coremium**). In synnema the conidiophores are fused at the base and free above near the tip, while in a coremium the conidiophores are only loosely aggregated. The conidiomata are of three types: pycnidial, acervular and sporodochial. The **pycnidial conidioma** is ostiolate, spherical or flask-shaped, whose inner wall is lined with short conidiophores. The **acervular conidioma** is a discshaped, flat stromatic mass of hyphae, formed beneath the cuticle, or the epidermis, of plants from which vertical, short conidiphores arise. The conidia are exposed when the overlying cuticle or epidermis is ruptured by pressure from within. **Sporodochial conidioma** consists of a cushion-shaped stroma bearing conidia externally. Frequently, fungi, forming acervular conidioma in nature, form sporodochia in laboratory cultures, but it does not occur *vice-versa*. The fruit bodies represent a progressive transition from imprisoned to free nature of conidiophores.

Sexual Reproduction

An important feature of sexual reproduction in *Ascomycota* is the separation of plasmogamy and karyogamy in space and time. The sexually compatible nuclei, that come to lie together during plasmogamy, do not fuse immediately but remain paired forming a **dikaryon**. The dikaryon gives rise to more pairs of such nuclei (dikaryons) by successive, **conjugate divisions**, until karyogamy takes place. The dikaryons are borne in specialized hyphae, called **ascogenous hyphae** (ascogenous = ascus-producing), originating from the female gametangium, the ascogonium. Thus, a dikaryotic phase intervenes between the haploid and diploid phases of the life cycle. It eventually terminates with the karyogamy which occurs in special cells, the **ascus initial cells**, formed at the tips of the ascogenous hyphae. Meiosis follows karyogamy. The four haploid nuclei formed, undergo mitosis to form eight haploid nuclei. The ascus initial cell transforms into an ascus containing eight ascospores, developed around the nuclei by **free-cell formation**. In yeasts, the zygote develops directly into the ascus. There is no dikaryophase. Plasmogamy is brought about by the following methods;

1. **Gametangial copulation** Two similar gametangia fuse by their tips and the fusion cell develops into an ascus. When two somatic cells fuse, as it happens in yeast (*Saccharomyces cerevisiae*), the gametangial copulation is termed **hologamy**'.
2. **Gametangial contact** Morphologically-distinct antheridia and ascognia (sing. ascogonium) are produced in some ascomycetes. The ascogonia are usually globular and bear a hair-like outgrowth, called **trichogyne**. Sometimes, as in *Aspergillus* and *Penicillium*, the ascogonium is cylindrical without a trichogyne.
3. **Spermatization** In this mode of plasmogamy, the male nucleus reaches the ascogonium through specialized **portable male cells** called **spermatia**. These are minute, uninucleate, spherical or rod-shaped cells, which are formed on short, erect **spermatiophores** and disseminated by wind, and insects. On coming in contact with an ascogonium, the male nucleus migrates into the ascogonium through a pore dissolved at the point of contact. The spermatia resemble minute conidia, but they cannot give rise to a new thallus. In some genera, microconidia, conidia and arthrospores serve as spermatia and bring about spermatization (e.g. *Neurospora*).
4. **Somatogamy** When two undifferentiated cells of same or different (but sexually compatible) hyphae fuse, plasmogamy takes place. This is called somatogamy. The nucleus from one

(the male) cell migrates into the other (female cell) and forms the dikaryon. Ascogenous hyphae that arise from the female cell, receive the daughter pairs of nuclei.

Development of Ascus

The male nuclei on reaching the ascogonium (Fig. 9.1) form dikaryons with the female nuclei. They don't fuse immediately to form diploid nuclei. Karyogamy is delayed. A dikaryophase intervenes between plasmogamy and karyogamy. Small protuberances appear on the ascogonium (or the somatic cell acting as the female cell), which grow and form ascogenous hyphae. The dikaryons multiply by conjugate (simultaneous) divisions and the daughter nuclear pairs migrate into the ascogenous hyphae. Later, septa are laid down and each cell of the ascogenous hypha, except the terminal cell, which contains a single nucleus, becomes dikaryotic. The dikaryophase in *Ascomycota* is represented by ascogenous hyphae; rest of the hyphae remain monokaryotic and take part in the formation of a thick protective covering around the asci (ascoma wall). Any dikaryotic cell of the ascogenous hyphae may elongate and bend to form a hook or **crozier**-like cell, called **crook cell**. Its two nuclei divide mitotically in such a way that their spindles lie vertically and parallel to each other. Two daughter nuclei, one from each spindle, lie at the bend of the hook; while one nucleus lies at the tip and one lies near the basal septum. Two septa are laid down to form three cells - the uninu-cleate basal and apical cells, and a binucleate penultimate cell. The penultimate cell acts as the **ascus initial cell**. The basal and apical cells later fuse and the nucleus of the basal cell migrates into the apical cell, through a pore formed at the point of contact. Thus, the apical cell becomes binucleate and repeats the above events to form a second ascus initial cell. This occurs repeatedly, and several penultimate cells are formed which give rise to asci.

The ascus initial cells elongate into a club-shaped structure. The two nuclei now fuse to form a diploid nucleus. Thus, the plasmogamy and karyogamy occur at different places; plasmogamy in ascogonium and karyogamy in the ascus initial cell. The diploid nucleus undergoes meiosis resulting in the formation of four haploid nuclei. Segregation of mating types and other characters takes places during the meiotic division. The four nuclei by a mitotic division form eight nuclei. Ascospores are organized around these eight nuclei by **free cell formation**. Some cytoplasm accumulates around each nucleus and, later, a wall is laid down. The cytoplasm left unused is known as the **epiplasm**, and probably serves for nutrition of the developing ascospores.

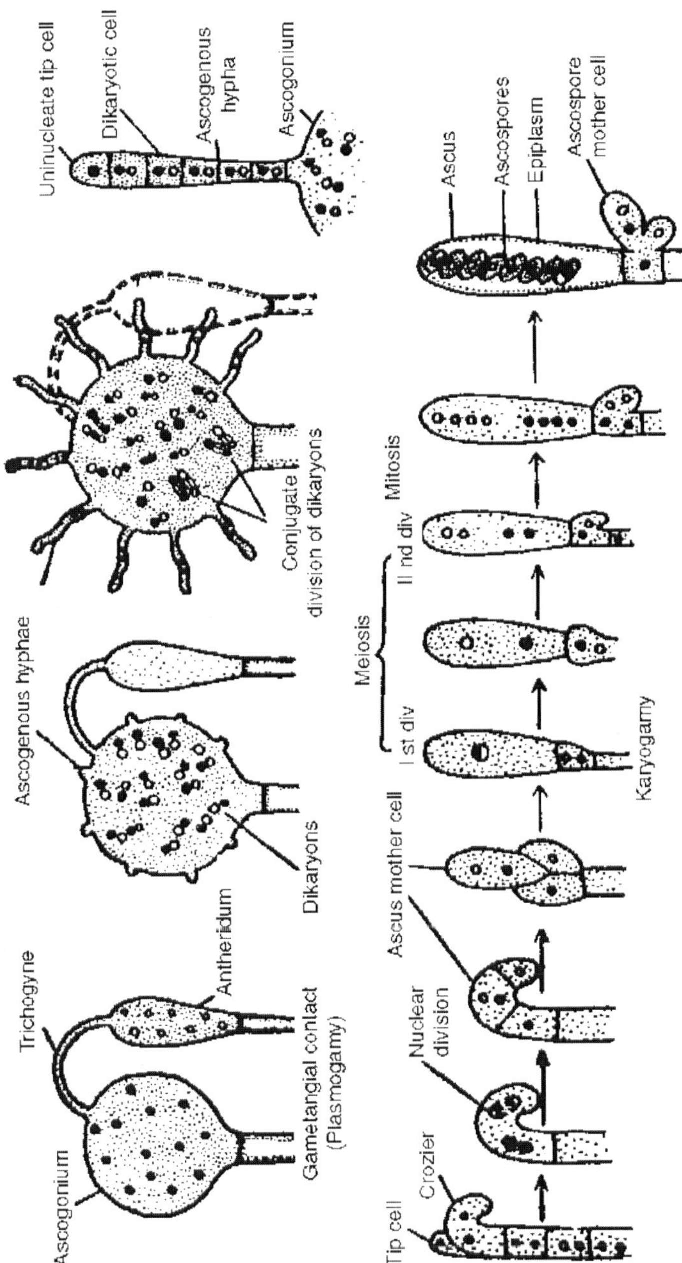

Fig. 9.1 Showing stages in ascus formation

The ascus has proved to be of great help to the geneticist in the analysis of segregation of characters during meiosis. The meiotic products are arranged in an ordered fashion and can be taken out for genetic analysis. The segregation of characters, whether it occurs during the first or the second meiotic division, can be known. Because the final division is mitotic, converting the tetrad into 8 nuclei, pairs of ascospores have the same genetical constitution. In all, there are six possible patterns of segregation of characters (Fig. 9.2); two patterns, if segregation of alleles occurs in the first division of meiosis, and four patterns if the segregation occurs during the second division of meiosis.

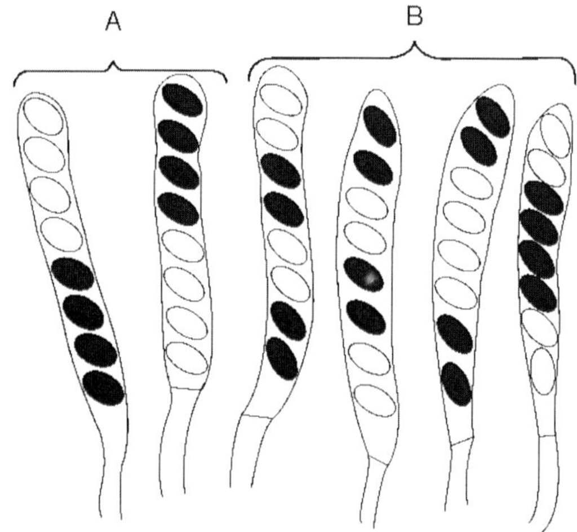

Fig. 9.2 Showing patterns of segregation of characters during Ist (A) or IInd (B) mieotic divisions.

The Ascus

Typically, asci are elongated, cylindrical or club-shaped, but globose asci are also common, especially in genera considered to be primitive. The asci may be stalked or sessile. These may be formed at various levels or arranged in one plane in a palisade fashion forming a hymenium. The asci are intermingled with sterile hyphal outgrowths — **the interascal tissue (hamathecium)**. There are three types of asci, based on ascus wall structure and mechanism of ascospore release. These are **prototunicate**, **unitunicate** and **bitunicate.** The wall in prototunicate asci is thin and fragile and breaks to release the spores passively. The unitunicate and bitunicate asci have two walls and discharge the ascospores forcibly. In the unitunicate asci both the walls are fused and rigid. Thus though bitunicate, it functions as unitunicate.

In the bitunicate asci, the outer wall (called exotunica, or exoascus) is rigid, and the inner (endotunica or endoascus) is extensible. During dehiscence of the bitunicate ascus, a transverse split occurs in the exotunica a little below the tip when the endotunica, carrying the ascospores, emerges out through it. The outer ruptured wall of the tip is thrown like a cap. Later, a pore is formed at the tip of the extruded endotunica through which the ascospores are discharged. This is called the **jack-in-a box** mechanism of dehiscence. The bitunicate asci showing "jack-in-a box" type of dehiscence are called **fissitunicate asci**. These are found in ascostroma (ascoma formed in locules in stroma). In unitunicate asci, the ascospores are ejected with force through a pore, a slit, and operculum (lid) or by breaking of the apical wall (Fig. 9.3). The unitunicate asci are formed in euthecial ascoma, cleistothecium, perithecium and apothecium (see below). There is one more type of ascus, the **rostrate** asci (found in *Lecanorales*), in which the inner wall, thickened at the apex, breaks through the outer wall and discharges he spores forcibly. This is sometimes referred as bitunicate, due to extrusion of the inner wall, but these are closer to unitunicate asci of ascohymenial forms.

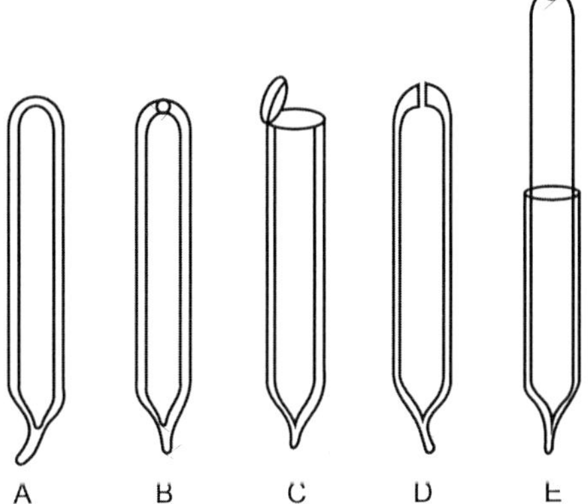

Fig. 9. 3 Types of ascus openings: A, no pore; B, aspical pore; C, operculum; D, slit; E, bitunicate ascus showing extension of inner wall (with an apical pore) through the rigid ruptured outer wall.

The apical apparatus of the persistent asci constitutes an important feature in identification and classification. It may have: (i) a single apical pore surrounded by thickened portion of the ascus apex, (ii) a similar thickening, as above, but it stains blue with cotton blue and is refractile, (iii) a funnel-shaped apex, (iv) have a simple apical pore and a swollen apex with a chitinous body inside, (v) an apical thickening that is stainable by iodine (vi), an apex stainable by iodine and also has a chitinous body inside, and (vi) several 'amyloid' rings i.e. staining blue and also chitinous body.

Interascal tissue (=**Hamathecium**) It includes all kinds of hyphae or other tissues between asci, or projecting into the locule or ostiole of ascomata. Seven categories of interascal tissues are known (Fig. 9.4).

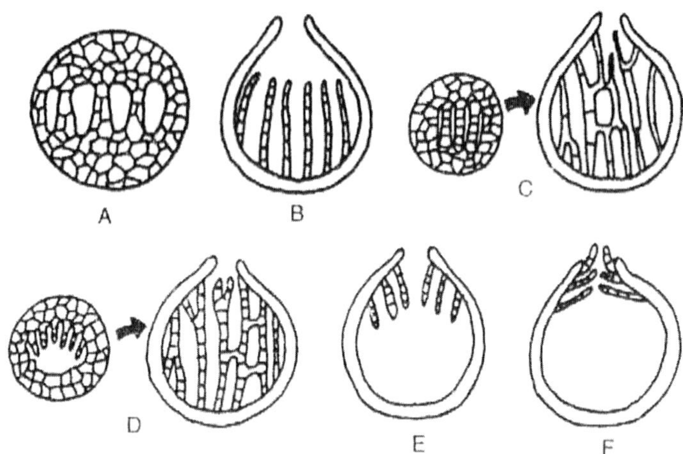

Fig. 9.4 Interascal tissue (=Hamathecium). A, interascal pseudopar-enchyma; B, paraphyses; C, paraphysoids; D, pseudoparaphyses; E, periphysoids; F, periphyses.

(A) **Interascal pseudoparenchyma**, carpocentral tissues unchanged or compressed between developing asci,

(B) **Paraphyses**, hyphae originating from the base of the cavity, usually unbranched and not anastomosed,

(C) **Pseudoparaphyses**, hyphae originating above the level of the asci and growing downwards between the developing asci, finally becoming attached to the base of the cavity and often also free in the upper part; often regularly septate, branched and anastomosing and broader,

(D) **Paraphysoids**, hyphae stretching and resembling pseudoparaphyses; often only remotely septate, anastomosing and very narrow,

(E) **Periphysoids**, short apical hyphae, originating above the level of the developing asci, growing downwards but not reaching the base of the cavity,

(F) **Periphyses**, hyphae confined to the ostiolar canal; unbranched and not anastomosing.

(G) **Hamathecial tissue absent.**

Formation of Ascoma (=Ascocarp)

During the development of asci, the surrounding monokaryotic hypahe form a thick protective wall, **excipulum**, around the developing asci. The hyphal tissues that form the excipulum (called **textura)** comprise of prosenchymatous or pseudoparenchymatous tissues. The exciplum is divisible into outer **ectal**- and inner, **medullary** excipulum. The wall and the asci are jointly called ascoma. **Ascomata** (plural) protect the asci and also regulate the release of the ascospores. Ascomata are of **four morphological types** (Fig. 9.5): **Cleistothecium**, which is completely closed and usually globular, **Perithecium**, which is ostiolate and globular or more commonly flask-shaped, the **Apothecium**, which is cup or saucer-shaped, bearing exposed hymenium, and **Ascostroma,** in which the asci are formed in preformed locules in stroma. The ascostroma, which often develop in flask-shaped or open cup-shaped stromatic tissue, superficially resemble perithecia and apothecia. Such ascostromatic 'perithecioid' or 'apothecioid' ascomata are called **pseudothecia** (sing. pseudothecium) to differentiate them from the **euthecia** (sing.euthecium) viz. cleistothecia, perithecia and apothecia.

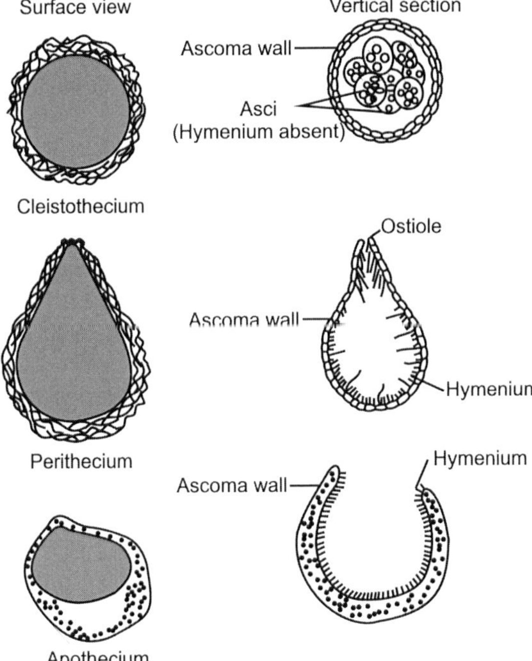

Fig. 9. 5 Showing types of ascocarps: whole fruiting bodies and their vertical sections.

The development of ascomata in stromatic locules is called **ascolocular** against **ascohymenial** ontogeny of the other three forms. Thus, there are two types of ascoma ontogeny, ascohymenial (in euthecia) and ascolocular (in pseudothecia). Besides ontogeny, the ascostroma (pseudothecia) differs from euthecia in that: (i) it starts development prior to fertilization of ascogonia (in ascohymenial development i.e. euthecia, the ascoma formation starts after ascogonial fertilization), (ii) the asci may be arranged in a fascicle, but not in a hymenium (in ascohymenial development, the asci are arranged in a hymenium), and (iii) there are no paraphyses in pseudothecia, which are present in euthecia, (iv) the asci are fissitunicate in pseudothecia, and unitunicate (operculate or inopecualte) in euthecia.

The ascogonia appear within a stromatic initial, which develops directly into the ascoma and the asci lie in unwalled locules in the stroma. This is so called **meristogenous** development. Unilocular ascomata with stroma reduced to a thin wall, closely resemble a perithecium or an apothecium.

The Centrum

All the tissues within the ascoma: the asci, the interascal tissue, and the inner portion of the wall, jointly constitute the centrum. Previously, the centrum was very important in defining the orders in former class, but recent 18S rDNA sequence data do not support the relationships based on centrum structure.

The Ascospores

Ascospores vary greatly in septation, number, shape, size, colour and arrangement. These characteristics are helpful in the identification of genera and species. Mostly, ascospores are unicellular but bi-celled or many-celled ascospores are also known. The ascospores are of varied shapes : globular, ovoid to long, thread-like e.g. *Claviceps*. In some genera, the ascospores are appendaged e.g. *Bombardia*, *Zygopleurage* (Fig. 9.6). The number of ascospores in an ascus too varies but is constant for a particular species. The most common number is eight. Some eight-spored asci have ascospores of two different sizes and different colours. Four-spored asci develop either by (1) degeneration of four nuclei, (2) inclusion of two nuclei in each ascospore e.g. *Neurospora tetrasperma*, (3) absence of the second mitotic division in the ascus e.g. *Sphaerotheca*, or (4) abortion of one daughter nucleus or its progeny e.g. *Bulgaria*. The asci of *Ptychoverpa bohemica* are consistently two spored. One-spored asci are known in some lichenized species. Multispored asci arise in two ways (1) by repeated nuclear divisions e.g.

Dipodascus, *Podospora*, and (2) by budding of the ascospores, while still within the ascus e.g. *Taphrina*. (In some cases the ascus is ultimately packed with innumerable, minute ascospores only a little bigger than bacteria). The ascospores may be hyaline or variously coloured.

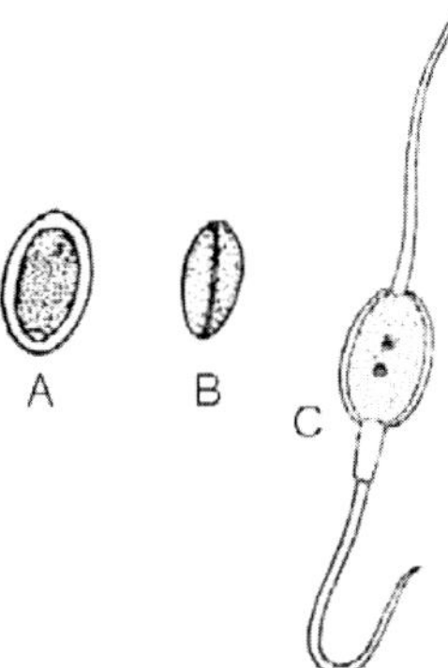

Fig. 9.6 Showing different forms of ascopores. A, *Sordaria* with a germ pore; B. *Xylaria* with a germ slit, and C, *Bombardia* with appendages.

The ascospores, when elongated and narrow, may be uniseriate (arranged in a single row), or biseriate (in two rows) or in parallel fascicles (=groups or bundles). The arrangement of the ascospores is a good taxonomic character.

Release of Ascospores

Asci contained in closed ascomata are exposed for release of the ascospores only after the disintegration of the ascoma. In perithecia, the asci elongate until their tips reach the ostiole (one at a time) for discharge of the ascospores. In some genera, especially those having long-beaked perithecia, the asci get detached and ooze out of the pore in succession. In other genera, the ascus wall dissolves and the ascospores are released within the cavity of the perithecium from where they are released in a mass of slime. Such ascospores are usually transmitted by insects. In apothecia, the asci are oriented in a way which facilitates easy discharge of the ascospores. In *Peziza* for example, the asci and paraphyses are positively phototropic and their tips bend towards the incident light. In unitunicate asci, the ascospores are released apically through a pore, slit, operculum or tear (formed on the lateral wall). The pore, before the discharge, is plugged with a substance that stains blue with iodine and is different from the wall material. In *Elaphomycetales*, the ascospores are freed only after the disintegration of the ascal wall. Operculate asci are encountered in *Pezizales*. The ascospores are, in most cases, discharged forcibly. During a period of quiescence the asci reach a super-stretched, unstable condition, and as soon as the ascomata are touched or blown upon in this condition, the asci burst, exploding the ascospores. The countless ascospores, puffed out by the numerous asci, form a visible 'cloud' that drifts away in the air like smoke. This 'puffing' of ascospores occurs frequently in the cup fungi – *Pezizales*. The puffing in some cases is accompanied by a hissing sound. Puffing occurs in installments, several times, until all the ascospores are discharged. Though the eight ascospores appear to be released simultaneously, they actually escape only one by one. Since the pore under stretched condition is narrower than the diameter of the ascospores, these are released only one at a time. As less than a fraction of a second elapses between two successive releases, it appears instant. The *Ascomycota* that lack the explosive discharge of ascospores, possess globose asci, with evanescent walls. The ascospores, after dissolution of the ascus wall, come to lie in the cavity of the ascoma and are released only after the disorganization of the ascoma wall.

Germination of Ascospores In yeasts and some other genera the ascospores multiply by budding before germination. During germination the germ tubes emerge from a germ pore, or germ slit, and gives rise to the somatic hyphae.

CLASSIFICATION

Ainsworth (1973) divided Ascomycota (then subdivision Ascomycotina) into six classes. The primitive genera lacking ascoma were segregated under class Hemiascomycetes. The rest, which formed ascomata, were divided under Loculoascomycetes, Plectomycetes, Laboulbeniomycetes, Pyrenomycetes and Discomycetes. But later, due to integration of lichenized and non-lichenized fungi within *Ascomycota*, it was difficult to determine diagnostic features for taxa above the rank of orders. Thus, the classes were derecognized. **Hawksworth** *et al.* (1983) in the 7th edition of *Dictionay of the Fungi* divided *Ascomycota* into 39 monophyletic orders. Later after of 12 years also the situation did not change and in the 8th edition of *The Dictionary* (1995) **Hawksworth** *et al.,* ruled out of grouping of orders into classes. The classification recognized 46 monophyletic orders arranged in alphabetical sequence due to lack of data regarding their phylogenetic relationships. A very important feature of the "46 order classification" was that it assigned a place (home!) to the lichen-forming (lichenizing) fungi, along with other ascomycetes. It is to be noted that the lichen-forming fungi constitute about 40% of the ascomycetes and these were not included in the "main stream" classification of ascomycetes. Instead, they were classified separately under Deuterolichens (now obsolete) or Mycophycophyta (phylum name used by Margulis, 1993) for lichen-forming fungi, as well as, the algal partners.

Relationships among the orders and designation of classes

Molecular phylogenetic data, based on 18S rDNA sequences, allowed grouping of the orders into classes. It became clear that there is a **basal group** which originated earlier than other ascomycetes (Fig. 9.7). **Nishida and Sugiyama** (1994) proposed creation of class, *Archiascomycetes* for this group. But they also considered the group to be non-monophyletic. Moving up, we find a dichotomy (Fig. 9.7) leading to two sister groups (i.e. two clades formed by splitting of a single lineage): *Saccharomycotina* and *Pezizomycotina* (an informal name for mycelial ascomycetes). **Kirk** *et al.*, in the 9th edition of *The Dictionary*, (2001) recognized **6 classes**: *Ascomycetes, Neolectomycetes, Pneumocistidomycetes, Saccharomycetes, Scizosacchar-omycetes* and *Taphrinomycetes*.

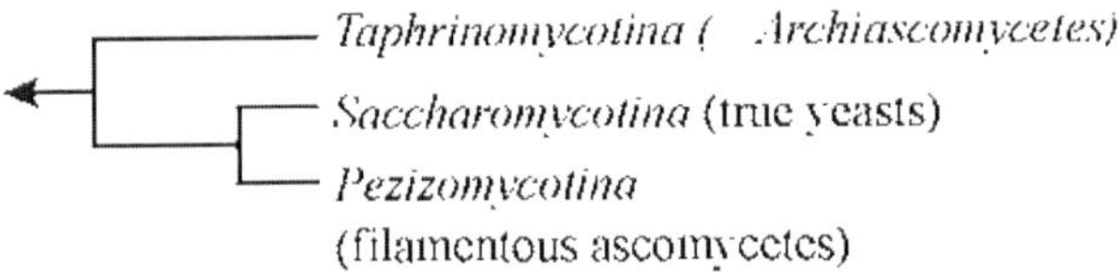

Fig. 9.7 Major groups of phylum *Ascomycota.*

The Present Classification

Kirk *et al.* (2008), in the 10th edition of *The Dictionary*, divided phylum *Ascomycota* into **3 monophyletic subphyla**: *Taphrinomycotina* (=*Archiascomycetes*), *Saccharomycotina* (true yeasts), and *Pezizomycotina* (mostly filamentous, sporocarp-producing and 'conidial ascomycetes'). The classification is based on recommendations of **Deep Hypha** and **AFTOL** (Assembling the Fungal Tree of Life) **projects*** sponsored by National Science Foundation (U.S.A.), as well as other resources like the ***Myconet***. An outline of the classification is as follows.

---

* The multi-gene sequence analysis undertaken in the Deep Hypha and AFTOL projects resulted in higher level phylogenetic classifications and better understanding of evolution of organisms in the kingdoms of fungi. The *Myconet* is an electronic and printed journal that publishes a continuously updated "Outline of *Ascomycota*" and "Notes on ascomycete systematics" informing on new concepts that may necessitate changes in classification. Contributions to the classification of other groups of fungi may be published in future.

Phylum *Ascomycota* (3 subphyla)

***Taphrinomycotina*:** 5 classes: *Archaeorhizomycetes* (1 order, 1 family, 1 genus), *Neolectomycetes* (1 order, 1 family, 1 genus), *Pneumocystidomycetes* (1 order, 1 family, 1 genus), *Schizosaccharomycetes* (1 order, 1 family, 2 genera), and *Taphrinomycetes* (1 order, 2 families, 8 genera)

***Saccharomycotina*:** 1 class, *Saccharomycetes*, 1 order, 13 families, 95 genera)

***Pezizomycotina*:** 11 classes: *Arthoniomycetes, Dothideomycetes, Eurotiomycetes, Geoglossomycetes, Laboulbeniomycetes, Lecanoromycetes, Leotiomycetes, Lichinomycetes, Orbiliomycetes, Pezizomycetes, Sordariomycetes*, and 3 "unplaced orders" (incertae sedis), *Lahmiales, Itchiclahmadion* and *Triblidiales.*

REFERENCES

Ainsworth, G.C. 1966. A general purpose classification of fungi. *Bibl. Syst. Mycol.* No. 1: 1–4.

Ainsworth, G.C. 1973. Introduction and keys to higher taxa. In: *The Fungi. An Advanced Treatise.* (G.C. Ainsworth, F.K. Sparrow and A.S. Sussman, eds.). Vol. IV A. 1–7. Academic Press, New York, London.

Hawkswoth, D.L. *et al.* 1983. *Ainsworth and Bisby's Dictionary of the Fungi*. C.A.B. International, Oxon, U.K.

Hawksworth, D.L. *et al.* 1995. *Ainsworth and Bisby's Dictionary of the Fungi* C.A.B. International, Oxon, U.K.

Kirk *et al.* 2008 *Ainsworth and Bisby's Dictionary of the Fungi* C.A.B. International, Oxon, U.K.

Margulis, L. 1993. *Symbiosis in Cell Evolution.* 2nd Ed. W.H. Freeman, N.Y.

Nishida, H. and Sugiyama, J. 1994. Archiascomycetes: detection of a major new lineage within the Ascomycota. *Mycoscience*, 35: 361–366.

Taylor, J.W. 1995. Making the Deuteromycota redundant: a practical integration of mitosporic and meiosporic fungi. *Can. J. Bot.* 73 (Suppl.): S754–SS759.

SUMMARY

The ascus and the two-layered cell wall are the diagnostic features of *Ascomycota*. The thin, electron dense outer layer and a thick, electron transparent inner layer of the wall enables the recognition of conidial ascomycetes (anamorphic) in absence of asci. The other characters, though not exclusive for these fungi are: the septate mycelium, absence of flagellated structures, formation of conidia as the main asexual spore, and the presence of a dikaryophase in the life cycle.

The *Ascomycota* is the largest phylum of fungi having almost 75% of all known fungi, including the 'conidial ascomycetes' and the lichen-forming fungi. Comparative studies of 18S rDNA sequences and the two-layered cell wall structure have made the deuteromycetes redundant as a separate taxonomic group. The enormous size makes it difficult to generalize the wide range of structures met within this phylum.

But for the yeasts, these fungi possess well-developed septate mycelium. The cell wall is made of chitin. In the yeasts, mannans and β 1-3 glucans are the principal cell wall polysaccharides and chitin is present in very small amounts. The cells of the hyphae may contain one to several nuclei. The septa have a central pore through which the cell contents, including the organelles, migrate freely from one cell to another. The septal pores may be blocked, when required, by Woronin bodies and more complex septal pore organelles. The somatic hyphae are often organized into somatic tissues like sclerotia, stroma and mycelial strands.

These fungi show extremes of behaviour in reproduction. While in the majority of genera, asexual spores, especially the conidia, play most important role, there are genera which multiply exclusively by sexual reproduction, i.e. through ascospores. The loss or degeneration of the sex organs, especially of the antheridium, is a conspicuous feature. Gametangial contact has been steadily taken over by spermatization and somatogamy, as the mode of plasmogamy. Thus, the sexual reproduction becomes subtle, and is certainly not lost. Asexual Reproduction is brought about by budding, fission, fragmentation, chlamydospores and conidia. Budding and fission are characteristic of the yeasts. Conidia are borne on conidiophores which may be scattered on the mycelium or formed in special structures, the conidioma (pl. conidiomata), which may be acervular-, pycnidial-, or sporodochial. The conidiophores, occasionally, aggregate and fuse up to varying lengths to form structures called synnemata and coremia.

Sexual Reproduction shows separation of plasmogamy and karyogamy in space and time. The sexually compatible nuclei, that come to lie together during plasmogamy, remain paired forming a dikaryon. The dikaryon gives rise to more dikaryons by successive, conjugate divisions in specialized hyphae, called ascogenous hyphae, until karyogamy takes place. A dikaryotic phase intervenes between the haploid and diploid phases of the life cycle. Karyogamy occurs in ascus initial cells, formed at the tips of the ascogenous hyphae. The diploid zygote nucleus undergoes meiosis followed by mitosis to form eight haploid nuclei. The ascus initial cell transforms into an ascus containing eight ascospores, developed around the haploid nuclei by free-cell formation. In yeasts, the zygote develops directly into the ascus. Therc is no dikaryophase.

Typically, asci are elongated, cylindrical or club-shaped, but globose asci are also common, especially in genera considered to be primitive. The asci may be stalked or sessile. These may be formed at various levels or arranged in one plane in a palisade fashion forming a hymenium. There are three types of asci based on ascus wall and mechanism of ascospore release. These are prototunicate, unitunicate or bitunicate. The bitunicate character becomes evident during the dehiscence of the ascus. The bitunicate asci showing "jack-in-a box" type of dehiscence are called fissitunicate asci and formed in pseudothecia. Unitunicate asci, opercualte or inoperculate, are found in euthecial ascomata.

Asci have a characteristic apical apparatus, which is of many types and constitute an important feature in identification and classification. The asci are intermingled with sterile hyphal outgrowths, the interascal tissue (= hamathecium), which may be interascal pseudoparenchyma, paraphyses, pseudoparaphyses, paraphysoids, periphysoids or periphyses. As the asci develop, the surrounding monokaryotic hypahe form a thick protective wall (prosenchymatous or pseudoparenchymatous) around the asci. The wall and the asci are jointly called ascoma. Ascomata are of four types: cleistothecium, perithecium, apothecium, and ascostroma. When the asci are formed in stroma, it is called ascostroma. These differ in ontogeny, the first three have ascohymenial ontogeny and called euthecia, while ascostroma has ascolocular ontogeny and is called pseudothecium. Besides, these differ in time of initiation of development, arrangement and nature of ascal wall and ascospore release. All the tissues within the ascoma: the asci, the interascal tissue, and the inner portion of the wall, jointly constitute the centrum.

Ainsworth (1973) divided *Ascomycota* (then Ascomycotina) into six classes, based on ascocarp characteristics. Later, when lichen-forming fungi were included among ascomycetes, the classification became difficult above the rank of order. It led to derecognition of classes in the later classifications of *Ascomycota*. Hawksworth *et al.* recognized 39 orders in the 7th edition of *The Dictionary* (1983), which was raised to 46 orders in the 8th edition (1995). A very important feature of these classifications was that it brought the lichen-forming fungi, in the main stream of fungal classification. However, by 2001, molecular phylogenetic data, based on 18S rDNA sequences, allowed grouping of the orders into classes. **Kirk** *et al.* (2001) in the 9th edition of *The Dictionary*, recognized 6 classes: *Ascomycetes, Neolectomycetes, Pneumocistidomycetes, Saccharomycetes, Scizosaccharomycetes* and *Taphrinomycetes*. Now, in the present classification, (Kirk *et al.*, 2008, 10th edition of *The Dictionary*), the phylum has 3 subphyla, which are further divided into classes, orders and families. The subphyla are *Taphrinomycotina* (= *Archiascomycetes*), *Saccharomycotina* (true yeasts), and *Pezizomycotina* (mostly filamentous, sporocarp-producing and 'conidial ascomycetes'). The classification is based on recommendations of Deep Hypha and AFTOL projects sponsored by National Science Foundation, as well as other resources like the ***Myconet***.

TEST QUESTIONS

1. Describe the diagnostic features of *Ascomycota*. Discuss the impact of inclusion of lichen-forming fungi on the classification.
2. Answer in brief:
   a. How are the ascomycetous conidial fungi recognized in absence of asci?
   b. Which structure represents the dikaryotic phase in *Ascomycota*?
   c. What are sites of karyogamy and meiosis in *Ascomycota*
   d. Where were the lichen-forming fungi classified before their inclusion in *Ascomycota?*
3. Write notes on any **four**: (a) Three types of conidiomata, (b) "Basal Group" in *Ascomycota* (c) Unitunicate asci, (d) Fissitunicate asci, (e) Ascomatal ontogeny
4. Distinguish between:
   a. Cell walls of yeasts and other ascomycetes.
   b. Woronin bodies and septal pore organelles.
   c. Classification of *Ascomycota* as given by Kirk *et al.,* in 2001 and 2008.
   d. Spermatia and conidia.
   e. Ascomycetous and Basidiomycetous yeasts.
   f. Synnemata and coremia.
   g. Prototunicate, unitunicate and bitunicate asci.
   h. The three types of ascomata.
   i. Euthecia and Pseudothecia.
   j. Paraphyses and Periphyses.
5. Fill in the blanks.
   a. There is no ----------------- phase in yeasts
   b. The Deep Hypha and AFTOL projects were financed by-------------------- of U.S.A.
   c. *Saccharomyotina* and -------------- form sister groups of *Ascomycota.*

d. The class ------------------------- was erected for the "Basal Group" of *Ascomycota.*

e. The 7th and 8th editions of *Dictionary of the Fungi* (Hawksworth *et al.,* 1983 and 1995.) restricted the classification of *Ascomycota* only up to-------------------

f. The number of ascospores increases by --------------- of ascospores.

g. The two basic types ascoma ontogeny are --------------- and --------------.

h. The ascospores are released passively by ------------- asci.

i. Ascospores are released forcibly by --------- and ----------- asci.

j. The bitunicate asci with 'jack in a box' mechanism of ascospore release are called ------------------- asci.

10

Subphylum *Taphrinomycotina*
The basal Group
(= former ARCHIASCOMYCETES)

The members of the subphylum *Taphrinomycotina* represent the 'basal group' of primitive ascomycetes, which appeared earlier than the other ascomycetes. Nishida and Sugiyama (1994) suggested the name *Archiascomycetes* (but not accepted formally) based on his study of the 18S rDNA sequences of several primitive genera *Neolecta*, *Pneumocystis*, *Protomyces*, *Saitoella*, *Schizosaccharomyces* and *Taphrina*. There are also some surprises in this group. For example, *Schizosaccharomyces* (a fission yeast) is not so closely related to *Saccharomyces* as previously understood. *Pneumocystis*, which causes pneumonia in mammals including human beings, was thought to be a protozoan until 1994. *Neolecta*, which forms a large ascoma (apothecium) is a puzzle, because ascomata are formed only by the euascomycetes, the highest evolved ascomycetes. The most recent addition to this basal group, class *Archaeorhizomycetes* (2011) represents a new class of fungi, which were earlier known as "Soil Clone Group 1" (SCG1) and identified only by DNA sequencing.

The subphylum *Taphrinomycotina* comprises mycelial fungi (*Archaeorhizomyces* and *Neolecta*), unicellular, yeast species, multiplying by fission (*Pneumocystis* and *Schizosaccharomyces*) and dimorphic, half mycelial and half yeasts (*Protomyces, Taphrina*). It is divided into 5 classes: *Archaeorhizomycetes, Neolectomycetes, Pneumocystidomycetes, Schizosaccharomycetes* and *Taphrinomycetes*. Kirk *et al.* (2008) in *The Dictionary of the Fungi* have mentioned only 4 classes. The class *Archaeorhizomycetes* was added by Rosling *et al.* (2011).

An outline of the classification up to families is given below.

Subphylum *Taphrinomycotina* (5 classes)

Archaeorhizomycetes (1 order, 1 family, 1 genus),

Neolectomycetes (1 order, 1 family, 1 genus),

Pneumocystidomycetes (1 order, 1 family, 1 genus),

Schizosaccharomycetes (1 order, 1 family, 2 genera), and

Taphrinomycetes (1 order, 2 families: *Protomycetaceae* and *Taphrinaceae*; 8 genera).

Fig. 10.1. Electron microphotogrph of *A. finlayi,* showing branched hyphae (1-2 μm in diameter) and chlamysdospores (credit Anna Rosling and Karelyn Cuz Martinez)

Class *Archaeorhizomycetes* (order *Arhaeorhizomycteales*, family *Archaeorhizomycetaceae*, genus *Archaeorhizomyces)*

The class was created by A. Rosling and T. James (2001) for the genus *Archaeorhizomyces.* These fungi were previously known as Soil Clone Group-1 (SCG-1), the most ubiquitous of several soil fungi which are known only by their DNA sequences (Environmental DNA) and have not been cultured (Environmental DNA is the genetic material of a living organism that is detected in bulk environmental samples, such as samples of soil or water in which the organism lives). Though Rosling *et al.* (2011) estimate the existence of as many 250 species in the class, only one has been described so far. The type species is *A. finlayi*, which forms branched hyphae with chlamydospores (Fig. 10.1), was isolated from a coniferous root. These organisms grow on simple synthetic media, but the growth is very slow.

The ecological niche of these organisms is not known but some species are found associated with ectomycorrhizal fungi that form a sheath on root tips of woody plants. There is no evidence that these organisms are mycorrhizal. Their apparent association with roots is possibly through their coexistence with mycorrhizal fungi.

Class *Neolectomycetes* (order *Neolectales*, family *Neolectaceae*, genus *Neolecta*)

Genus *Neolecta*

Neolecta is a filamentous, ascoma-producing soil fungus. It has a unique position in modern mycology. It is a puzzle because ascomata are, otherwise, formed only by highest evolved ascomycetes, the *Pezizomycotina*. *Neolecta*, lies near the very bottom of the evolutionary tree of ascomycetes, as indicated by rRNA, RPB2 and β-tubulin gene sequences. Landvik *et al.* (2001) referred to *Neolecta* as **fungal dinosaurs** and **living fungal relic of the past** (= a **living fossil**). No one expected such a large ascoma-bearing fungus to lie at the bottom of the evolutionary tree. LSU rDNA sequences also confirm that *Neolacta* forms part of a sister group to other fruiting body-forming ascomycetes and the "budding yeasts" i.e. *Pezizomycotina* and *Saccharomycotina*. *Neolecta* is suggested as a model organism for studies of the regulation of ascomal ontogeny and physiology.

Neolecta species are called **earth tongues** because of the characteristic ascoma they form (Fig. 10.2). The genus has 3 species: *N. flavovirescens, N. irregularis* and *N. vitellina* (= *N. aurantia*).

Fig. 10.2 Ascomata of *Neolecta vitellina.*

All the species live in association with trees, and at least one, *N. vitellina* grows from the rootlets of its host. But, it is not known whether the fungus is parasitic, mutualistic or saprobic. So far, the organism has not been cultured, suggesting it to be obligately-parasitic or symbiotic. The ascoma is edible. The hymenium lacks paraphyses, and the ascogenous hyphae do not form croziers. These are primitive characters which show the rudimentary nature of the ascoma. This makes it distinct from other "earth tongues", formed by *Geoglossum,* belonging to *Pezizomycotina*. The ascospores germinate to form yeast-like conidia.

Class *Pneumocystidomycetes* (order *Pneumocystidales*, family *Pneumocystidaceae*, genus *Pneumocystis*)

Genus *Pneumocystis*

Pneumocystis is a yeast which has so far not been cultured. It has 7 species. *P. jirovecii*[*] causes pneumocystis-pneumonia (PCP) in patients with weak immune system, especially in people suffering with AIDS or cancer. This pathogen is specific to humans and does not infect other animals. *P. carinii* infects animals like rats, mice, horses, pigs, rabbits etc., but not humans (Aliouat-Denis *et al.,* 2008; Stringer *et al.,* 2002). Earlier, *P. carinii*, was described as the causal agent of pneumocystis pneumonia in humans and the disease was known as PCP (pneumocystis carinii pneumonuia). But now the correct name of the disease is **PJP** (**P**neumocystis **J**irovecii **P**neumonia). However the acronym **PCP** can be continued, because of long usage but understood as **P**neumo **C**ystis **P**neumonia.

The organism was earlier believed to be a protozoan (*Pneumo-cystis jirovecii*) but, because of yeast-like morphology and life cycle, as well as DNA sequences, it is placed here. The somatic cells are thin-walled, irregularly shaped, uninucleate, haploid and divide by fission (not budding). The life cycle (Fig. 10.3) involves fusion of two somatic cells and later of their nuclei, leading to the formation of the zygote. It grows bigger in size, develops a thick, two-layered wall and transforms into an ascus. Eventually, meiosis, followed by mitosis, leads to formation of 8 nuclei, which develop into ascospores. These are released after rupture of the ascus wall. Due to difficulty in taking *P. carinii* in culture, some stages in the life cycle are not very clear.

---

* Named after **Otto Jirovec**, who first observed pneumocystis pneumonia in humans in 1952

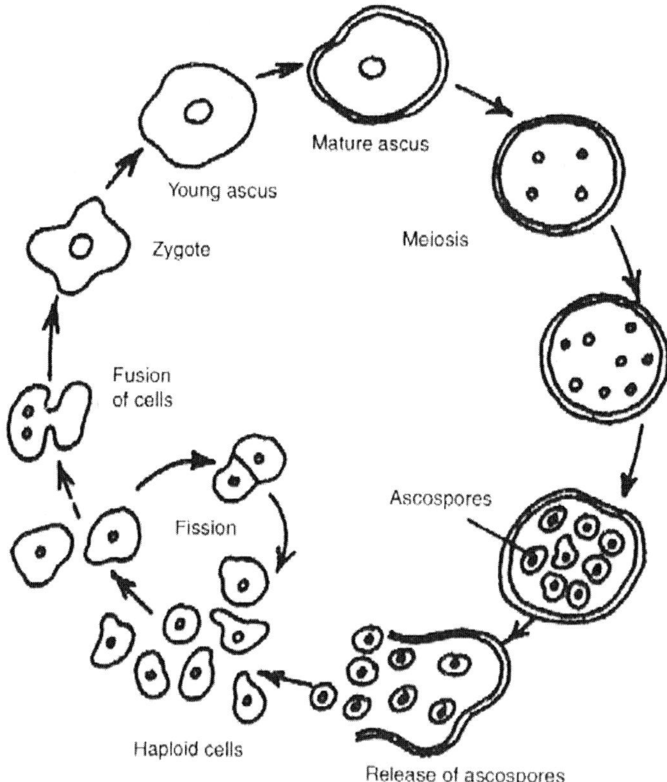

Fig. 10.3. Life cycle of *Pneumocystis jirovecii*

Class *Schizosaccharomycetes* (order *Schizosaccharomycetales*, family *Schizosaccharomycetaceae*)

There are 2 genera, *Schizosaccharomyces* (4 species) and *Hasegawea* (1 species).

Genus *Schizosaccharomyces*

It has been traditionally placed with *Saccharomyces* and other yeasts, though the differences in cell division and wall carbohydrates were well-known. The cell wall of *Schizosaccharomyces* is made of chitin, while in *Saccharomyces* and other yeasts it is made of β 1-3 glucan. Kurtzman (1993) shifted *Schizosaccharomyces* to an independent order due to differences in rDNA sequences.

The genus reproduces asexually by fission (a fission yeast). It has four species (*S. cryophilus, S. japonicus, S. octosporus,* and *S. pombe*), which can be isolated from high sugar substrata like mollases, cane sugar, grapes etc. The fungus exists mainly as haploid cells; the diploid phase is very short, represented only by the zygote. In culture, the fungus grows as single cells or as a pseudomycelium, which soon breaks into arthrospores. The cells of *S. octosporus* are rectangular and haploid. During fission (Fig. 10.4), the cell is divided into two equal halves, which are later separated. A peripheral ring of plasmamembrane invaginates and grows centripetally, to form a double-membraned septum after deposition of wall material in between the two membranes. The matrix between the newly-synthesized walls dissolves and the two daughter cells separate.

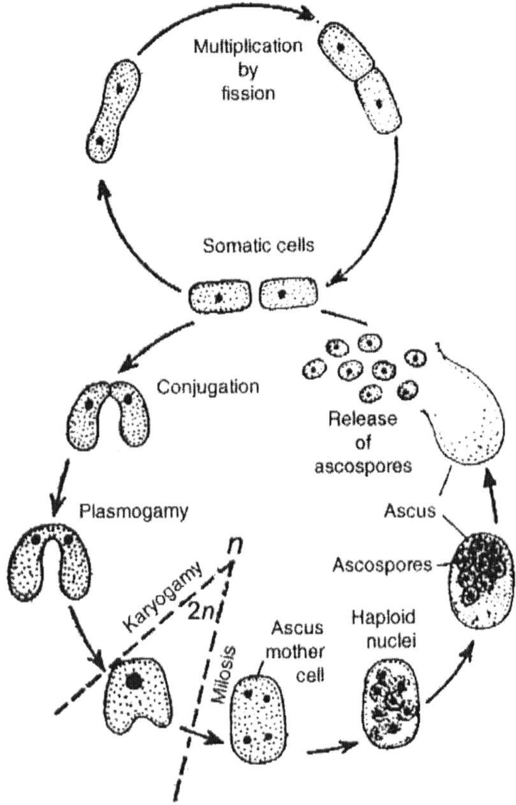

Fig. 10.4. Life cycle of *Schizosaccharomyces octosporus.*

Sexual reproduction involves copulation between two similar haploid cells. All species are homothallic. The two cells fuse at the tip; their nuclei migrate into the narrow isthumus (conjugation tube). Karyogamy takes place and the fused **zygote cell** enlarges to become the ascus. The diploid nucleus undergoes meiosis, followed by mitosis, and 8 haploid nuclei are produced, around which eight ascospores are then formed. The cell wall dissolves by autolysis liberating the ascospores, which grow into haploid cells.

S. pombe (Fig. 10.5), like the distantly related *Saccharomyces cerevisiae*, is used as a model organism in the study of eukaryotic cell biology. For cell-cycle studies, it is better than *S. cerevisiae*. It is particularly useful in evolutionary studies because it is thought to have diverged from *S. cerevisiae* lineage 300 million to 1 billion years ago *S. pombe* is also used in biotechnological production of citrulline and also for production of protein from kerosene and diesel oil.

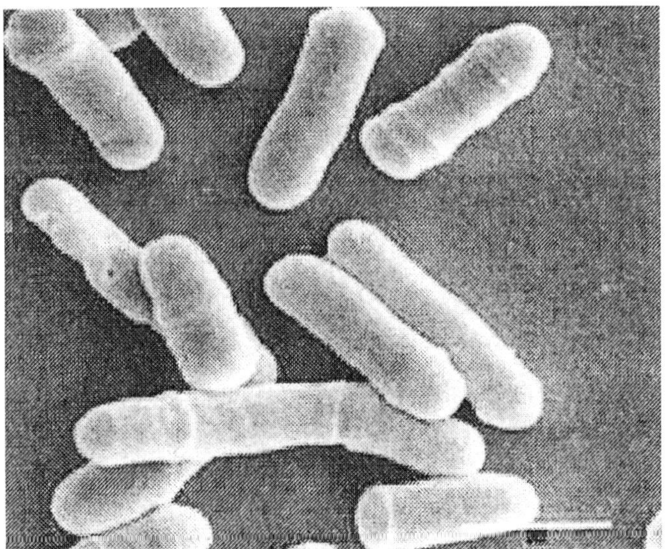

Fig. 10.5. Cells of *Schizosaccharomyces pombe*

Class *Taphrinomycetes* (1 order, *Taphrinales*, 2 families: *Protomycetaceae* and *Taphrinaceae*; 8 genera)

Family *Protomycetaceae*

There are 6 genera, which attack flowering plants and cause lesions and galls, accompanied by excessive discolouration. The apparently diploid mycelium spreads intercellularly in the host and produces thick-walled resting spores. These germinate and form a vesicle called **synascus**, which is considered to be a compound ascus, containing numerous ascospores. The ascospores are discharged in a single mass.

The genus *Protomyces* has 10 species, *Protomyces macrosporus* being the most widespread, causing galls and discolouration on coriander and other members of the family *Umbelliferae*. *P. inundatus* and *P. pachydermus* attack leaves of *Apium nodiflourm* and *Taraxacum officinale* (dandelion), respectively.

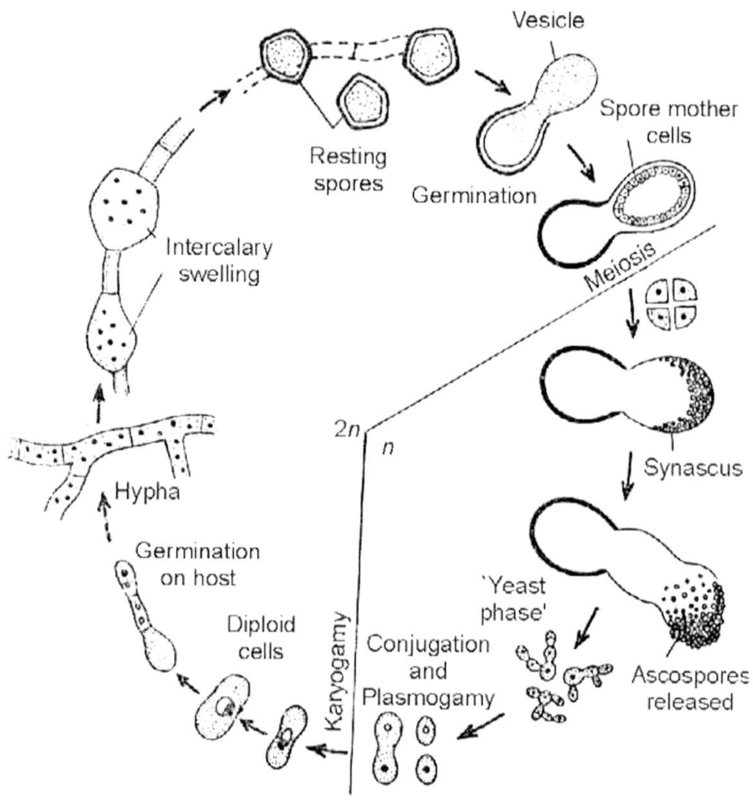

Fig. 10.6. Life cycle of *Protomyces macrosporus*

The mycelium (Fig. 10.6), which is filamentous and septate, spreads intercellularly in the host. The galls develop as a result of uncontrolled division and growth of cells. Intercalary **hyphal swellings**, which develop into **resting bodies**, appear in plenty in the galls. The resting bodies of *P. macrosporus* undergo a period of dormancy before germination. At the time of germination, the outer wall splits and the contents move out into a sac-like vesicle. A large vacuole develops in this vesicle pushing the cytoplasm and the nuclei (about 100 in number) to the periphery. The nuclei undergo meiosis; the resulting haploid nuclei later become incorporated in large number of ascospores (= endospores). The vesicle is regarded as a *synascus* or compound ascus. The DNA content of the spores is half of the amount recorded for the hyphae which suggests the occurrence of meiosis before spore formation. The ascospores are discharged forcibly in a single mass. These multiply by budding. In laboratory cultures, the fungus exists only in the budding or 'yeast state'. The ascospores or 'bud cells' (blastospores) apparently conjugate to form diploid spores (the DNA content is restored to level of the hyphae). These fused spores germinate only on the host and cause infection. The unfused ones fail to infect.

Family *Taphrinaceae*

The family is characterized by the development of asci from dikaryotic **ascogenous cells**, formed by breaking up of the cells of the dikaryotic mycelium.

The genus ***Taphrina***, has 95 species, all parasitic on higher plants and ferns. All the species are **dimorphic**, forming mycelium and asci in the parasitic phase and 'yeast' colonies of budding cells in the saprobic phase. The ascospore budding is **enteroblastic**, which is a characteristic of basidiomycetous yeasts, like red yeasts of *Exobasidiales* and *Ustilaginales*. In this, the wall of the bud is derived from the inner wall of the parent cell or is formed anew.

The species of *Taphrina* cause necrotic leaf spots, galls and deformations of various types and, in a few cases, form "witches broom". Some important species and the diseases they cause are: *T. deformans* (peach leaf curl); *T. pruni* (pocket plum disease of plum fruits) and *T. epiphylla* (witches broom of cherries). The species have narrow host range, and are named on their hosts.

The blastospores (Fig. 10.7), formed by budding of ascospores germinate on young peach leaves. The nucleus divides and the two nuclei form a dikaryon, which, by subsequent conjugate divisions, gives rise to dikaryotic cells. The hyphae grow intercellularly, taking irregular shape of the intercellular spaces, which they fill. The adjoining cells show neoplastic (tumorous) growth resulting in curling and thickening of the leaves. The hyphae eventually aggregate below the cuticle and break up into dikaryotic, ascogenous cells. Each ascogenous cell develops into a stalked ascus. Karyogamy occurs in the elongating ascogenous cell. The diploid nucleus divides mitotically: one daughter nucleus migrates to the apical portion while the other remains at the base of the elongating cell. A septum is laid down between the two nuclei, separating a basal 'stalk cell' from an upper 'ascus mother cell'. The nucleus and the cytoplasm of the stalk cell disintegrate making the cell empty. The diploid nucleus, in the ascus initial cell, divides by meiotic and mitotic divisions and forms 8 haploid nuclei, which later transform into 8 ascospores. The ascospores are forcibly discharged by a rupturing of the ascus tip. The ascospores (outside or when still inside the ascus) produce small, round blastospores by budding, which, in turn, produce secondary blastospores, and in this way, 'yeast' colonies are formed. This yeast phase represents the conidial or asexual phase and is designated as *Lalaria* (anamorph genus, asexual state).

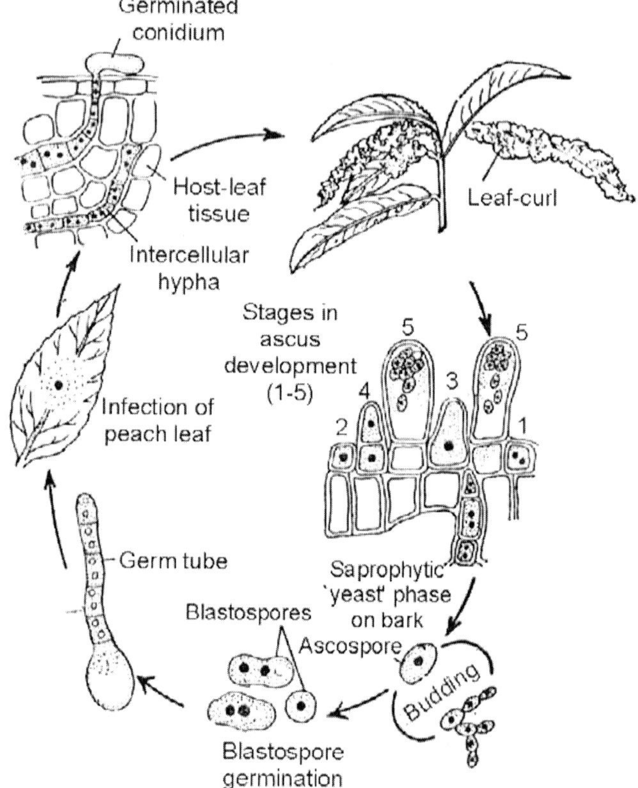

Fig. 10.7 Life Cycle of *T. deformans*

REFEENCES

Aliouat-Denis, C.M. *et al.,* 2008 Pneumocystis species, coevolution and pathogenic power. *Infection, Genetics and Evolution.* 8(5): 708-726.

Kurtzman, C.P. 1993. Systematics of the ascomycetous yeasts assessed from ribosomal RNA sequence divergence. *Ant. v. Leeuw*. 63:165-174.

Landvik, S., Erikson, O.E. and Berbee, M.I. 2001 *Neolecta* – a fungal dinosaur? Evidence from β-tubulin amino acid sequences. *Mycologia*. 93(6): 1151-1165.

Nishida, H. and Sugiyama, J. 1994. *Archiascomycetes*: detection of a major new linage within the Ascomycota. *Mycoscience*, 35: 361-366.

Rosling, A. *et al.* 2011. *Archaeorhizomycetes*: unearthing an ancient class of ubiquitous soil fungi. *Science*, 333 (6044) 876.

Stringer, J.R. *et al.* 2002 A new name (*Pneumocystis jirovecii*) for *Pneum-ocystis* from humans. Emerging Infectious *Diseases.* 8 (9): 891-896.

SUMMARY

Taphrinomycotina includes the most primitive ascomycetes, and, as revealed by the 18S rDNA sequences, constitute the 'basal group' of phylum *Ascomycota*. It is divided into 5 classes: *Archaeorhizomycetes, Neolectomycetes, Pneumocystidomycetes, Schizosaccharomycetes* and *Taphrinomycetes*. The most recent addition to this basal group, genus *Archaeorhizomyces* (2011) represents a new class of fungi, which were earlier known as "Soil Clone Group 1" (SCG1) and identified only by DNA sequencing. They are found in the ectomycorrhiza but independently can not cause mycorrhizal association. The organism which lived only as an "environmental DNA" so far, is now available in culture showing their branched mycelial nature. The other fungi in this subphylum are equally interesting. For example, *Neolecta*, forms a large ascoma (apothecium) similar to those formed by most highly-evolved ascomycetes. The genus thus shares characters of both primitive and highly evolved ascomycetes. It has been called a living "fungal relic of the past". *Pneumocystis*, which is always unicellular, causes pneumonia in mammals including human beings. It was thought to be a protozoan until 1994. These three fungi are the lone genera of the classes named after them. The 18S rDNA sequences have shown that *Schizosaccharomycetes* and *Saccharomyces* are not as closely related as previously understood, and thus, they are now separated into two different subphyla. These too, like *Pneumocystis*, exist only as unicellular organisms. *Protomyces* and *Taphrina*, which are half yeasts and half filamentous, induce tumors in the plant hosts.

TEST QUESTIONS

1. Describe the important features of subphylum *Taphrinomycotina* and give its classification up to families.
2. Write notes on any **two** of the following: (a) *Archaeorhizomyces*, (b) *Neolecta*, (c) *Pneumocystis*, and (d) *Schizosaccharomyces*.
3. Briefly describe any **three** the following: (a) Septum formation in *Schizosaccharomyces*, (b) Environmental DNA, (c) *Schizosaccharomyces pombe*, (d) Resting bodies in *Protomyces*.

4. Explain:
   (a) The basis of designating *Taphrinomycotina* as the "basal group" of *Ascomycota*.
   (b) The differences in the chemical composition of the cell wall of *Schizosaccharomyces* and *Saccharomyces*.
   (c) The role of *Archaeorhizomyces* in nature.
   (d) Dimorphic nature of *Taphrina*.
   (e) The difference between the ascomata of *Neolect*a and those formed by *Pezizomycotina*.
5. Answer in brief:
   (a) Which genus in *Taphrinomycotina* is unique in showing primitive and advanced character?
   (b) What is environmental DNA? From which environmental DNA was *Archaeorhizomyces* isolated?
   (c) Which type of soil fungi and *Archaeorhizomyces* live together?
   (d) Which genera in *Taphrinomycotina* live only as unicellular yeasts?
   (e) Which classes in *Taphrinomycotina* have only one genus?
   (f) Which genera in *Taphrinomycotina* live only in mycelial form?
   (g) What is synascus in *Protomyces*?
   (h) Which is the pathogen causing PJP.
   (i) What are the two expansions of PCP,
   (j) Which are the primitive characters of ascoma of *Neolecta*.
6. Fill in the blanks in **any ten** of the following:
   (a) The designation of *Taphrinomycotina* as "basal group" was given in 1994 by -------------and--------------.
   (b) The genus *Archaeorhizomyces* was discovered by ---------- in 2011.
   (c) The ascospore budding in *Taphrina* is -----------, which is typical of basidiomycetous yeasts.
   (d) The two genera in *Taphrinomycotina* that multiply by fission are -----------and --------------.
   (e) *Archaeorhizomyces* coexists with ---------------- fungi, but they themselves do not form -------------.
   (f) The *Neolecta* species are also called as -----------------.

(g) The ascogenous hyphae of *Neolecta* do not form ----------, and their hymenium lacks -------------.

(h) *Neolecta* and *Pneumocystis* have so far not been ------------, and therefore, much is not known about their biology and life cycle.

(i) *Archaeorhizomyces* was isolated from soil sample earlier known as -------.

(j) *Pneumocystis* divides by -----------, like -------------------.

(k) In culture, *Schizosaccharomyces* forms -----------------, which breaks into -------------.

(l) In *Taphrina*, ascogenous cells are formed by breaking of ---------- hyphae.

(m) *Protomyces* forms galls on plants belonging to family ------------.

(n) Neoplastic growth is induced in the host plants by ------ and ------.

(o) *Taphrina* forms mycelium and asci in the ---------phase, and ----------- in ------------ phase.

(p) *Taphrina* and *Protomyces* exist in ----------- phase in culture.

11

Subphylum *Saccharomycotina*

(True yeasts, Ascomycetous yeasts)

Moving upwards from the 'basal group', towards filamentous ascomycetes (informally Euascomycetes), we find the *Saccharomycotina* comprising mostly of unicellular yeasts. These 'ascomycetous yeasts* comprise a monophyletic group. These are cosmopolitan and are found in wide range of habitats. The yeasts are:

(1) **Unicellular organisms** that multiply by budding or fission. The daughter cells form chains (fragile filaments), called pseudomycelium, which breaks into individual yeast cells.

(2) **Mycelium absent or poorly developed**; if present have septa, which have several minute pores instead of a single pore.

(3) **Cell wall made up of β 1-3 glucans**. Chitin is found only around the bud scars and that too in minute amounts.

(4) **Direct development of asci from the zygote**. Asci are formed singly or in chains, sometimes look like somatic cells.

(5) **Absence of ascomata.**

The subphylum comprises 1 class, ***Sacharomycetes***, 1 order ***Saccharomycetales*** and 13 **families**. The families are:

Ascoideaceae, Cephaloascaceae, **Dipodascaceae**, Endomycetaceae, Ermotheciaceae, Lipomycetaceae, Metschnikowiaceae, Phaffomycetaceae, Pichiaceae, **Saccharomycetaceae**, **Saccharomycodaceae**, Saccharomycops-

---

* Yeasts are defined as fungi "which in a stage of their life cycle occur as single cells, reproducing by budding or fission." The name 'yeast' in its broader sense, is applicable to a heterogeneous group of fungi. In addition to common ascomycetous yeasts, there are basidiomycetous yeasts.

idaceae, and Trichomonascaceae. We shall study only a few of these families (printed in bold).

Family *Ascoideaceae* (contains a single genus, *Ascoidea*)

Genus *Ascoidea*

Ascoidea is represented by 4 species. *A. rubescens*, the type species, was obtained from exudates of peach. The hyphae are made of uni- or multinucleate cells. Conidia are produced singly at the hyphal tips, but due to resumed growth of the tip, the hypha appears to bear one apical and several lateral conidia (Fig. 11.1). Oval asci, bearing hat-shaped ascospores, develop parthenogenetically on the hyphae. The asci proliferate by forming new asci successively within the old ruptured ascus (ascal proliferation).

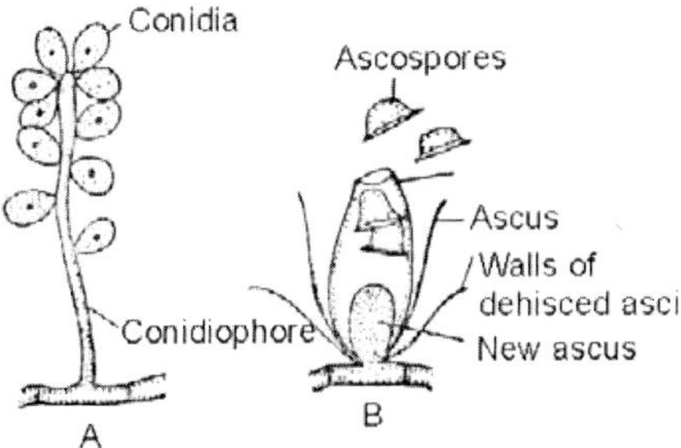

Fig. 11.1. *Ascoidea rubescens*; A, conidia and conidiophore; B, hat-shaped ascospores and ascal proliferation.

Family *Dipodascaceae*

The family has four genera, *Dipodascus*, *Galactomyces*, *Sporopachydermia* and *Yarrowia*, which have well developed mycelium, with uni- or multinucleate cells. Fragmentation gives rise to conidia (arthrospores). Asci are formed by fusion of gametangia formed on adjacent cells or separate mycelia. There are one to several ascospores. The species are widespread and found in decaying plant tissue or as spoilage organisms in food industry.

Genus *Dipodascus*

There are 16 species. *D. albidus*, *D. uninucleatus* and *D. aggregatus* live saprobically on plant exudates. The mycelium consists of multinucleate cells in *D. albidus*, and of uninucleate cells in the other two species. These are homothallic or heterothallic.

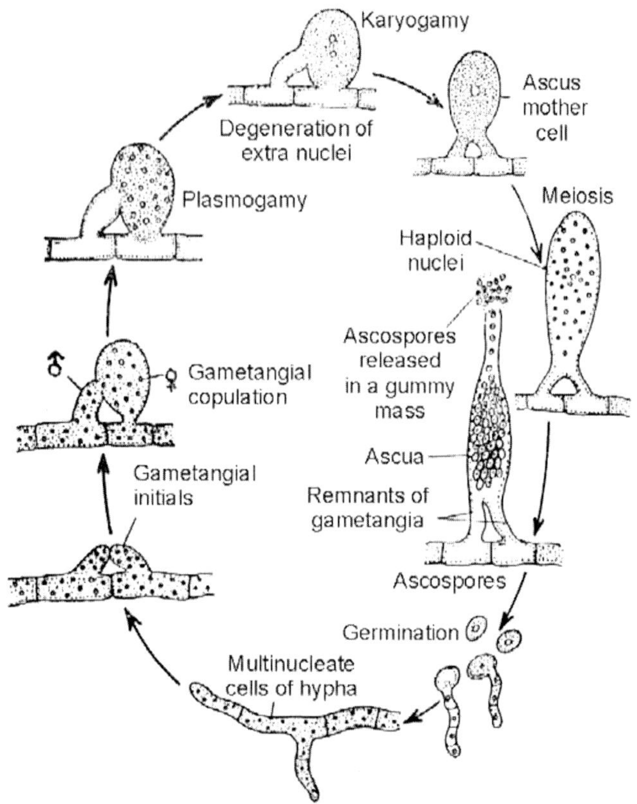

Fig. 11.2 Life cycle of *Dipodascus albidus*

The multinucleate, septate hyphae (Fig. 11.2) form arthrospores for asexual multiplication. Multinucleate gametangia are formed, which are initially similar, but later one becomes larger than the other. The gametangia fuse and the nuclei from the smaller (male) gametangium migrate into the bigger (female) gametangium, but except two, the rest degenerate. After karyogamy, the zygote cell enlarges to form an elongated, tapering ascus. The lower persisting portions of the gametangia look like two 'feet' of the ascus. The zygote nucleus by one meiotic and repeated mitotic divisions, forms several nuclei which later give rise to numerous round or oval ascospores. The ascus tip dissolves and the ascospores are released in a gummy mass.

Family *Saccharomycetaceae*

The family includes majority of the ascomycetous yeasts. There are 14 genera and 121 species (Kirk *et al.,* 2008), which exist as single cells in a wide range of habitats. The mycelium is more or less absent. The somatic cells proliferate characteristically by multiple budding. Asci are morphologically similar to the somatic cells and contain 1 to 4 globose ascospores.

Most of the yeasts are saprobes growing on sugary substrata, like flower-nectar, ripe fruits, and also occur in milk, soil, animal excreta and plant parts. The organisms ferment sugars under anaerobic and partially-aerobic conditions, which in general favour pseudomycelial formation. Even some strains of *Saccharomyces* do not always exist as descrete cells. In addition to anaerobic condition, some other factors like deficient nutrition, lower than optimum temperature, chemicals (like camphor and penicillin) and metal complexes (especially of cobalt and boron), enhance production of the pseudomycelium. Reversal to the unicellular budding stage is brought about by the addition of readily-assimilated nutrients like sulfhydral compounds (cysteine, glutathione and sodium thioglycolate) and certain chemicals, like potassium tellurite or sodium selenite.

Asexual reproduction occurs by budding. Asci develop directly from the zygote, formed by conjugation between two cells, or parthenogenetically from single somatic cell. Ascospores, which are of various shapes (but not needle, or spindle-shaped), are used in the identification of the genera. Assimilation of nitrate is another important characterise for identification. The genera may be homo-, or heterothallic. The phenomena of **agglutination** and **flocculence** are common in yeasts.

Genus *Saccharomyces*

The genus has 26 species, the most important being ***S. cerevisiae***, (with 25 races), called as the ''yeast of commerce'' or "bakers and brewers yeast". It is perhaps most useful yeast which has been used for baking and brewing since ancient times. It was originally isolated from the skin of grapes, showing a thin white film. Due to its extensive use in production of alcohol (in brewing industry) and carbon dioxide (to render the bread spongy in baking industry), as well as a **probiotic**, the commercial production of yeast tablets is an industry in itself. *S. cerevisiae* var. *ellipsoides* is the strain used in the brewing industry.

It is also used as a **probiotic**. (Probiotic is defined by FAO/ WHO as "Live organisms which when administered in adequate amounts, confer benefit on the host''). Commercially-produced 'nutritional yeast', if taken 2 table spoons daily, provides 52% of the **Recommended Daily Amount** (RDA) of protein. It is high in fiber, B-vitamins and folic acid. Besides, the β 1–3 glucans, present in nutritional yeast, stimulates the immune system of the body, lowers LDL cholesterol level which helps in healing of wounds and prevents infection.

S. cerevisiae is one of the most extensively studied eukaryotic organisms in biology, much like *Escherichia coli* among prokaryotes. *S. cerevisiae*, was the first eukaryotic organism whose genome was completely sequenced. More than 600 scientists from Europe, North America and Japan participated in the **Yeast Genome Project** which was initiated in 1989 and completed by 1995 end. The genome has 12,156,677 base pairs and 66257 genes, compactly organized in 16 chromosomes. But only 5800 are functional genes. *S. cerevisiae* shares 23% of its genome with the human genome.

Another species, ***S. boulardii***, first isolated from 'Litchi' (*Litchi chinensis*) and mangosteen (*Garcinia mangostana*), is used as a probiotic to maintain and restore the natural flora of the alim entary canal in humans. It is marketed in lyophilized form (referred as *S. boulardii lyo*) for treatment of gastroenteritis, acute diarrhea, irritable bowel syndrome and antibiotic-associated diarrhea.

The somatic cells may be haploid or diploid; both phases are equally common and alternate with each other. The cells of both phases are capable of multiplication by budding (Fig. 11.3). The haploid cells are smaller, about half the size of the diploid cells. On solid nutrient-agar medium, the cells form typical, slimy-white, slow-growing colonies.

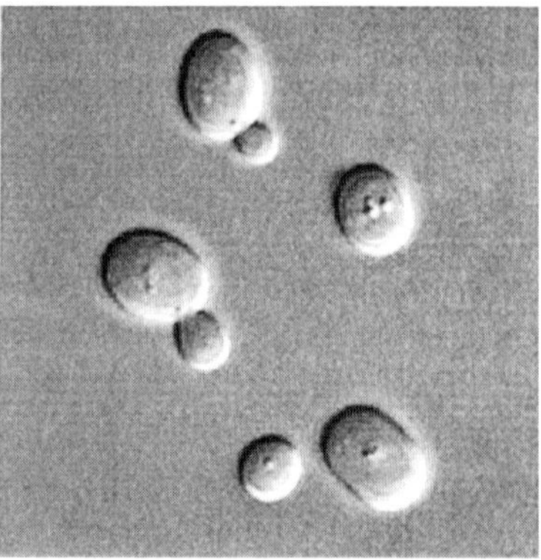

Fig 11.3. Budding in *Saccharomyces cerevisiae*

The cells are round to oval (Fig. 11.4), and 5-10 μm in diameter. The cell wall is made up of β1-3 glucans. Small amount of chitin is found around the bud-scars. The bud-scars (up to 23 in number) are remnants of the detached daughter bud cells that persist on the external surface of the parent cell wall. There is an unusually big central vacuole pushing the protoplasm to the periphery. The cytoplasm contains the usual cell organelles like nucleus, mitochondria and reserve food granules (glycogen, oil and refractile 'volutin' granules which are inorganic metaphosphate polymers). The tiny nucleus lies in the cytoplasm, appressed to the vacuole.

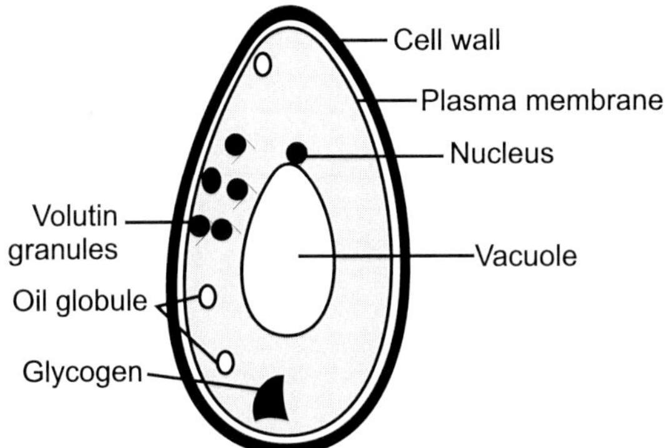

Fig. 11.4. A typical cell of *Saccharomyces cerevisiae*

The cells (both haploid and diploid) reproduce asexually by multipolar 'budding'. A small zone of the cell wall softens and bulges out like a bud, which grows and is eventually pinched off by constriction. The nucleus divides by constriction and the daughter nucleus enters the bud along with other cell organelles. The bud frequently gives rise to a bud cell itself before its separation from the parent cell, and in this way a chain of cells is formed. The bud formation is **exogenous** i.e. the wall of the bud is derived from the wall of the parent cell. (In basidiomycetous yeasts it is **enterogenous** i.e. the bud wall is synthesized afresh).

S. cerevisiae shows a haplo-diplobiontic life cycle in which both haploid and diploid phases are equally extensive and important (Fig. 11.5).

Both types of somatic cells i.e. haploid and diploid, multiply by budding and eventually, transform into the other type. What triggers this change is not precisely known. The haploid cells, which are of two mating types, a and α (alfa), after fusion (plasmogamy and karyogamy) give rise to diploid cells, and, thus, initiate the diploid phase. At the end of the diploid phase, the diploid nucleus undergoes meiosis and forms four haploid nuclei around which four ascospores are developed. In laboratory, the diploid cells can be induced to form ascospores by adding 5% sodium acetate to the growth medium. Thus, the diploid cells are potential asci. On release after lysis of the ascus wall, these grow into somatic cells establishing the haploid phase.

Sexual compatibility is controlled by a single, 2-allele (a and α) gene, which segregates during meiosis. Conjugation may take place between somatic cells or ascospores immediately after release. Agglutination (sexual aggregation) greatly increases the chances of zygote formation. **Agglutination** is the aggregation of cells of opposite mating types and, therefore, is also called "sexual aggregation". It enhances zygote formation but has no effect on ascospore formation. Agglutination is brought about by hydrogen binding between protein-like substances of the cell wall of one mating type and a polysaccharide on the other. There are evidences suggesting role of sex hormones in bringing about the mating of *S. cerevisiae* (*See Chapter 31*).

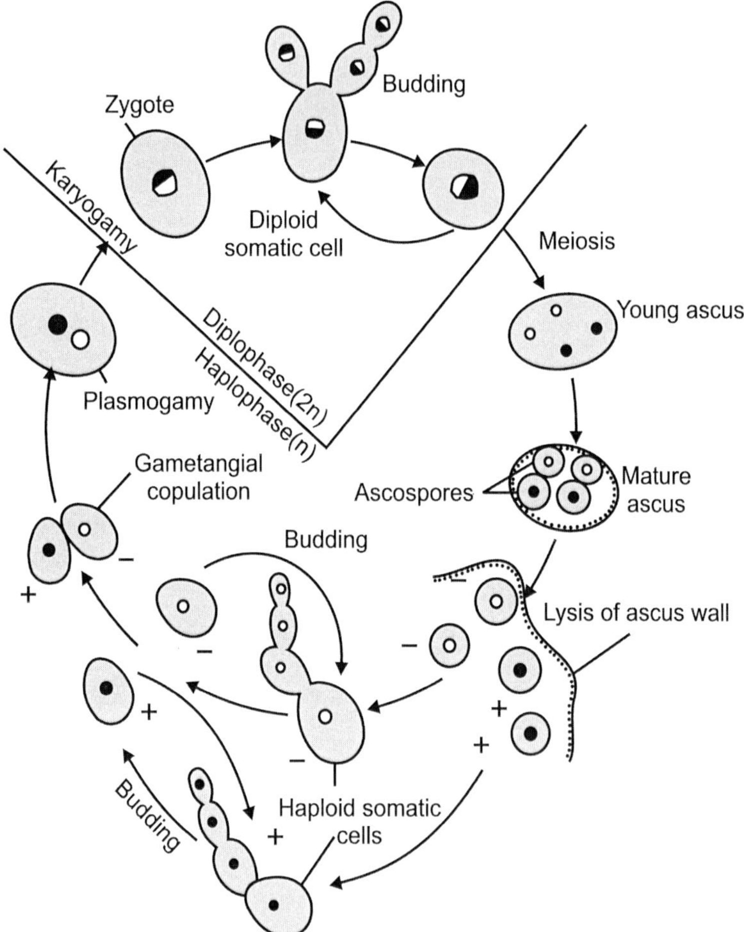

Fig. 11.5. Life cycle of *Saccharomyces cerevisiae*

Flocculence (or autoaggregation) has been defined as a reversible aggregation of similar cells to form discrete groups. It is known in *Saccharomyces cerevisiae* since Pasteur first observed the phenomenon in 1876. With the onset of alcoholic fermentation under anaerobic conditions, the yeast cells show tendency to flocculate. **Flocs** are formed either at the top

(**top yeasts**) or at the bottom of the vessel (**bottom yeasts**), depending on the brewing process. In 'top yeasts', during fermentation, the hydrophobic surface of the yeast cells causes the flocks to adhere to CO_2 and rise to the top of the fermentation vessel. Flocculent yeasts are of value in continuous fermentation systems since the cells may be removed from the effluent system and recycled.

Flocculation, if not properly adjusted, may be harmful to fermentation. It has been observed that some ions can **enhance or inhibit** flocculation. Calcium and magnesium are essential for floc formation, while zinc, cadmium and ferrous ions enhance the flocculation. Anions like carbonate, phosphate and fluoride inhibit flocculation. Carbohydrates also inhibit flocculation and disperse the cells of the flocs. All strains do not behave similarly towards the different carbohydrates. It is also known that the ability to flocculate develops only when the cells have ceased reproducing. Reproducing yeasts are non-flocculent. When fermentation starts and sugar level falls to a critical level, flocculation starts.

Flocculence is an inherent character of the yeast cell itself and controlled by three gene pairs. Hybridization between flocculent and non-flocculent strains results in segregation of this character, the flocculent character being dominant. It is also controlled, at least in part, by inheritable cytoplasmic elements, and the unequal distribution of these elements in the daughter cells is the cause of the varying degree of flocculence shown by different strains. The inheritable character finds expression in the cell wall. A difference in the cell wall of flocculent and non-flocculent yeasts has been reported. Isolated cell walls of mature cells retain the characteristic of flocculence. A **phosphate-mannan-protein complex** of the cell wall is involved in the process.

Family *Saccharomycodaceae*

There are 3 genera in this family which show feeble or no mycelium. Somatic cells, which are lemon-shaped, multiply by bipolar budding. Asci are formed either directly from an undifferentiated somatic cell or by conjugation between a mother cell and bud.

Genus *Saccharomycodes*

It reproduces asexually by both budding and fission. The ascospores conjugate while still inside the ascus. There is a single species, *S. ludwigii*, which is diploid and heterothallic. The life cycle (Fig. 11.6) shows a long diploid and a short haploid phase i.e. **diplobiontic life cycle**. The haploid stage is represented by a single generation of ascospores.

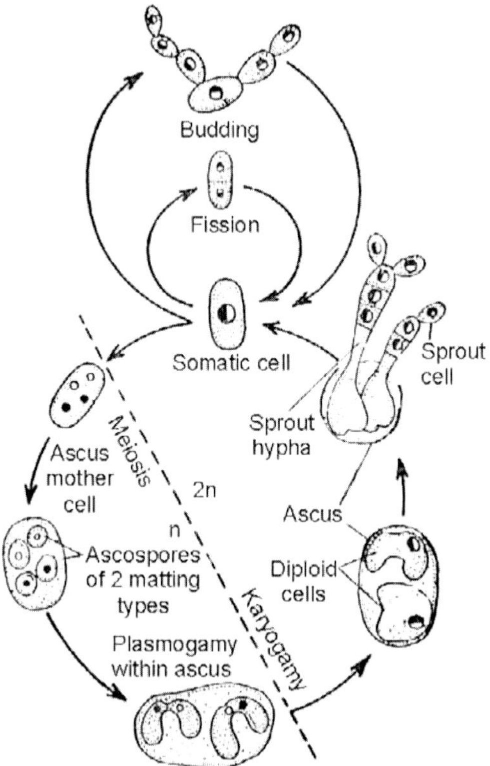

Fig. 11. 6. Life cycle of *Saccharomycodes ludwigii*.

The diploid cells multiply by budding as well as fission. Each cell is a potential ascus. When conditions become unfavourable for growth and multiplication, it behaves as an ascus. The diploid nucleus undergoes meiosis and forms four haploid nuclei of two mating types, which develop into four ascospores. The ascospores of the opposite mating types undergo conjugation while still inside the ascus. The zygote cells, formed within the ascus, form a germ tube, which breaks through the ascus wall and forms a 'sprout mycelium' from which are budded off several diploid somatic cells.

REFERENCE

Kirk, P.M. *et al.* 2008 *Ainsworth and Bisby's Dictionary of the Fungi* (10th ed.) CAB International, Oxon, U.K.

SUMMARY

Subphylum *Saccharomycotina* comprises majority of 'yeasts' and forms a monophyletic group. These (1) are unicellular and multiply by budding or fission, (ii) mycelium is absent or poorly developed; if present have septa, which have several minute pores instead of a single pore, (iii) cell wall is made up of β 1–3 glucans, (iv) asci develop directly from the zygote, and (5) there is no ascoma. These are 'true yeasts' and placed in class, *Sacharomycetes*, order *Saccharomycetales*, and 13 families. The genera *Ascoidea* and *Dipodascus* represent mycelial yeasts, while *Saccharomyces* and *Saccharomycodes* exist only as 'yeasts'. They do form pseudomycelium which breaks into component cells. *Ascoida* bears one terminal and several lateral conidia on the conidiophore. The asci develop parthenogenetically which form hat-shaped ascospores. Ascal proliferation is a characteristic feature of this genus. *Dipodascus* multiplies asexually through arthrospors. Sexual reproduction is by fusion of gametangia and the zygote directly develops into the ascus. The cylindrical ascus appears as standing on 2 legs. The ascospores are numerous and oval to round in shape.

Saccharomyces and *Saccharomycodes* live exclusively as unicellular yeasts. The former multiplies by budding, while the latter by budding and fission both. They also differ in their genetic mode of life cycle. While *Saccharomyces* has haplo-diplobiontic life cycle, *Saccharomycodes* has diplobiontic type of life cycle.

Saccharomyces is historically important as the organism used since ages for baking and brewing. It has been used as the model eukaryotic organism for biochemical and molecular studies, in the same way as *E. coli* has been used as prokaryotic organism. *S. cerevisiae* is the first organism whose genome was sequenced in a global scientific project. *S. cerevisiae* is commercially produced as 'nutritional yeast' for its use as a probiotic. It provides protein, fiber, folic acid, B-vitamins and β 1–3 glucans. *S. boulardii* is also used as a probiotic for maintaining and restoring the normal flora of the alimentary canal of humans. Agglutination and flocculation have been studied for academic and practical purposes. Agglutination is sexual aggregation; while floccualtion is aggregation between similar cells. *S. cerevisiae* cells are of two mating types (a and α) and their conjugation is under control of sex hormones.

With the onset of alcoholic fermentation, the yeast cells form flocs, either at the top (top yeasts) or at the bottom of the vessel (bottom yeasts). Flocculation starts when sugar level falls below a critical level during fermentation. Flocculation, if not properly adjusted, may be harmful to fermentation. It can be done through use of ions that enhance or inhibit flocculation. Flocculence is inherited as a dominant gene. It is controlled, at least in part, by inheritable cytoplasmic elements. The unequal distribution of these cytoplasmic elements in the daughter cells is the cause of the varying degree of flocculence shown by different strains.

TEST QUESTIONS

1. Write notes on: (1) Characteristic features of *Saccharomycotina,* (2) *Saccharomyces cerevisiae* as a probiotic, (3) Flocculence in yeast.
2. Distinguish between any **three**: (1) Agglutination and Flocculence, (2) Exogenous and Enterogenous budding in yeasts. (3) Flocculent and Non-flocculent yeast cell walls, (4) Top yeast and bottom yeast, (5) life cycle of *Saccharomyces* and *Saccharomycodes*, (6) *S. boulardii* and *S. boulrdii lyo*.
3. Fill in the blanks:
   (a) Yeast cells are of two mating types, ---- and ----.
   (b) The yeast, ------------ multiplies by budding as well as by fission.
   (c) *S. boulardii* is used as a --------- for maintaining and restoring natural flora of the alimentary canal of humans.
   (d) In *S. boulardii lyo*, *lyo* stands for -----------------.
   (e) Chitin in yeast is present only around ------ ---------.
   (f) Volutin granules in yeast cells are ---------------------.
   (g) Diploid yeast cells can be induced to form ascospores by adding.------------ to the medium.
   (h) Asci develop parthenogenetically in the genus --------------.
   (i) Ascal proliferation and hat-shaped ascospores are found in the genus-----
   (j) Numerous ascospores are formed by -----------------.
   (k) *S. boulardii* was first isolated from ----------------.
   (l) *S. cerevisiae* was first isolated from ----------------.
   (m) Agglutination is also called as --------------.
   (n) Flocculence in yeast is controlled also by ----------- --------.
   (o) Multinucleate gametangia are formed in the genus ------ .

12

Subphylum *Pezizomycotina*
(the 'Crown Group' of *Ascomycota*)
(Filamentous Ascomycetes, Euascomycetes)

These filamentous, ascoma-forming fungi (with the exception of *Neolecta* of *Taphrinomycotina*) are the biggest and highest evolved members of the phylum and regardeds as the **crown group** of *Ascomycota*. Traditionally, these have been segregated into classes based on the morphology of ascomata, but multigene phylogenetic analyses rejected this classification and supported the existence of at least 11 clades i.e. monophyletic groups (Schoch *et al.,* 2009; Spatafora *et al.,* 2006), which are now designated as classes. Studies on the evolution of these fungi, based on rDNA sequences (Berbee and Taylor, 1999) reveal that apothecial ascomycetes with exposed asci lie at the base of the evolutionary tree of *Pezizomycotina*. The closing of the ascomata resulted in the development of perithecia (with an opening) and cleistothecia (closed ascomata). Studies on ascomatal ontogeny reveal that ascohymenial taxa are ancestral and the ascolocular taxa with fissitunicate asci have evolved from within ascohymenial taxa (Liu and Hall, 2004).

Several lines of evidence suggest that the common ancestor of ascomycetes had mycelial growth. *Zygomycota* (in RPB1 and RPB2 studies) occupies a position immediately basal to the divergence of asco- and basidiomycetes. Thus, it appears that the yeast forms of *Taphrinomycotina* and *Saccharomycotina* result from independent loss of filmentous growth. In the two **sister groups** (the two clades are formed by splitting of a single lineage) *Saccharomycotina* and *Pezizomycotina*. *Saccharomycotina* have a simple mycelial and/or yeast dimorphic growth, where as the pezizomycotan branch evolved an array of different ascoma structures (ascohymenial and ascolocular).

Classification

In the current classification (Eriksson, 2006, Kirk *et al.,* 2008, Schoch *et al.,* 2009), based largely on rDNA phylogenies, *Pezizomycoina* is divided into 11 classes. The classes with their ascomatal and ascal characters are as follows.

Subphylum *Pezizomycotina* (11 classes)

Class

Arthoniomycetes (ascoma apothecia; asci bitunicatre) (1 order)

Dothideomycetes (ascostroma; asci bitunicate)

2 sub classes:

Dothideomycetidae (3 orders)

Pleosporomycetidae (4 orders)

Eurotiomycetes (perithecia, cleistothecia, ascostroma; bitunicate, prototunicate)

3 sub classes: *Chaetothyriomycetidae,*

Eurotiomycetidae,

Mycocaliciomycetidae.

Geoglossomycetes (apothecia, cleistothecia; inoperculate, prototunicate) (1 order)

Laboulbeniomycetes (perithecia; prototunicate)

Lecanoromycetes (apothecia, perithecia; bitunicate, inoperculae, prototunicate)

3 sub classes: *Acaurosporomycetidae,*

Lecanoromycetidae and

Ostropomycetidae

Leotiomycetes (apothecia, cleistothecia; inoperculate, prototunicate)

Lichinomycetes (apothecia; bitunicate, inoperculate, prototunicate)

Orbiliomycetes (apothecia; inoperculate)

Pezizomycetes (apothecia, operculate) and

Sordariomycetes (perithecia, cleistothecia; inopercualte, prototunicate)

Incertae sedis: *Lahmiales, Medeolariales, Triblidiales* (not assigned any class)

Phylogenetic Realtionships

Orbiliomycetes and *Pezizomycetes* are the two **basal classes** of *Pezizomycotina* (Schoch *et al.*, 2009). Both taxa produce apothecia but differ in the asci. *Pezizomycetes* produce operculate asci, while *Orbiliomycetes* produce inoperculate asci. The remaing 9 classes with inoperculate asci, form a monophyletic group that is designstd by Spatafora *et al.* (2006) as a superclass taxon, informally referred to as **Leotiomyceta**. While the super class relationship has not been resolved, the 'sister group' relationship of *Arthoniomycetes* and *Dothideomycetes*, as well as of *Leotiomycetes* and *Sordariomycetes* is well accepted.

Ascomyceteous Lichens and their Evolution

Lichens, like mycorrhiza, are a biological and not a systematic group. It is important to remember that 20% of all fungi and 40% of all ascomycetes are **lichen-forming fungi** (Kirk *et al.*, 2008). All the ascomycetous lichen-forming fungi belong to *Pezizomycotina*. Most recent estimate of global biodiversity suggest that 17500-20000 lichen-forming species are known and there are additional 1500 **lichenicolous fungi**, growing on lichens as parasites, commensals, or saprobes.

RPB2 pylogenetic studies indicate the lichen-forming fungi have arisen independently several times, losing and forming the lichen association. These occur in almost all of the major lineages and found in both ascohymenial and ascolocular ascomycetes. Majority of the lichenized fungi are found in *Arthoniomycetes, Eurotiomycetes, Lecanoromycetes* and *Lichinomycetes*. Only 4 orders are exclusively lichen-forming. However, while many families are exclusively lichen-forming, there are several families and even genera which have both lichen-forming and non-lichen-forming fungi (e. g. *Arthonia, Arthothelium, Mycomicrothelia, Toninia*). It indicates that **lichen-mode of nutrition** is at various stages of evolution. Some are evolving towards stronger lichen-foroming species, while some others towards less-strong lichen-forming species. These fungi can be grown in axenic cultures (with approximately 42% success) from ascospores or from the thallus by fragmentation. The cultures are slow-growing and have little organized structure and do not form ascomata (although some form conidiomata).

The Basal Classes: Orbiliomycetes and Pezizomycetes

Class *Orbiliomycetes* (apothecial ascoma; inoperculate asci)

The class has 1 order, *Orbiliales* and 1 family *Orbiliaceae*. There are 3 teleomorphic and 12 anamorphic genera. The teleomorphic genera are

Orbilia, *Hyalorbilia* and *Pseudorbilia*. These are widespread but more prevelant in temperate regions. They grow on wood and form small, brightly-coloured to translucent, disk-shaped (typically convex) apothecia. The asci are inoperculate, having oval or ellipsoidal ascospores. The anamorphic genera, which were earlier classified as hyphomycetous genera (belonging to imperfect fungi, now obsolete) represent the asexual stage. For example *Orbilia* has the former anamorph genera *Arthrobotrys, Dactylella, Dicranidion, Dwayaangam, Helicon, Monaacrosporium, Trinacrium* and many others as the asexual stage.

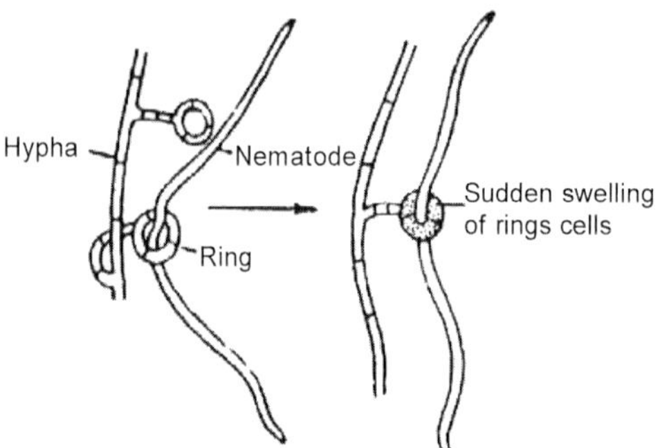

Fig. 12.1 *Dactylella bembicoides*: a nematode is trapped in hyphal ring acting as a trap.

The family is well known for its many nematophagus species. Several of the anamorphic genera are nematophagus and form several types sticky or mechanical loops and branches as traps which help in capturing the prey (nematode) and devour it. *Dactylella bembicoides* (= *Orbilia bembicoides*) forms rings, whose cells swell when a nematode gets into the trap (Fig. 12.1). When the prey is held tightly in the ring, a penetration tube grows out of a hypha and enters the body of the prey. It forms haustoria which secrete enzymes that kill and digest the nematode.

Class *Pezizomycetes* (apothecial ascoma; operculate asci)

The class has 1 order, *Pezizales*, and 16 families

Order *Pezizales* (cup fungi)

The order includes operculate ascomycetes that form apothecia (rarely cleistothecia), which are minute to large, discoid, cupulate (cup like) or more or less, globose, and often brightly coloured. These are epigeous i.e., formed above ground, (rarely hypogeous). The excipulum is usually thick-walled, fleshly, sometimes brittle or leathery, and made up of pseudoparenchy-matous cells. The asci, arranged in a hymenium with paraphyses, are elongated and persistent. The ascus apex is thinner than the side-wall, and has an apical or sub-apical lid (operculum), which is pushed out during ascospore release. Rarely, there is an apical longitudinal slit. It contains a number of species of economic importance, such as morels, black and white truffles and the desert truffles.

This is a large order comprising 16 families and genera, which are saprobic growing in soil, dung, rotten wood etc. These form ascomata in habitats with high pH and low organic content, including disturbed grounds. Some hypogeous forms are mycorrhizal. The families and the genera which we shall study are: *Ascobolaceae* (*Ascobolus*), *Helvellaceae* (*Helvella*), *Morchellaceae* (*Morchella*), *Pyronemataceae* (*Genea*), *Pezizaceae* (*Peziza*), and *Tuberaceae* (*Tuber*).

Family *Ascobolaceae*

The apothecia are minute and occur on dung of herbivore animals, burnt grounds, soil or debris. The asci are phototropic, and, when mature, protrude above the paraphyses. The ascospores are arranged in 2 or 3 irregular rows. There are 6 genera.

Genus *Ascobolus*

Ascobolus (Fig. 12.2) is the type genus, having 61 species, most of which are coprophilous forming apothecia on dung of herbivore animals. *A. carbonarius* grows on old burnt grounds. Most of the species are hetero-thallic. The female element consists of a coiled hypha (Fig. 12.3) with a stalk of few cells, a swollen ascogonial region and an apical trichogyne. All the cells are multi-nucleate.

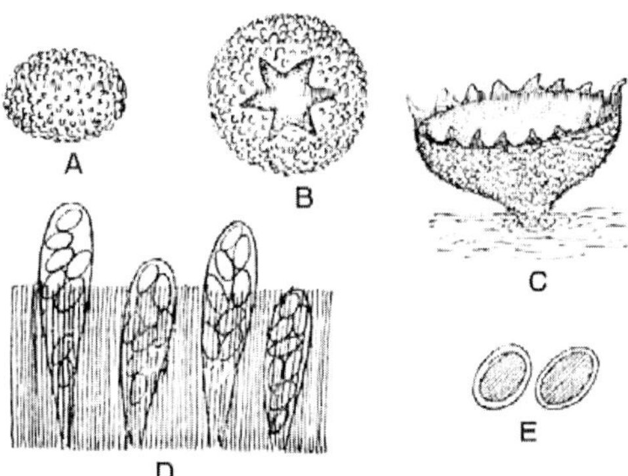

Fig. 12.2 *Ascobolus crenulatus*. A young developing apothecium; B, excipulum ruptured exposing the hymenium; C, a large mature apothecium; D, asci projecting above the surface of the hymenium; E, ascospores with straiations.

Oidia (arthrospores), borne in chains on short hypha, serve both as asexual spores and male cells. (*Rhizostilbella* is recognized as the asexual state). These are transported to the trichogyne through insects. The apothecium is usually 1 mm in diameter. The spaces between asci and the paraphyses are filled with mucilage, which also forms a layer over the hymenium. The mature asci elongate, push out of the mucilage surface and bend towards light. Ascospores are thick-walled, free from one another, and have a mucilaginous epispore, which helps in sticking to the herbage. The ascospores are discharged explosively upto a horizontal distance of 30 cm.

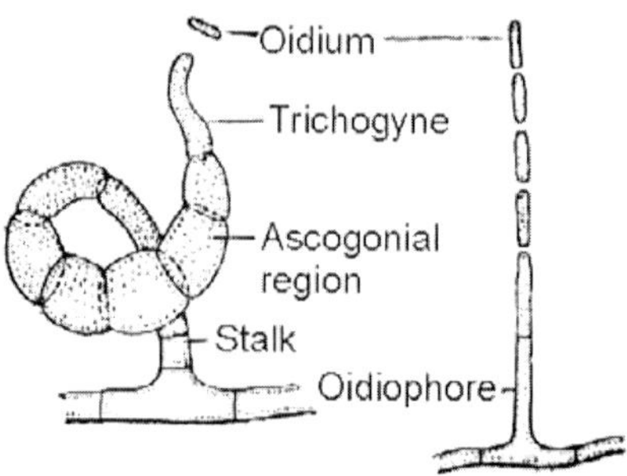

Fig. 12.3. Sexual apparatus of *Ascobolus stercorarius*. A, coiled ascogonium with stalk; B, chain of oidia which also act as male cells.

Stimulation for ascospore germination is provided in the alimentary canal of the herbivore animal; otherwise, the spores fail to germinate on nutrient media. *A. crenulatus*, however, is an exception whose ascospores germinate without any special treatment. It can be grown on RDA (Rabbit-Dung-Agar) with filter paper in a Petri dish. Apothecia appear in about 10 days. The asci, which are formed a few days later, are visible to the naked eye, protruding above the apothecia. When ripe, the ascospores are shot on to the lid of the Petri dish.

Family *Helvellaceae*

The family includes 6 genera, producing fairly large, stalked, cup or saddle-shaped apothecia on soil or on wood. The asci don't stain blue with iodine. The ascospores are elliptical to fusiform, and contain 1-3 conspicuous oil-globules and are invariably tetranucleate.

Genus *Helvella* (saddle fungi, false morels)

The species of *Helvella* are popularly known as the 'saddle fungi' or 'false morels', on account of their saddle-shaped caps (Fig. 12.4). The stalk is long and heavily ribbed or furrowed. *H. crispa*, *H. gigas*, and *H. underwoodii* are the common species. *H. gigas* forms the largest fruit body, 15 cm long and about 8 cm in diameter. The false morels are delicious, but should not be eaten raw, as some species contain a poison, called 'helvelic acid'. On cooking, the poison is destroyed and, thus, cooked helvellas can be enjoyed without fear. The false morels (and also true morels), as evidenced by molecular data, are closely related to ascomycetous truffles (O'Donnell, 1997).

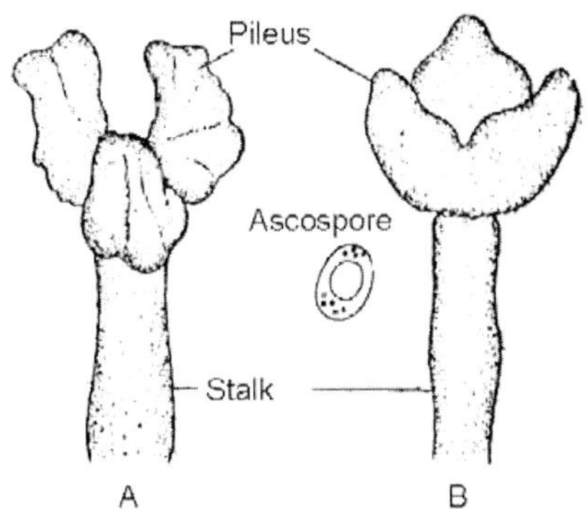

Fig. 12.4. Saddle-shaped ascoma of *Helvella crispa*

Family *Morchellaceae* (morels and bell morels)

The family, with 4 genera, is characterized by the apothecia that are large and usually stalked. The cap (pileus) is sponge-like or piltted in *Morchella*, and campanulate, like a bell, in *Verpa* (hence called bell morel). The ascospores are multinucleate and have numerous tiny granules adhering at the two ends.

Genus *Morchella* (true morel or sponge mushroom)

Morchella is edible and is next only to the truffles as a table delicacy. Like false morels, true morels are closely related to ascomycetous truffles In India, it is popularly known as 'gucchi'. The various edible species are *M. conica*, *M. delicosa*, *M. hybrida*, and *M. crassipes*.

The ascomata of *Morchella*, grow on humus-rich soils and consist of a sponge-like pileus and a stalk (Fig. 12.5). The pileus has broad, shallow depressions and ridges (Fig. 12.6). While the depressions are lined with hymenium, the ridges are sterile (Fig. 12.7). Both pileus and the stalk are hollow. There are 36 species. The two best known species are *M. esculenta* and *M. elata*. The ascoma of *M. esculenta* is bigger (upto 20 cm high); *M. elata* is much smaller (10 cm high) and also narrower. *Morchella* is now grown in culture up to fruiting stage, but it grows better in natural soils and is cultivated like a vegetable.

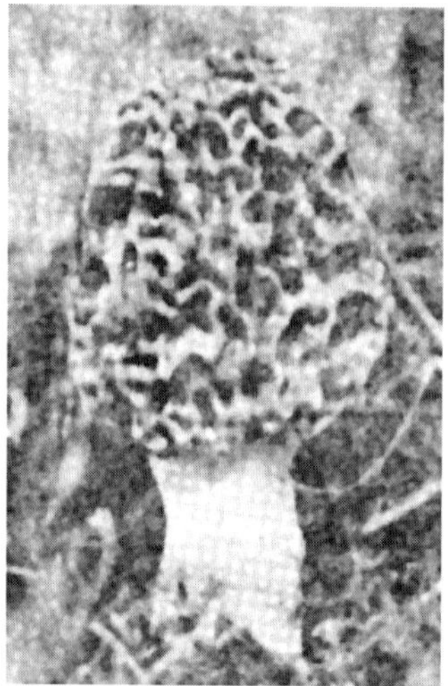

Fig. 12.5. Fruit body of *Morchella esculenta.*

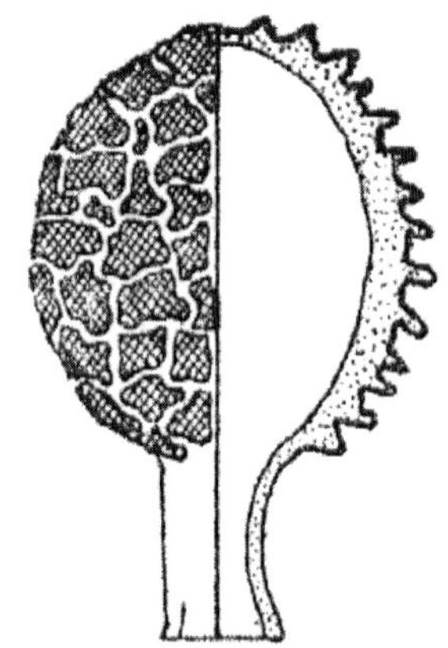

Fig. 12.6. *M. esculenta.* The left half shows the surface view and right half is a median longitudinal section.

Family *Pyrenomataceae* (= *Aleuriaceae, Humariaceae, Otideaceae*)

This is the largest and least studied family of the order. Recent studies suggest that the family is not monophyletic (Perry *et al.,* 2007). Its break-up is imminent. *Glaziellaceae* forms asister group to *Pyrenomataceae.*

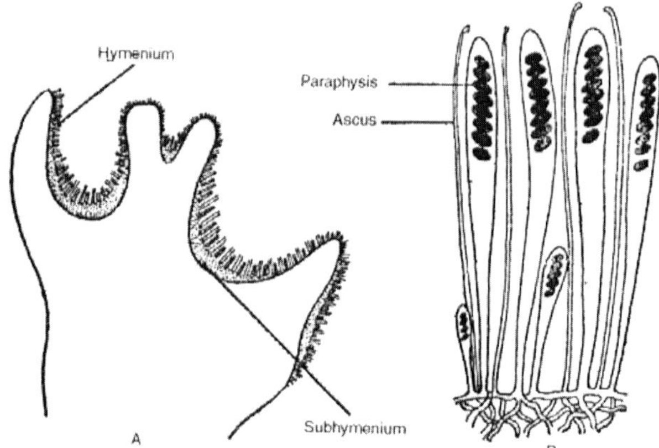

Fig. 12.7. *M. esculenta.* A, a section through the pileus showing hymenium in furrows; B, magnified view of a portion

Genus *Genea* (truffle-like fungus)

Genea (Fig. 12.8), with about 32 species, is of great academic importance. Trappe (1979) placed the genus in family *Geneacea*e because it is apothecial while *Pyrenomataceae* are cleistothecial. However, molecular data do not support this separation. The genus is hypogeous and mycorrhizal. The ascoma resembles the apothecium of *Peziza*, and, thus, shows similar origin. A vertical section of the ascoma shows its apothecial nature. The hymenium is convoluted and consists of elongate asci, each containing 8 sculptured ascospores and a non-functional ascus pore. The paraphyses grow above the asci and fuse to form a compact pseudoparenchymatous epithecium. The asci never burst and the rodents, which locate them by their scent, bring about the dispersal of the ascospores.

Family *Pezizaceae*

Apothecia are sessile or feebly stalked, and usually cup-or disc-shaped. These are found growing on dung, soil, rotting wood etc. The ascus apex shows distinct amyloid reaction i.e. stains blue with iodine. Slowly, the whole wall takes the stain. Ascospores are thin-walled, hyaline or brown, rarely greenish-yellow and always uninucleate. There are 31 genera and 1230 species (Kirk *et al.,* 2008).

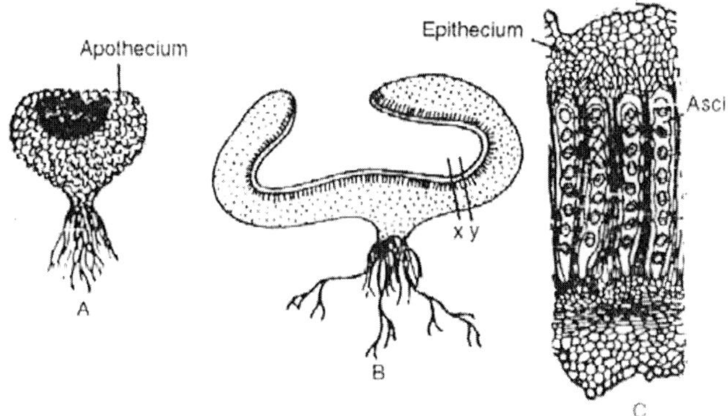

Fig. 12.8. *Genea* sp. A, mature fruit body showing apical opening and more complex infolding of hymenium; B, vertical section of the apothecium; C, microscopic details of a portion of the apothecium between the lines X and Y

Genus *Peziza*

Peziza, with 80 species, is the most common genus of the family. Its cupulate apothecia (flattened in some species) are found on rotten wood, manure heaps and soils rich in organic matter. However, the apothecia of *P.*

ammophila (Fig. 12.9) lie semi-immersed in sand. The size of the cup of different species varies from one to several centimeters. *P. aurantia* (Fig. 12.10.) is highly attractive due to its bright orange colour and looks like an orange peel.

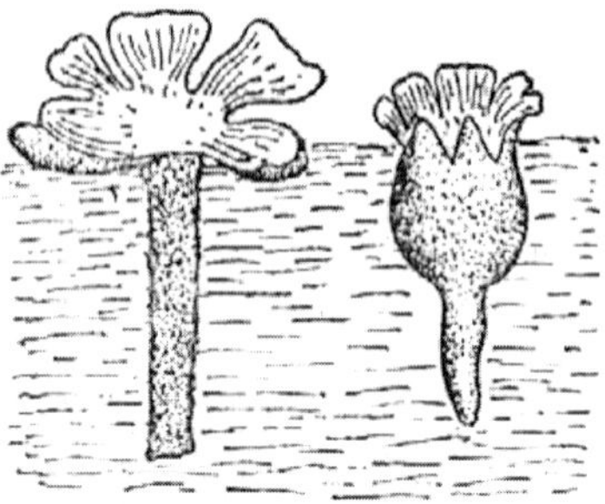

Fig. 12.9. Showing habit of *Peziza ammophila.*

Fig. 12.10 *Peziza aurantia*

The exposed hymenium, which lines the interior of the cup, consists of elongated asci (exhibiting various stages of development) and paraphyses. Both the asci and the paraphyses are phototropic, and their tips are always bent towards incident light (Fig. 12.11). In apothecia, receiving light from above, the asci at the base of the cup are straight, while those on the sides are curved towards the light. This orientation of the asci ensures smooth upward discharge of the ascospores. The asci have a distinct amyloid ring at the apex, which stains blue with iodine. Ascospores are large, usually, elliptical and are discharged violently and in one lot. On the basis of ascospore characteristics, the species of *Peziza* fall into two series: (1) those producing smooth-walled spores without oil drops, and (2) those producing spores ornamented by warts or ridges and containing two large oil drops. Some species of the second series yield a coloured fluid from the flesh of their ascoma. The spore ornamentation, which is visible when stained with cotton blue, is used in identification of the species. The habitat of the species too is characteristic. The ascospores, on germination at the suitable substratum, give rise to a haploid mycelium of uninucleate cells.

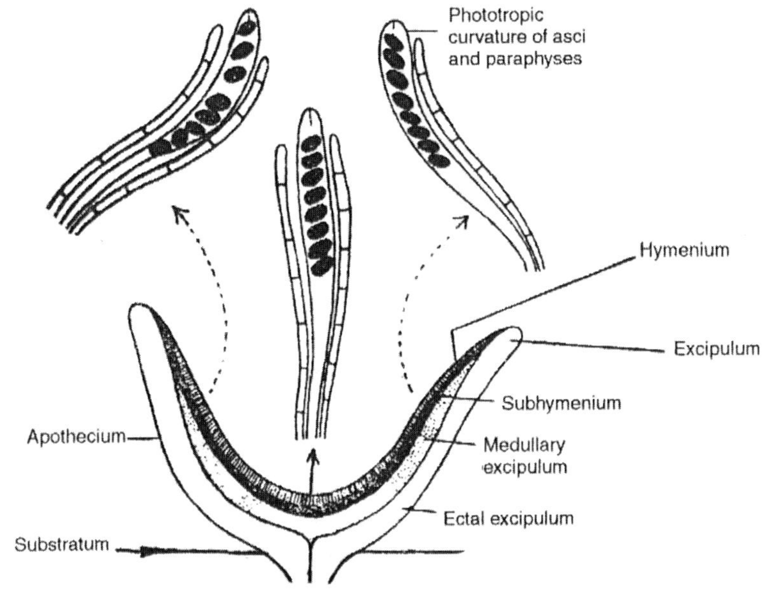

Fig. 12.11. Vertical median section of a *Peziza* cup showing phototropic curvature of asci and paraphyses at three places in the hymenium.

Several species are known to have *Oedocephalum* (Fig. 12.12) or *Chromelosporium* as the conidial stage. Their efficacy in disseminating the fungus is doubtful.

Sex organs are lacking and replaced by undifferentiated cells. Plasmogamy involves somatogamy between hyphae of opposite strains (in heterothallic species) or between two cells of the same hypha (in homothallic species). The cell that receives the male nucleus and becomes dikaryotic, acts as the female cell. Ascogenous hyphae arise from this cell and develop asci at the apices. The somatic hyphae weave a peridium surrounding the asci, resulting in the formation of apothecium. After liberation of the ascospores, the apothecium disintegrates, but the underground mycelium persists and forms apothecia again the next season. This is repeated innumerable times until the mycelium is destroyed.

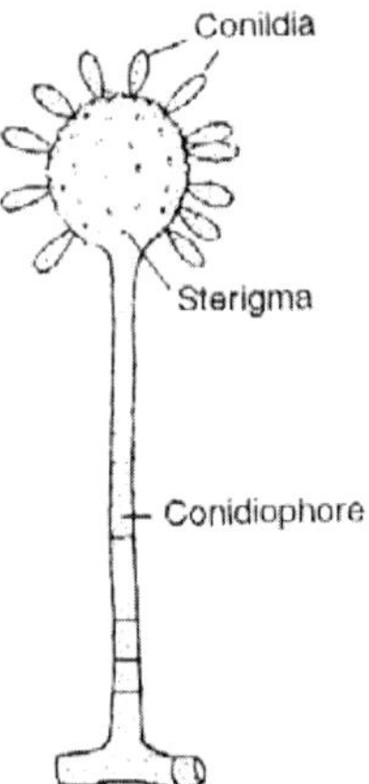

Fig. 12.12 *Oedocephalum*, the coniial state of *Peziza*.

Family *Tuberaceae*

It is a family of mycorrhizal fungi. There are 7 genera in this family. ***Tuber*** forms hypogeous, closed ascomata i.e., clestiothecia, which are globular, fleshy or waxy, of various colours and ornamented. The asci are globose, thick-walled and indehiscent. They have lost explosive 'puffing' mode of ascospore release. The species are mycorrhizal with angiosperm trees.

Genus *Tuber* (true truffles)

This widespread and important genus of the family is highly prized as a delicious food ever since man came to know the use of fungi as food. Truffles, which were known to the Romans 2000 years ago, have been the most valued hunt of the mycophagists. There are 86 species. The popular ones are: *Tuber melanosporum* (black or Perigord truffe), *T. magnatum* (white Piedmont truffle), *T. aestivum* (summer truffle), *T. uncinatum* (desert truffle) and comparatively recent one, *T. indicum* and *T. himalayensis* (Chinese truffles). *T. aestivum* (summer truffle) is considered the best, but its over-ripe ascoma emits such a foul smell that if left overnight in the house, the stink persists for several days.

The fungus grows in ectomycorrhizal association with specific trees, especially oaks. The ascomata (Fig. 12.13) lie buried, but occasionally come above ground. These may be globose, ovoid or irregular in shape, 2–3 cm in diameter and with hard, black and warted exterior. Sections of the fruit body (Fig. 12.14) show an outer peridium or rind, made of thick-walled cells surrounding a pseudoparenchymatous central fertile portion. Dark, sterile veins, studded with asci traverse the central mass, and, in some species, the veins communicate with the exterior through pores.

Fig. 12.13 *Tuber* (the ascoma)

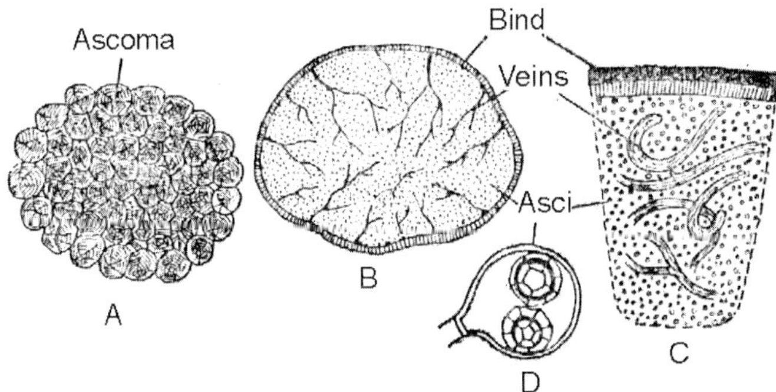

Fig. 12.14. *Tuber aestivum*. A, hypogean fruit body; B, same in section showing sterile veins and asci which look like dots; C, magnified view of a small portion of B; D, ascus with two ascospores.

The asci, which are not organized in a hymenium, contain 2–4, thick-walled ascospores with reticulate or spiny surface. These are released after rupture of the ascus wall. The germination of ascospores has not been seen.

Collection and Cultivation of Truffles

Since the ascomata grow underground, at a depth of 1–6 inches, their collection is difficult. They can be located by the scent they emit, which is perceived only by some insects and animals (like rodents, rabbits, dogs and pigs), but not man. Professional truffle-collectors prefer pigs for the job. A good truffle-hunting pig is treated very tenderly and taken to the woods in arm or in a wheel chair, so that it may not get tired. As soon as it locates the spot and starts digging the soil, the clue is taken by the master. The pig is tied to the fartherest tree and the collector himself completes the digging to make the gain sure (Fig. 12.15). The pig and the man are competitors on such occasions!

Fig. 12.15. Mycophagists raking for truffles.

The cultivation of truffle is traditionally done in France, but now it is popular in Australia and New Zealand. It is a cash crop, different from agriculture of the familiar sort. First, oak or beech trees are planted and as soon as they are established, the land is tilled, mixed with pieces of truffles and covered by soil. The land is left as such. The first harvest comes only after 6–10 years, but, after that, an annual crop of truffles is ensured for two decades or even more, without further investment. In addition to their use as food, truffles have been used in liquor making, perfumery and for scenting tobacco.

Classes under Superclass Leotiomyceta

The remaining nine classes with inoperculate asci form a monophyletic superclass informally referred to as **Leotiomyceta**. This is well supported by molecular data (Spatafora *et al.,* 2006). *Dothideomycetes* is basal to all other Leotiomyceta and is sister group to *Arthoniomycetes*. Similarly, *Leotiomycetes* and *Sordariomycetes* are form another sister group.

Class *Arthoniomycetes* (1 order, *Arthoniales*, 4 families)

The order *Arthoniales* includes 4 families of lichen-forming fungi, which are prevalent in tropical and subtropical regions. These form cup or saucer-shaped apothecia. The asci are fissitunicate. The four families are: *Arthoniaceae* (14 genera), *Chrysotrichaceae* (2 genera), *Melaspileaceae* (1 genus), and *Roccellaceae* (46 genera). All these fungi form crustose lichens with green algae or grow on other lichens (i.e. lichenicolous). Asexual stages are pycnidial. The mostly lichenized *Arthoniomycetes* is the **sister group** to mostly non-lichenized *Dothideomycetes*. The presence of fissitunicate asci with ascohymenial development is unusual and suggests it to be an **intermediate form** between ascohymenial and ascolocular fungi.

Class *Dothideomycetes* (ascostroma; asci bitunicate)
2 subclasses

Dothideomycetes is basal to all other Leotiomyceta. It includes most **ascolocular ascomycetes** with fissitunicate asci. Earlier most of its members were placed in Loculoascomycetes, now derecognized. DNA sequence data strongly suggest that another group that shares ascolocular character is the subclass *Chaetothyriales* (class *Eurotiomycetes).*

It is the largest and most diverse class of ascomycetes comprising 11 orders, 90 families, 1300 genera and 19000 species. The class includes mostly non-lichenized genera, which are saprobes, coprophilous, parasites on plants. A few lichenized taxa, have been recently included based on molecular evidence (Poggeler and Wostemeyer, 2011). These are the *Arthopyreniaceae* and the *Trypetheliaceae* and the vegetatively-reproducing species *Cystocoleus ebeneus* and *Racodium rupestre*. The ascoma ontogeny has not been studied in these lichenized members, except in *Arthopyrenia* which has shown ascolocular development.

The *Dothideomycetes* mostly produce perithecioid ascostroma. The asci are clavate and fissitunicate. The interascal tissue comprises branched or anastomosed paraphysoids or pseudoparaphyses, initially attached to the base. Ascospores are almost always septate and longitudinally asymmetrical. The two sub-classes and orders are given below. We shall study two genera, one from each subclass (shown in bold).

Dothideomycetidae: 3 orders

Capnodiales: 9 families, *Mycospharellaceae* (***Mycospharella***)

Dothideales: 4 families, *Dothioraceae*

Myriangiales: 3 families, *Cookellaceae, Elsinoaceae, Myriangiaceae*

Pleosporomycetidae: 1 order

Pleosporales: 23 families, *Arthopyreniaceae* and *Trypetheliaceae* (both comprise lichen-forming fungi); *Venturiaceae*, plant parasites (***Venturia***)

Family *Mycospharellaceae*

The family includes 53 plant parasitic genera which form ostiolate, perithecioid pseudothecia, immersed in or erumpent from the plant tissues or stroma. Asci are ovoid to saccate and fissitunicate. The ascospores are transversely septate. Conidial stage is enormously varied.

Genus *Mycospharella*

It is the largest (647 species) and most widespread genus with enormously varied anamorphs, like *Cercospora, Cercosporella, Cercosporidium, Passalora, Pseudocercospora, Ramularia* and *Septoria*. It includes some important plant parasites causing leaf spot diseases. *M. musicola* causes the most destructive disease of banana, the 'Sigatoka' disease. *M. arachidicola* and *M. berkeleyii* (with *Cercospora arachidicola* and *C. personata* as the conidial stage) cause the leaf-spot disease of groundnut (*Arachis hypogea*). The conidial stages of the various species are represented by several genera (Fig. 12.16), which may be pycnidial, acervular or hyphomycetous. But the perfect stage is same. The conidia are formed on living leaves, while the pseudothecia are formed on detached leaves. Plasmogamy occurs by spermatization. The spermatia are formed in flask-shaped spermogonia, which resemble pycnidia and therefore, the spermatia are frequently confused with conidia. But, the spermatia never germinate to give rise to a thallus, and can easily be distinguished from the conidia. Ascogenous hyphae are not formed in this species or in *M. tassiana* and *M.*

typhae. Some multinucleate cells, lying at the base of locule, form small protrusions into which two nuclei migrate. These protrusions form asci directly or by crozier formation. The asci, which contain 8 bi-celled ascospores, are arranged in a fascicle (fan-shaped tuft) at the base of the ascostroma.

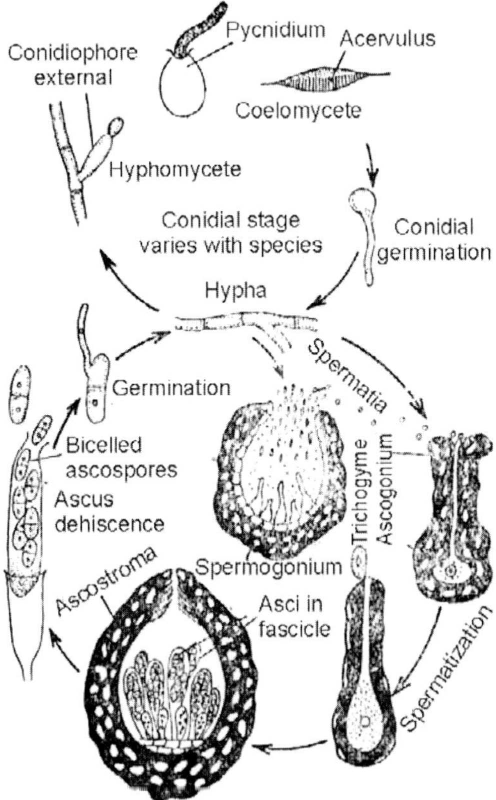

Fig. 12.16. Generalized life cycle of *Mycospharella.*

Family *Venturiaceae*

Genus *Venturia*

The mycelium is primarily internal, forming stroma under the epidermis or cuticle. Conidiophores arise from the stroma and form conidia. Later, erumpent or immersed (rarely superficial) perithecioid pseudothecia are formed on the stroma. The ascospores are bicelled (one cell longer than the other), ovoid or ellipsoid; yellowish, greenish or brownish, never opaque. The pseudothecial wall consists of distinct dark-brown cells.

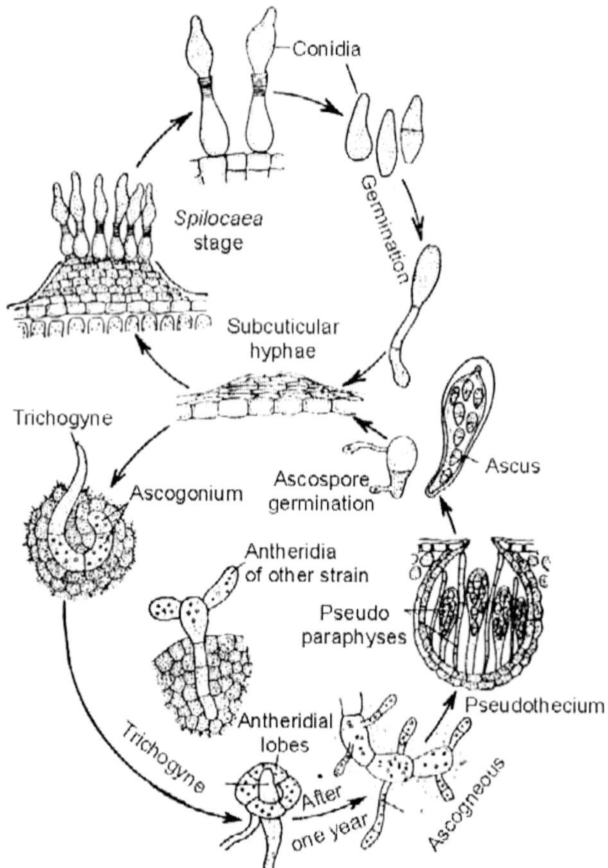

Fig. 12.17. Life cycle of *Venturia inaequalis*

Venturia is the largest and most important genus of this family, which includes several destructive plant pathogenic species like *V. inaequalis*, (apple scab) and *V. pyrina* (pear scab). The perithecioid pseudothecia, which are separate, remain immersed in the host tissue and open to the exterior through an erumpent ostiole. The conidial state is *Spilocaea pomi* a hyphomycetous anamorphic acomycete.

Venturia inaequalis (Fig. 12.17) causes apple-scab. The fungus attacks leaves, stems, and fruits. It overwinters in the form of ascostroma buried in dead apple leaves on the ground. In early spring, when leaf-buds of apple start opening, the ascostromata, buried in the dead leaves, release ascospores explosively. The ascospores are wafted by wind to the young leaves. These are yellowish, unequally bi-celled (hence the name *inaequalis*). The upper cell is broader and smaller than the lower. The ascospores germinate on the surface of young leaves, penetrate the cuticle, and form hyphae which eventually aggregate and form sub-cuticular stromata. In about ten days, short, brown conidiophores appear all over the stroma, which break through the cuticle and form one-celled (occasionally bicelled), flame-shaped, brown conidia of *Spilocaea pomi*. The conidia cause secondary infections on leaves and fruits of all ages.

At the end of the season, sexual reproduction sets in. The hyphae penetrate deeper into the tissues. A small coil of hypha of uninucleate cells initiates formation of a stroma around it. Soon, a coiled, multinucleate ascogonium, with a trichogyne, appears in the growing stroma. The trichogyne pushes through the stroma and protrudes over the surface. The fungus being heterothallic, plasmogamy occurs only when the trichogyne, contacts the antheridium borne by hypha of a complementary mating type. The antheridium is lobed. The lobes, formed by the contact stimulus, surround the trichogyne. The trichogyne becomes enclosed as if in a box. The male nuclei pass into the ascogonium through a pore, dissolved at the point of contact, and pair with the female nuclei. During the following year, ascogenous hyphae arise and form asci in the usual way by crozier formation in the young fructification. Perithecioid ascostroma is formed in the meantime. The asci grow in-between the pseudoparaphyses that are attached to both the roof and floor of the locule.

Class *Eurotiomycetes*

It is a monophyletic but morphologically heterogenous group including the well-known *Penicillium, Aspergillus* and related organisms. It includes saprobes, parasites pathogens many of the **black yeasts, sooty molds** and **lichen-forming** families. The ascomata are apothecia, perithecia, cleistothecia and ascostroma (pseudothecia). It is classified as follows:

Subclasses : three

Chaetothyriomycetidae : 3 orders

Chaetothyriales: 3 families: ascolocular; lichen-forming and saprobes. (*Chaetothyriaceae*, *Coccodiniaceae*, *Herpotrichilliaceae*.)

Pyrenulales: 6 families, ascohymenial; lichen-forming fungi

Verrucariales: 2 families, ascohymenial; lichen-forming fungi

Eurotiomycetidae 3 orders

Coryneliales : 1 family

Eurotiales: 3 families

Onygenales: 4 families

Mycocaliciomycetidae:1 order *Mycocaliciales*: 2 families

Subclass *Chaetothyriomycetidae*

The subclass includes saprobic, parasitic and lichen-forming fungi. There are three orders, *Chaetothyriales*, *Pyrenulales* and *Verrucariales*. The order *Chaetothyriales* typically form ascolocular perithecioid ascomstroma with bitunicate (fissitunicate) asci. The other two orders *Pyrenulales* and *Verrucariales* form ascohymenial ascoma with unitunicate asci.comprise predominantly of lichen-forming fungi.

In *Chaetothyriales*, the family *Chaetothyriaceae* (13 genera) includes **sooty molds**, fungi that grow on stems and leaves of living trees forming dense dark mat of black hyphae, and hence the name 'sooty mold'. These occur in warm temperate or topical countries, where there is abundant moisture. The perithecioid ascostroma, covered with sharp spines and with fissitunicate asci, lie under the mycelial mat, which looks like a stromatic covering. These are epitrophic or biotrophic on plants.

The family *Herpotrichiellaceae* (13 genera) often grow on bark of tree trunks or dead rotting wood forming minute perithecioid ascostroma covered with spines. The anamorph *Exophiala* (of *Carponia*) forms black, slimy colonies, usually without perithecia, called **black yeasts.** The black yeasts are sometimes found in unusual places like unsterilized steroid perepa-rations, utility poles, sewage sludge and wood pulp samples. Their ability to digest industrial chemicals makes them useful in filtering out organic pollutants. Some cause disease in marine fishes and are reported causing mild to serious diseases in humans, some even leading to death. They are frequently found infesting dish washers and become source of infection.

The orders ***Pyrenulales*** (6 families) and ***Verrucariales*** (2 families), which form ascohymenial ascoma with unitunicate asci, comprise mostly lichen-forming fungi. Their asexual stages are pycnidial fungi. They form crustose lichens with green algae and some are lichenicolous. *Verrucaria maura* is one of the most abundant perithecial lichens in the world.

Subclass *Eurotiomycetidae* (3 orders)

It is characterized by: (1) small cleistothecia (occasionally absent, as in *Byssochlamys*) with thin, membranous peridium, usually brightly coloured, (2) absence of interascal tissues, (3) asci prototunicate, clavate or saccate, and evanescent which dissolve releasing ascospores passively, (4) unseptate ascospores, often ornamented and with equatorial thickening, and (5) prominent asexual phase.

The organisms, many of which are highly important, are mostly saprobic in soil or on decaying plant and food material. There are 3 orders detailed below. We shall study some members of families in bold type.

Order *Coryneliales*: 1 family, *Coryneliaceae*

Order *Eurotiales* 3 families: *Elaphomycetaceae*, ***Trichocomaceae*** and *Thermoascaceae*.

Order *Onygenales*: 4 families: ***Ajellomycetaceae, Arthodermataceae, Gymnoascaceae***, and ***Onygenaceae.***

Order *Eurotiales*

Family *Trichocomaceae*

The family contains 39 genera which have *Aspergillus, Penicillium* and *Paecilomyces* as the common anamorph genera. The members are ubiquitous in soil and decaying plant and food material. We shall study *Byssochlamys, Aspergillus* and *Penicillium*.

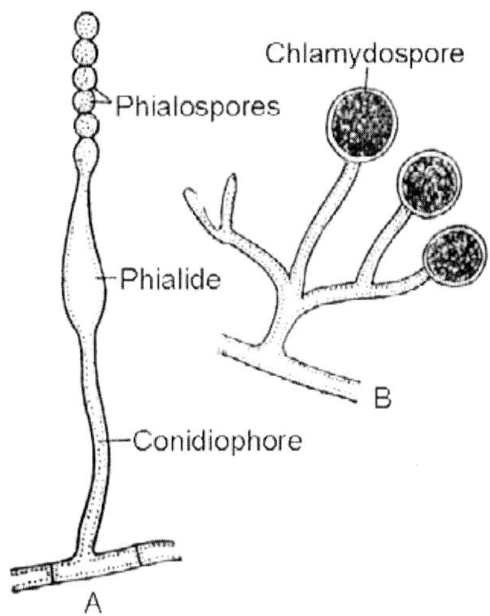

Fig. 12.18. *Byssochlamys*: A, phialospores and B, chlamydospores.

Genus *Byssochlamys*

This is the simplest genus of this family in which the asci are exposed as there is no ascoma. There are 4 species. *Byssochlamys fulva* causes spoilage of tinned fruits. Its ascospores are heat-resistant and survive fruit canning. *B. nivea* occurs in soil. The genus (Fig. 12.18) has *Paecilomyces* as the anamorph. The phialospores are formed in a chain at the tapering end of a long phialide. In addition, "terminal chlamydospores" (aleuriospores) are also formed. These are thick-walled spores cut off at the apices of hyphae. The slender ascogonium, coils around the club-shaped antheridium. Migration of the male nucleus into the ascogonium is not certain. The ascogonium becomes septate; ascogenous hyphae arise from it and form 8-spored, globular asci terminally or laterally.

GENERA WITH ASPERGILLUS AS THE ANAMORPH

(*Emericella, Eurotium* and *Neosartorya*)

All the three teleomorph genera are similar in having *Aspergillus* as the anamorph and are distinguished by their perfect state i.e. ascoma, as described below.

Emericella (Fig. 12.19) does not form sex organs and the ascoma originates from a loose hyphal knot. The cleistothecium is reddish to purple, and the peridium is composed of several layers of cells. It is surrounded by a hyphal investiture bearing globose, thick-walled, 'hulle' cells, whose function is not known.

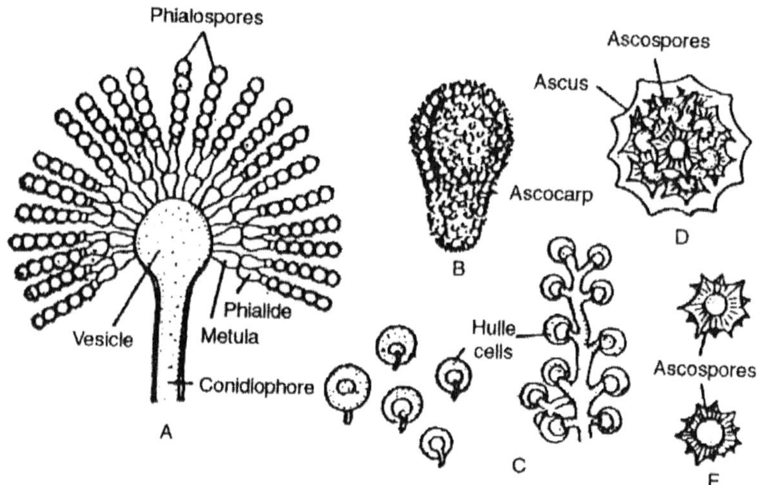

Fig. 12.19. *Emericella variecolar* (= *Aspergillus stellatus*)
A, apex of conidiophore bearing metulae, philiades and conidia; B; an ascoma with its investment of Hulle cells; C, Hulle cells; D an ascus with ascospores showing their orientations; E ascospores with paired stellate flanges in plane view.

Eurotium (Fig. 12.20) forms antheridia and ascogonia and both are functional (see later). The cleistothesium is minute, globose, bright-yellow lying superficially on the mycelium. The peridium is pseudoparenchymatous, made of a single layer of polygonal cells. There is no investiture of hyphae around the cleistothecium.

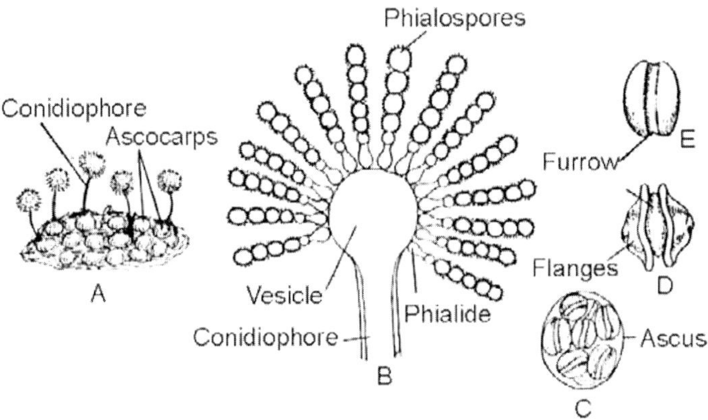

Fig. 12.20. *Eurotium;* A, habit, showing superficial spherical ascomata and conidiophores; B, apex of conidiophore showing phialides and conidia; C, an ascuss with ascospores; D, ascospores with flanged furrow and warty surface; E, ascospore with shallow furrow, no flanges, and smooth surface.

Neosartorya. The ascoma originates from the ascogonium alone: antheridium is not formed. Cleistothecia are white to slightly yellowish and surrounded by loose mass of sterile hyphae.

The anamorph *Aspergillus*

Aspergillus, which is the conidial state of the above genera, since long, has been treated as an independent genus under class Deuteromycetes (now derecognized). Since *Aspergillus* is very important in biodegradation, biodeterioration, genetical studies, environmental pollution, toxins of food-feed stuffs and in industry, we shall study it in some detail.

Aspergillus maintains its individual status also, because many (two-third) of its species have not been assigned to any ascomycete, and in some cases the sexual stage just does not exist. The species of *Aspergillus* have been placed under groups where perfect stages are known. There are:

A. nidulans group	*Emericella*
A. glaucus group	*Eurotium*
A. fumigatus group	*Neosartorya*

Distribution and Importance

The aspergilli, though world wide in distribution, are more prevalent in tropical countries. Conidia of *Aspergillus* are always present in the air and cause contamination in laboratory cultures of bacteria and fungi. *A. niger* is called 'weed of the laboratory'. These are important as pathogens of humans and animals, producers of toxins, in industries for production of chemicals, and as an organism of choice in genetical research.

Disease in humans and animals, caused by *Aspergillus* spp. (especially *A. fumigatus*) is called aspergillosis. Among animals it is especially common in birds. In humans *Aspergillus* infects lungs causing pulmonary aspergillosis. It appears in four forms: (1) *invasive aspergillosis*, which spreads to surrounding tissues, more common in immunocompromised patients such as AIDS, and has high mortality rate, (2) non-invasive aspergillosis (3) *aspergilloma* which is a 'fungus ball' composed of fungal hyphae formed in preexisting cavities in upper lobe of the lung (4) severe asthama with fungal sensitization. Aspergillosis of the ear canal and sinuses causes pain, congestion and fluid discharge.

Toxins in food-feed stuffs In England, in 1960, about 100,000 Turkey poults died by eating groundnut cake infested with *Aspergillus flavus*. A toxin, called aflatoxin, produced by *A. flavus*, was responsible for the poultry deaths. Aspergilli are now known to produce several deadly toxins on various food/feed-stuffs, which cause mycotoxicoses in animals and human beings.

Industrial production of chemicals Strains of *Aspergillus niger* are used in the manufacture of several acids like citric, gluconic and itaconic acids, and enzymes, like glucose oxidase and lysozume. Over 99% of total global production of citric acid is through fermentation by *A. niger*. Species of *Aspergillus* (*A. oryzae*) are used for hydrolysis of starch into glucose which is then fermented into alcohol by yeast.

As Research organism *Aspergillus nidulans* is the historical fungus in which the parasexual phenomenon was discovered in 1952 by Pontecorvo and Roper. *A. nidulans* was more recently used as the pioneering filamentous organism for genome sequencing. In 2008, the genomes of seven more *Aspergillus* species were sequenced. The species: *A. niger, A. oryzae, A. terrus, A. clavatus, A. fischerianus, A. flavus* and *A. fumigatus*. **The genome size** varied from 29.3 Mb for *A. fumigatus* to 37.1 Mb for *A. oryzae. A niger* had intermediate size of 33.9Mb. The **number of genes** varied from 9926 for *A. fumigatus* to 12071 for *A. oryzae.*

Growth in varied environment Species of *Aspergillus* are found in any conceivable habitat because of the varied enzymes that they produce. They cause spoilage of stored products, such as grains, food stuffs, jams and jellies, salted meat, fish and leather. They are the 'green molds' that cover the shoes or leather goods in humid weather. Some species are weak parasites causing storage diseases of fruits during storage and transit. Members of the *Aspergillus glaucus*, *A. halophilicus* and *A. restrictus* groups are the most xerophytic fungi known, and can grow where other fungi are excluded. The grains stored commercially below 13% of moisture content (equivalent to R.H. below 70%), are attacked by these three groups of *Aspergillus* and are thus, called 'storage fungi' (Christensen, 1965). Other fungi cannot grow at this low relative humidity. As a result of their activities, the percentage germination of grains falls; they are rendered unfit for food, as they contain toxins which can cause health hazard.

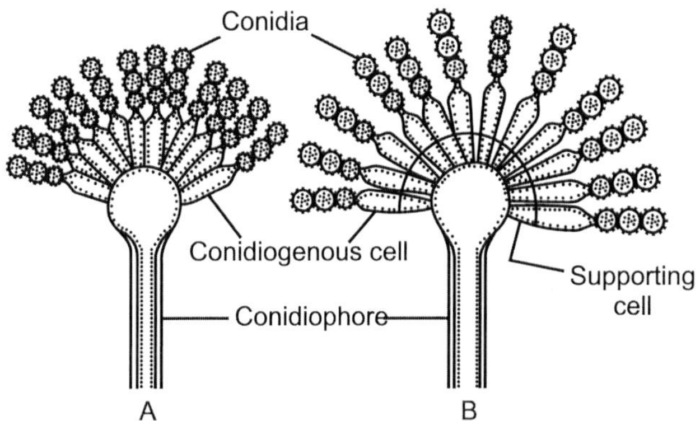

Fig. 12.21. *Aspergillus* sp: A, conidiophore bearing conidia on phialides; B conidiophore with metulae, phialides and conidia.

Asexual Reproduction

The multinucleate hyphae are hyaline, septate and branched. Conidia are produced in enormous quantity. The mycelium starts forming conidiophores very early. The long and erect conidiophores arise from a particular thick-walled **foot cell.** It forms a terminal swollen vesicle, on which bottle-shaped structures called sterigmata or phialides (conidiogenous cell is now a preferred term for phialide), arise all over the surface and produce chains of conidia at the tips. When there are two layers of sterigmata, the lower is called primary sterigmata or metulae (better term is supporting cell) and the upper, as secondary sterigmata or phialides (conidiogenous cells; Fig. 12.20). The neck of the phialide is called the spore-producing tube as the conidia are formed inside this tube (Fig. 12.21). A portion of protoplasm with a nucleus inside the neck is delimited by a septum. The protoplasm rounds off, develops a wall, which fuses with the wall of the tube and becomes a conidium. The basal portion of the tube elongates and cuts off another conidium, and in this way conidia are produced at the tip of the phialide. The conidia, which remain attached in chains, are circular, multinucleate and have a thick, rough wall.

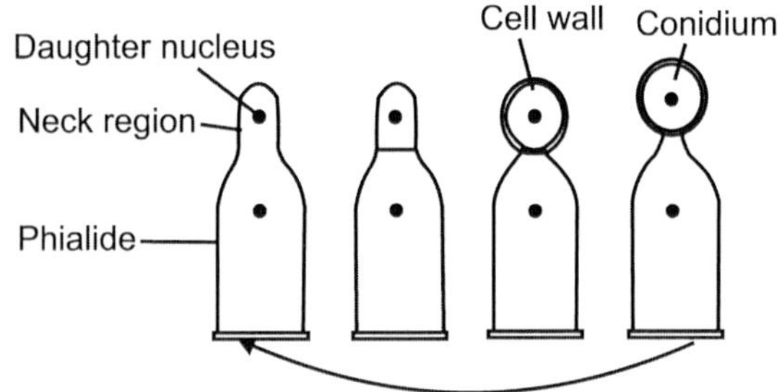

12.22. *Aspergillus*. Formation of conidia by phialides.

The colour of the conidiophores and the conidia impart colour to the colony, which may be green (*A. flavus*), yellow (*A. ochraceus*), brown (*A. tamari*), black (*A. niger*), white (*A. candidus*) etc. The colony colour is an important criterion in the identification of species. The conidia, which are 'dry', are easily detached and disseminated by wind.

Life Cycle of *Eurotium* sp.

The fungus has hyaline, septate and, branched hyphae with multinucleate cells. The asexual phase is the anamorph *Aspergillus*, described above. The sexual reproduction involves distinct sex organs, the antheridia and ascogonia, which are formed closely on the same hypha (Fig. 12.23).

Both are multinucleate and elongated, which coil around each other. Antheridia may be functional or defunct. Pairing of nuclei, derived from antheridia and ascognia (or more commonly only from ascognium), occurs in the ascogonium. Ascogenous hyphae arise from the ascogonium and pairs of nuclei migrate into them. The ascogenous hyphae grow to unequal lengths and produce asci at different levels. Simultaneous with the development of the asci, hyphae arise from the base of the ascogonium and collect around the sex organs to form a characteristic simple cleistothecium. The cleistothecia, which are spherical and small (100-200 μm), contain large number of minute, 8-spored, globose asci. These are evanescent, and dissolve, leaving the ascospores free inside the cleistothecium. The clestothecia open by natural weathering or rupture due to internal pressure, created by maturing ascogenous hyphae. The ascospores have a distinct central peripheral groove, which divides the ascospore into two equal halves. The ascospores resemble pulley wheels. The wall of the ascospore is sculptured. On germination, the ascospores produce germ tubes which give rise to the mycelium.

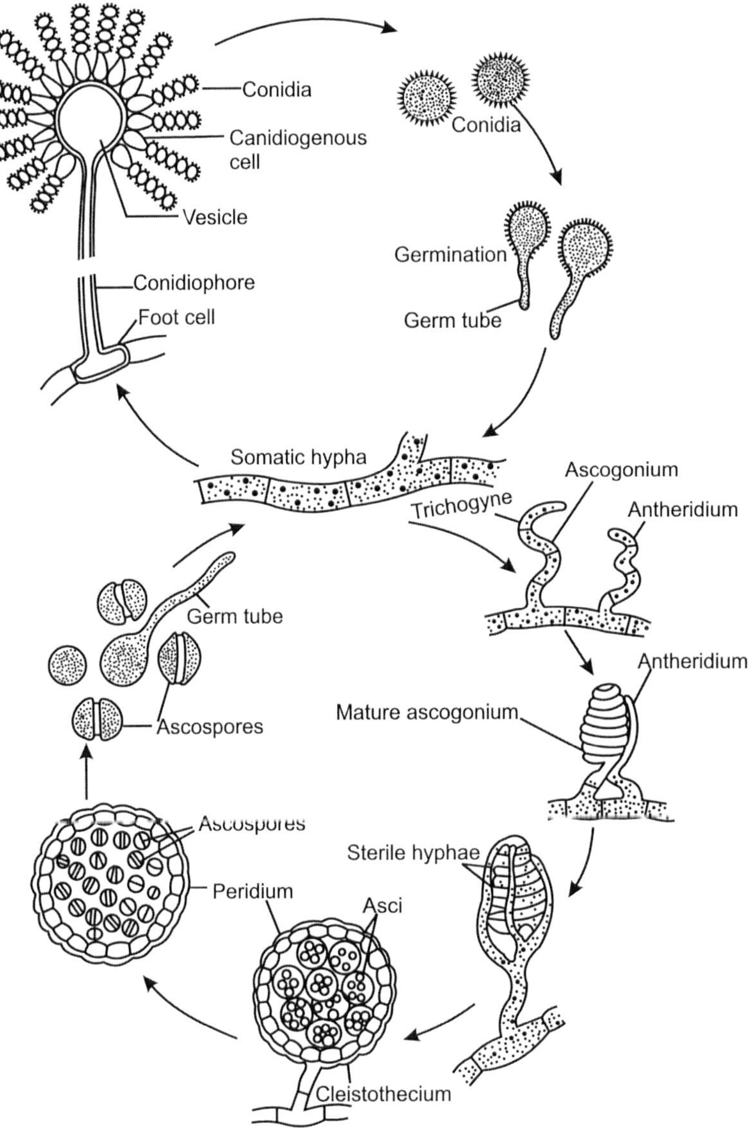

Fig. 12.23. Life cycle of *Eurotium* sp.

GENERA WITH PENICILLIUM AS THE ANAMORPH (*Eupenicillium* and *Talaromyces*)

Both the genera *Eupenicillium* and *Talaromyces* have *Penicillium* as the conidial state, but differ in their sexual reproduction and the ascomata. While *Talaromyces* forms the asci in clestothecia, in *Eupenicillium*, the asci are formed in **ascostroma**

Distribution and Importance

Though cosmopolitan, species of *Penicillium* are more prevalent in temperate countries. (This is the opposite of *Aspergillus*). Because of its importance in industry, especially in antibiotic and cheese manufacture, extensive work has been done on various aspects of *Penicillium*. The important activities of *Penicillium* are as given below.

Antibiotic production The antibiotic penicillin is manufactured from *P. chryrsogenum* whose several high antibiotic producing strains are available. *P. notatum*, from which Alexander Fleming discovered penicillin, is, however, now not used in the industrial production. The 'wonder drug', though discovered by Fleming in 1927, was not put to commercial production until 1945. It is active (as a bacteriostatic agent) against Gram +ve bacteria and also against some of the larger viruses and rickettsia. Penicillin is now a generic term used for an entire group of antibiotics. The antibiotic griseofulvin, produced from the fungus *P. griseofulvum*, is used in the treatment of dermatophytic infections of skin, nails, hair and feathers, like ringworms, atheletes' foot and dermatomycoses. It is fungistatic and not fungicidal. It is active only against fungi having chitin in cell walls and, thus, ineffective on `Chromistan fungi'.

Cheese Industry *P. roquefortii* and *P. camembertii* are used in the production of cheese having special flavours called Roquefort and Camembert cheese.

Plant Parasites Blue mold of citrus fruits (*P. italicum*), green mold of citrus fruits (*P. digitatum*) and rot of apples (*P. expansum*) are some of the important diseases caused by *Penicillium*. Several species of *Penicillium* are associated with spoilage of food, leather and clothes.

Mycotoxins Several species of *Penicillium* produce toxins on food/feed stuffs, which causes toxaemia in man and animals. Examples: yellowed rice toxins, penicillic acid and patulin

Asexual reproduction The characteristic conidial apparatus (Fig. 12.24), which resembles a brush or broom, is called 'penicillus' (Penicillium = small brush). Long, septate conidiophores can arise from any cell of the hypha, and not from a specific 'foot cell' as in *Aspergillus*. The conidiophores branch once or twice at two-third of its total length. These branches are called primary sterigmata, or rammi (sing. ramus) secondary sterigmata or metulae (sing.metula), which finally bear the bottle-shaped phialides (= conidiogenous cell). Conidia (phialospores) are produced in the neck region of the phialide as described for *Aspergillus*. Long chains of conidia are formed with the youngest conidium at the base. These are globose to ovoid and look like glass beads under the microscope. The coloured conidia, green, blue or yellow, give the characteristic colour to the colony, which aid in identification of species. In some species the conidiophores unite to form coremia (sing. coremium).

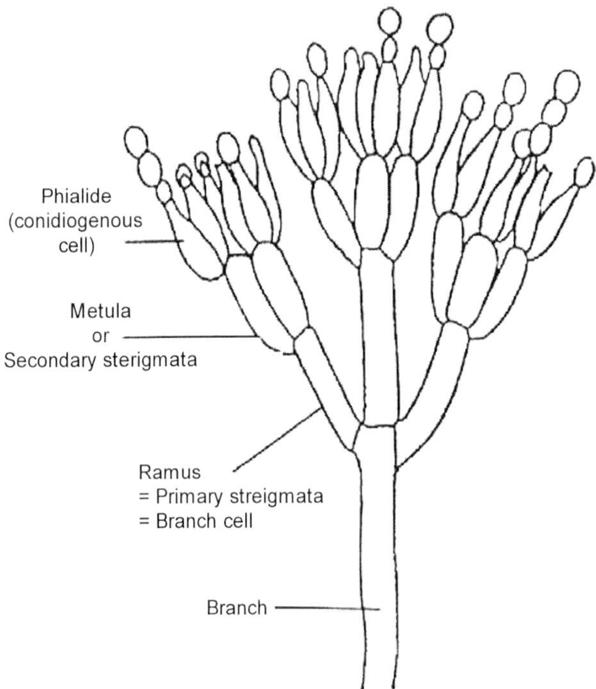

Fig. 12.24. *Penicillium*: conidial apparatus.

Life Cycle of *Talaromyces favus*

The fungus reproduces asexually by the method described above for anamorph *Penicillium*. Sexual reproduction in most species does not occur at all, and conidia are the only method of reproduction. Even the species in which sexual reproduction is known, it is a rare event. The uninucleate mycelium forms a swollen cylindrical ascogonium (Fig. 12.25), which is uninucleate in the beginning, but by repeated nuclear divisions, comes to have 64 nuclei. An antheridial branch, which originates on a separate hypha, coils around the ascogonium and cuts a terminal antheridium.

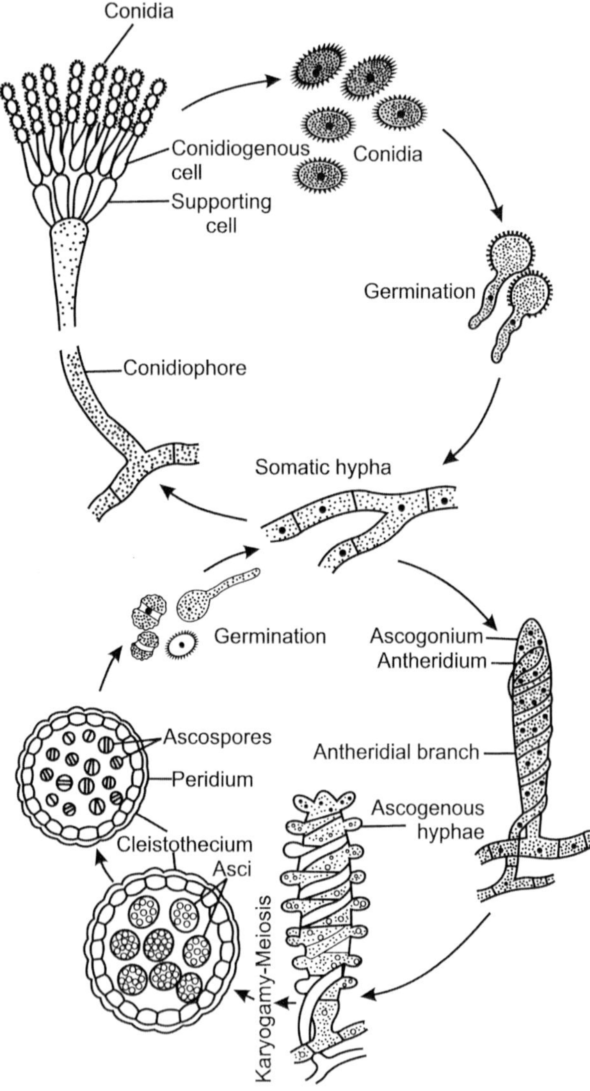

Fig. 12.25. Life cycle of *Talaromyces* sp.

The wall at the place of contact of the two sex organs dissolves but nuclear migration from the antheridium has never been observed. The male nuclei can be seen in the antheridium even after the formation of ascogenous hyphae. Sometimes, the antheridial branch appears when the ascogonium is in advanced stage of development. Thus, the female nuclei in the ascogonium themselves pair and form the dikaryons, which by conjugate divisions multiply and form more dikaryons. Septa are formed in the ascogonium. Ascogenous hyphae arise from the binucleate cells of the ascogonium and produce asci at different levels. The somatic hyphae, in the meantime, grow and surround the ascogenous hyphae so as to form a two-layered peridium. Ultimately, a cleistothecium is formed.

The asci, which are globose and evanescent, liberate the ascospores in the cleistothecium. The ascospores are freed after decay of the cleistothecial wall. The ascospores, as in *Aspergillus*, resemble pulley-wheels, and on germination, form a germ tube which develops finally into the somatic mycelium.

Eupenicillium

The asci in this genus are formed in ascostroma (pseudothecium), which is made up of thick-walled pseudoparenchymatous cells. Multinuleate ascogonia are formed in the centre of stroma. Information about antheridum is lacking. Even if present, it is non-functional and the nuclei of the ascogonium come to lie in pairs. Ascogenus hyphae are formed on the ascogonium which branch dichotomously and radiate in all directions. Asci are formed singly or in helical chains. In mature stroma, the cells lyse and form a cavity which contains the asci and later, the ascospores, after rupture of the asci.

Order *Onygenales* (4 families)

The order is characterized by asci, which are exposed, or formed inside cleistothecium. The peridium may be pseudoparenchymatous or loose aggregation of intertwined hyphae, with complex appendages. Asci are globose, evanescent and contain 8, brightly-coloured unicellular ascospores. The asexual stage is represented by hyphomycetous aleurio-, or arthro-conidial fungi.

The order is distinct in having several keratinophilic fungi growing on skin, hair, nail and feather, as parasites or saprobes. Some are coprophilous, while others are cellulolytic. The order has 4 families: *Ajellomycetaceae*, *Arthrodermataceae*, *Gymnoascaceae*, and *Onygenaceae*. Family *Eremascaceae* is placed here as incertae sedis.

Family *Ajellomycetaceae*

There are seven non-keratinophylic genera, *Ajellomyces* being the type genus.

Genus *Ajellomyces*

Ajellomyces has two species, *A. dermatitidis* and *A. caspulata*, both medically very important, causing deep-seated mycoses which are, occasionally, fatal. Both the species are dimorphic and have a **saprobic mycelial phase** in soil, and a **parasitic yeast phase** in animals, including human beings, causing mycoses. In the soil, both the species grow on keratinaceous substrata and multiply by forming conidia represented by hyphomycetous anamorphic fungi *Blastomyces* and *Histoplasma*. They form asci in cleistothecia with peridium made of intertwined mass of hyphae having helical appendages.

The conidia, however, when inhaled reach the lungs and give rise to the yeast phase in the lung tissues. Depending on the yeast anamorph, *Blastomyces* or *Histoplasma*, the disease is called **blastomycosis** or **histoplasmosis** (Fig. 12.26). *A. dermatitidis* which has *Blastomyces zygonema* (syn. *B. dermatitidis*) as the yeast phase, causes blastomycosis, while *A. capsulata*, whose yeast phase is called *Histoplasma capsulatam*, causes histoplasmosis. The infection of lung resembles pulmonary tuberculosis. In histoplasmosis, the yeast also infects the histocytes and macrophages. In both, blastomycosis and histoplasmosis, the infection can spread to other organs of the body and may become fatal.

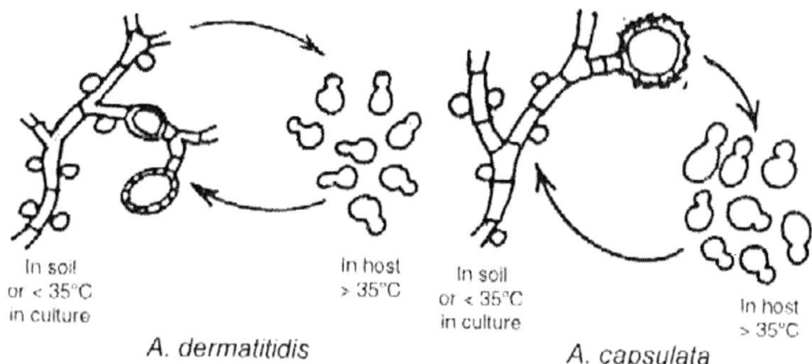

Fig. 12.26. The parasitic and saprobic phases of *Ajellomyces* species.

In culture, the growth phase, yeast or mycelial, is determined by temperature. At high temperatures, more than 35°C, resembling animal body temperature, the fungus grows in yeast phase and multiplies by budding. At lower temperatures (below 35°C), the fungus forms mycelium and then forms conidia.

Family *Arthrodermataceae*

There are 5 keratinophylic genera. The two more important genera, *Arthroderma* and *Ctenomyces*, are kerationophilic, growing on hair, skin, nail, feather etc. as parasites or saprobes. The genera are characterized by non-stipitate cleistothecia with peridium made up of loose hyphal aggregation with complex appendages. Ascospores are obovate and smooth. The asexual stage is represented by anamorphic genera *Chrysosporium*, *Microsporum* and *Trichophyton*.

Genus *Arthroderma* (ring-worm organisms)

Arthroderma is kerationophilic; some species cause dermatomycoses (tinea or ringworm). The asexual reproduction is brought about by conidia (Fig. 12.27) that are designated as *Microsporum* and *Trichophyton*. On the basis of the habitat, these organisms are called zoophilic (growing on animals), anthrophilic (on human beings) or geophilic (saprobic in soil). The dermatomycoses (called tinea or ringworm) are designated by the part of the body infected viz. ringworm of scalp (tinea capits), of beard (tinea barbae), of nails (tinea ungium), of body (tinea corporus), of feet (tinea pedis). Before the discovery of fungi as causal agents of "ring worm" by David Grab in 1841, these infections were thought to be caused by worms (tinea in latin means worms). The 'hangover' continues!

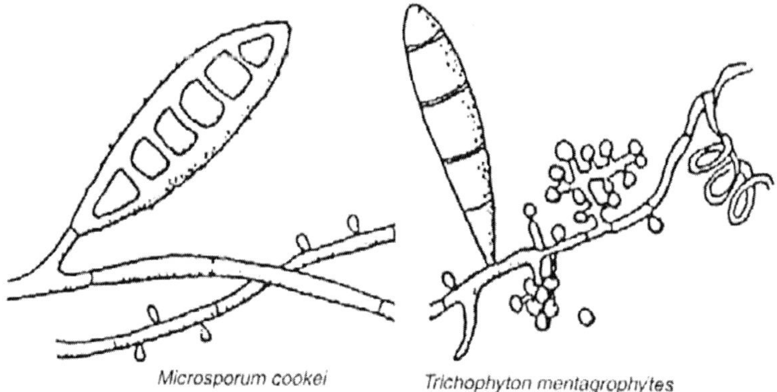

Fig. 12.27. Anamorphs of *Arthroderma*

Genus *Ctenomyces*

The monotypic genus, *C. serratus*, grows saprobically in soil or dung and especially on feathers. A thin prosenchymatous layer of thin-walled hyphae surrounds the asci (Fig. 12.28). Serrated, comb-like appendages arise from the weft of orange-brown, thick-walled peridial hyphae, which lie external to the prosenchymatous layer. Anamorph is *Chrysosporium*-like.

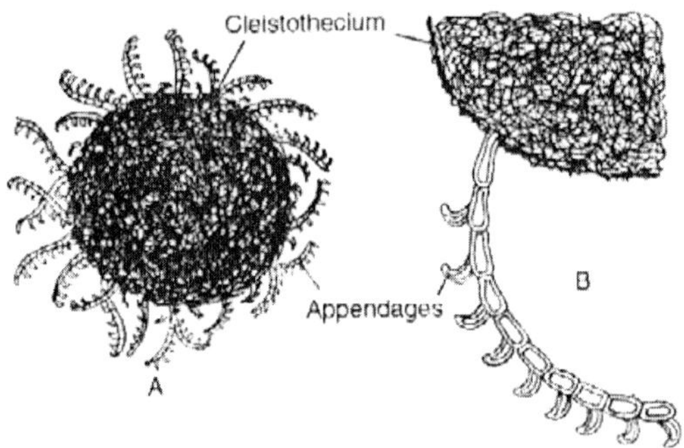

Fig. 12.28. *Ctenomyces serratus*; A, cleistothecium; B, enlarged view of the peridium and appendage.

Family *Gymnoascaeae*

The family, with 12 genera and 38 species, is characterized by the ascoma which is called **gymnothecium**, as it does not have a definite peridum. The ascogenous hyphae and asci are surrounded by a loose mass of interwoven, little differentiated hyphae. The outer hyphae of this protective hyphal covering are modified to form characteristic appendages which aid in identification of the genera. Asexual reproduction is by arthrospores and aleuriospores. It includes dermatophytes and forms that grow on dung, soil and feathers.

Genus *Gymnoascus*

The genus has 8 species (Fig. 12.29) which grow in soil and animal substrata. The peridium is a lax network of hyphae. The free ends of the peripheral hyphae are spine-like or elongate with short, straight or curved lateral branches. The gametangia, arising on same or different hyphae, consist of a club-shaped antheridium and a cylindrical ascogonium, coiling around the antheridium. The migration of the male nuclei is not certain. The ascogonium becomes septate; ascogenous hyphae arise from its cells, which bear croziers at the apex and form asci. The somatic hyphae, in vicinity of the gametangia, form a loose envelope, which represents the peridium. The anamorph genus *Malbranchea* represents he asexual phase.

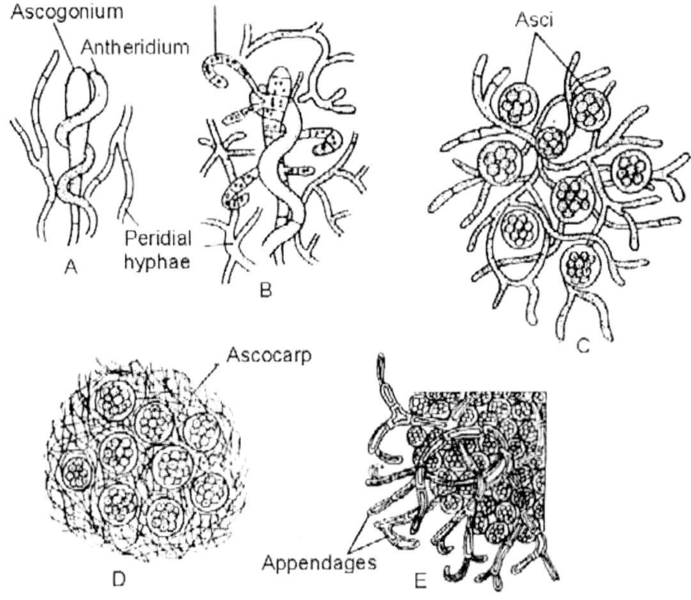

Fig. 12.29. *Gymnoascus reesi*: A, sex organs and peridial hyphae; B, ascogenous hyphae; C, Asci among peridial hyphae; D, mature ascocarp; E, enlarged view of appendages

Family *Onygenaceae*

The family with 23 genera and 134 species is characterized by large sessile or stalked cleistothecia and pseudoparenchymatous peridium. The 23 genera grow on keratinaceous animal remains like shed horns, rotting hooves, hair and feathers, while some are coprophilous or saprobic in soil. Asexual reproduction is by conidia, designated as *Chrysosporium* or *Malbranchea*.

Genus *Onygena*

In this genus (Fig. 12.30), which grows on feathers and bones, the ascoma is a stalked cleistothecium, 1–2 cm high. The cleistothecium is globose and brown; the peridium is pseudoparenchymatous, made of polygonal cells. The stalk is cylindrical, smooth and white. The globose asci, which lie scattered, dehisce to liberate a powdery mass of ascospores. Some of the disintegrated tissues form a sort of capillitium (thread-like structures), which helps in the dispersal of ascospores.

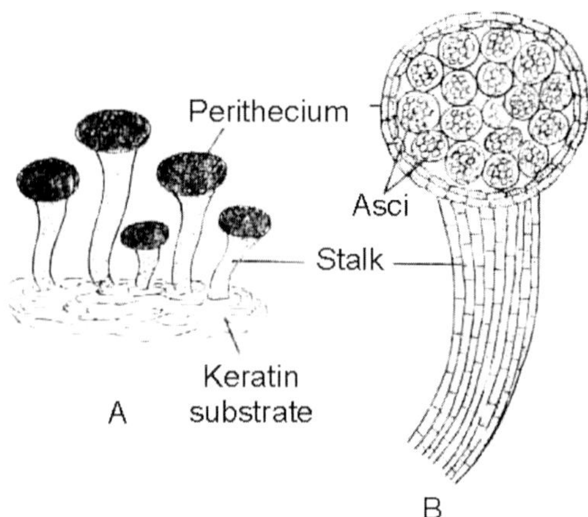

Fig. 12.30. *Onygena equina*: A, habit stipitate ascomata; B, longitudinal section through ascoma.

Subclass *Mycocaliciomycetidae* consists of parasite or commensals on lichens. Some are saprobes. These have unitunicate asci with active ascospore dispersal, rarely passive.

Class *Geoglossomycetes*

The class, created in 2009 (Schoch *et al.*) has 1 order *Geoglossales*, and 1 family *Geoglossaceae*. There are three genera: *Geoglossum, Sarcoleotia* and, *Trichoglossum* which form ascomata called the 'earth tongue' (Fig. 12.31). These grow in soil or rotting vegetation forming black, 2-8 cm high, tongue-shaped ascoma (apothecia or cleistothecia) containing prototunicate inoperculate asci.

Fig. 12.31 'Earth tongue' ascomata of *Geoglossum umbratile*

Class *Laboulbeniomycetes* (perithecia; prototunicate)

These have evolved numerous unique morphological features which for a long time made their classification and phylogenetic affinities obscure. These produce perithecia with unitunicate, clavate, early-dissolving asci. These appear to be an independent lineage of *Ascomycota*.

Laboulbeniomycetes are minute obligate ecto-parasites of insects and other arthropods. They produce thalli, which are not made of hyphae but are cellular structures of definite morphology. They remain attached to the host by a stromatic black haustorium without entering the body. Though not of any economic importance, they are interesting for their host specificity and strict developmental control. The host specificity is to the extent that they grow on a particular sex of a particular species, and only on a particular portion of the body. The class is divided into 2 orders *Laboulbeniales* and *Pyxidiophorales*. The *Laboulbeniales* has 5 families. We shall study 2 members from the family *Laboulbeniaceae*.

Family *Laboulbeniaceae*

The family includes 126 genera, which are monoecious or dioecious. We shall study one example of each; *Stigmatomyces* and *Amorphomyces*, respectively.

Genus *Stigmatomyces*

The genus (Fig. 12.32) includes species, which grow on free-living flies (Diptera) and on a few species of *Coleoptera*. A short, pointed, black haustorium (foot) penetrates the chitinous covering of the host and derives nutrition, possibly by degrading the chitin. The receptacle consists of two superposed cells, and in monoecious forms, bears appendages terminally and laterally. The lateral appendage forms the enclosed ascogonium with protruding trichogyne, while the terminal appengdage bears divergent simple antheridia. Spermatia, formed endogenously, are extruded through the exit tube of the antheridium, that fuse with the trichogyne. The fertilized ascogonium develops into a perithecium.

Genus *Amorphomyces*

It is a dioecious genus (Fig. 12.33) in which the antheridia and the ascogonia are borne on separate individuals, which, of course, grow attached to each other. The male thalli are much smaller than the female. The ascospores, from which they arise, also show dimorphism. The one that gives rise to the male thallus is much smaller than that which gives rise to the female thallus. The two ascospores, as a rule, are liberated together in pairs to ensure their proximal growth. The male thallus consists of two superposed cells and a single terminal antheridium, forming endogenous spermatia. The female thallusl bears only a terminal ascogonium surrounded by a single layer of protective cells. The trichogyne is external and branched. The perithecium has 5 tiers of cells in its wall. The ascospores are two-celled, but the upper small cell is not easily seen.

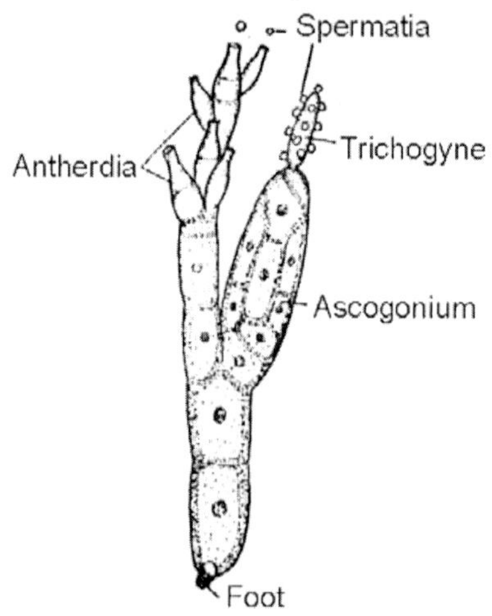

Fig. 12.32. *Stigmatomyces baeri*, mature thallus showing at left the appendage with several antheridia and at right the enclosed acogonium with its papillate trichogyne.

Class *Lecanoromycetes*

The class includes the largest group of lichen-forming fungi. There are 630 genera and 14190 species distribued in 12 orders and 77 families. The anamorphs are pycnidial or absent. The class is divided into 3 subclasses: *Acaurosporomycetidae*, *Lecanoromycetidae* and *Ostropomycitidae.*

Acaurosporomycetidae is the smallest group, having 11 genera and 183 species, placed in the family *Acaurosporaceae*, order *Acaurosporales*. These form apothecia. The asci are bitunicate and contain more than 100 unicellular spores.

Lecanoromycetidae also forms apothecia, with unitunicate, evanescent asci and highly variable ascospores. There are 3 orders, 119 families, 119 genera and over thousand species. The well-known lichenized fungi *Lecanora* (552 spp), *Peltigera* (91 spp.), *Lobaria* (67 spp.) belong here.

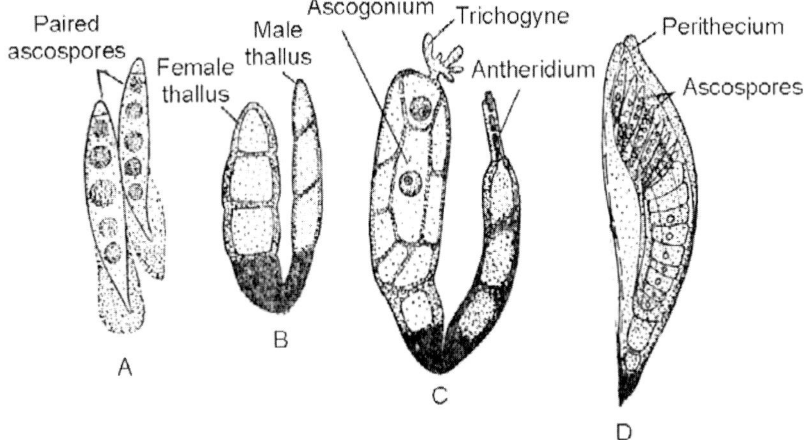

Fig. 12.33. *Amorphomyces falagriae*. A. paired ascospores with small terminal cells; B. young adjacent male and female thalli; C. mature thalli; D. mature female with perithecium containing ascospores.

Ostropomycitidae comprise 4 orders, 11 families. The members form apothecia or perithecia. Ascospores are varied, often filamentous. It includes some non-lichenized species, which are considered 'secondarily de-lichenized' (Lutzoni *et al.,* 2001). Sedin *et al.* (2004) unraveled the ***Conotrema-Stictis* connection,** to be a unique case of ecological adaptation. It is an interesting and the lone example of **optional lichenization**. *Conotrema* (lichen-foming) and *Stictis* (non-lichen-forming) can undergo their whole life cycle (both asexual and sexual stages) either as lichens or as saprobes under different environmental conditions. This optional-lichenization allows separate individuals to exploit different niches. This challenges the obligatory nature of lichen symbiosis.

Class *Leotiomycetes* (apothecia, cleistothecia; inoperculate, prototunicate)

It includes the majority of non-lihenized apotheccioid ascomycota, with inoperculate unitunicate asci. Now, the cleistothecial powdery mildew genera with prototunicate asci are also included in the class. Molecular data suggest that the class is a sister group to the perithecial class *Sordariomycetes*. There are 5 orders: *Cyttariales* (1 family), *Erysiphales* (1 family), *Helotiales,* (10 families), *Leotiales* (2 families) and *Rhytismatales* (3 families).

Order *Cyttariales*

The order has a single family *Cyttariaceae* and 2 genera.

Genus *Cyttaria*

The type genus *Cyttaria*, has 10 species, with *Cyttariella* as the anamorph. All species are parasitic on *Nothofagus* (*Fagaceae*) and confined to the Southern hemisphere. Brightly coloured stromata (Fig. 12.34 a,b), each consisting of about 200 embedded apothecia, appear on the stems of infected plants. Asci in all apothecia are formed simultaneously. The asci are cylindrical, 8-spored.

Fig 12.34 (a) Edible fruit bodies of *Cyttaria* sp.

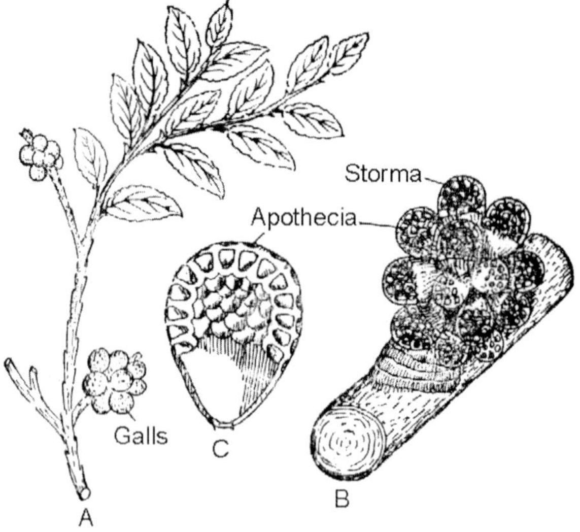

Fig. 12.34 (b) *Cyttaria gunnii.* A, a twig of *Nothofagus cunninghami* bearing the fungus; B, a group of stromata; C, ongitudinal section of a stroma.

In some species pycnidia are formed in the stroma prior to apothecial formation. The brightly coloured, fleshy ascostromata of *Cyttaria* are **edible** and have been, since centuries the **food of the rich** because of their high price. They cannot be produced commercially as their fruiting is not certain. In laboratory cultures also fruiting never occurs.

Order *Erysiphales* (powdery mildew fungi)

The order comprises well defined obligate parasites causing 'powdery mildew' disease of plants. There is one family *Erysiphaceae* having 19 genera. Mycelium is superficial and hyaline. Exceptions to the superficial, epiphytic growth are two genera *Phyllactinia* and *Levielulla*, especially the latter, which establishes an intensive intercellular mycelium. The ascomata are minute cleistothecia with external appendages, and contain globose or ovoid, evanescent asci, arranged in a basal layer. There are no paraphyses. There is one family and 19 genera.

A

B

Fig. 12.35. Powdery mildew of A, *cucurbit* and B, grapes.

The family ***Erysiphaceae*** comprises well-defined genera which are all obligate ectoparasites (except above two) of angiosperms, causing 'powdery mildew'. The superficial mycelium produces enormous number of conidia, usually on the leaf surfaces which appear like a mass of white powder; hence the name 'powdery mildew' (Fig. 12.35). In fact, both the fungi and the disease are called powdery mildew. Conidia are the main source of spread of the disease. Of the 40,000 host species, which are all angiosperms, and spread world over, 90% are dicotyledons. The monocotyledons are infected only by one species, *Erysiphe graminis*. Powdery mildews are more common on cultivated plants and grow luxuriantly in dry, cool seasons. However, other powdery mildew diseases are of minor significance, and inspite of being chronic, are seldom destructive.

The Genera

There are eight important genera (teleomorphic) whose identification is primarily based on the nature of the appendages on the cleistothecia, nature of mycelium-superficial or internal, and the number of asci (one or many). The appendages are of four types (Fig. 12.36): (1) simple hypha-like (2) rigid and dichotomously branched at tips, (3) rigid and swollen at the base, and (4) rigid and coiled at tips.

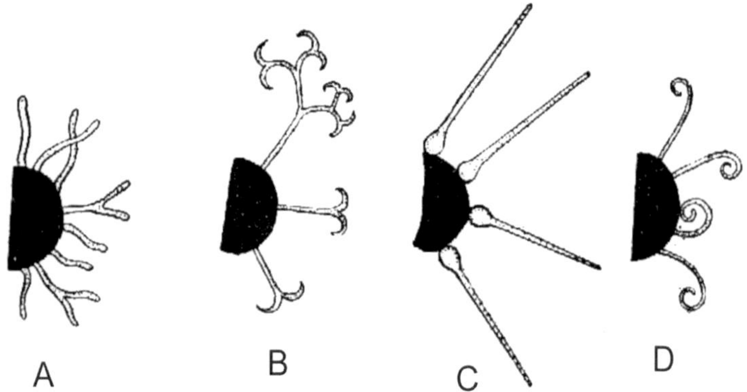

Fig. 12.36. Showing various types of appendages on the cleistothecia. A, simple, hypha-like; B, apically branched dichotomously; C, swollen at base; D, with coiled tips.

Now, in the current scheme (Kirk *et al.*, 2008) identification of the genera includes atributes of the 'whole fungus' i.e. anamoroph and teleomorph both. Now the genus is the whole fungus (holomoorph). This has resulted in the merger of some of the previous genera and addition of new genera. For example *Microsphaera, Uncinula* and *Erysiphe* are now one genus *Erysiphe*, placed under three sections (which is not a formal, recognized taxon). Similarly, *Sphaerotheca* and *Podosphaera* are merged and placed under 2 sections of *Podosphaera*. Two new genera added are *Blumeria* and *Golvinomyces*. Finally, the common genera now are: ***Erysiphe*** (anamorph *Oidium* subgenus *Pseudoidium*), ***Blumeria*** (anamorph *Oidium* subgenus *Oidium*), ***Golvinomyces*** (anamorph *Oidium* subgenus

Reticoloidium), ***Leveilulla*** (anamorph *Oidiopsis*), ***Phyllactinia*** (anamorph *Ovulariopsis*), ***Podosphaera*** (anamorph *Oidium* subgenus *Fibroidium*). So, with this change, the pathogen of 'powdery mildew of grape' is *Erysiphe necator* (formerly *Uncinula necator*), and for 'powdery mildew of wheat' the pathogen is *Blumeria graminis* (formerly *Erysiphe graminis tritci*).

Identification of the Genera

This requires information nature of the appendages on the cleistothecia, nature of mycelium—superficial or internal, and the number of asci (one or many), and the anamorph genus /subgenus.

Mycelium internal

Clelistothecium with several asci,

Appendages with bulbous base; anamorph *Ovulariopsis* ***Phyllactinia***

Appendages hypha-like; anamorph *Oidiopsis* ***Leveilulla***

Mycelium superficial

Cleistothecium with several asci;

Anamorph *Oidium* subgenus *Pseudoidium*

Appendages hypha-like ***Erysiphe*** section ***Erysiphe***

Appendages dichotomously-branched at tip ***Erysiphe*** section ***Microsphaera***

Appendaes with coiled tips ***Erysiphe*** section ***Uncinula***

Anamorph *Oidium* subgenus *Oidium*

Appendages hypha-like ***Blumeria***

Anamorph *Oidium* subgenus *Reticoloidium*

Appendages branched, hypha-like ***Golvinomyces***

Cleistothecium with single ascus

Annamorph *Oidium* subgenus *Fibroidium*

Appeendages dichotomously-branced a tip ***Podosphaera*** section ***Podosphaera***

Appendages hypha-like ***Podosphaera*** section ***Sphaerootheca***

Conidia, produced in enormous numbers on vertical conidiophores (Fig. 12.37), are the chief, rather exclusive, source of dissemination of these fungi. Secondary infections are caused repeatedly by the conidia. The conidiophores comprise a chain of cells, which are shorter, but wider than the hyphal cells.

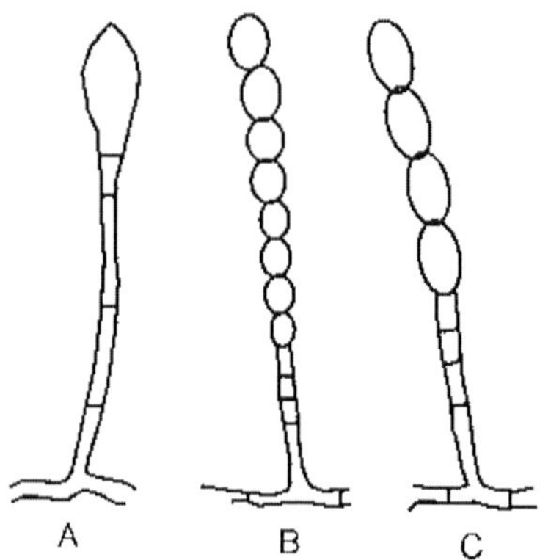

Fig. 12.37 Conidiophore types. A, single clavate conidium; B, conidia in true chains C, conidia in pseudochains.

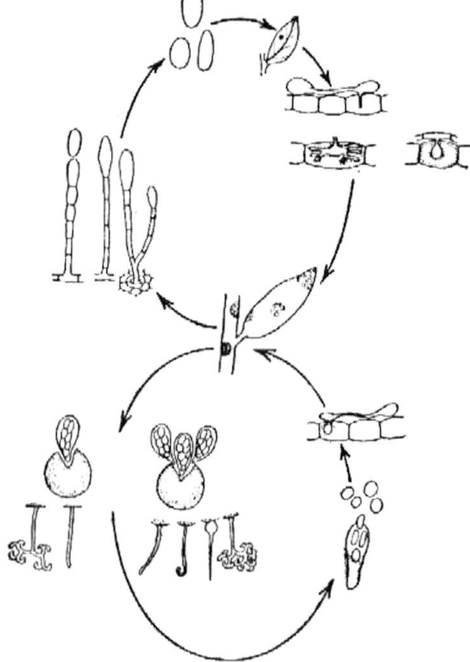

Fig. 12.38 A generalized life cycle of powdery mildews.

The conidia of the powdery mildew fungi are characteristically of **large size** and **high water content**, 70% against 10% water content of other air-borne conidia. It explains their abundance in dry areas. The conidia are largest in size among all fungi. These are released passively, and, when they fall on the proper host, germinate in about 2 hours. Only vacuolated conidia germinate, and there is no need of external supply of liquid water; humid air is enough. In fact liquid water is toxic to the germ tubes. Cleistothecia are formed at the end of the season though in some species they are formed regularly. The role of clesitothecia as primary inoculum is variable. The fungus can survive as mycelium in host debris and on collateral hosts. Conidia on collateral hosts can cause primary infection. A generalized life cycle of powdery mildew fungi is depicted in Fig. 12.38.

Order *Helotiales*

It is a large and diverse order containing 10 families; we shall study family *Sclerotiniaceae*

Family *Sclerotiniaceae*

This family of 47 genera and 284 sppecies is of great economic importance. The cup-shaped apothecia, which are usually yellowish-brown, are borne on long stalks arising from sclerotia or stromata formed in association with the host tissue. The ascus pore stains blue with iodine. The genera are distinguished by the origin of the apothecia from sclerotia or stromatized host tissue, and their conidial stages. *Sclerotinia* and *Monilinia* are two important genera. In *Sclerotinia* the apothecia, arise from a seed-like sclerotium (Fig. 12.39) which, though formed in host tissue, fall off to lie in the soil. The conidial stage is represented by *Myrioconium* and *Sclerotium*. *Monilinia* forms apothecia on mummified and stromatized fruits; the conidial stage is *Monilia*.

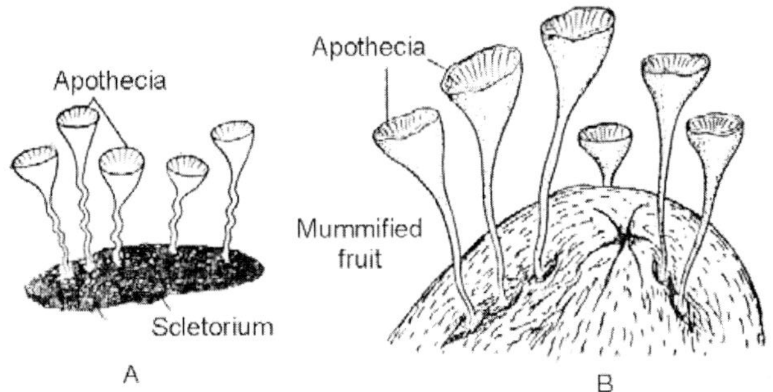

Fig. 12.39 A, *Sclerotinia*, apothecia formed on sclerotia; B, *Monilinia*, apothecia formed on mummified fruit.

Genus *Sclerotinia*

Three species of *Sclerotinia* are parasites, with a narrow host range. The important diseases they cuse are cottony rot of vegetables (*S. sclerotiorum*) 'crown-rot' of legumes, 'pink-rot' or celery, 'drop' of lettuce, 'stalk-rot' of potato, 'timber-rot' of tomato, 'watery soft rot', 'sclerotiniose' etc. Cottony mycelia, studded with irregular sclerotia, lie over watery, soft rot formed on the infected host. The sclerotium (see Fig. 12.39) serves to perennate the fungus between two successive host seasons. At the return of favourable conditions, it germinates to form stalked, cupulate apothecia. The ascospores that are explosively discharged infect and cause the disease. At the end of the season again the sclerotia are formed to repeat the cycle next growth season. *Myrioconium* and *Sclerotium* are the anamorphs but they play no role in disease causation.

Genus *Monilinia*

Monilinia causes fruit-rot, leaf-spot and 'dieback' diseases of economically important plants, belonging to the family *Rosaceae* and *Ericaceae*. *M. fructicola*, *M. laxa* and *M. fructigena* are responsible for the 'brown-rot' on stone fruits, like plums, peaches, cherries, apricots, almonds etc. Twigs and leaves are also attacked. The life cycle (Fig. 12.40) starts with germination of the conidia or ascospores, which infect the flower, leaf or twig of the susceptible host, causing blossom blight, leaf-blight or twig-blight, respectively.

The mycelium grows in host tissues and forms conidia (*Monilia* stage). Branched conidiophores form chains of oval conidia, which are easily separated and dispersed by wind. If the weather is humid, the infection spreads in epiphytotic proportions, causing severe damage. Brown, soft rots appear on the fruits, covered with conidial masses of *Monilia*. The infected fruits later dry up and shrivel. The mummified fruits may remain attached to the tree or fall on the ground. The fungus perennates in mummified fruits or at the site of the twig-cankers. On the return of favourable season, the conidia start reappearing on the mummified fruits, swinging on the twigs, which serve as an excellent source of dissemination of the conidia.

The mummified fruits, lying buried in the soil, give rise to long-stalked, cupulate apothecia. Globular spermatia (or microconidia) are formed on bottle-shaped spermatiophores. The asci discharge the ascospores explosively by puffing. The ascospores are hyaline, one-celled or oval. Till date only conidia, and not ascospores, are known to spread the disease. The reason why apothecia are formed only on the buried mummified fruits and not on those attached to trees is also not known.

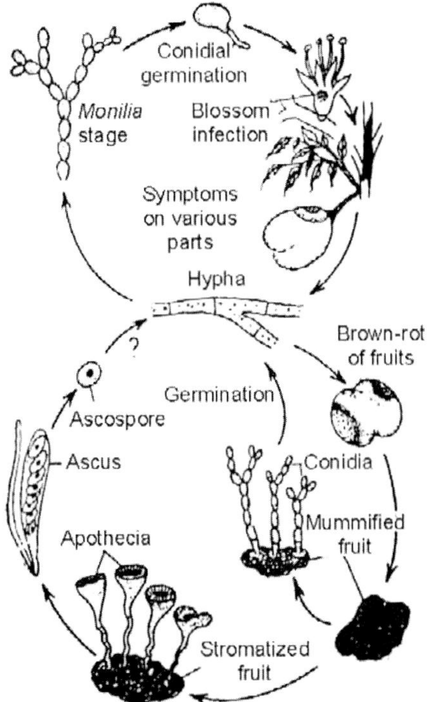

Fig. 12.40 Life cycle of *Monilinia fructicola.*

Class *Lichinomycetes*

This is an exclusive class of fungi lichenized with cyanobacteria. The thallus is fruticose, foliose and sometimes peltate. The apthecia may be sessile or immersed. The apothecia are initially perithecia formed from a pycnidium. Asci are bitunicate or prototunicate and sometimes evanescent. The ascospores are hyaline. Anamorphs are pycnidial. There is 1 order, ***Lichinales*** and 3 families, *Gloeoheppiaceae, Lichinaceae and Peltulaceae. Lichinaceae* has 47 genera, including, *Lichina* and *Lichinella*.

Class *Sordariomycetes*

Sordariomycetes is one of the largest classes in the *Ascomycota* with more than 600 genera and 3000 species. It includes most of the non-lichenized perithecial ascomycetes, and also some cleistothecial forms. Asci are inoperculate, unitunicate or prototunicate. Stroma is present in some (*Xylariales*). The class is phylogenetically related to *Leotiomycetes* as sister groups. Memberas of the class are ubiquitous and well known as plant patogens (causing chestnut blight, Dutch elm disease, blast diseae of rice and most ancient 'ergot' of rye), endophytes (*Xylaria*), arthropod and mammalian pathogens (*Sporothrix, Trichoderma*), and as saprobes (*Neurospora, Chaetomium*).

In the current classification (Eriksson, 2006; Kirk *et al.*, 2008), it has 3 subclasses: *Sordariomycetidae, Hypocreomycetidae*, and *Xylariomycetidae*. The further classification is as follows (all orders and all families are not mentioned). We shall study the genera in bold type.

Subclasses: 3

Sordariomycetidae: 7 orders

Sordariales: 5 families.

Chaetomiaceae: ***Chaetomium***

Sordariaceae: ***Sordaria***, ***Neurospora***

Diaporthales: 10 families

Cryphonectriaceae: ***Cryphonectria***

Ophiostomatales: 1 family: *Ophiostomataceae:* ***Ophiostoma***

Hypocreomycetidae: 4 orders:

Hypocreales: 7 families

Clavicipitaceae: ***Claviceps, Metacordyceps***

Cordycipitaceae: ***Cordyceps***

Hypocreaceae: ***Hypocrea***

Nectriaceae: ***Necrtria***

Microascales: Ceratcystidaceae: ***Ceratocystis***

Xylariomycetidae: 1 order

Xylariales: 9 families

Xylariaceae: ***Xylaria***

Incertae sedis: *Meliolales (Meliola), Phyllachorales (Phyllachora), Magnaporthales (Magnaporthe)*

Subclass *Sordariomycetidae*: 7 orders

Order Sordariales

Ascospores are dark with, a germ pore, and often have a gelatinous sheath or appendage. Conidial stage is usually absent or spermatial. Most of the species are saprobes on rotting wood or in soil, coprophilous, cellulolytic and fungicolus. The order has 5 families. We examine *Chaetomiaceae* (*Chaetomium*), and *Sordariaceae* (*Sordaria* & *Neurospora*).

Family *Chaetomiaceae*

The family includes genera, which often produce beaked and ostiolate perithecia usually with complex ornamental hairs. The asci are spherical or globose, evanescent and lack apical structures. The ascospores are dark and may, or may not, have germ pores. After dissolution of the ascus, the ascospores, which are brown, fill the cavity of the perithecium and are released in a mucilaginous matrix, like a tooth paste. Anamorphs are absent i.e. there is no conidial stage. The members are mostly saprobes growing in soil, on cellulose rich substrata.

Genus *Chaetomium*

It is an important cellulose-degrading fungus, growing on wet paper, fabrics, and other cellulose-rich materials. Some species are thermophilic and coprophilous, frequently inhabiting manure heaps. *C. thermophile* is thermophilic, cellulolytic, and coprophilous, growing on composting manure heaps. *C. globosum* (Fig. 12.41) is another important cellulose-degrading species.

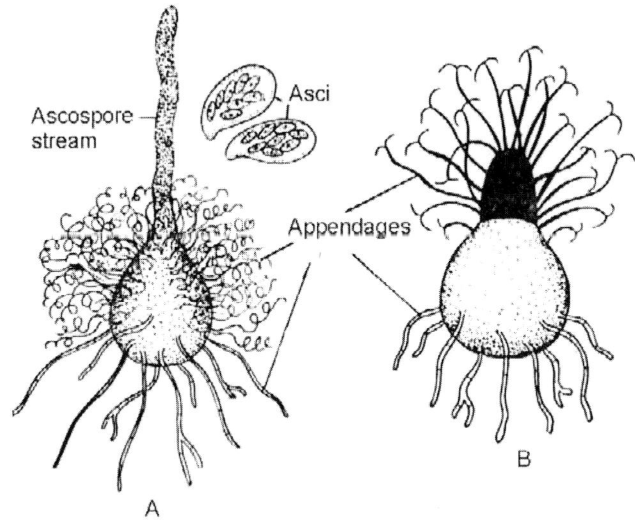

Fig. 12.41 *Chaetomium*. A, *C. globosum*; B, *C. thermophile*

The genus has 95 species and is easily identified by the perithecium, which has a long beak and covered with hairs. The fungus is named on this character (gr. chaite-long hair). The hair is of two types: in *C. globosum*, curly at top and straight at the base, and in *C. thermophile*, dichotomously branched above and mycelioid below. The asci are stalked, club-shaped, cylindrical or oval and produced in a basal tuft. These gelatinize leaving the ascospores free in the cavity of the perithecium. The ascospores are released passively in an oozing mass of mucilage. The anamorph *stage is Botry-trichum* Dissemination is brought about mostly by ascospores.

Family *Sordariaceae*

The family is characterized by dark, thick-walled perithecia; ascospores are mostly dark when mature and have germ pores (or germ slits) and a mucous sheath, or appendages that are sometimes ornamented. There are 8 genera. We shall study the genera *Sordaria* and *Neurospora*.

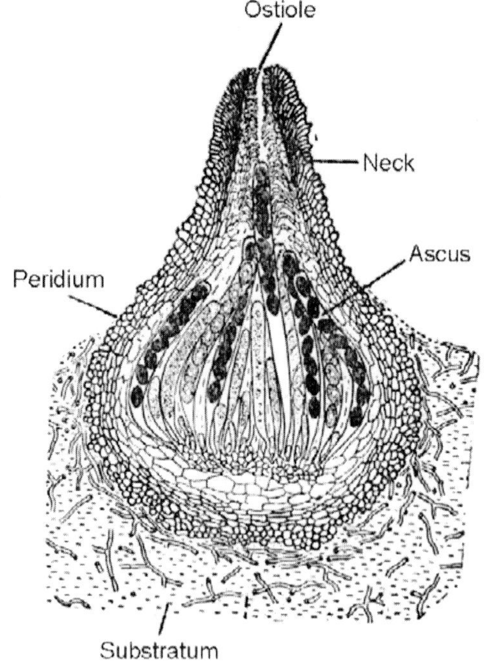

Fig. 12.42 *Sordaria fimicola,* vertical section of perithecium.

Genus *Sordaria*

There are 12 species. *Sordaria fimicola* (Fig. 12.42) is a common coprophilous fungus, growing on dung of herbivore animals. The fungus has been widely used in the studies on physiology and genetics of fungi. The reproduction is solely by means of ascospores; asexual reproduction is not known. The fungus is homothallic. The dark, pear-shaped perithecia, about 0.5 mm high, lie more or less sunken in the dung. The ascoma shows positive phototropism and the spores are discharged violently towards light.

The neck of the mature perithecium projects out, and bends towards light. The asci of all stages arise from the base in a tuft. Paraphyses deliquesce early and, therefore, are absent in mature perithecia. The ostiole is lined by periphyses. One ascus, which is ahead in growth from the rest, elongates and pushes through the ostiole. On reaching the ostioler opening, it ejects the ascospores violently into the air upto a distance of about 10 cm. The empty ascus shrinks and retracts into the perithecium. This is repeated by the second longest ascus, and in this way all the asci discharge their ascospores one by one. There is no puffing out of ascospores in instalments.

The ascospores are provided with a minute germ pore at the lower end, and surrounded by a gelatinous sheath. The sheath helps the ascospores to stick to objects like grass and find their way into the alimentary canal of animals, where they recive the stimulation for germination.

Genus *Neurospora*

Neurospora made history in 1947 when Tatum and Beadle initiated a new era in genetics, called the 'biochemical genetics'. Neurospora 'dethroned' *Drosophila* as the organism of choice in genetical studies. *Neurospora* gets its name from the characteristic striations present on the ascospores. The ascospores have one or two germ pores. There are 51 species. The three best studied species are: *N. sitophila*, *N. crassa*, (both heterothallic, with *Chrysonilia* (=*Monilia*) as the conidial stage) and *N. tetrasperma* (secondarily homothallic and lacking conidia).

By nature, *Neurospora* occurs saprobically on bread and burnt grounds in tropical countries. *N. sitophila* is called the 'red bread-mold', because it frequently infests bakeries and causes trouble. It attacks bread and very rapidly covers it with a pink growth. It is not advisable to grow *Neurospora* in Petri dishes as it is a notorious 'weed of the laboratory'. Dry conidia escape as pink clouds, when the dish is opened and contaminate cultures.

The mycelium (Fig. 12.43) consists of highly branched, pigmented hyphae with multinucleate cells. The fungus reproduces asexually by two types of conidia: the macroconidia, 6-8 μm in a diameter, which are multinucleate (1-8 nuclei, usually 2 or 3), and microconidia, 3-4 μm in diameter and uninucleate. The macroconidia are pink and borne in branched chains in large numbers on short conidiophores. These are designated as the *Chrysonilia* stage. Erect, unspecialized hyphae serve as conidiophores, which cut chains of multinucleate, pink conidia apically. Later, conidia are formed by budding of the terminal conidium. If more than one bud is formed, the chains become branched. The conidia are disseminated by wind. The microconidia are also formed in similar way but are less frequent. Both types of conidia, on germination, give rise to somatic hyphae, which quickly (within 24 hours) start forming conidia.

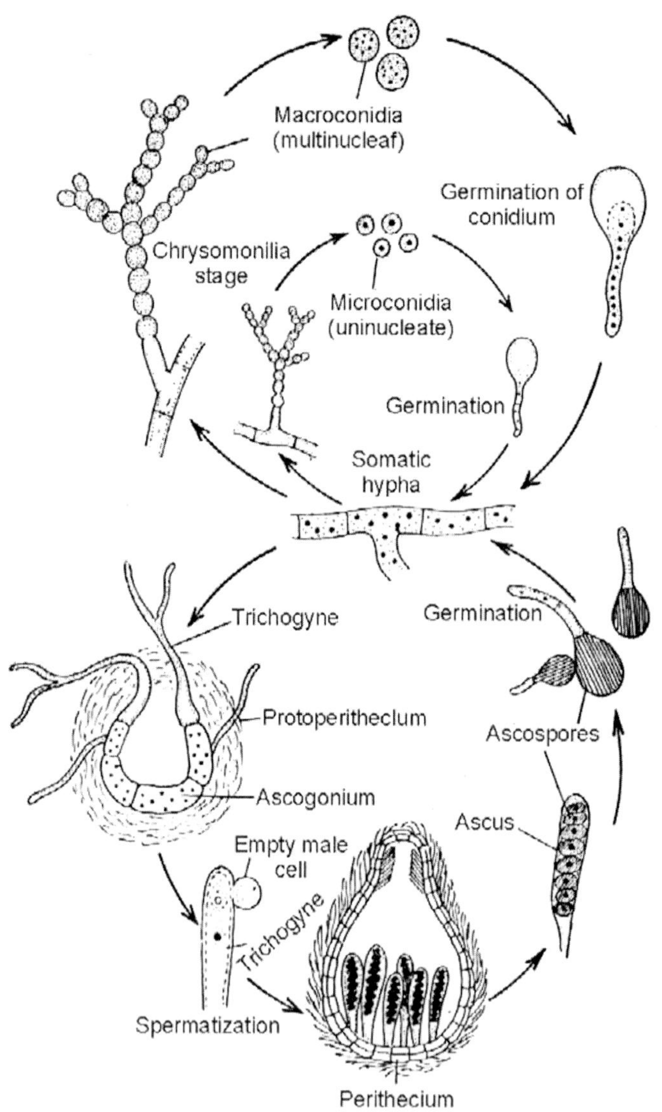

Fig. 12.43 Life cycle of *Neurospora sitophila.*

N. crassa and *N. sitophila*, the two important species are heterothallic. There is no antheridium. The female sex organ is also degenerate. The ascogonium lies embedded in a knot of hyphae called protoperithecium. It consists of a thick coil of few multinucleate cells, from which branched, long (2 mm or so) receptive hyphae protrude out and serve as trichogynes. There are no specialized male cells like spermatia. However, both macro-, and microconidia, and somatic cells of the compatible mating types can, on contacting the trichogyne, supply the male nucleus. The male nuclei reach the ascogonium and pair with the female nuclei. Ascogenous hyphae arise from the ascogonium and bear asci apically on characteristic croziers. The hyphae of the protoperithecium give rise to the peridium of the ascoma.

The mature perithecium is dark, pyriform and beaked, bearing numerous, 8-spored asci but no paraphyses. The ascospores are brown to black, with characteristic ridges or striations.

N. tetrasperma, as the name suggests, produces only four ascospores in each ascus, which are binucleate; the nuclei being of the opposite mating types. On germination, they produce heterokaryotic, (secondarily homothallic) mycelium, which bears perithecia. However, sometimes a few uninucleate 'dwarf' ascospores are produced which give rise to homokaryotic thalli that do not produce perithecia unless mated with the opposite strain.

Order *Diaporthales*

Perithecia often aggregated into a pseudostroma, usually long-necked, paraphyses dissolve early.

There are 10 families. We shall study the genus *Cryphonectria* of the family *Cryphonectriaceae*.

Genus *Cryphonectria* (formerly *Endothia*)

The fungus *Cryphonectria parasitica* (formerly *Endothia parasitica*) is of great historical importance, which devastated the chestnut trees to extinction in USA. The disease called chestnut blight (or canker), was introduced into U.S.A. around the beginning of 1900 through infected lumber or trees and by 1940 the once plentiful trees were wiped out. This suggests that pathogens, when introduced to new areas, become more destructive than in their natural habitat. The conidial state is the pycnidial anamorph genus, *Endothiella*. The conidia or the ascospores enter the tree through wounds and establish the infection under the bark. The conidia and ascospores are also spread by beetles (*Agrilus* spp.) and birds through wounds produced by insect vectors. The fungus forms yellowish-orange stromata that break through the bark of the canker. Conidomata (pycnidia) are formed on the stroma, which release hyaline conidia in form of yellowish tendril (cirrus). Perithecia in groups of 10-20 are formed on the stroma. Ascospores are hyaline, 2-celled, and constricted at the septum.

Order *Ophiostomatales*: 1 family: *Ophiostomataceae*

There are 12 genera which form perithecia (except in 1 genus), with a long-neck and ostioler setae. Inteascal tissue is absent. The asci are small, evanescent and formed in chains. Ascospores are unicellular. The anamorphs are hyphomycetous. Many species growing on wood are associated with bark-beetles which help in their dispersal. Some species cause important diseases of conifers, while some are mycoparasites. We shall examine genus *Ophiostoma* belonging to the lone family *Ophiostomataceae*.

Genus *Ophiostoma*

The 2 species of *Ophiostoma* (*O. ulmi* and *O. novoulmi*) made history by devastating the elm trees in two pandemics of Dutch elm disease, in 1920s and in 1940s, respectively. The first was caused by *O. ulmi*, and the second by its more aggressive form, which later was designated as a new species *O. novoulmi* (Brasier, 1991). Both the species cause wilt disease of elms and depend on insects, the bark-beetles, for their dispersal.

The Dutch elm disease exemplifies, like chestnut blight, that pathogens, when introduced to new areas, become more destructive than in their natural habitat. The disease was reported in 1919 from Holland, and soon it spread to Central Europe. By 1935, one-third of the elm trees in Britain had to be felled because of the wilt. In USA, the disease appeared in 1930 and within 8 years, 25,000 trees were killed. The second pandemic originated in Romania and the Ukraine in 1940s, and soon spread to North America.

Life Cycle of *O. ulmi* (= *O. novoulmi*)

The fungus overwinters on elm logs in tunnels made by beetles for laying eggs (Fig. 12.44). The fungus grows saprobically and forms conidia (anamorph *Graphium*), which are carried by beetles to elm trees on whose sap they feed. The insect vectors are two species of *Scolytus*. In xylem vessels, the fungus produces small conidial 'buds' (anamorph *Sporothrix*), which move up to petioles and veins in the leaves, and cause extensive invasion of the vessels. In this way, multiple infections result and susceptible trees die.

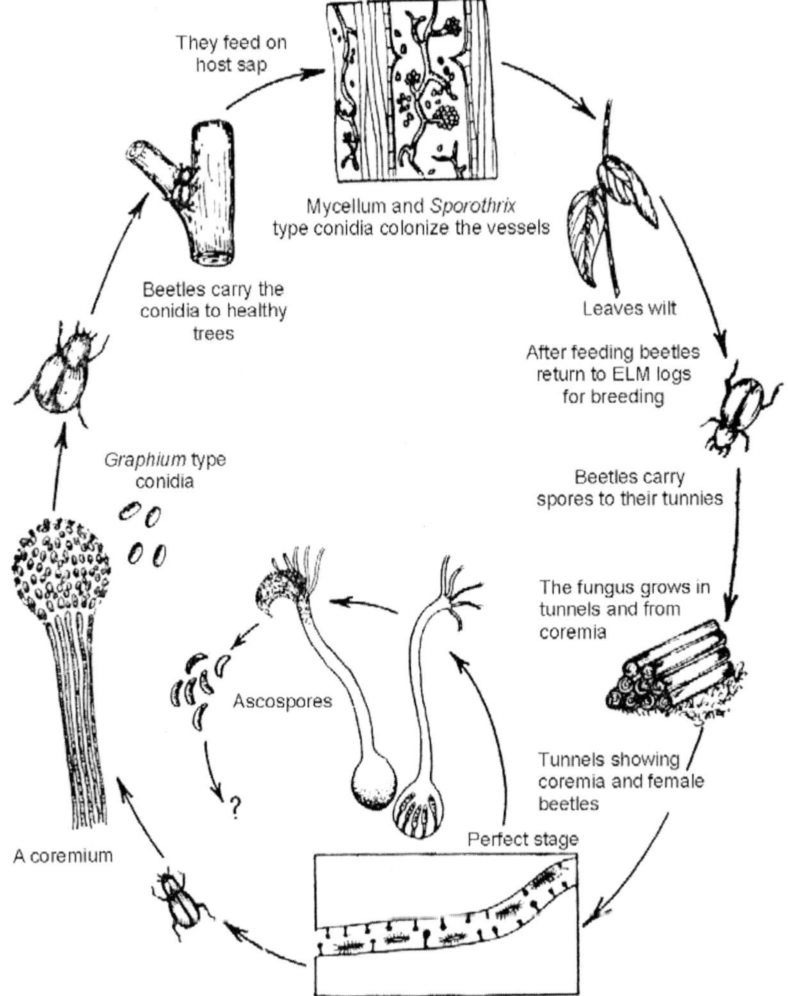

Fig. 12.44 Life cycle of *Ophiostoma ulmi*

Subclass *Hypocreomycetidae*: 4 orders

Order *Hypocreales*: 7 families

The genera included in this order produce brightly coloured, fleshy and ostiolate perithecia. The stromatic genera produce soft, brightly coloured stromata bearing perithecia, either superficially or immersed in the stromatic tissues. The perithecia lack paraphyses but pseudoparaphyses or 'apical praphyses' are present. Ascospores are spherical to needle-like, septate and sometimes breaking into 'part spores'. These are usually forcibly discharged. The anamorphs are represented by several hyphomycetous genera. There are 7 families and 43 genera, found in uncommon habitats as parasites or saprobes.

Family *Clavicipitaceae: Claviceps, Metacordyceps*

The family is distinct by the long, narrow asci and the filiform (thread-like) ascospores. The apex of the ascus is slightly thickened and a narrow canal passes through the thickened ascus tip, which opens outside through a pore. The perithecia, with a definite peridium, are formed in stroma.

Genus *Claviceps*

There are 41 species of *Clavicep*s, which are chiefly parasites of grasses infecting the inflorescence and causing the disease, called **ergot**. The best-known species are *C. purpurea* (causing ergot of rye, *Secale cereale*) and *C. microcephala* (causing ergot of 'bajra', *Pennisetum typhoides*). The infected 'ears' of the rye plants develop several, hard, black sclerotia projecting among the healthy grains. The sclerotia are of the same shape as the grains but much larger in size. These are highly toxic to animals and human beings, and cause the diseased condition called **ergotism** (see Chapter 28). In addition to ergotoxins, the sclerotia also contain some useful alkaloids, which are used in medicine. The well-known hallucinogenic chemical LSD (d-lysergic acid diethyl amide) is obtained from the sclerotia. The sclerotium is, therefore, of great industrial importance and is known as the 'ergot of commerce'.

Life Cycle of *C. purpurea*

The filiform ascospores, released forcibly from the long and narrow asci are disseminated by wind (Fig. 12.45). These germinate on rye flowers and infect the ovaries. The infection occurs through base of the young ovaries and not from stigma or the style. The mycelium that develops soon fills the ovary, and in less than a week, numerous condiophores arise from the peripheral hyphae. This conidial state is called *Sphacelia* .These form onely-celled, hyaline conidia in enormous numbers. A viscous sugary substance oozes out of the fungus-stuffed ovaries in form of 'drops', called 'honey dew'. During humid weather, the honey dew drips on the leaves and the soil. It contains glucose, fructose and sucrose. Insects visit the infected plants for the honey dew, and in doing this, disseminate the conidia to healthy plants. The conidia multiply by budding. At the end of the rye season, the mycelia in the ovaries develop into pink or purplish sclerotia, which project out of the glumes. The grain-shaped, but much larger, sclerotia appear conspicuous in the ears.

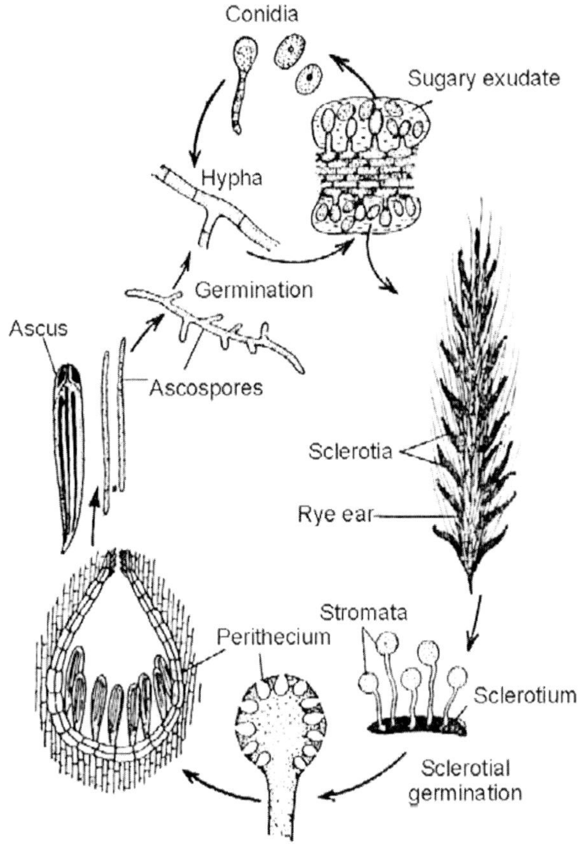

Fig. 12. 45. Life cycle of *Claviceps purpurea.*

Sclerotia, knocked off during harvesting, reach the soil where they over winter. These germinate during spring, which is also the time for flowering of the rye plants. A cold pre-treatment is essential for the sclerotial germination. During germination, several stromatic masses arise from within the sclerotia rupturing the rind. These elongate and become globular at the apex. The stromata finally become "mushroom-shaped" and attain a length of about 2 cm. The stalk of the stroma is made of prosenchymatous tissue, while the globular head consists of loose mass of hyphae. Perithecia develop in the peripheral zone of the head, at right angles to its surface. The perithecia possess periphyses in the ostiole, paraphyses at the lateral wall and asci at the base. The ascospores are violently released which pick-up the life cycle.

Family *Cordycipitaceae*

The family has 14 genera and several anamorphs.

Genus *Cordyceps* (plant worms, vegeaive wasps)

The genus *Cordyceps*, which has 90 species, grows mostly on insects, caterpillar fungi and vegetable caterpillars. The genus comprises about 90 species, most of which are parasites of insects, though some grow on subterranean ascoma of *Elaphomyces* or on sclerotia of *Claviceps*. This is the oldest entomogenous fungus, recorded in literature as 'plant worms', 'vegetative wasps' or 'trees growing on insects'. In China the dead caterpillars, containing the fungal sclerotium, are eaten as a delicacy and tonic. *C. militaris* is the best-known entomogenous species; *C. canadenensis* and *C. ophioglossoides* grow on *Elaphomyces* ascoma. *C. sinensis* is highly valued in traditional Chinese medicine. **Cyclosporin**, a drug used in transplant surgery, is derived from *Tolypocladium inflatum*, the anamorph of *C. subsessilis*. Recently, the genus has been split and 3 new genera have been added (Sung and Spatafora, 2007). These are *Ophiocordyceps*, *Elaphocordyceps* (both belonging to *Ophiocordycipitaceae, Hypocreales*), and *Metacordyceps* (*Clavicipitaceae*).

Life Cycle of *C. militaris*

The species gets its name *militaris* because of erect, soldier-like, stromata (Fig. 12.46). The infection is initiated by the filiform ascospores or their fragments landing on the proper insect host. The fragments become globular and put out a germ tube, which penetrates the integument and gives rise to hyphae. The fungus produces a cutinase enzyme *in vitro*. The hyphae break-up into fragments that spread in the haemocoel. The fragments bud off more propagules, which give rise to more hyphae and soon the insect gets stuffed with the mycelium. The insects become sluggish and then die. The mycelium modifies and develops into a sclerotium, which remains covered by the integument of the dead insect. The corpse of the insect enlarges in size and becomes resistant to decay due to antibiotics (e.g. cordycepin) produced by the fungus. The dead insect with endosclerotium, lies buried underground until the conditions favourable for sclerotial germination are available. Club-shaped, orange-coloured stromata, with a stalk and head, arise above ground and bear perithecia in the periphery of the head region.

Development of perithecia, which has been studied in *C. canadenensis*, is similar to that of *Claviceps*.

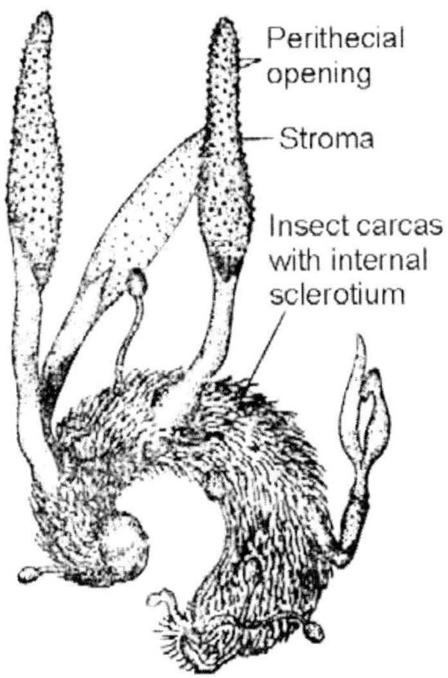

Fig. 12.46 *Cordyceps militaris* sclerotium inside insect corpse gives rise to stromatic masses bearing perithecia.

The asci, with a characteristic apical apparatus, contain 8 filiform ascospores, which are released violently in succession. The ascospores multiply by division into small segments and get dispersed by wind. The fungus has many anamorphs, *Beauveria, Lecanocillium Metarhizium* and *Isaia*. *Beauveria* and *Metarhizium* are used in biological control of insects.

Family *Hypocreaceae*

The family has 22 genera, which are mostly saprobic and grow on plant debris. The members are characterized by (1) perithecia, which usually lie embedded in brightly coloured, flattened or cushion-shaped, sessile stromata, and (2) the colourless ascospores that divide into two at maturity. Thus, sixteen ascospores are released from each ascus.

Genus *Hypocrea*

The genus, with 171 species, forms different types of conidia which are all phialospores and represent well-known anamorph genus *Trichoderma*.

H. rufa forms a cushion-shaped stroma (upto 10 mm across) on rotting wood, bark, old fructifications of fungi and other debris (Fig. 12.47). The surface is studded with the openings of the ostioles of thc prithccia. The asci contain 16 colourless spores (part-spores) which are globose or broadly elliptical. The conidial state is the anamorph, *Trichoderma viride*, which is a well-known soil inhabitant, showing antagonism against several plant pathogenic fungi. The branched conidiophores bear apical phialides, which produce phialospores (conidia) held in a mucilage mass at the tip.

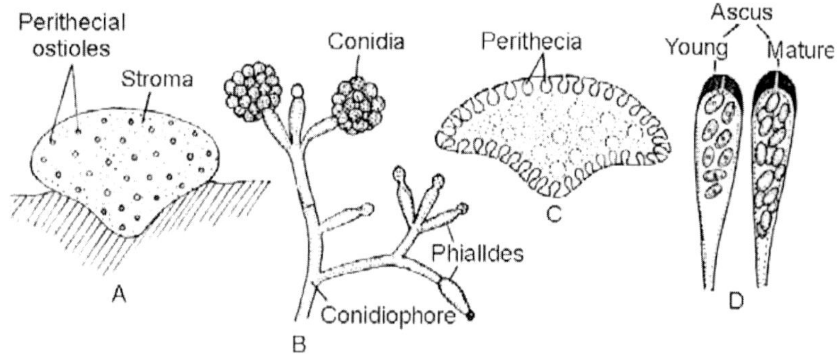

Fig. 12.47 *Hypocrea rufa*, A, habit – perithecial stroma with distinct ostioles; B, the conidial stage (*Trichoderma viride*); C, perithecia in section; D, young and mature asci.

Family *Nectriaceae*

Genus *Nectria*

Nectria species (82) are saprobes on decaying wood, or parasites on trees causing cankers. *Nectira, Calonectria, Haematonectria* are some of the common genera These are economically important in forestry and forest products. Anamorph states are *Tubercularia* and *Acremonium*-like. *N. galligena* and *N. cinnabarina* are the important species causing cankers on many hardwood trees like apple, elm, birch and oak. *Fusarium solani* is the anamorph of *N. haematococca.*Cushion-shaped hyphal masses formed on the infected areas bear the asexual conidiomata and later bright-coloured perithecia which produce one- or bicelled ascopores. The ascospores cause infection when they reach wounds or natural openings on the trees.

Order *Microascales*

The order is characterized by absence of stroma, formation of perithecia or cleistothecia, which are usually black, thin-walled, and, sometimes covered with well-developed setae. There is no ascal tissue.The asci are spherical, ovoid, and evanescent which contain 8 small hyaline ascospores. The ascospores, after the break-down of the asci, are released in a mucilaginous mass. The anamorphs are hyphomycetous. The order has 2 families *Ceratocystidaceae* (6 genera) and *Microascaceae* (12 genera) which are saprobes in soil or on rotting vegetation. A few are opportunistic human and animal pathogens.

Family Ceratcystidaceae

Genus *Ceratocystis*

This genus is almost indistinguishable from *Ophiostoma* and both were earlier placed in *Ophiostomatales*, but because of different asexual form (and also several biochemical characteristics, including rDNA sequences) it is now placed in *Microascales*. However, the species *Ceratocystis ulmi* has been retained under *Ophiostomales* as *Ophiostoma ulmi*.

The genus includes several important plant-pathogenic species, as well as, some wood-attacking saprobes. The important pathogenic species are *C. fagacearum* (causing oak wilt) and *C. fimbriata* (causing black rot of sweet potato). Several species (*C. piliferum*, *C. minor* and *C. piceae*) attack logs and sawn timber and produce blue stains which reduce their commercial value. The anamorph varies with the species. Sexual reproduction involves spermatization through conidia of opposite strain. There are no ascogenous hyphae, and the asci develop directly from the ascogonia. Perithecia (Fig. 12.48) with typically long necks are formed on mycelial mats, which lie hidden in crevices of the bark. The anamorph is *Chalara*.

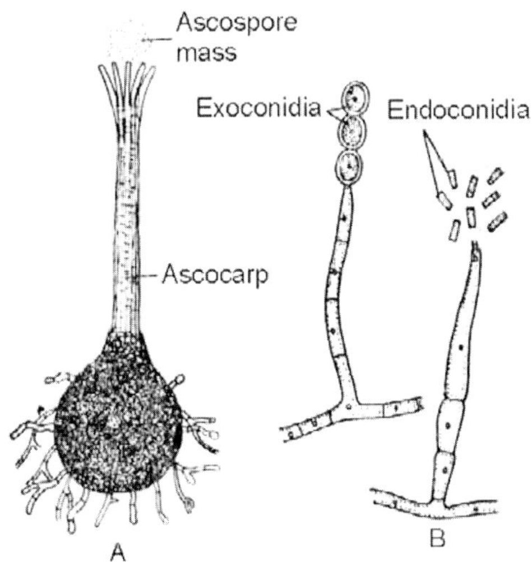

Fig. 12.48 *Ceratocystis fimbriata*. A, perithecium; B, exo-and endoconidium.

Subclass *Xylariomycetidae*

The subclass is characterized by well-developed stromata, mostly made of fungal tissue. Perithecia, rarely cleistothecia, which are black and thick-walled, are globose, spherical or immersed in the stroma. The ostiole is papillate, and periphysate. Paraphyses well-developed. Asci cylindrical and persistent with ascospores that are sometimes septate. The anamorphs, are hyphomycetous or pycnidial. The members are saprobes and plant parasites. There is 1 order, ***Xylariales*** and 9 families, including the family ***Xylariaceae*** which has of 85 genera. We shall study genus *Xylaria*.

Genus *Xylaria*

Xylaria has 300 species which are widespread, growing on dead wood. The common species are *X. hypoxylon* (Fig.12.49, Candle-snuff fungus, root rot of apple), *X. polymorpha* (dead man's fingers), *X. digitata* (root rot of heart woods), *X. longipes* (dead Moll's fingers) and *X. vaporaria* (invader of mushroom beds). Some species are associated with termites but the nature of the realationship is not known. The fungus forms erect stroma, several centimeters tall, at the base of the tree or from wood just below the ground. The stroma may be, depending on the species, firm or hollow internally. *X. hypoxylon* (Fig. 12.48) forms erect, 3-7 cm tall candle-like stromata that bear perithecia at the base and conidiomata in the upper portion at the top. The conidiomata form a compact palisade layer of conidiophores that cut off conidia.The anamorphs are *Molleroclavus, Geniculosporium-like, Xylocoremium.* Perithecia are of bright colours. The ascospores are dark and septate (bicelled).

Fig. 12.49 Stromata of *X .hypoxylon*

X. longipes improves the quality of the wood used in string instruments, in lightening its structure. In Indian traditional medicine (Ayurved) the powedered stroma, mixed with sugar is used to promoe lactation after chilld birth.

Incertae Sedis Orders of *Sordariomycetes*

Meliolales (*Meliola*), *Phyllachorales* (*Phyllachora*), *Magnaporthales* (*Magnaporthe*) are yet to be firmly assigned a subclass in the classification of *Sordariomycetes*.

Order *Meliolales (Meliola)* (black or sooty mildews)

The members of this group show resemblance to the powdery mildews, and several authors, including Yarwood (1973), have placed these fungi under *Erysiphales*. But, since the differences outweigh the resemblances, the two are placed under different orders. The order *Meliolales* consists of a single family *Meliolaceae*, including 22 genera and 1980 species. The genera are based on the number of cells constituting the ascospores, shape of the ascoma, presence or absence of the mycelium etc.

The members are parasitic on higher plants and grow on leaves. The superficial hyphae are dark-coloured (hence the name, 'black mildew'), and lie attached to the leaves by means of **hyphopodia**. Stiff erect bristles, called setae, are present on the hyphae. The hyphopodia are lobed structures formed at the tips of short lateral branches of the hyphae. Fine hyphae arise from the hyphopodia that penetrate the epidermal cells and form globular haustoria. Spherical, thin-walled cleistothecia, which are borne on the superficial hyphae, form a large apical pore or rupture irregularly. The asci are unitunicate thin walled, clavate or globose and evanescent. The asci contain 2–4 (rarely 8) ascospores, which are dark brown, thick-walled, septate (2–5 celled) and unitunicate. Paraphyses are often present.

Genus *Meliola*

This is the largest genus with about 1300 species widespread in tropical countries. The genus (Fig. 12.50) is characterized by presence of black setae on the hyphae. The thick-walled, dark-brown hyphae lie in small, irregular patches on both surfaces of the infected leaves. Numerous short hyphopodia originate from the alternate cells of the hyphae in opposite directions. Each hyphopodium consists of 2 cells, a short basal cell and a swollen, club-shaped or lobed head, which serves as appresorium. The hyphae also bear stiff erect, black setae, about 1 mm long. Globose black cleistothecia, which lie attached to the mycelium, contain oblong evanescent asci. Each ascus has 2-4, brown, elliptical, fusiform or sightly-curved, 5-celled ascospores.

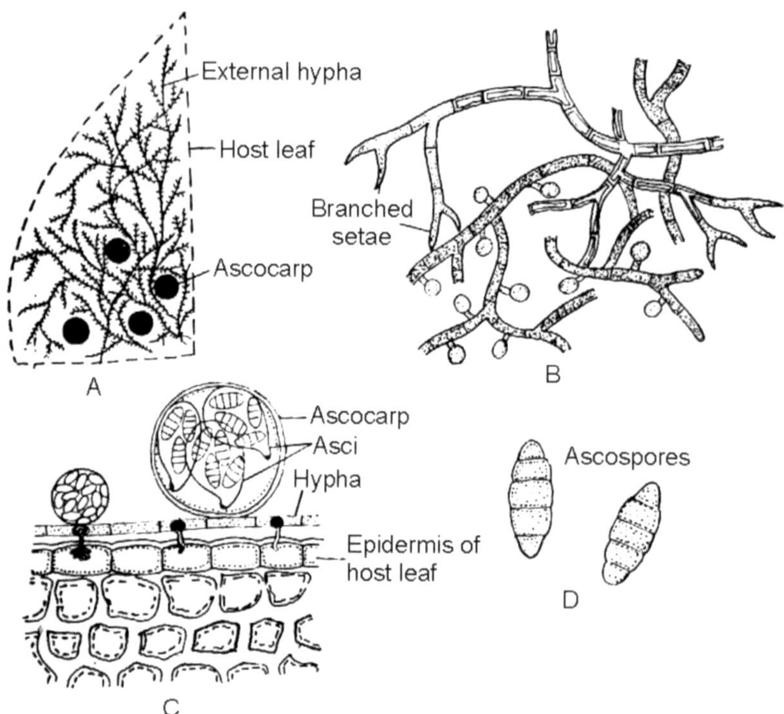

Fig. 12.50 *Meliola*. A, showing superficial hyphae with hyphopodia and ascomata; B, branched pointed mycellial setae and dark hyphae with alternate and opposite hyphopodia; C, young and mature ascomata on host leaf (in section), the hyphopodia forming haustoria; D, 5-celled ascospores.

Order *Phyllachorales*

The order has 2 families, *Phaeochoraceae and Phyllachoraceae*, and 63 genera, which form well-developed black stromata immersed in plant tissues. Perithecia are formed in the stroma with a shield-shaped structure (clypeus) of differentiated stromatic tissue around the ostiole, which is periphysate (i.e., lined with periphyses). Asci, intermingled with paraphyses, are cylindrical, persistent and with apical pores encircled by rings. Ascospores are usually hyaline and one-celled. The anamorphs are coelomycetous anamorphic fungi. We shall examine 2 genera, *Phyllachora* and *Glomerella*.

Genus *Phyllachora*

The genus has 944 species, which are widespread. True perithecia with definite walls lie sunken in endophyllous stromata, formed on both sides of infected leaves (Fig. 12.51). The stromata, which resemble rust pustules, are firm and well-defined near the leaf surface. The perithecia are globose with distinct erumpent ostiole and possess paraphyses. The asci have a uniformly thick wall and a conspicuous apical pore. Young stromata contain pycnidia bearing hyaline, thread-like spores. *P. graminis* and several other species parasitize grasses, as well as, *Ficus* and many other plants. The anamorph is *Linochora*.

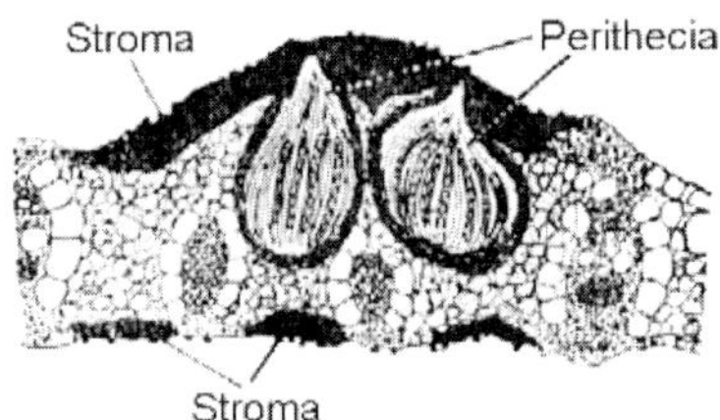

Fig. 12.51 *Phyllachora graminis*, section of infected leaf showing perithecia.

Order *Glomerellales*, Family *Glomerellaceae*

Genus *Glomerella*

Glomerella (11species) produces beaked perithecia in stroma or in host tissues. The asci are short-stalked, 8-spored, and lie intermingled with paraphyses. Ascospores are hyaline, one-celled and curved. *Colletotrichum* species are the anomorphs. *Colletotrichum gloeosporioides*, causing bitter rot of apples, and anthracnose of several plants, is the anamorph of *G. cingulata*, while *Colletotrichum graminis* (= *C. falcatum*), which causes the 'red-rot' disease of sugarcane, is the anamorph of *G. tucamanensis*.

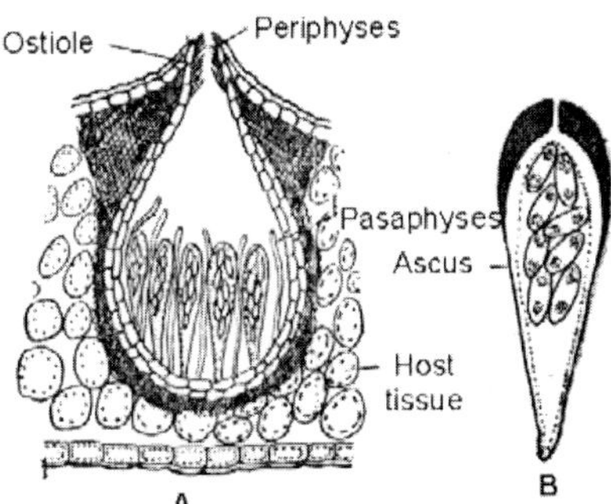

Fig. 12.52 *Glomerella cingulata*. A, perithecium in section; B, an ascus.

G. cingulata (Fig. 12.52) has been the organism of choice for fungal geneticists ever since Edgerton (1914) reported heterothallism in this fungus. The sexuality of this genus is complex and has been a subject of intense research.

REFERENCES

Alexopoulos, C.J., Mims, C.W. and Blackwell, M. 1996. *Introductory Mycology*. 4th ed. John Wiley, N.Y.

Berbee, M.L. and Taylor, J.W. 1999. Fungal phylogeny. In: *Molecular Fungal Biology* (R.P. Oliver and M. Schweizer eds.). pp. 21–77. Cambridge Univ., Press, Cambridge.

Brasier, C.M. 1991. *Ophiostoma novo-ulmi* sp. nov., Causative Agent of the Current Dutch Elm Disease Pandemics. *Mycopathologia*. 115: 151-161.

Christensen, C.M. 1965. *The Molds and Man*. 3rd ed. Univ. of Minnesota Press.

Edgerton, C.W. 1914. Plus and minus strains in the genus *Glomerella. Amer. J. Bot.,* 1: 224–254.

Liu YJ and Hall BD 2004 Body plan evolution of ascomycetes, as inferred from an RNA polymerase II phylogeny. PNAS 101(13) 4507-4512

Lutzoni F et al., 2001 Major fungal lineages are derived from lichen-symbiotic ancestors. *Nature* 411: 937-940

O'Donnell K 1997 Phylogenetic relationships among ascomycetous truffles and the true and false morels inferred from 18S and 28S rDNA sequence analyses. *Mycologia* 89(1):48-65

Perry BA et al., 2007 A phylogenetic overview of family *Pyrenomycetaceae. Mycol. Res.* 111(5) 549

Poggeler S and Wostemeyer Eds. 2011 *Evolution of Fungi and Fungal-Like Organisms* Vol XIV. Springer Verlag

Pontecorvo, G. and Roper, J. A. 1952. Genetic analysis without sexual reproduction by means of polyploidy in *Aspergillus nidulans. J. Gen. Microbiol*, 6: VII (Abstr.).

Sedin M et al., 2004 Saprotrophy and lichenization as options for the same fungal species in different substrata: environmental plasticity and fungal life styles in the *Stictis-Conotrema* complex. New Phytologist. 164:459-465

Schoch, C.L. et al., 2009 *Geoglossomycetes* cl. nov., *Geoglossales* ord. nov and taxa above class rank in the *Ascomycota* tree of life. *Persoonia* 22:129-138

Spatafora, J.W. et al., 2006 A five gene-phylogeny of Pezizomycotina. *Mycologia* 98 (6) 1018-1028.

Yarwood, C.E. 1973. Pyrenomycetes: Erysiphales. In: *The Fungi, An Advanced Treatise*. Vol. IV A. pp. 71–87 (G.C. Ainsworth, F.K. Sparrow and A.S. Sussman eds.). Academic Press, N.Y. and London.

SUMMARY

The subphylum *Pezizomycotina* comprises the highlest-evolved ascomycetes, forming the 'crown group' of *Ascomycota*. These are filamentous, and form ascomata (also by *Neolecta* of *Taphrinomycotina*). The earlier classifications based on the morphology of ascomata are obsolete in light of multigene phylogenetic analyses. There exist 11 clades i.e. monophyletic groups, which now form the classes. Studies on the evolution of these fungi, based on rDNA sequences reveal that apothecial ascomycetes are the initial members of *Pezizomycotina*. The perithecial and cleistothecial ascomycetes evolved from them by closing of the ascomata. Studies on ascomatal ontogeny reveal that ascohymenial taxa are ancestral and the ascolocular taxa with fissitunicate asci evolved from within ascohymenial taxa.

Molecular phylogenetic data also suggest that *Zygomycota* is the common ancestor of ascomycetes, which occupies a position immediately below the point of divergence of asco- and basidiomycetes. Thus, the yeast forms of *Taphrinomycotina* and *Saccharomycotina* appear to be result of independent loss of filmentous growth. In the two sister groups, *Saccharomycotina* and *Pezizomycotina*, *Saccharomycotina* have a simple mycelial and/ or yeast dimorphic growth, where as the *Pezizomycotina* evolved different types of ascomata.

The current classification (Eriksson 2006, Kirk *et al.,* 2008, Schoch *et al.,* 2009) is based largely on rDNA phylogenies. *Pezizomycoina* is divided into 11 classes: *Arthoniomycetes*, *Dothideomycetes, Eurotiomycetes, Geoglossomycetes, Laboulbeniomycetes Lecanoromycete, Leotiomycetes*, *Lichinomycetes*, *Orbiliomycetes, Pezizomycetes* and *Sordariomycetes.*

Orbiliomycetes and *Pezizomycetes* are the two **basal classes** of *Pezizomycotina*. Both taxa produce apothecia but differ in the asci. The remaing 9 classes with inoperculate asci, form a monophyletic group that is designstd as a superclass taxon, informally referred to as **Leotiomyceta**.

Lichens, like mycorrhiza, are a biological and not a systematic group. It is important to remember that 20% of all fungi and 40% of all ascomycetes are 'lichen-forming' or 'lichenizing' fungi. All the ascomycetous lichen-forming fungi belong to *Pezizomycotina*.

RPB2 pylogenetic studies indicate the lichen-forming fungi have arisen independently several times, losing and forming the lichen association. Majority of the lichenized fungi are found in *Arthoniomycetes, Eurotiomycetes, Lecanoromycetes* and *Lichinomycetes*. Only 4 orders are exclusively lichen-forming. However, while many families are exclusively lichen-forming, there are several families and even genera which have both lichen-forming and non-lichen-forming fungi (e. g. *Arthonia, Arthothelium, Myco-*

microthelia, Toninia). It indicates that 'lichen-mode of nutrition' is at various stages of evolution. Some are evolving towards stronger lichen-forming species, while some others towards less-strong lichen-forming species.

Class *Orbiliomycetes* (apothecia with inoperculate asci) has 1 order, *Orbiliales* and 1 family *Orbiliaceae.* The family is well known for its many nematophagus species. Several of the anamorphic genera are nematophagous form several types sticky or mechanical loops and branches as traps which help in capturing the prey (nematode) and devour it. *Dactylella bembicoides* (= *Orbilia bembicoides*) forms rings, whose cells swell when a nematode gets into the trap

Class *Pezizomycetes* (apothecial ascoma; operculate asci) has 1 order, *Pezizales*, and 16 families, containing "false morels (*Helvella*), 'true morel' (*Morchella*), 'truffle-like *Genea*, *Peziza*, 'true truffles (*Tuber*), which are edible and highly prized.

Dothideomycetes is basal to all other Leotiomyceta and is sister group to *Arthoniomycetes*. Similarly, *Leotiomycetes* and *Sordariomycetes* form another sister group.

Class *Arthoniomycetes* (1 order, *Arthoniales*, 4 families) comprise of lichen-forming fungi.

It is the 'sister group' to primarily non-lichenized *Dothideomycetes*. The presence of fissitunicate asci with ascohymenial development is unusual and suggests it to be an intermediate form between ascohymenial and ascolocular fungi.

Class *Dothideomycetes* (ascostroma; asci bitunicate, 2 subclasses) is basal to all other Leotiomyceta. It includes most ascolocular ascomycetes with fissitunicate asci.

It is the largest and most diverse class of ascomycetes. The class includes mostly non-lichenized genera, which are saprobes, coprophilous, parasites on plants. A few lichenized taxa, have been recently included. These are the *Arthopyreniaceae* and the *Trypetheliaceae* and the vegetatively-reproducing species *Cystocoleus ebeneus and Racodium rupestre*. The *Dothideomycetes* mostly produce perithecioid ascostroma. The two subclasses are: *Dothideomycetidae* and *Pleosporomycetidae*. The important genera are: *Mycospharella* and *Venturia*.

Class *Eurotiomycetes* form apothecia, perithecia, cleistothecia and ascostroma. It has 3 subclasses: *Chaetothyriomycetidae* (lichen-forming fungi, sooty molds, black yeasts, and *Verrucaria*) *Eurotiomycetidae* (*Aspergillus*- and *Penicillium*-related organisms, *Ajellomyces*, *Arthroderma*) and *Mycocaliciomycetidae* (parasites or commensals on lichens).

Class *Geoglossomycetes* (created in 2009) has 1 order *Geoglossales*, and 1 family *Geoglossaceae*. There are three genera: *Geoglossum, Sarcoleotia* and, *Trichoglossum* which form ascomata called the 'earth tongue'.

Class *Laboulbeniomycetes* (perithecia with prototunicate asci) are minute obligate ectoparasites of insects and other arthropods. These fungi with unique morphological features appear to be an independent lineage of *Ascomycota*. The thallus is not hyphal but cellular structures with definite morphology, under strcit developmental control. *Stigmatmyces* (monoecious) and *Amorphomyces* (dioecious) are the representative genera.

Class *Lecanoromycetes* includes the largest group of lichen-forming fungi. It is sister group to primarily perithecial *Sordariomycetes*. The class is divided into 3 subclasses: *Acaurosporomycetidae*, *Lecanoromycetidae* (*Lecanora*, *Peltigera, Lobaria*) and *Ostropomycitidae* (includes some 'secondarily de-lichenized' species, and show property of optional lichenization).

Class *Leotiomycetes* includes majority of non-lihenized apotheccioid ascomycota, with inoperculate unitunicate asci. The cleistothecial powdery mildew genera with prototunicate asci are also included in the class. The class is a sister group to the perithecial *Sordariomycetes*. There are 5 orders: *Cyttariales* (*Cyttaria*), *Erysiphales* (powdery mildews), *Helotiales*, (*Sclerotinia* and *Monilinia*), *Leotiales* (2 families) and *Rhytismatales* (3 families).

Lichinomycetes is an exclusive class of fungi lichenized with cyanobacteria. There is 1 order, *Lichinales* and 3 families, *Gloeoheppiaceae, Lichinaceae and Peltulaceae.*

Class *Sordariomycetes* is one of the largest classes in the *Ascomycota*. It includes most of the non-lichenized, perithecial ascomycetes, and also some cleistothecial forms. Asci are inoperculate unitunicate or prototunicate. Stroma is present in some (*Xylariales*). The class is phylogenetically related to *Leotiomycetes* as sister groups. Members of the class are ubiquitous and well known as plant pathogens (causing chestnut blight, Dutch elm disease, blast diseae of rice), endophytes (*Xylaria*), arthropod and mammalian pathogens (*Sporothrix, Trichoderma*), and as saprobes (*Neurospora, Chaetomium*). It has 3 subclasses: *Sordariomycetidae*, *Hypocreomycetidae* and *Xylariomycetidae.*

Incertae sedis orders of *Sordariomycetes*

Order *Meliolales (Meliola,* black or sooty mildews) shows resemblance to the powdery mildews. The members are parasitic on higher plants and grow on leaves. The superficial hyphae are dark-coloured (hence the name, 'black mildew'), and lie attached to the leaves by means of hyphopodia.

Order *Phyllachorales* form well-developed black stromata immersed in plant tissues. Perithecia are formed in the stroma. *Phyllachora* and *Glomerella* are the important genera. The latter is much studied organism in genetics.

TEST QUESTIONS

1. Why is *Pezizomycotina* called the 'crown grioup' of phylum *Ascomycota*? Give its present classification up to classes, and indicate the basal classes. Name the classes that form the two sister groups.
2. Distinguish between:

   (a) Asci of *Orbiliomycetes* and *Pezizomycetes*, (b) *Ceratocystis* and *Ophiostoma*, (c) Lichenizing and lichenicolous fungi, (d) *Stigmatomyces* and *Amorphomyces*, (e) *Ophiostoma ulmi* and *O. novoulmi*, (f) Blastomycosis and Histoplasmosis
3. Answer in brief:
   (a) Which is the common mycelial ancestor of asco- and basidiomycetes
   (b) Which type of asci (operculate of inoperculate) is foumd in classes under superclass Leotiomyceta?
   (c) Which classes are placed under Leotiomyceta?
   (d) Name three classes of lichen-forming (or lichenicolous) fungi.
   (e) Name any two nematogenous fungi from the class *Orbiliomycetes*.
   (f) Name the species of *Aspergillus* whose genome has been sequenced?
   (g) Which is the basal class of superclass Leotiomyceta
4. Write note on any **three**: (a) Ergot of commerce, (b) Aspergillosis, (c) Ringworms, (d) *Xylaria.*
5. Describe any **three** of the following genera giving their class and order: (a) *Ajellomyces*, (b) *Tuber*, (c) *Cyttaria*, (d) *Nectria,* (e) *Chaetomium.*
6. Describe any **two** of the following diseases: (a) Chestnut blight, (b) Dutch elm disease, and (c) Ergot of rye.
7. Describe some edible fungi from *Pezizomycetes*, giving their common names and family.
8. Name the more important powdery mildew genera. What are the characteristic features of conidia of powdery mildews? Describe the genera *Erysiphe* and *Podosphaera*.

9. Fill in the blank.
   - (a) Perithecia and cleistothecia have evoved from --------------
   - (b) Lichenicolous fungi grow on------------.
   - (c) The two basal classes of *Pezizomycotina* are ------------- and---------------.
   - (d) *Genea* is truffle-like, while ------------- is true truffle.
   - (e) Trained dogs and pigs are used for digging up --------------
   - (f) *Arthoniomycetes* and ------------------ form sister groups.
   - (g) *Verrucaria maura* is most abundant perithecial ------------.
   - (h) *Penicillium* is the anamorph of------------ and --------------.
   - (i) Fungi that grow on skin, hair, nails are called ------ fungi.
   - (j) Fungi of the classes of Leotiomyceta have ------------- asci (operclate/inoperculate)
   - (k) Members of class *Lichinomycetes* form lichen association with --------------.
   - (l) Members of order *Meliolales* are called ------------------ mildew.
   - (m) In *Ajellomyces*, the -------- phase is saprobic, while the -------- phase is parasitic.
   - (n) Ringworm of beard is called ---------.
   - (o) The ascomata of------------------ is called 'earth tongue'.
   - (p) The powdery genera ---------- and------------ are endophytic, while others are epiphytic.
   - (q) In *Monilinia*, the disease is spread by -------------- , and not by ----------------.
   - (r) In *Sclerotinia*, the ------------- play no role in disease causation.
   - (s) *Chaetomium* is spread by ------------.
   - (t) Powdery mildews are spread mostly by ---------------.
   - (u) *Neurospora* is called ---------- mold and --------- of the lab.
   - (v) *Cryphonectria* ----------- causes the chestnut blight.
   - (w) The anamorph of *Hypocrea rufa* is ------------------.
   - (x) *Xylaria hypoxylon* is also called as -------------------, while *Xylaria* ---------is called 'dead man's fingers'.
   - (y) The ascolocular ascomata have ---------------------- asci.
   - (z) The ---------- ascomata have evolved from ----------- ascomata. (use ascohymenuial and ascolocular).

13

Anamorphic Fungi

(Residual Fungi, Former Deueromycetes, Deuteromycotina, Fungi imperfecti, Mitosporic fungi)

The septate, mycelial fungi which reproduce exclusively by asexual means, usually by conidia (mitospores), and which are not the asexual stages of any ascomycete or basidiomycete, are called **anamorphic fungi**. Previously, these fungi were placed under class Deuteromycetes (Saccardo, 1866), and sub-division Deuteromycotina (Ainsworth, 1973), but their being an artificial assemblage was always known. They were never considered equal in status to sexually-reproducing fungi. They were called deuteromycetes (deuteron = secondary or second), or imperfect fungi, because they lack the sexual phase.

The genera (and species) of the imperfect fungi, which have been connected to any ascomycetous or basidiomycetous genera have been accordingly shifted to *Asco*- and *Basidiomycota* (Table 13.1). But for the majority (more than 95%) of the deuteromycetes, the perfect states are still not known. These unconnected "residual" fungi have either lost their sexuality, or adopted alternatives to sex, like parasexuality. They are not a monophyleic unit. The organisms which have no sexual cycle can nevertheless, have the status of a formal genus and species. *Aspergillus niger,* which has no sexual stage, is officially and formally designated as *A. niger*. But once the sexual state is discovered, the species will have to be designated as species of the sexual (teleomorph) genus. Thus, *A. nidulans* is called *Emericella nidulans* and not *A. nidulans*. This 'dual nomenclature' (separate names for different species of pleomorphic fungi) is permitted by 'rules of nomenclature'.

DNA sequencing now makes it possible to place these residual organisms within the groups of teleomorphic fungi (*Asco*-, or *Basidiomycota*) from which they are derived. By this comparative analysis of DNA sequences (molecular systematics) many asexual fungi have been placed in the existing classes.

Table 13.1 Anamorphic (imperfect) genera and their teleomorphs (perfect stage)

Anamorph	Teleomorph
A. Yeast genera	
Candida allbicans	*Saccharomycetales*
Cryptococcus neoformans	*Filobasidiella neoformans**
Sporobolomyces salmonicolor	*Aessospora salmonicolor*
B. Hyphomycetous genera	
Alternaria	*Lewia, Comoclathris, Cochliobolus*
Bipoloris sorokiniana	*Cochliobolus sativus*
Botrytis	*Botryotinia*
Cercospora	*Mycospharella, Sphaerulina*
Curvularia	*Cochlobiolus*
Drechslera	*Pyrenophora*
Fusarium	*Gibberella, Nectria, Calonectria*
Monilia	*Monilinia*
Nigrospora	*Khuskia*
Pyricularia	*Magnaporthe*
Rhizoctonia	*Thanetophorus**
C. Acervular genera	
Colletotrichum	*Glomerella*
Pestalotiopsis	*Pestalosphaeria*
Sphaceloma	*Elsinoe*
D. Pycnidial genera	
Ascochyta	*Didymella*
Diplodia	*Botoryosphaeria*
Phoma	*Didymella*
Phomopsis	*Diaporthe*
Phyllosticta	*Guignardia*
Septoria	*Mycospharella*

Sign * indicates *Basidiomycota*; rest *Ascomycota*.

CLASSIFICATION

The classification which has been commonly used for nearly a hundred years was proposed by **P.A. Saccardo** (1880-1886). He divided the 'fungi imperfecti' into three orders: *Moniliales* (or *Hyphomycetales*), for members which formed conidiophores or conidia directly on the hyphae, *Melanconiales*, for members forming conidia in saucer or disk-shaped structures, called acervuli (sing. acervulus), and *Spheropsidales*, for members producing conidia in flask-shaped receptacles, called pycnidia (sing. pycnidium).

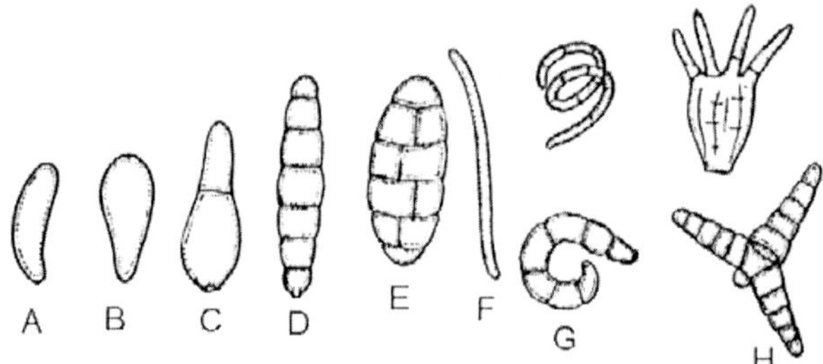

Fig. 13.1 Saccardo's spore types. A, B amerospore; C, didymospore; D, phragmospore, E, dictyospore; F, scoleospore (filiform); G, helicospore, H, staurospore (star-shaped).

Later, a fourth order '*Mycelia Sterilia*' was added. This included members which were sterile, (i.e. lacked conidia). The orders were divided into families and sections. Moniliales into four families: Melanconiales into one, and Sphaeropsidales into four families. Saccardo divided each family into 7 sections based on septation and shape of the mature conidia (Fig. 13.1). Remember that 'section' is not a recognized taxon but an arbitrary grouping. The seven sections are Amerosporae (spores 1-celled), Didymosporae (spores 2-celled), Phragmosporae (spores with 2 - or many celled with septa), Dictyosporae (muriform spores, i.e. possess both transverse and longitudinal septa), Staurosporae (star-shaped spores), Scolecosporae (filiform spores), Helicosporae (helical spores). Each section was further divided into two groups: Micro-nemeae (conidiophores-short and inconspicuous) and Macro-nemeae (distinct conidiophores). The genera and species were identified by the colour (hyalo -, i.e., hyaline or bright, or phaeospores i.e., pigmented and dark), shape and septation of their conidia.

Ainsworth (1973) divided the above fungi into 3 classes: Blastomycetes, Hyphomycetes, and Coelomycetes. The class Blastomycetes included the 'imperfect' yeasts, while the Hyphomycetes included genera with sterile hyphae (mycelia sterilia) or bearing conidia on the hyphae. In Coelomycetes, the spores were formed in pycnidia and acervuli (i.e. conidiomata, sing. conidioma). The conidiomata are specialized conidia-bearing structures and includes acervulus, pycnidium, sporodochium and synnema.

Classification Based on Conidial Origin (Hughesian classification)

Vuillemin, a French mycologist (1910, 1911) mooted out conidial development as the basis of classification of these fungi. He studied the various modes of origin of conidia and stressed that the conidia should be distinguished by their origin and not by their morphology.

S.J. Hughes 1953 proposed a scheme of classification for hyphomycetes based on conidial ontogeny i.e. their origin. He divided these fungi into 8 sections, each representing one particular mode of conidial development. The developmental morphology as the taxonomic criterion attracted the leading hyphomycetologists like E.W. Mason, C.V. Subramanian, K. Tubaki, M.B. Ellis, E.S. Luttrell, W.B. Kendrick, S. Nilson and G.L. Barron. The system of classification based on conidial origin is designated as the 'Hughesian' classification. The *International Conference on Criteria and Terminology in the Classification of Fungi Imperfecti*, held at Kanaraskis, Canada, in 1969 endorsed that Saccardoan classification must be replaced by a system based on conidial ontogeny.

Origin of spore types The five types of spores and their ontogeny is described below (Fig. 13.2).

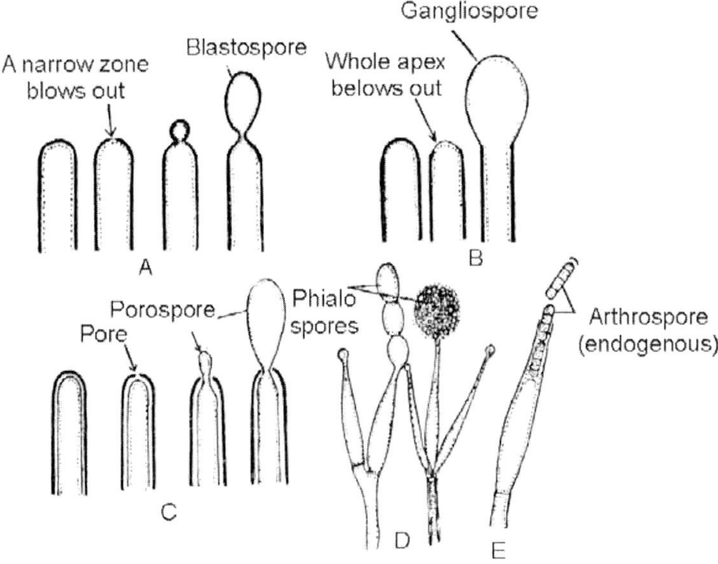

Fig. 13.2 Conidial ontogeny in Hyphomycetes. A. blastospore; B. gangliospore; C. porospore; D. phialospore (slimy phialospore form a ball; E. arthrospore (endogenous).

Blastospores A blastospore is formed by simple budding as in yeasts. A narrow zone of the conidiogenous cell blows out, enlarges and develops into a conidium. A constriction is conspicuous between the conidium and the conidiogenous cell, before separation. Typical blastospores are formed in *Torula*, *Cladosporium*, etc.

Porospore A pore is enzymatically dissolved in the outer layer of the wall, at the tip of the conidiophore. The inner wall blows out through this pore like a bud and develops into the conidium. Electron microphotographs have confirmed this mode of development in *Alternaria, Helminthosporium, Curvularia, Alternaria*, etc.

Phialospores The **phialide** is a unicellular structure, oval, sub-cylindrical or flask-shaped, with a distinct basal swelling and a narrow distal neck. It is borne terminally on simple or branched conidiophores. The neck may show annellations left by detachment of successive conidia. From the tip of the phialide arise, exo- or endogenously, two types of conidia — **dry spores**, which persist as chains (e.g. *Aspergillus, Penicillium*) or **slimy solitary spores**, which occasionally adhere in a slimy ball, (e.g. *Trichoderma, Stachybotrys*, etc.). Phialosopores are thin-walled. The species which form persistant chains of dry spores are clearly different in their mode of development from those that form solitary slimy spores.

Gangliospores The term gangliospore refers to a conidium developed by the transformation of the swollen tip of a hypha into a spore. A conidium initial may or may not be delimited. Blastospores and gangliospores are basically similar. Blowing out of a limited zone, less than the width of the conidiophore apex, gives rise to a blastospore, whereas, in gangliospore, the whole apex blows out. Consequently, while a constriction appears between the blastospore and the conidiophore, no such constriction forms in gangliospore development. Nevertheless, the demarcating line between a blastospore and a gangliospore is narrow and it becomes difficult to distinguish between the two types when the conidiophore tip is thin.

Arthrospores The term arthrospore has been used for spores resulting from the breaking up of a hypha into cells. Obviously, there is no conidiogenous cell, but only a conidiogenous hypha of determinate growth. Conidial initiation begins only after growth of the hypha has ceased.

Present Classification

Kendrick and Carmichael (1973); Hawksworth *et al.* (1995) and Kirk *et al.* (2008) discarded the traditional hierarchical divisions into orders and families as it is meaningless in terms of natural or phylogenitic classification. This is now universally accepted and we shall study the genera only in the alphabetical sequence under the following heads: **Imperfect yeasts, Hyphomycetous Anamorphic Fungi**, and **Acervular-**, and **Pycnidial Anamorphic Fungi**.

Three categories of information employed for description of an asexual fungus are: conidiomatal types (1-9 listed in Table 13.2), Saccardo's groups and conidiogenus events, (which includes 43 types of conidiogenesis – conidial outogeny).

Table 13.2(a) Conidiomatal types and conidial spore groups used in description of mitosporic fungi.

	Conidiomata
1	hyphal
2	synnematal
3	sporodochial
4	pycnidial
5	pycnothyrial
6	acervular
7	cupulate
8	stromatic (complex conidiomata excluded from other categories)
9	sclerotia

Table 13.2(b) Conidial shape and septation (Saccardo's groups)

			1 Conidia hyaline or bright (hyalo-)	**2** Conidia pigmented or dark (phaeo-)
A	1-celled (not E, F or G)	amerosporae	Hyalosporae	Phaeosporae
B	2-celled	didymosporae	Hyalodidymae	Phaeodidymae
C	2-many-celled (multiseptate)	phragmosporae	Hyalophragmae	Phaeophragmae
D	muriform	dictyosporae	Hyalodictyae	Phaeodictyae
E	filiform	scolecosporae		
F	helical	helicosporae		
G	branched	staurosporae		

IMPERFECT YEASTS (Former Blastomycetes)

The 'Imperfect' yeasts exist as budding cells or as pseudomycelium by end-to-end attachment of cells, due to incomplete wall separation. The bud cells may be discharged explosively, like ballistospores of some basidiomycetes, or detached passively.

Genus *Candida*

The genus, having 355 species, is anamorph of *Saccharomycetales*. The conidial origin is not confirmed. The species are based on physiological characteristics. The yeast phase, comprising budding cells, is dominant over the pseudomycelial phase, which is formed due to inadequate nutrition or other adverse factors. Only few species, under rare circumstances, from true mycelium. The mode of life is correlated with the growth phase. When the fungus grows as a commensal, it exists in the yeast phase, and when it is pathogenic, it becomes pseudomycelial.

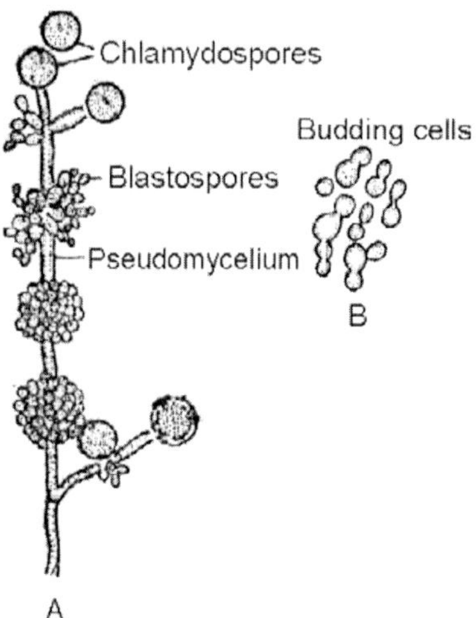

Fig. 13.3 *Candida albicans*. A, pseudomycelium bearing blastospores and chlamydospores; B, 'yeast' phase comprising of budding cells.

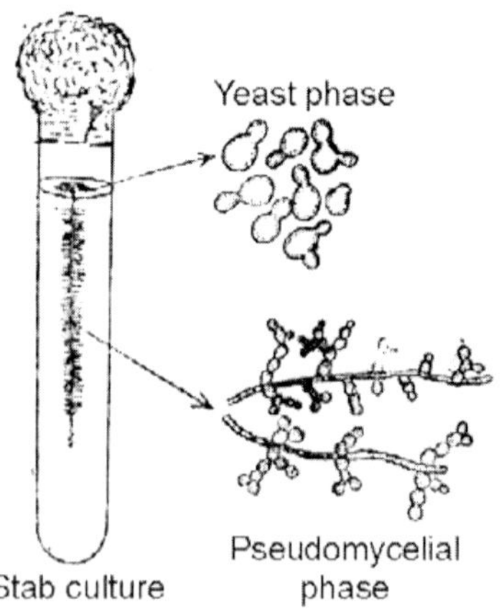

Fig. 13.4 Stab-culture of *Candida albicans*

In laboratory cultures, it forms white or creamy, opaque paste like colonies which consist of sub-globose to oval budding cells (3.6 μm × 3.5–

12.5 μm). The cells frequently show *bipolar* or *tetrapolar* budding i.e. buds arise from two or four places. As the colony becomes older, it forms pseudomycelium or branched true mycelium, bearing blastospores and chlamydospores (Fig. 13.3), which is a characteristic feature of this species. The mycelium can be visibly demonstrated by a "stab culture" (Fig. 13.4) which is also used as a method for differentiating a pathogenic species of *Canidida* from non-pathogenic true yeasts.

Candidiasis

The infection by *Candida albicans* is called candidiasis, or 'thrush'. It was the first fungus reported to cause disease in man (Langeubeck, 1839). **Superficial candidiasis,** affecting the skin and mucous membranes, is more common than **visceral or systemic candidiasis**. In 'oral thrush' (a superficial candidiasis), the mucous membrane of mouth, tongue and palate, develop soft white or creamy, curd-like patches. The white material consists of pseudomycelium and blastospores of *C. albicans*, mixed with dead epithecial cells, leucocytes and particles of food. Similar infections of the oesophagus and the intestines are called 'oesophagal thrush'. Other examples of superficial candidiasis include ulcers on the cornea of eyes, maceration of finger clefts (more common in persons engaged in washing work), infections of the sub-mammary folds, especially in obese women, infections of genitocrural folds or perianal skin. Systemic candidiasis involves infection of the respiratory, circulatory, urinary and the central nervous system. Like many other fungal diseases, candidiasis also is common in immunocompromized patients.

Genus *Cryptococcus*

Cryptococcus, which has 75 species, exists exclusively in the yeast phase. The conidial ontogeny is holoblastic. All the species assimilate inositol and produce the enzyme urease. The cells are hyaline, capsulated and produce mucus. *Cryptococcus* is polyphyletic and shows basidiomycetous affinity. For example *C. neoformans* is anamorph of *Filobasidiella* (*Filobasidiella neoformans*) family *Filobasidiaceae* of the order *Tremellales*. *C. neoformans* (Fig. 13.5) is pathogenic to human beings, causing **cryptococcosis**. It grows on bird excreta, which supplies the specific requirement of nitrogen in form of creatinine. The fungus forms glistening, mucoid, yellow-brown colonies, consisting of round, capsulated cells, 4 mm – 6 mm in diameter, excluding the capsule. The capsules are best visible in Indian ink. The daughter cells which appear as buds, are surrounded by the capsule. The varietal differentiation of *C. neoformans* is done by D-proline assimilation.

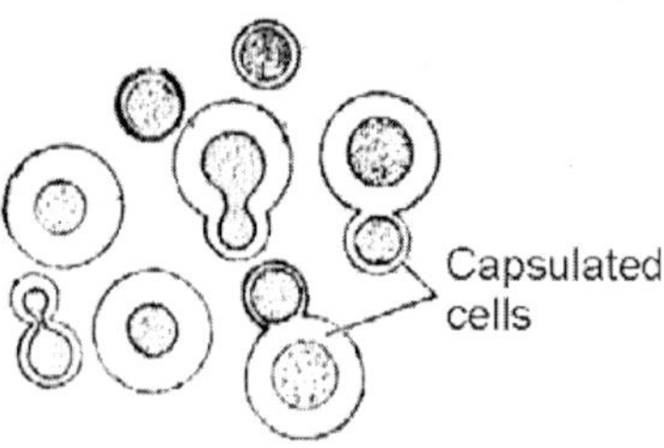

Fig. 13.5 *Cryptococcus neoformans*. Round capsulated cells multiplying by budding

Cryptococcosis *Cryptococcus* can cause infections in any part of the body, but is most dangerous when it infects the central nervous system, causing cryptococcal meningitis, which is usually fatal. Workers involved in cleaning places contaminated with bird droppings may contract *Crytococcus* infections, which first appears in lungs and later encroaches on and paralyses the central nervous system. Similar infections are reported in animals also.

Genus *Sporobolomyces* (mirror picture yeasts)

Set up a monument to him
Say on a stab of schist;
He was an ardent **Spor-obol-**
O-my-cet-o-log-ist!

This genus represents ballistosporous yeasts and called "mirror picture yeasts" or "shadow yeasts". It is anamorph of genus *Aessospora* (*Sporidiobolales; Microobotryomycetes, Pucciniomycotina*). The conidial ontogeny is holoblastic.

A.H.R. Buller

The genus has 25 species that grow saprobically on ripe fruits, senescent leaves, and rust-infected leaves. If a senescent grass leaf is glued to the inner surface of the lid of a Petri dish, containing nutrient agar, in a few days tiny "yeast" colonies appear on the agar forming a mirror image of the leaf outline and its veins. The *Sporobolomyces* species, hence, are called "mirror-picture yeasts". *S. salmonicolor* (= *Aessospora salmonicolor*) (Fig. 13.6) frequently comes in culture if a dead leaf is placed on an agar medium. Yeast colonies consisting of budding cells develop in about 3 days, and after 6 – 7 days, ballistospores start shooting out.

Sporobolomyces forms oval cells, which secrete mucilage in copious amounts and contain pigments that impart colour to the colonies. In the initial stages, the cells multiply by budding. Later, the peripheral cells of the colony form ballistospores on sterigmata. Each cell bears 1–2 (rarely 3) kidney-shaped ballistospores with distinct hilar appendix, borne asymmetrically on pointed sterigmata which are occasionally forked. In old cultures, the fungus forms pseudomycelium (short, branched hyphae) bearing ballistospores.

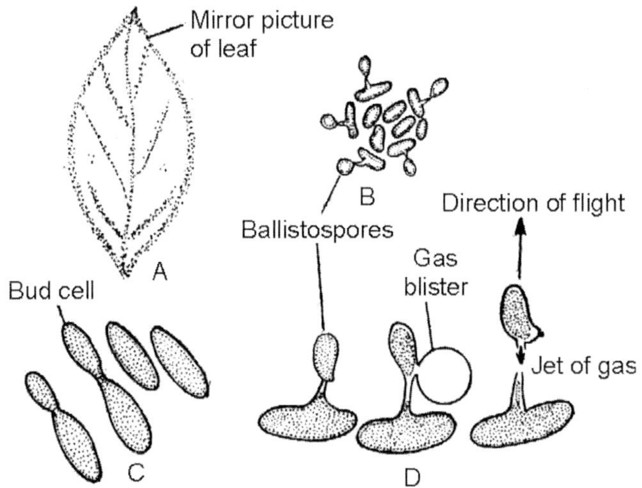

Fig. 13.6. *Sporobolomyces salmonicolor.*, A, mirror image of a leaf by basidiospore discharge in a Petri-dish; B, magnified view of a portion of one yeast colony; C, budding; D, ballistospore formation and shooting.

The violent discharge of the basidiospores intrigued Buller. It has been studied in detail. A gas filled bubble, or water bubble (not water drop), appears on the hilar appendix. The gas pressure built inside the bubble, jerks off the spore from the sterigma. At the moment of the break off, the gas erupts out from near the hilum, like a jet, and propels the spore at an outward angle from the axis of the spore.

THE HYPHOMYCETOUS ANAMORPHIC FUNGI (Former Hyphomycetes)

In this, we shall examine 'imperfect' fungi that form spores or sporophores directly on the hyphae, free or aggregated (synnema or sporodochia). It is a large group thriving on a variety of substrata, mostly terrestrial. Some aquatic forms are also known. Interesting predacious members have been described by Drechsler (1941).

Genus *Alternaria*

The genus (Fig.13.7) stands distinct from other genera by its transversely, and longitudinally-septate (muriform) conidia. The conidia show a distinct beak which may be short or very long (equal to the conidia). The conidia are usually formed in chains. These are dark brown, typically obclavate, with a distinct oval body and a beak. The beak is light in colour, which may be short or long, and serve as conidiophore. The conidiophore is dark, broader than the hyphae and is a sympodula*, bearing geniculations marked with scars of detached conidia.

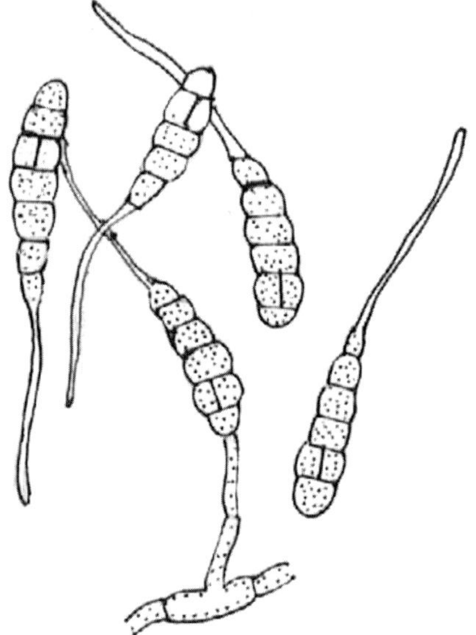

Fig. 13.7 *Alternaria solani*

The species are placed in three sections on the basis of the size of conidial chain or absence of the chain. The sections are **Longicatenatae**, with 10 or more conidia in the chain. e.g. *A. tenuis*; **Brevicatenatae,** with shorter chain of 3-5 conidia, e.g. *A. citri*, and **Noncatenatae**, solitary conidia with long, filiform beak, e.g. *A. solani*.

---

*Sympodula describes a conidiophore which produces a single terminal conidium, then bends just below the conidium to produce a very short branch on which a second spore is formed, and as a result of a terminal position of the conidiophore looks like a highly compressed sympodium.

Most of the species are parasitic causing leaf-spot diseases. *A. solani* causes "early-blight" of potato and other members of family Solanaceae. The 'target-board'–like leaf spots show distinct concentric bands of light and dark colours. The perfect stages are *Lewia* (*Pleosporaceae, Pleospoorales*).

Genus *Botrytis*
(Gr. botrys = a bunch of grapes)

The conidia look like grape bunches (Fig. 13.8) on the conidiophores, which are brown, erect, simple, variable in length, and become hyaline and irregularly-branched towards the apex. The ultimate branches bear the densely-aggregated conidia; each conidium being borne on a minute sterigma. The conidia are unicellular, oval, apiculate at base, dry and easily blown away. In culture, sclerotia appear after a few days, which are irregular in shape and serve to perennate the fungus.

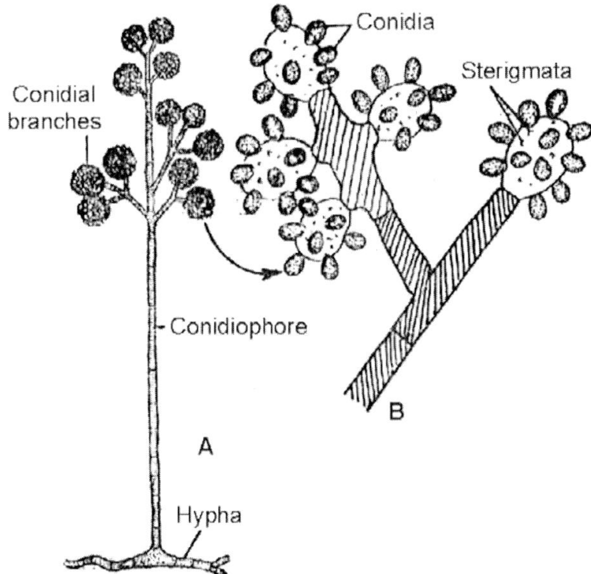

Fig. 13.8 *Botrytis cinerea*. A. whole conidiophore bearing conidia like grape bunches; B, a lateral branch magnified showing conidia borne on sterigmata.

There are 54 species, which are anamorphs of *Botryotinia* (*Sclerotiniaceae*). *B. cinerea*, the "grey-mold", is the best known species. It commonly attacks strawberries, covering it with a gray moldy mass.

B. cinerea was first described by **de Bary** from vine leaves. He studied the parasitism of this fungus and initiated Physiological Plant Pathology. He showed that the fungus killed host cells ahead of the growing hyphae. **Ward**

(1888) elucidated the production of enzymes by this fungus in 'lilies', which caused swelling and disintegration of host cells. Later, **Brown** (1915 – 1922) used this organism in his studies on the physiology of parasitism and described the production of enzymes **pectinase** and **protopectinase** in host, during disease development, which brought about tissue disintegration.

Genus *Cercospora*

The genus, with 1416 species, is anamorph of *Mycosphaerella* (*Mycosphaerellaceae*). It is characterized by long cylindrical obclavate, multi-septate, hyaline or brownish conidia (Fig. 13.9). The conidia vary in size but always have a breadth, length ratio of 1:10–1:150. The species cause leaf-spot diseases of several important crops.

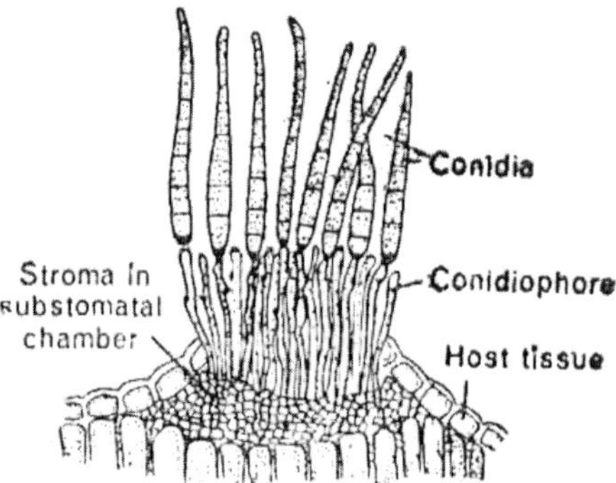

Fig. 13.9 *Cercospora* sp. Conidiophores bearing long, septate conidia arising from stroma formed in substomatal chamber of leaf.

The hyphae, when young, are hyaline but later turn brown. Some of the hyphae aggregate in the stomatal cavity and form a stroma. Conidiophores arise in tufts from the stroma and emerge through the stromata or ruputured epidermis. Conidiophores are hyaline to dark brown, septate or non-septate, sometimes branched, straight or flexuous and show distinct geniculate (knee-like) bends. These form single conidia acrogenously i.e. at the tip. The mature conidium is pushed aside due to sub-apical growth of the conidio-phore. This results in geniculations bearing scars of the detached conidia.

The species are more common in the tropics, causing leaf-spot diseases. Some important species are: *C. arachidicola* and *C. personata*, both causing leaf-spot (tikka-disease) of groundnut. Their perfect stages, respectively, are *Mycosphaerella arachidicola* and *M. berkeleyii*. *C. musae* (= *Mycorph-aerella musae*) causes the destructive, "sigatoka" leaf-spot disease of banana; *C. necotinae* (*M. necotinae*) causes 'frog-eye' leaf-spot of tobacco.

Species of *Cercospora* do not readily grow in culture and if they grow, usually don't sporulate.

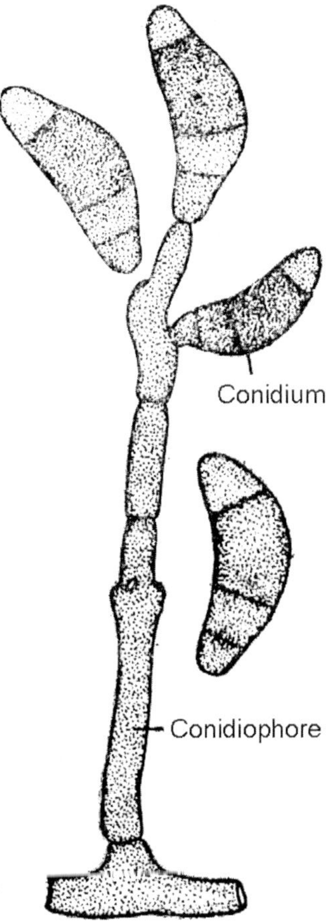

Fig. 13.10 *Curvularia lunata*. Conidia curved about the third cell from the base which is also larger and darker.

Genus *Curvularia*

The genus is an anamorph of *Cochliobolus* (*Pleosporaceae*). Both conidia and the hyphae are dark-brown (Fig. 13.10). The conidia are 3-4 septate and mostly curved about the third cell from the base (hence the name *Curvularia*), which is also much larger and darker in colour than rest of the cells. However, the conidia may be straight also. *C. lunata*, the type species, is crescent (half-moon) shaped. The conidiophore forms a sympodula bearing a cluster of conidia, which may be compact or lax. The detached conidium leaves a knob-like scar on the conidiophore on the 'bends' or geniculations, formed due to the characteristic growth as a sympodula. Perfect stage of *C. lunata* is *Cochliobolus lunata*.

There are 54 species of *Curvularia* which are common as leaf-spot fungi; some occur in the soil. Some are active degraders of cellulose and damage textile. The species are segregated into three groups on the basis of septation and shape. These are *Maculans*, *Lunata* and *Geniculata*, as described below.

Conidia 3 septate

Conidia mostly straight or feebly-curved with the central two cells larger and darker than the end cells, broadest in the middle. **Maculans group**

Conidia distinctly curved, rarely straight with one of the central cells larger and darker than the other cells, broadest above middle. **Lunata group**

Conidia 4 or more-septate **Geniculata group**

Genus *Fusarium*

The genus is an important plant pathogen causing wilt and root-rot diseases of several economically important plants, especially in tropical countries. The hypahe are hyaline (Fig. 13.11), branched and septate. Conidia are formed in slimy, effuse sporodochia, called **pionnotes**, or sometimes, scattered on the mycelium. Conidiophores are quite short, simple or branched, septate, bearing a terminal phialide. The phialides are **subulate** i.e. widest at the base, narrowing to a point like an 'awl'. The conidia, which are slimy, are of two types: **microconidia** (one-celled, oval or comma-shaped, pyriform or elongate) and **macroconidia** (which are hyaline or pale, spindle-shaped with pointed ends; hence the name *fusus–spindle*), 3 to many septate, with a distinct pedicellate base. The macroconidia with a distinct "foot cell", bearing some kind of 'heel', is the most distinct feature of *Fusarium*. Chlamydospores, if present, are formed on the hyphae or on the macroconidia. Microconidia may be present or lacking. The conidia are formed in succession and become enveloped in a slimy mass. In addition to conidia and chlamydospores, some species form brightly coloured sclerotia. The perfect stages of the various species belong to *Gibberella* and *Haemonectria*. *Fusarium* is a problematic genus in identification due to its great variability.

There are two main classifications of the species: one by **Wollenweber and Reinking** (1935), and other by **Snyder and Hansen** (1940). Wollenweber and Reinking grouped the 1000 or more isolates into 16 *sections* (or groups), 6 sub-groups and 142 species. Snyder and Hansen simplified it and designated all the species belonging to the section *Elegans* of Wollenweber and Reinking as one species, *F. oxysporum*. The erstwhile species, which were merged under *F. oxysporum* were all vascular wilt pathogens. These were distinguished by host specificity and designated as formae speciales in a trinomial nomenclature e.g. *F. oxysporum* f. sp. *lycopersici* (causing wilt of tomato); *F. oxysporum* f.sp. *vasinfectum* (on cotton), *F. oxysporum* f. sp. *cubense* (on banana) etc. In the same way, the species placed under the section *Martiella* in Wollenweber's classification were merged and named as *F. solani*.

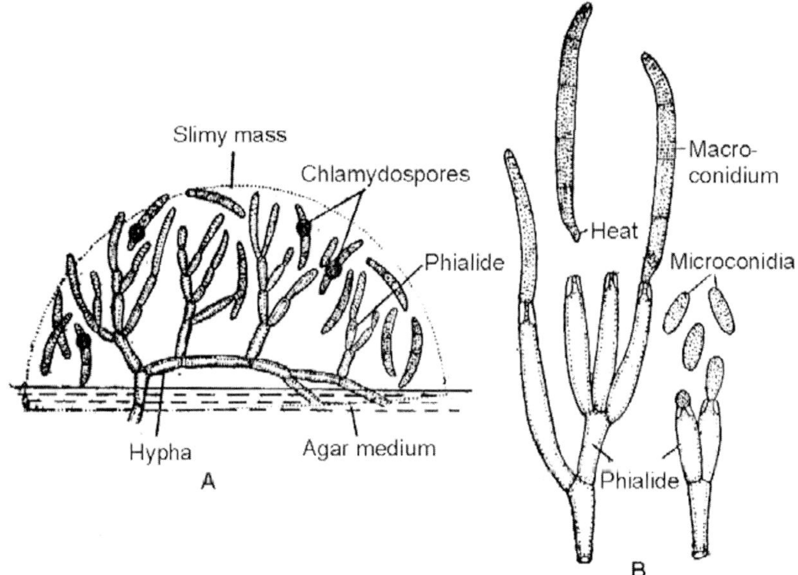

Fig. 13.11. *Fusarium* sp. A, conidiophores producing macro-and microconidia in a slimy mass; B, conidia (= phialospores) phialides and conidiophore.

Later, **Synder**, and **Hansen** (1941, 1945) and Snyder (1965) reduced the 145 species into nine species. The "nine-species concept", as it is called, has been adapted by **Toussoun** and **Nelson** (1968) in their monograph of this genus. But it is by no means universally accepted. **Subramanian** (1971), and also **Booth** (1971), basically still accept the sections of Wollenweber and Reinking.

Genus *Helminthosporium*

(Gr. helminthos – a worm; hence worm-like spores)

The hyphae are typically dark, bearing conidia on simple, erect, septate conidiophores. The conidia are obclavate, brown, thick-walled with 8-10 transverse septa and a basal scar, indicating the point of attachment on the conidiophore (Fig. 13.12). The conidia are produced singly from a pore at the tip and on lateral sides in the upper part of the conidiophore; often forming verticels (whorls) above or below the septa. *H. velutinum*, the type species, was reported from wood.

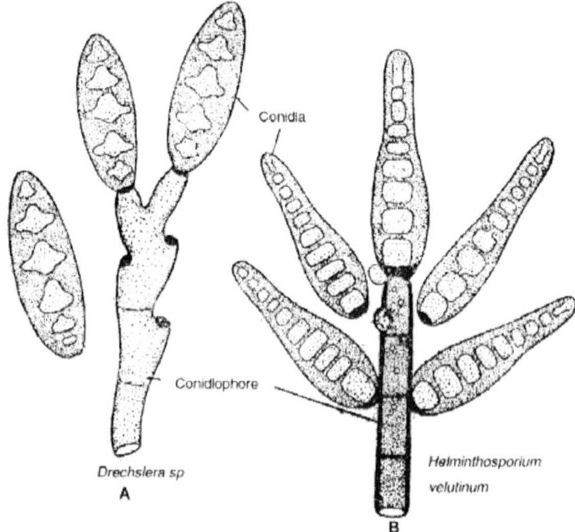

Fig. 13.12 A, *Drechslera* sp; B, *Helminthosporium velutinum*

The genus was an assemblage of heterogenous species. It also included plant parasites which differed from the lignicolous 'type species' in the method of spore production. The parasitic species were segregated and placed in two new genera *Bipolaris and Drechslera. Helminthosporium* included only the lignicolous species, resembling *H. velutinum* in producing terminal and lateral conidia. *Bipolaris* and *Drechslera* were characterized by formation of clusters of conidia on the conidiophore i.e. a typical sympodula. The genera *Bipolaris* and *Drechslera* differed in their mode of germination; the former germinated only by the two end cells (hence the name *Bipolaris*), while the latter germinated from all the cells. The genus *Exerohilum* is another offshoot of *Helminthosporium*, with *Septosphaeria* as the teleomorph.

Perfect stages of several species have been discovered, which suggest that these fungi are the asexual stages of *Pleosporaceae*. *Bipolaris* is anamorph of *Cochliobolus* and some of the species are being designated by their perfect stage viz. *C. sativus*. *Drechslera* is the anamorph of *Pyrenophora*.

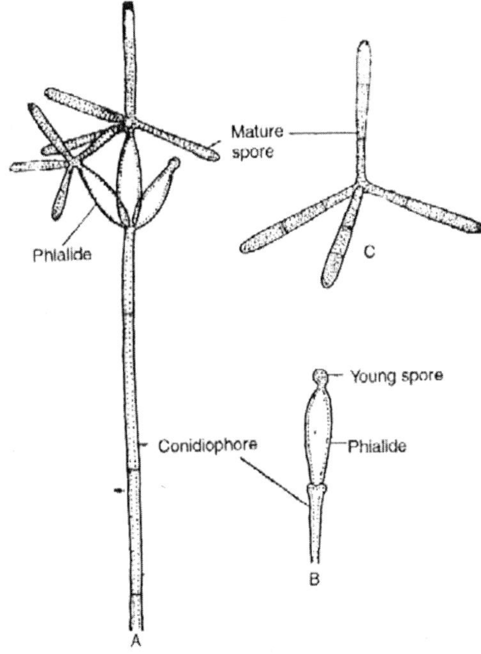

Fig. 13.13. *Lemonniera aquatica*, an aquatic hyphomycete. A, conidiophore with three phialides each bearing a spore at a different stage of maturity; B, developing spore; C, mature spore.

Genus *Lemonniera*

Lemonniera is one of the several interesting 'aquatic hyphomycetes' described by Ingold (Fig. 13.13) in 1942. The most striking feature of these fungi is the hyaline and usually tetra-radiate nature of their conidia. There are 6 species that are found on decaying leaves in water.

C.T. Ingold

The branched, septate hyphae ramify inside the leaf tissues, especially in the petiole and larger veins. The somatic hyphae, emerge into water and form conidia. In *L. aquatica*, the conidiophore is branched apically and bears phialides. The conidium starts as a small spherical swelling, which soon becomes tetrahedral with long septate arms. When the mature conidium is detached, another begins to develop.

Genus *Monilia*
(Lat. *monile*, a necklace of beads)

The genus (Fig. 13.14) is characterized by long, branched chains of bead-like, oval or lemon-shaped conidia which are often united by isthmus. These are produced in basipetalous succession (i.e. the youngest spore lying at the outermost end of the chain) and easily break off from the chain. The conidiophores arise laterally or terminally from the hyphae and are variously-branched.

The teleomorphs belongs to *Monilinia* (*Sclerotiniaceae*). *M. fructigena*, *M. fruticola* and *M. laxa* are the asexual conidial stages of *Monilinia*, a well-known pathogen, causing brown-rot of several tree fruits.

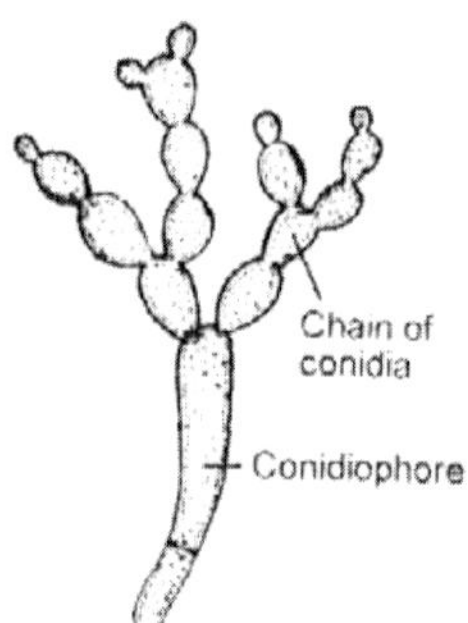

Fig. 13.14. *Monilia fructigena* (conidial stage of *Monilinia*)

Genus *Nigrospora*

The genus, which has 6 species, is easy to identify. The black, one–celled globular conidia, which may be flattened or slightly bulged apically, lie studded on the hyphae in enormous numbers (Fig. 13.15). Mature conidia completely obstruct the short conidiophore from view.

The hyphae are septate, branched, hyaline, later dark-brown. The conidiophores are typically small, ampulliform (flask-shaped) forming a single conidium at the tip. Three species are reported from India causing leaf-spots, or growing saprobically on plant debris or soil. *Khuskia*, belonging to *Trichosphaeriaceae* (*Sordariomycetes*), is the perfect stage of *N. oryzae*.

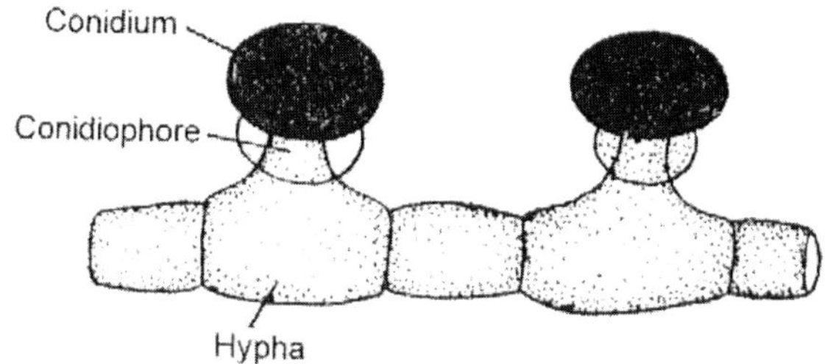

Fig. 13.15 *Nigrospora oryzae*, jet black conidia formed singly on ampuliform conidiophore.

Genus *Pyricularia*

Pyricularia forms hyaline or pale-gray, bi-celled, pyriform conidia (with a round base and basal papilla) singly at the tip of the conidiophore and its successive growing points. The conidiophores are simple, erect, septate and hyaline or gray.

There are 4 species with *Magnaporthe griseae* (*Magnaporthaceae, Sordariomycetes*) as the teleomarph of *P. oryzae* (= *P. griseae*). It causes the destructive 'blast' disease of rice (Fig. 13.16). Spindle-shaped spots, which are ash-coloured in the center, appear on all parts (leaves, culms, glumes, stalks of the panicle). The stalks are rotted (neck-rot) and the ears bent downwards. Diseased plants form poorly-filled grains.

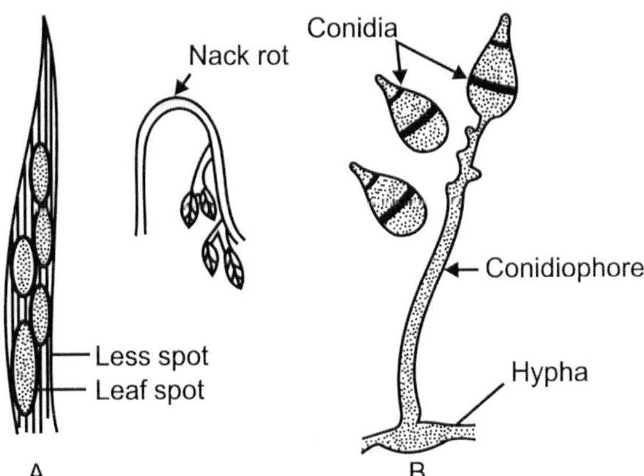

Fig. 13.16 *Pyricularia* (= *Magnaporthe*) *oryzae.* A,'blast' symptom on paddy B, conidia and conidiophore.

The conidia, disseminated by wind, germinate on rice plants. The germ tubes enter through the stomata and establish the septate, much branched mycelium. Conidiophores emerge through stomata and form conidia. The fungus also grows on several collateral hosts and is reported to perennate in the rice seeds.

Genus *Rhizoctonia*

There are some imperfect fungi which do not form even asexual spores, and are, therefore, called mycelia sterilia. But they do form sclerotia, which perpetuate and disseminate the fungus. There are two genera: *Rhizoctonia* and *Sclerotium* which include several plant pathogenic species.

Rhizoctonia is commonly found in soils, causing root-rot and "damping off" disease of several economically important plants. The sclerotia are variable in shape and lack anatomical differentiation. *R. solani* (Fig. 13.17) is the most important pathogenic species which also grows as a vigorous saprobe in soil. It is a strong cellulose decomposer and, during the saprobic growth in soil, derives its nutrition from cellulose. The hyphae, which are initially hyaline, but later brown, are characteristically branched. The branches arise at right angles (90°) from below the septa and show distinct constrictions at the point of origin. The sclerotia are formed in great numbers on the surface of infected roots and are clearly visible.

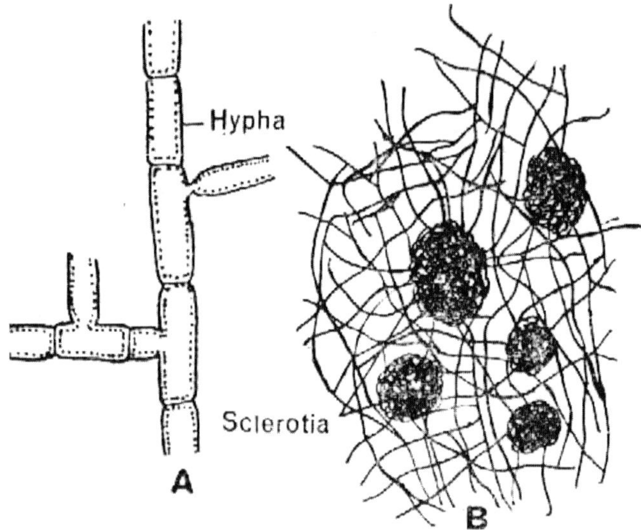

Fig. 13.17 *Rhizoctonia solani*. A, the hyphae showing perpendicular branching below septa; B, sclerotia.

The perfect stage of *R. solani* is *Thanatephorus cucumeris*, which belongs to *Ceratobasidiaceae* (*Basidiomycota*).

Genus *Trichothecium*

Trichothecium (Fig. 13.18) forms conidia in basipetalous succession due to sympodial growth of the conidiophore and repeated formation of spores. The conidia form a chain. Each conidium is so arranged that its long axis is horizontal with respect to the chain and the alternate spores are placed in opposite directions. The conidia are attached to one another through slight thickenings on each side of the base.

There are 5 widespread species. *T. roseum*, the type species, causes pink rot of apple, brinjal, banana, etc. The colonies are initially white, but later turn pink, hence named *roseum*. The hyphae are septate and highly branched. Conidiophores are erect, unseptate or sparsely-septate, hyaline and bear cluster of conidia at the apex, which is broader than the remainder portion. Conidia are acrogenous (i.e. formed terminally), bi-celled, ovate with a small projection at the point of attachment.

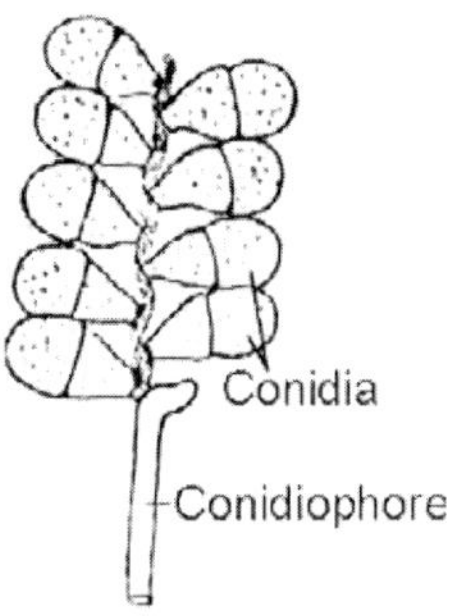

Fig. 13.18. *Trichothecium roseum*

Genus *Verticillium*

Verticillium is an important vascular wilt pathogen, but, unlike *Fusarium*, it is more prevalent in temperate countries. The genus is characterized by verticillate branching of the conidiophores i.e. the branches arise in whorls (*verticillus* = a whorl of branches). The phialides are also formed in whorls on the upper portions of the conidiophore, below the septa.

The genus has 51 species, comprising both saprobic and plant pathogenic species. *V. albo-altrum* and V. *dahliae* are the two wilt-causing species, that attack a large number of plants (Fig. 13.19). *V. theobromae* causes the "cigar-end" disease of banana. The pathogenic species lack saprobic ability and survive in soil through their perennating structures in host debris. The two less important pathogenic species are *V. nigriscens* and *V. tricorpus* (reported only from U.K.). The pathogenic species are identified by the characteristics of their resting bodies: *V. albo-atrum* forms **resting mycelium**, *V. dahliae* **microsclerotia**; *V. nigriscens,* **chlamydospores**, while *V. tricorpus* forms all the three. *V. intertextum* and *V. lateritium* are saprobic species.

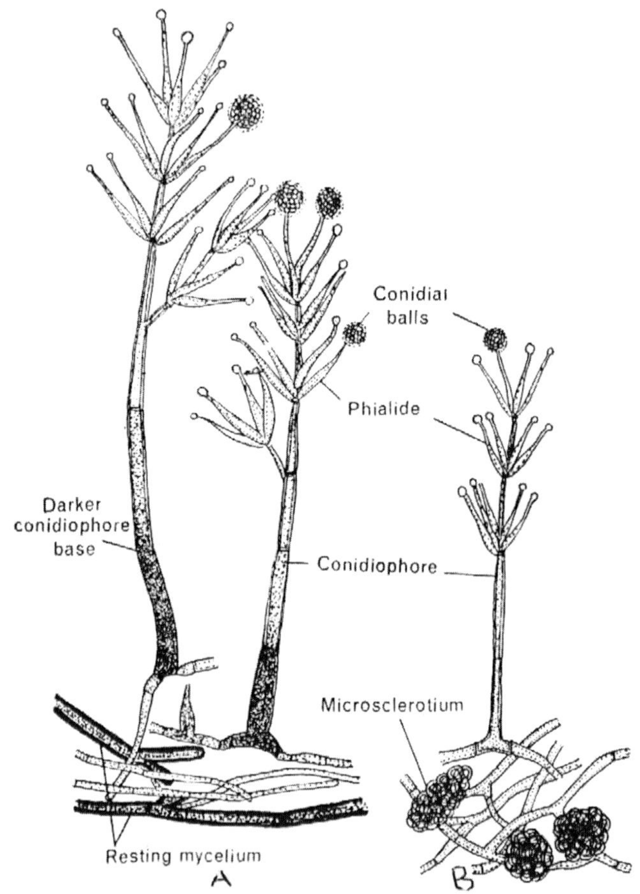

Fig. 13.19 *Verticillium. V. albo-atrum* (note the thickened dark base of conidiophore and the resting mycelium); *V. dahliae* characterized by the microsclerotia.

In *V. albo-atrum* the hyphae are hyaline in the beginning, but later turn black; and hence the name: (Lat. *albus*, white; *ater*, black). It is more virulent than *V. dahliae*, especially in temperate countries. The conidiophores are simpler than other species: unbranched, with swollen and coloured basal portion. The verticels of phialides are restricted to the upper hyaline portion of the conidiophore. It forms resting mycelium by blackening of the hyphae, which become closely septate. The short cells so formed, increase in size, become spheroid and give a torulose appearance to the mycelium. The pathogen perennates in the host debris in this stage. It grows at lower temperatures of 22-25°C and shows no growth at 30°C.

V. dahliae differs from *V. albo-altrum* in: (1) forming branched conidiophores, with no swelling or colouration of the basal portion, (2) forming microsclerotia as the resting stage, and (3) showing fairly good growth at 30°C. The fungus was first reported to cause serious wilt of cotton from Tamil Nadu and Rajasthan.

The North-American workers usually refer to both the species as *V. albo-atrum*, and mention their distinction by adding the suffix, 'resting mycelial' or 'microsclerotial; British workers designate them as two species.

ACERVULAR AND PYCNIDIAL IMPERFECT FUNGI (Former Coelomycetes)

The term Coelomycetes (and also Hyphomycetes) was coined by W.B. Groove (1919) whose monumental 2-volume treatise 'British Coelomycetes' is the biggest source of information on these fungi. These are saprobes, living on dead plant materials in soil; some are destructive plant parasites.

The studies on conidial ontogeny and conidiophore characteristics, which picked up so fast in the hyphomycetous imperfect fungi, have not kept the same pace here. It is because of the small size of the conidiophore and limitations imposed by their formation inside conidioma. The differentiation of the conidiogenous cells and conidiophores from the somatic cells of the conidioma wall is rather difficult. As a result, most of the information on conidiophore structure and development is presumptive and conjectural. The conidia are formed by budding and are referred to as blastospores. The conidial shape is not as diverse as in hyphomycetous members, again because of the limitations of space and other mechanical restrictions imposed by the conidiomata. However, the ornamentation and appendages present on the conidia are highly characteristic and diverse. The perfect states, wherever known, belong to *Ascomycota*.

ACERVULAR IMPERFECT FUNGI (Former Melanconiales)

An acervulus is a typically flat, open bed of short conidiophores, arising from the upper layers of a pseudoparenchymatous stromatic mass of hyphae. There is no wall layer. However, in culture, these fungi often form conidia and conidiophores and are indistinguishable from hyphomycetous genera. The acervuli may be sub-cuticular, sub-epidermal or epidermal. In some genera, the acervuli form large, brown or black setae (sing.= seta) intermingled with the conidiophores. Setae are external appendages with no cytoplasmic continuity. The developing conidial mass ruptures the overlying host tissues and gets exposed for dissemination. There is no special mechanism for dehiscence.

Genus *Colletotrichum*

The genus represents the anamorphic state of *Glomerella* (*Sordariomycetes*). There are about 60 species of *Colletotrichum*, that cause **anthracnose** (coal-like) leaf-spot diseases of several crops, and are, therefore, known as the **anthracnose fungi**. *Sphaceloma* is the other anthracnose fungus. The former genus Gloeoporium is now merged into *Colletotrichum*.

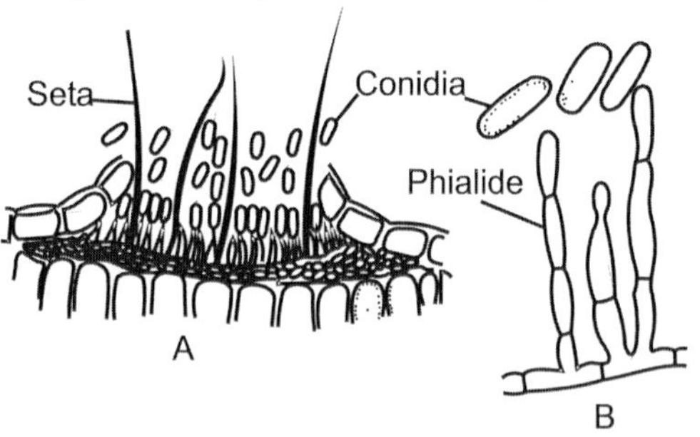

Fig. 13.20 *Colletotrichum*, A, acervulus in section; B, showing origin of conidiophore and conidia

Acervuli (Fig. 13.20) may be sub-cuticular, sub-epidermal or epidermal, with a characteristic peripheral ring of black, long and stiff setae. Conidia are hyaline, one-celled, cylindrical or falacate (sickle-shaped). Conidiogenous cells are cylindrical phialides (monophialidic). The erstwhile genus Gloeosporium differed only in lacking the setae, but this difference was untenable as formation of setae is highly variable. Some important pathogenic species are *C. falcatum* (*Glomerella tucamanensis*), the cause of 'red-rot' of sugarcane; *C. corchorum*, causing anthracnose of jute; *C. gloeosporioides* (*Glomerella cingulata*), causing anthracnose of citrus and *C. graminicola*, anthracnose of 'jowar'.

Genus *Pestalotiopsis*

The genus, with 162 widespread species, is unmistakable, because of the characteristic shape of its conidia (Fig. 13.21). The conidia are spindle-shaped or clavate, five-celled with three central coloured cells and two (upper and lower) hyaline cells. The upper hyaline cell bears two or three cellular appendages called **setulae**, which, in some species, are branched.

There is a short hyaline pedicel below the lower hyaline cell. The species are frequently isolated from necrotic zones of infected leaves. Some are destructive pathogens e.g. *P. theae*, causing 'grey-blight' of tea leaves, and *T. mangiferae*, causing leaf-spot of mango. The teleomorph is *Pestalosphaeria* (*Amphisphaeriaceae*, *Xylariales*).

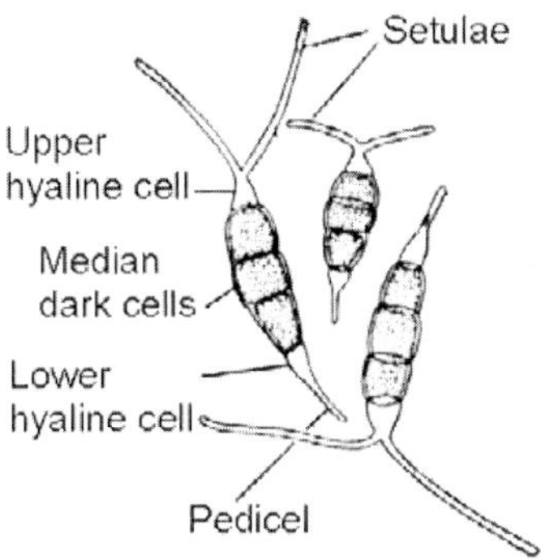

Fig. 13.21 *Pestalotiopsis*

Genus *Sphaceloma*

It is anamorph of *Elsinoe* (*Elsinoaceae*, *Dothidiales*), which differs from *Colletotrichum* in: (1) production of conidiophores from a stroma, but it is an inconsistent character, and (2) the polyphialedic conidiophore (it is monophialidic in *Colletotrichum*). There are 52 species which are widespread, causing anthracnose and scab diseases e.g. *S. rosarum* (rose leaf-spot).

PYCNIDIAL IMPERFECT FUNGI (former Sphaeropsidales)

These fungi produce the conidia in globular or flask-shaped pycnidia, that appear as minute black dots on the bleached, dead plant tissue. These may be superficial, sub-immersed or immersed in the substratum. Mostly, they are **glabrous** (smooth), but some are **setose** i.e. bear brown setae. The wall of the pycnidium is made up of pseudoparenchymatous tissue and the conidiophores arise from the cells of the inner wall. The cavity is usually undivided but multilocular pycnidia are also known. The pycnidia open by a single ostiole, but in culture these are usually multi-ostiolate. The pycnidia superficially resemble perithecia, but while the pycnidia are asexual, perithecia are sexual structures bearing asci and ascospores.

The pycnidia may be separate or gregarious; in the latter case aggregated in groups, connected by prosenchymatous fungal tissue. In some genera, the conidia are formed in cavities developed in stroma. In the simplest form, it is indistinguishable from a pycnidium, but usually the cavities are multiloculate. Characteristics of some common pycnidial genera are given below:

(a) Conidia hyaline, single-celled, globose to avoid – *Phyllosticta*, *Phoma*, *Phomopsis*.

(b) Conidia hyaline, bi-celled – *Ascochyta*.

(c) Conidia dark, bi-celled – *Diplodia*, *Lasodiplodia*.

(d) Conidia hyaline, filiform – *Septoria*.

Genus *Ascochyta*

It has 388 species, which are worldwide in distribution and represent the anamorph of *Didymella* (*Pleosporaceae*). The pycnidia are brown, thick-walled and immersed in leaf-spots. Conidia (Fig. 13.22) are bi-celled, obovoid, hyaline to pale-brown. The *Ascochyta* leaf-spot of pea is caused by 3 species: *A. pisi, A. pinodella* and *A. pinodes* (= *D. pinodes*), which differ in their spore size and appearance in culture. These fungi perennate in pea seeds and, if such seeds are used, they give rise to infected plants. Thus, the disease is internally-seed borne.

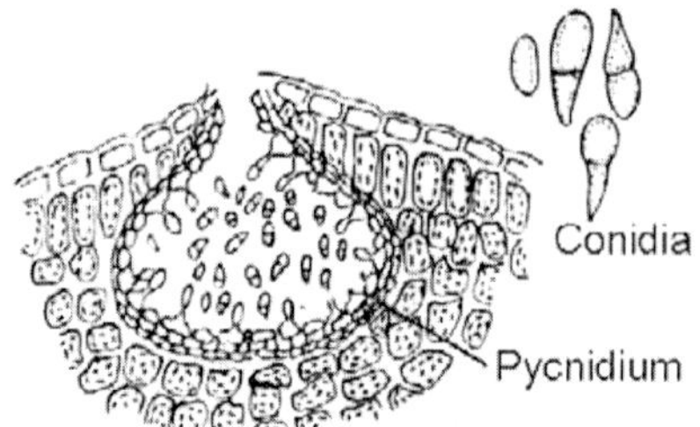

Fig. 13.22 *Ascochyta*. section through pycnia, and spores

Genus *Lasiodiplodia* (former Botryodiplodia)

The pycnidia lie aggregated in 'botryose' (grape-like) clusters; hence the name Botryodiplodia. The pycnidia are thick-walled, dark-brown and ostiolate. Conidiophores are cylindrical, hyaline, and show distinct apical annellations. Conidia are ellipsoid or ovoid, bi-celled, dark-brown, with characteristic striations at maturity. Young conidia are single-celled and hyaline. *L. theobromae,* the most common species, causing fruit rot of mango, cacao etc. It is anamorph of *Botryosphaeriaceae*.

Genus *Diplodia*

Diplodia (Fig. 13.23) with nearly 960 species, forms solitary, black, globose pycnidium bearing two-celled, dark-brown, ovoid or ellipsoid conidia. The species may be parasitic or saprobic. *D. zeae* colonizes dead maize stalks while *D. macrospora* causes dry-rot of maize ears and stalks. It is anamorph of *Botryosphaeria* (*Botryosphaeriaceae*, *Dothidiales*).

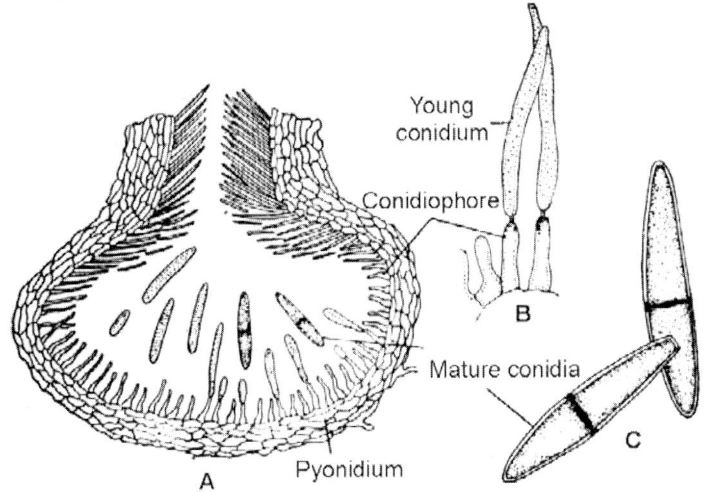

Fig. 13.23 *Diplodia macrospora* A, section through pycnidium; B, a portion of wall showing conidiophores; C, conidia.

Genus *Phyllosticta*

The genus has 92 species, which are anamorphs of *Guignardia* (*Botryosphaeriaceae*). *Phyllosticta* is a common leaf-spot fungus. The pycnidia are thin-walled, dark-brown, immersed in host tissues, and look like black dots on the leaf-spots (Fig. 13.24). If these leaf-spots are teased with a needle, and examined under the microscope, an unending stream of minute conidia starts flowing out of the pycnidia. The conidia are hyaline, 1-celled, globose to oval and guttulate i.e. have one or more oil drops inside. An apical mucilaginous appendage, stainable by modified Leifson flagellar stain, is the characteristic feature of the conidia. It may vary in size from minute to bigger than the size of the spore. The conidia are blastospores. The elongated, cylindrical, conidiogenous cells are distinct from the cells of the inner wall. The species are usually named after the host.

Genus *Phoma*

Phoma is similar to *Phyllosticta*. However, the conidia of *Phoma* invariably lack the apical mucilaginous appendage. The conidia in *Phoma* are phialospores. The genus has 140 species, which represent the conidial stage of *Didymella* (*Pleosporaceae*).

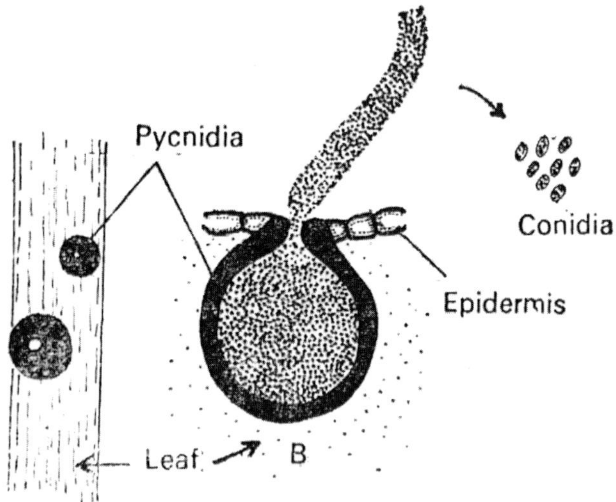

Fig. 13.24 *Phyllosticta* sp. A. pycnidia on leaf (habit): B, showing release of spores in a stream (as seen under microscope).

Genus *Phomopsis*

It is a big genus with 100 species. The conidioma is a stroma (Fig. 13.25) which is thick-walled brown, simple or convoluted, but with a common ostiole. Conidiogenous cells are monophialidic, borne on branched, septate conidiophores. Conidia are hyaline and of 2 types: α **conidia** – ovoid to fusoid, and β **conidia** (or stylospores) – filiform, curved at the apex, like a walking-stick, and biguttulate (i.e. have 2 oil globules). Since the β conidia do not germinate, their function remains obscure. *Phomopsis* has been found to be the conidial state of *Diaporthe* (*Diaporthaceae*, *Diaporthales*), causing leaf-spot and 'die-back' diseases of a large number of plants. *D. citri* (=*P. citri*) causes 'Phomopsis stem rot' of citrus fruits.

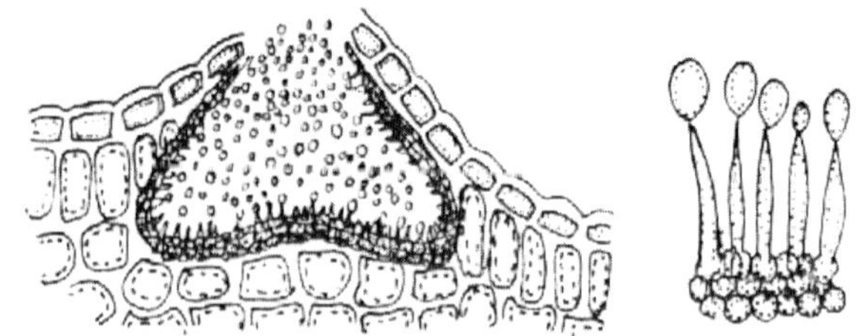

Fig. 13.25 *Diaporthe* (*Phomopsis*) *citri*.
A section through pycnidium; portion of wall showing α and β conidia.

Genus *Septoria*

The conidia (Fig.13.26) are hyaline, filiform and multiseptate, hence easily recognized. The pycnidia are dark, separate, globose, immersed or papillate. There are 1072 species, causing important leaf-spot diseases. *S. apicola* causes leaf-spot of celery (*Apium*); *S. nodorum* attacks several plants of *Graminae*; *S. lycopersici* causes tomato leaf-spot and *S. chrysanthemella* produces leaf spots on *Chrysanthemum*. The species are anamorphs of *Mycosphaerella* (*Mycosphaerellaceae*).

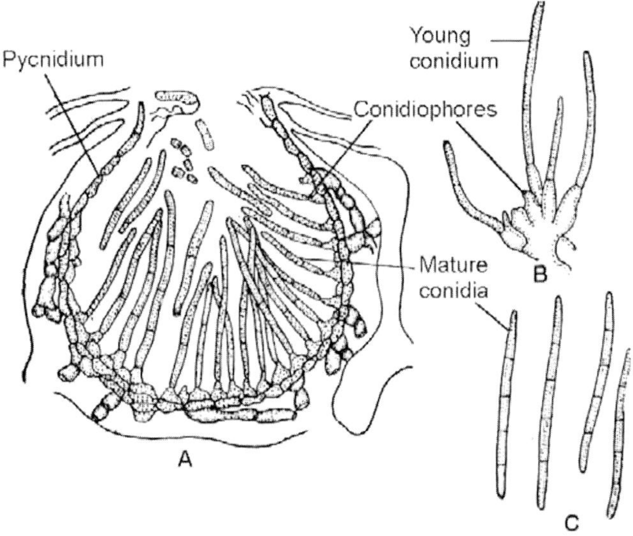

Fig. 13.26 *Septoria chrysanthemella*. A, vertical section of a pycnidrum; B, conidiophores and conidia; C, mature septate conidia.

REFERENCES

Ainsworth, G.C. 1973. Introduction and keys to higher taxa. In: *The Fungi. An Advanced Treatise*. (G.C. Ainsworth, F.K. Sparrow and A.S. Sussman, eds.). Vol. IV A. 1–7. Academic Press, New York, London.

Booth, C. 1971. *The genus Fusarium*. Commonwealth *Agric.* Bureau, England.

Brown, W. 1915. Studies on the physiology of parasitism, I. The action of *Botrytis cinerea. Ann. Bot.,* **29:** 318–348.

Hawksworth, D.L. *et. al.,* 1995. *Ainsworth and Bisby's Dictionary of the Fungi* C.A.B. International, Oxon, U.K.

Hughes, S.J. 1953. Conidiophores, conidia and classification. *Can. J. Bot.*, **31:** 577–659.

Ingold, C.T. 1942. Aquatic hyphomycetes of decaying alder leaves. *Trans. Brit. Mycol. Soc.,* **25:** 339–417.

Kendrick, W.B. 1962. The *Leptographium* complex. *Verticladiella* Hughes. *Canad. J. Bot.,* **40:** 771–797.

Kendrick, W.B. and Carmichael, J.W. 1973. Hyphomycetes. In: *The Fungi, An Advanced Treatise*. Vol **IV A** (G.C. Ainsworth, F.K. Sparrow and A.S. Sussman, Eds.). pp. 325–509. Academic Press, N.Y. and London.

Kirk PM. et al., 2008 *Ainsworth and Bisby's Dictionary of the Fungi* (10th ed.) C.A.B. International, Oxon, U.K.

Langeubeck, B. 1839. *Neue. Mot. Geb. Natur-u-Heilk.* (Frorip), **12:** 145–147.

Saccardo, P.A. 1886. Sylloge Fungorum Omnium Huscusque Cognitorum, 4: 1–807. Pavia.

Snyder, W.C. 1965. Current status of taxonomy in *Fusarium* species and their perfect stages. *Phytopathology*, **55**(8) : 833–837.

Snyder, W.C. and Hansen, H.N. 1940. The species concept in *Fusarium. Amer. J. Bot.*, **27:** 64–67.

Snyder, W.C. and Hansen, H.N. 1941. The species concept in *Fusarium* with reference to section *Martiella. Amer. J. Bot.*, **28:** 738–742.

Snyder, W.C. and Hansen, H.N. 1945. The species concept in *Fusarium* with reference to *Discolor* and other sections. *Amer. J. Bot.*, **32:** 657–666.

Subramanian, C.V. 1962. A classification of the Hyphomycetes. *Curr. Sci.*, **31:** 409–411.

Subramanian, C.V. 1971. *Hyphomycetes. Indian Counc. Agri. Res.,* New Delhi.

Subramanian, C.V. 1975. Hyphomycetology in retrospect and prospect. Presidential address. Indian Science Congress. 62nd session held at Delhi.

Toussoun, T.A. and Nelson, P.E. 1968. *A Pictorial Guide to the Identification of Fusarium* species. 1968 Pennsylvania State Univ. Press.

Ward, H.M. 1888. *A lily disease, Ann. Bot.,* **2:** 319–382.

Wollenweber, H.W. and Reinking, O.A. 1938. *Die Fasarien*, Paul Parey, Berlin.

SUMMARY

The septate, mycelial fungi which reproduce exclusively by asexual means, usually by conidia (mitospores), and which are not the asexual stages of any ascomycete or basidiomycete, are called **anamorphic fungi**. They are not a monophyletic unit but are fungi which have either lost a sexual phase or adapted alternatives to sex, like parasexuality.

The organisms can nevertheless, have the status of a formal genus and species e.g. *Aspergillus niger*. But once the sexual state is discovered, the species will have to be designated as species of the sexual (teleomorph) genus. Thus, *A. nidulans* is called *Emericella nidulans* and not *A. nidulans*. This **dual nomenclature** (separate names for different species of pleomorphic fungi) is permitted by 'rules of nomenclature'.

DNA sequencing now makes it possible to place these residual organisms within the groups of teleomorphic fungi (*Asco-*, or *Basidiomycota*) from which they are derived. By this comparative analysis of DNA sequences (molecular systematics) many asexual fungi have been placed in the existing classes.

P.A. Saccardo (1880-1886) divided the 'fungi imperfecti' into three orders: Moniliales (= Hyphomycetales), Melanconiales, and Spheropsidales. Later, a fourth order Mycelia Sterilia was added. This included members which were sterile, (i.e. lacked conidia). **Ainsworth** (1973) divided the above fungi into 3 classes: Blastomycetes (imperfect yeasts), Hyphomycetes (conidia on the hyphae), and Coelomycetes (conidia in pycnidia and acervuli). **Hughes** (1953) proposed a scheme of classification for hyphomycetes based on conidial ontogeny i.e. their origin. He divided these fungi into 8 sections, each representing one particular mode of conidial development. The spore types based on conidial ontogeny were: blastospore, porospore, phialospore, gangliospore, and arthrospores.

Kendrick and Carmichael (1973), Hawksworth *et al.* (1995), and Kirk *et al.* (2008) discarded the traditional hierarchical divisions into orders and families as it is meaningless in terms of natural or phylogenitic classification. This is now universally accepted and the genera are grouped under the following heads: **Imperfect yeasts** (e.g. *Candida, Cryptococcus, Sporobolomyces*) **Hyphomycetous Anamorphic Fungi** (*Alternaria, Botrytis, Cercospora, Curvularia, Fusarium, Helminthosporium, Bipolaris, Drechslera, Limnoniera, Monilia, Nigrospora, Pyricularia, Rhizoctonia, Trichothecium, Verticillium*), **Acervular-** (*Colletotrichum, Pestalotiopsis, Sphaceloma*), and **Pycnidial** (*Ascochyta, Lasiodiplodia, Diplodia, Phyllosticta, Phoma, Phomopsis, Septoria*) **Anamorphic Fungi**. All are very important as human or plant pathogens.

TEST QUESTIONS

1. What are anamorphic fungi? Describe in brief the classifications given by Saccardo, Ainsworth, and the currently followed as given by Kirk *et al.* (2008).
2. Distinguish between any **four**:
   (a) *Aspergillus niger* and *Aspergillus nidulans*
   (b) *Helminthosporium, Bipolaris* and *Drechslera*
   (c) *Verticillium albo-atrum* and *V. dahliae*
   (d) *Phoma and Phyllosticta*
   (e) *Fusarium* and *Verticillium,* and
   (f) *Lasiodiplodia* and *Diplodia.*

3. Describe the conidia of the following genera: *Alternaria, Curvularia, Helminthosporium, Lemnoniera, Pestalotiopsis* and *Pyricularia.*

4. Explain any **four:**
   (a) Why Hughes' classification was limited to Hyphomycetes.
   (b) What are the three sections of *Alternaria* species?
   (c) What are the three groups of *Curvularia* species?
   (d) What are the diagnostic features of *Rhizoctonia*?
   (e) What is the purpose of 'stab culture' of *Candida*?
   (f) Why is *Sporobolomyces* called the "mirror picture yeast"
   (g) How are bird droppings (excreta) responsible for cryptococosis?
   (h) Which two genera are anamorphs of basidiomycetous genera?

5. Write names of the teleomorph of any **four**: *Pyricualria, Cercospora, Monilia, Botrytis, Nigrospora, Rhizoctonia,* and *Colletotrichum.*

6. Write notes on any **three**: Candidiasis, Cryptococosis, Classification of *Fusarium* species, *Verticillium* species, Spore types based on conidial ontogeny.

7. Fill in the blanks, any **ten**;
   (a) Hughes' classification was based on ------------------.
   (b) *Candida* is pathogenic in ------------- phase.
   (c) The white material in oral thrush consists of -------------.
   (d) ------------------ is an aquatic hyphomycete.
   (e) Cryptococosis is caused by -------- ----------.
   (f) Candidiasis is caused by --------- ------------.
   (g) *Cryptococcus* grows on bird excreta which provides -------- as nitrogrn source.
   (h) *Didymella* is the perfect stage of pycnidial genera ---------- and ------------.
   (i) Conidia are hyaline and filiform in the pycnidial genus ----------.
   (j) Two types of conidia α and β are formed by the genus -------.
   (k) The conidioma is a stroma in the genus ------------------.
   (l) The perfect state of *Phyllosticta* is -------------------.
   (m) 'Grey blight' of tea is caused by --------------- ----------.
   (n) *Colletotrichum* and ---------------- are called anthracose fungi because they form --------- like leaf spots.
   (o) 'Red rot' of sugarcane is caused by ------------ --------------.

14

Phylum *Basidiomycota*

(formerly Basidiomycotina, Basidiomycetes)

The beautiful, umbrella-shaped mushrooms and toadstools with which mycology started, the rusts and the smuts that have been devastating our crops since man started agriculture, the 'shelf' and 'bracket' fungi that lie attached to the trees and sap their vitality, and the less known 'puff balls', 'stink horns' and the 'jelly fungi' – all belong to *Basidiomycota*.

The *Baidiomycota* contains about 30,000 species which is 37% of true fungi These, recognized as the highest evolved group of fungi, are characterized by several distinct features, the most important being the production of **basidiospores**. Like ascopores, these are also sexual spores, or meiospores, formed after karyogamy and meiosis. But, instead of being internal, the basidiospores lie externally on the **basidium**. Because of the similar mode of development of basidiospores and ascospores, mycologists consider them homologous and conclude that the *Basidiomycota* and *Ascomycota* are **sister groups** with a common ancestor. The informal name **dikarya** or **Dikaryomycotima** is used for the combined *Asco-*, and *Basidiomycota* highlighting the shared dikaryophase (Berbee & Taylor, 1999).

The other distinguishing features include: the highly developed dikaryophase of the thallus, the formation of **clamp connections** on the dikaryotic (secondary) hyphae, the presence of the **dolipore septa** in some, and a **double-layered cell wall**, that appears lamellate and electrondense in electron micro-photographs.

The Basidium

A typical basidium is a unicellular, club-shaped structure formed terminally on dikaryotic hyphae. It bears four basidiospores on projections called sterigmata (sing. sterigma). Such a hollow, single-celled basidium is called **holobasidium**. There are basidia which are septate (transversly or vertically); these are called **phragmobasidia** (sing. **phragmobasidium**). In rusts and smuts, however, the basidia develop from special thick-walled resting spores, formed by the dikaryotic hyphae, called the **teleuto-** or **teliospores**.

The basidium is the site of sexual reproduction i.e. karyogamy and meiosis, resulting in the formation of four haploid recombinant nuclei. Four sterigmata develop as protrusions of the basidial wall, each of which develops an apical swelling, the basidiospore initial. A single haploid nucleus migrates into each of these swellings. A wall is formed around the nucleus to delimit basidiospore. No cross wall separating the basidispore from the basidium is formed. The basidiospore is thus double-walled; the outer **perispore**, being the wall of the basidium and the inner **epispore**, being the true spore wall (formed aound the nucleus). Then, the basidiopores can be interpreted as internally-produced spores, much in the way as the ascopores develop within an ascus by 'free-cell formation.'

The Thallus

Basidiomycota are typically mycelial, but there are also some yeasts, (or have a yeast phase). Such yeasts are distinguished from ascomycetous yeasts by (1) the morphology of the bud scars, (2) formation of a red colour with diazonium blue B, (3) urease +ve reaction, and (4) high G:C percentage. The molecular sequence data can also clearly place other fungi in the phylum.

Three types of mycelia are involved in the life cycle of *Basidiomycota*. These are called the **primary**, **secondary**, and **tertiary** mycelium.

The primary mycelium is formed by germination of the basidiospore and consists of haploid, monokaryotic, septate hyphae. Since all the nuclei are descendants of a single nucleus, the monokaryotic primary mycelium is homokaryotic. The life cycle in the *Basidiomycota* starts with the formation of primary mycelium, which may produce oidia as asexual spore. It is of short duration and soon gives rise to the dikaryotic secondary mycelium.

The secondary mycelium consists of dikaryotic hyphae, the cells of which contain two nuclei of opposite mating types, brought together during plasmogamy and maintained by conjugate divisions. The greater part of the life cycle, and in some genera the whole life cycle, is lived in this stage. The secondary mycelium becomes long lived by virtue of **urediniospores** (conidia), which are formed successively in rusts during the dikaryotic hyphae. However, it also ends and the secondary mycelium winds up with the formation of basidia (teliospores, in rusts and smuts).

Since dikaryotization is achieved by plasmogamy, the various mechanisms by which it is brought about constitute an important event in the life cycle. In *Ascomycota*, the dikaryophase is represented by the ascogenous hyphae or the ascogenous cells, which are not free but dependent on the monokaryotic hyphae.

The tertiary mycelium consists of specialized dikaryotic hyphae which participate in the formation of the **basidioma** (pl. basidiomata).

The dikaryotic secondary and tertiary hyphae are physiologically equivalent to diploid cells, but are mentioned as (n + n) and not (2n) to indicate the separate existence of their two nuclei.

Clamp Connections

These are 'curious' structures formed during cell division in dikaryotic secondary hyphae having dolipore septa. Clamp connections indicate the dikaryotic nature of the hyphae, although those lacking them cannot be excluded from being dikaryotic hyphae. These are not formed on the tertiary hyphae. A small outgrowth called "clamp connection" arises (Fig. 14.1) between the two nuclei of a binucleate cell (a and b) and forms a curved hook. The two nuclei divide simultaneously with their spindles oriented in such a way that the daughter nucleus of the upper nucleus goes into the clamp, while that of the lower nucleus comes close to the upper nucleus. The clamp bends and touches the cell near the lower nucleus. The intervening walls dissolve. The nucleus of the clamp migrates into the cell and comes to lie close to the lower nucleus. Two septa are now laid down, one longitudinal, at the place of origin of the clamp (so as to cut off its original connection with the cell), and the other transversely at right angles to the first longitudinal septum. A new daughter cell is formed. The clamp remains permanently attached to the hypha. The clamp-like "croziers' formed in *Ascomycoa* at the base of the asci (where meiosis occurs) may be homologous to clamp connections (Hibbett *et al.*, 2007)

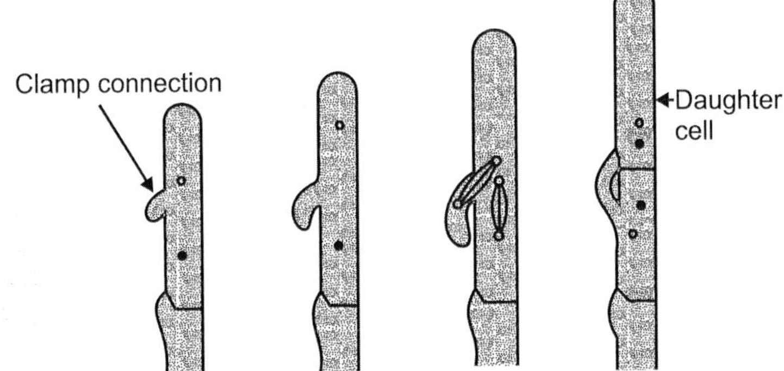

Fig. 14.1 Cell division and clamp conncection

The clamp connections do not occur uniformly all over the mycelium. While in some species these occur at every septum, in others they occur at long intervals. Several environmental factors are known to affect their formation. In liquid cultures the submerged hyphae fail to form clamp connections but the aerial hyphae form them abundantly. The function of clamp connections is to ensure the dikaryotic condition. The separation of the daughter nuclei, brought about by the clamp, may not be possible through other mechanisms.

Dolipore Septum

The dolipore septum is equal to, or even more important, than the basidium, as a characteristic feature of *Basidiomycota*. However, most of the smuts do not possess dolipore septa. The structure of the septum has been elucidated by electron microscopy. The septum (Fig. 14.2) flares up in the middle of the hyphae to form a barrel-shaped pore, called 'dolipore'.

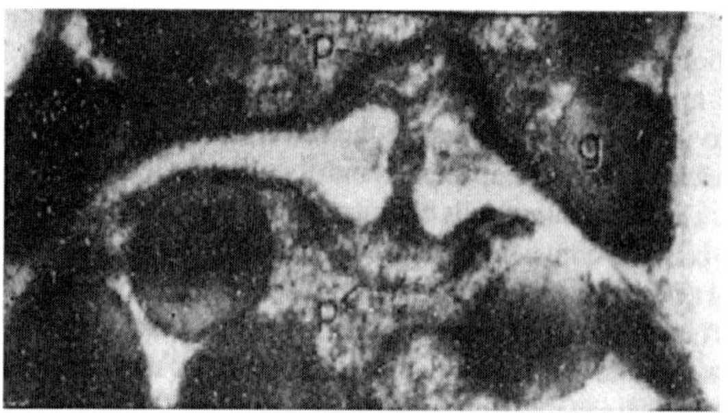

Fig. 14.2 Dolipore septum, electronmicrograph

Both the open ends of the pore are covered by a double-membraned 'pore cap', called **parenthesome**. Thus, the complete name for the septum is **dolipore-parenthesome septum**. The parenthesome is absent in rusts and smuts. The dolipore septa of monokaryotic and dikaryotic hyphae are not similar. They differ in the following points:

(1) The material composing the septa in the dikaryotic hyphae is the same as the wall material, while in the monokaryotic hyphae it is a different material.

(2) The parenthesomes in the primary hyphae are derived from the endoplasmic reticulum, while in dikaryotic hyphae, the parenthesome is isolated and unconnected with the endoplasmic reliculum; nevertheless, it is double-membraned.

(3) The parenthesomes are perforated in the primary hyphae but in the dikaryotic hyphae, these are not only entire (without pores) but the two ends of the dolipore are also blocked by plugs.

Thus, the karyotic state of the hyphae (mono or dikaryotic) greatly influences the nature of the dolipore septum. The function of this septum is to prevent movement of nuclei from one cell to another, while permitting cytoplasmic continuity. The dolipore septum is, thus, another mechanism to sustain the dikaryotic nature of the hyphae (Clamp connections also perform the same function). The perforated parenthesome of monokaryotic hyphae offer little or no obstruction to the movement of nuclei. There is free migration of nuclei even in the dikaryotic hyphae of *Coprinus lagopus*. How? The electron photomicrographs indicated that the dolipore septum had changed into a simple septum, like that of *Ascomycota*.

Reproduction

Asexual reproduction Unlike *Ascomycota*, asexual reproduction is not common in the *Basidiomycota*, and the multiplication is brought about mostly by the basidiospores. Nevertheless, budding, fragmentation, conidia, arthrospores and oidia do contribute their share in reproduction. Budding of basidiosphores is quite common in smuts. Fragmentation of secondary mycelium is used in the preparation of inoculum in mushroom cultivation. Oidia are formed by fragmentation of the monokoryotic hyphae or on special hyphal branches, called the oidiophores. The oidia adhere in a drop of mucilage at the tip of the oidiophores and are set free and dispersed when moistened with water. These are hyaline, thin-walled, unicellular and uninucleate cells which serve as conidia or spermatia. The urediniospores of *Puccinia* represent conidia, which repeat the dikaryophase for several successive generations, until the environment becomes hostile for their production.

Sexual reproduction Distinguishable sex organs are not formed in *Basidiomycota* except in *Puccinia*, where the **spermatia** and **the receptive hyphae** represent the male and female sex organs, respectively. However, the disappearance of definite sex organs has not done away with the three fundamental events of sexual reproduction viz. plasmogamy, karyogamy and meiosis. The mating competence too has not diminished. The **incompatibility systems** regulate out-breeding and in-breeding.

Sexual reproduction is a long-drawn process here. The separation of plasmogamy and karyogamy, which first appeared in the *Ascomycota*, has reached its full development here. The dikaryophase starts with the coming together of two compatible nuclei during plasmogamy, extends through the greater part of the life cycle and is perpetuated by the urediniospores in rusts. It ends with the occurrence of karyogamy and meiosis in the basidium, which is usually the terminal cell, subtended by a clamp connection (Fig. 14.3). The resulting four recombinant haploid nuclei are segregated in the four basidiospores. In rusts and smuts, karyogamy and meiosis occur in specialized, thick-walled spores called the **teliospores** (Fig. 14.4). These germinate after a resting period and give rise to a short hypha, which bears the basidiospores. *The telio-, and the basidiospore-bearing structure jointly constitute the basidium.*

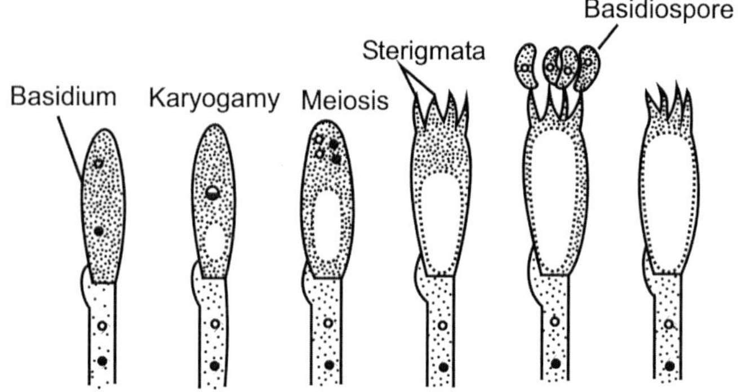

Fig. 14.3 Stages in basidium formation.

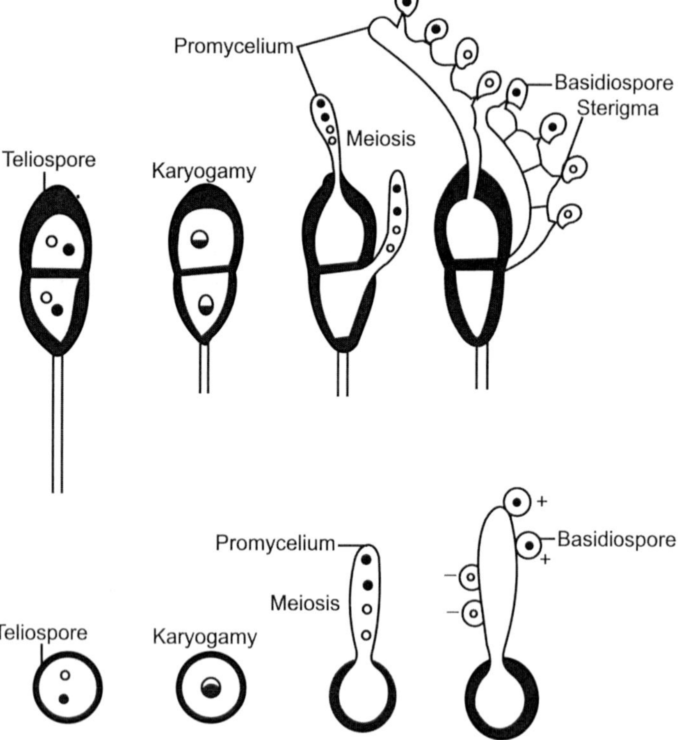

Fig. 14.4 Basidial apparatus in rusts and sumts

Plasmogamy (dikaryotization) is brought about by (1) somatogamy between cells of monokaryotic hyphae, (2) fusion between an oidium and a monokaryotic hypha and (3) spermatization of receptive hyphae (in rusts only).

Dikaryons may be formed directly after meiosis, by inclusion of two haploid nuclei of opposite mating types in the basidiospores. Such genera skip over the haploid, primary mycelial stage, as the dikaryotic basidiospores directly give rise to dikaryotic mycelium. Another common type of plasmogamy involves the dikaryotization of a monokaryotic mycelium by a dikaryotic hypha (Fig. 14.5), first noted by Buller (1931) in *Coprinus*. This is known as the **Buller phenomenon**. It is also called **di-mon mating**.

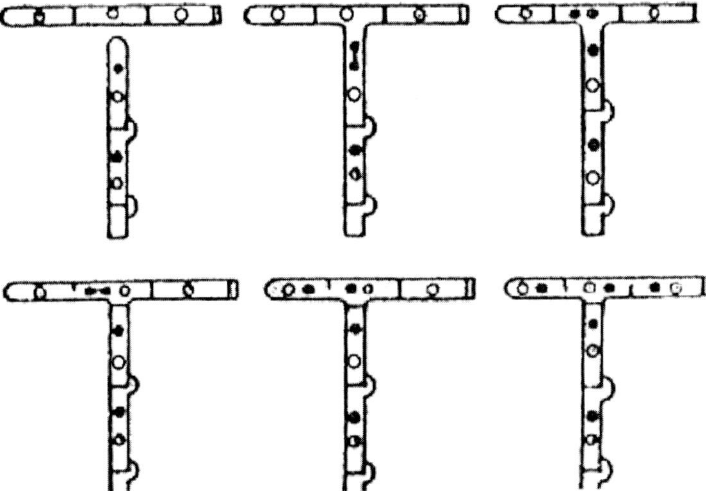

Fig. 14.5 Showing steps in dikaryotization of monokaryotic mycelium by a dikaryotic hypha (Buller phenomenon)

The dikaryotic hypha is formed by three methods:

(1) The binucleate cell, formed by fusion of two cells, by repeated cell divisions may give rise to a dikaryotic, secondary mycelium. The two nuclei divide conjugately and the new nuclear pair migrates into daughter cell through clamp connections.

(2) As it happens in *Puccinia*, the entrant nucleus divides and the daughter nucleus migrates into the adjacent cell. This is repeated and in this way the entire monokaryotic mycelium is dikaryotized.

(3) When the terminal cell of a monokaryotic mycelium is dikaryotized, the two nuclei divide conjugately and the daughter pair of nuclei are separated to form a new cell. A new dikaryotic hypha,

having clamp connections and dolipore septa, grows out from the terminal cell of a monokaryotic mycelium.

The Basidioma (= basidiocarp)

The basidiomata are highly attractive and beautiful, hence called **fungus flowers** (rusts and smuts, mostly do not form basidioma). These are tremendously variable in colour, size, shape and texture. Their numerous shapes are evident from the names given to some of them e.g. mushrooms, bracket fungi, puff balls, stink horns, birds' nest fungi etc. These may be microscopic or large, over one meter in size. In texture, they may be crust-like, gelatinous, fleshy, papery, spongy, cartilaginous or woody. The basidiomata may be open from the beginning (gymnocarpous) or only at a later stage (hemiangicarpic). In gasteromycetes, they are permanently closed (angiocarpic) and the spores are released after disintegration of the wall, the peridium.

Hyphal Types and Hyphal Systems

The hyphal construction of the basidioma is now an important aspect of study in the taxonomy of *Basidiomycota.* **Corner** (1932 a,b) introduced **hyphal analysis** and coined the term **hyphal system** in taxonomy. Corner's has been the most powerful voice in exposing the artificiality of the classification that gave too little attention to the problems of **homoplasy** (i.e. similarity in structure and development due to parallel or convergent evolution).

The Hyphal Types

The hyphae that compose the flesh or context of the basidiomata are of three basic types: **generative, skeletal** and **binding hyphae** (Fig.14.6).

Generative hyphae are the main and first-formed hyphae produced directly from secondary hyphae. They are thin-walled (rarely thick), branched, possess clamp connections and are cyanophilous (i.e. take stain of cotton blue or methylene blue). The basidia and, directly or indirectly, all sterile structures of the hymenium, context and trama arise from the generative hyphae. The other types of hyphae viz. the skeletal and binding hyphae originate as specialized branches of the generative hyphae. These, when they gelatinize, give rise to gelatinous basidiomata. Occasionally, they become strongly thick-walled, forming sclerified generative hyphae or pseudoskeletals.

Skeletal hyphae are thick-walled to solid, unbranched or sparsely-branched, non-septate or pseudoseptate hyphae, without clamp-connections and usually non-cyanophilous. They serve as strengthening structures lending rigidity to the basidioma.

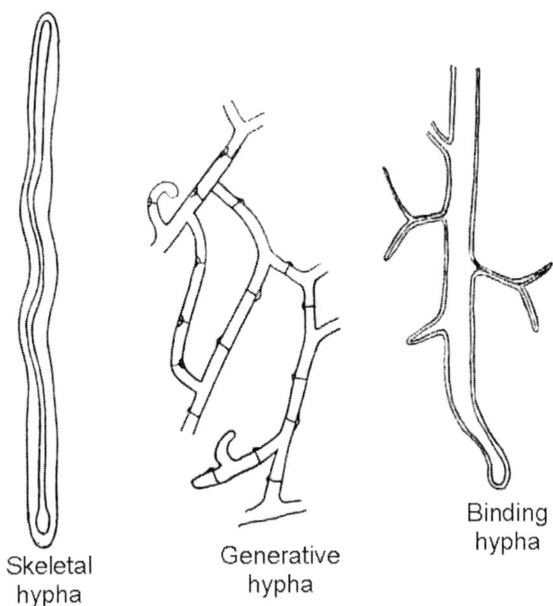

Fig. 14.6 Hyphal types: skeletal, generative and binding hyphae

Binding hyphae, also called *ligative hyphae*, (like the skeletal hyphae) are thick-walled to solid, unseptate, non-cyanophilous and lack clamp-connections. However, these differ in being of limited growth and highly and irregularly branched. Their function, as evident by the name, is to weave and bind the different hyphae.

Hyphal Systems (= the mitic system)

There are three types of hyphal systems: **monomitic**, **dimitic** and **trimitic**. When the basidioma consists exclusively of generative hyphae, the hyphal system is called **monomitic**. When in addition to the generative hyphae, there is any one of the other type of hypha, it is **dimitic**,. When all the three types of hyphae are present, the hyphal system is **trimitic**.

The hyphal systems are now important in the identification, especially at the generic level. Gross morphology of basidiomata is of limited value, as different species in the same genus produce basidiomata of varying shapes.

Hymenophore and the Hymenium

These terms have been used in literature ambiguously and, therefore, need a clear understanding. ***Hymenophore*** is the surface that bears the hymenium. It may be smooth, toothed, poroid, lamellate etc. The ***hymenium*** is the row of basidia and the sterile companions, the basidioles, cystidia, hyphoides, setae etc.

Basidia bear the basidiospores and are the most important element of the basidioma.

Basidioles are young, aborted or sterile basidia, which support the basidia in the palisade. In *Coprinus* and several hymenogastraceous species, the basidioles inflate markedly to form a kind of pavement supporting the basidia. When they inflate to attain a diameter, twice or more of the young basidia, the basidioles are termed **brachybasidioles**. Earlier, the basidioles were called as **paraphyses**, but when the karyological difference between them and the paraphyses of *Ascomycota* was realized, they were renamed as basidioles (In *Ascomycota* the paraphyses arise from monokaryotic hyphae, whereas the basidioles arise from the dikaryotic secondary hyphae).

***Cystidia* (*sing. cystidium*)** are of several types and vary in number from abundant to nil. Usually, these are vesiculate to clavate, projecting beyond the level of the basidia and, in addition to serving as air traps, provide mechanical strength to the hymenium. The cystidia are classified in different ways, based on their origin location, shape, content or function. They may be simple or branched, coloured or colourless.

Hyphidia (sing. Hyphidium) are hyphal end cells which function as air traps around the developing basidia. These are more common in *Poriales* than in the *Agaricales*. These may be unbranched, (halophysis) or branched. On the basis of branching, three types of hyphidia are known; viz. *Dendrohyphidium* (with irregularly strongly branched), *Acanthohyphidium* (with bottle-brush like projections), and *Dichohyphidium* (repeatedly, dichotomously-branched).

Setae (sing. seta) are dark-brown, stiff, thick-walled, smooth, pointed and fusiform to ventricose (inflated in the middle or at the side) structures found only in family *Hymenochaetaceae*. Setae are found in the hymenium, trama and the context. Those in the hymenium are smaller and called **haplosetae**, while others that occur in the trama and context are bigger and called **macrosetae**. The setae may be in level with other elements or longer, protruding above the hymenium.

Chemical Reactions

Some chemical tests provide useful information which is helpful in defining families and genera of poroid families. The important reactions are the following.

Amyloidity (amyloid reaction). Hyphal or spore walls turn blue in Melzer's reagent.

Dextrinoidity (dextrinoid reaction). Walls turn brown in Melzer's reagent.

Cyanophilly. The walls take intense blue colour in lactic acid containing cotton blue.

Xanthochroic reaction (darkening or non-darkening of the brown context or hyphae in aqueous KOH solution). If the brown hyphae become darker to black, the reaction is termed positive.

Discharge of Basidiospores

The active (violent or forcible) and passive discharge of basidiospores occur in almost equal proportions in the *Basidiomycota*. The basidiospores, which are violently discharged, are called **ballistospores***, while passively discharged basidiospores are called **statismospores**.

Our knowledge of the mechanism of ballistospore discharge is largely derived from the research of Buller. The mature basidiospores lie asymmetrically perched on the tip of a fine curved sterigma. The spore has a minute projection, the **hilar appendix**, near the point of attachment with the sterigma (hilum). Just before a spore is shot out, a minute bubble, limited by an aqueous film (not a water drop), appears at the hilar appendix, which grows in a few seconds to attain a critical size. Its expansion results in the explosive throwing off of the basidiospores. The distance of discharge is short, 0.1-0.2 mm and occasionally more than 1 mm (Fig. 14.7). The sterigma soon collapses and undergoes lysis. The **bubble theory** as outlined above, envisages a surface tension mechanism. According to Ingold (1971), the surface energy available to affect the discharge is double for a water bubble, as compared to a water drop. This propels the spore in an outward direction. What is visible at the hilar appendix, for long, was considered to be a water 'drop' and the mechanism of basidiospore discharge was called 'water drop' method. It was originally described by Buller. The question of bubble versus drop is now decided in favour of the bubble.

The four basidiospores, as a rule, are discharged one after the other, and not simultaneously. The trajectory (direction of discharge) of the released spore, takes an outward angle from the long axis of the basidium (Ingold, 1966).

---

* Ballistospores may be sexual or asexual, and may be produced by basidia, hyphae, yeast cells or even other ballistospores

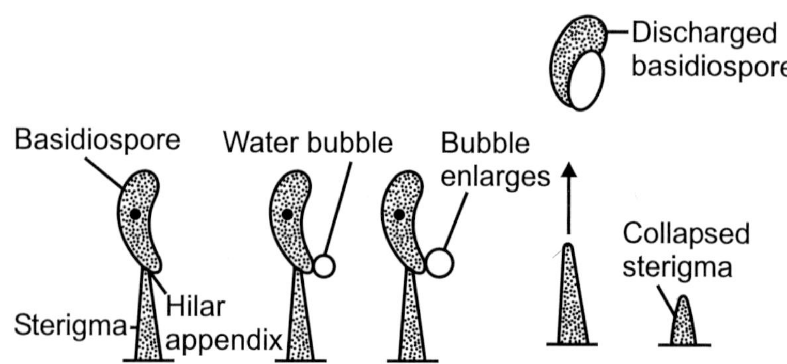

Fig. 14.7 Explosive discharge of basidiospores

Classification

The classification of *Basidiomycota*, like that of any other group, has been changing with the addition of information, especially regarding evolutionary relationships. The morphological similarities are being replaced by information on small sub-unit (18S) rDNA sequences as the basis of the classification. The important classifications are tabled below:

Older Classifications

Ainsworth (1973) classified Basidiomycotina into **3 classes:** Gasteromycetes, Hymenomycetes, and Teliomycetes, based on: (1) the presence or absence of the basidioma, (2) angiocarpic, hemiangiocarpic or gymnocarpic nature of the basidioma, (3) the septate or unseptate nature of the basidia, and (4) explosive or passive discharge of the basidiospores. Later, **Hawksworth** *et al.* (1983) split the Teliomycetes into two classes, Urediniomycetes and Ustilaginomycetes, respectively, for rusts and smuts. The other two classes remained unchanged. However, in 1995, **Hawksworth** *et al.* merged the Hymenomycetes and Gasteromycetes into one class, the Basidiomycetes. Swann and Taylor (1993) preferred the name Hymenomycetes over Basidiomycetes). The rusts and smuts, despite their forming teliospores, were put in separate classes, Teliomycetes (mostly rusts) and Ustomycetes (mostly smuts). The classifications of Swann and Taylor (1993) and Hawksworth *et al.* (1995) are based on comparisons of SSU rDNA sequences, which suggest three lineages, two without basidiomata (Teliomycetes and Ustomycetes) and one with basidiomata (Basidiomycetes). Kirk *et al.* (2001) divided phylum Baidiomycota into 3 classes : Urediniomycetes, Ustilaginomycetes and Basidiomycetes.

Ainsworth 1973	Hawksworth 1983
Sub division Basidiomycotina Class Gasteromycetes Hymenomycetes Teliomycetes	Sub division Basidiomycotina Class Gasteromycetes Hymenomycetes Urediniomycetes Ustilaginomycetes
Swann and Taylor 1993	**Hawksworth *et al.*, 1995**
Phylum Basidiomycota Class Hymeomycetes Ustilaginomycetes Urediniomycetes	Phylum Basidiomycota Class Basidiomycetes Teliomycetes Ustomycetes
Kirk *et al.*, 2001	**Kirk *et al.*, 2008**
Phylum Basidiomycota Class Baidiomycetes Urediniomycetes Ustilaginomycetes	Phylum *Basidiomycota* (3 subphyla) *Agaricomycotina* (3 classes) *Pucciniomycotina* (8 classes) *Ustilaginomycotina* (1 class)

The Current Classification

In the 10th edition of *The Dictionary of the Fungi* (Kirk *et al.*, 2008), the phylum *Basidiomycota* is divided into 3 subphyla: *Agaricomycotina, Pucciniomycotina* and *Ustilaginomycotina*. There are 6 classes which are Incertae sedis i.e. not assigned subphylum. The 3 subphyla are further divided as shown below (up to orders). All orders are not given.

Phylum *Basidiomycota* (3 Subphyla)

Subphylum *Agaricomycotina* (3 classes)

Class *Agaricomycetes* (17 orders), *Agaricales, Atheliales, Boletales*

Class *Dacrymycetes* : (1 order), *Dacrymycetales*

Class *Tremellomycetes* (3 orders), *Cystofilobasidiales, Filobasidiales, Tremellales*

Subphylum *Pucciniomycotina* (8 classes)

Class *Pucciniomycetes* (5 orders), *Pucciniales, Septobasidiales*

Subphylum *Ustilaginomycotina* (2 classes)

Class *Ustilaginomycetes* (3 orders), *Urocystidales, Ustilaginales*

Class *Exobasidiomycetes* (6 orders), *Exobasidiales, Tilletiales, Entylomatales*

Phylogenetic relationships

There is strong evidence that the *Basidiomycota* is monophyletic. Ballistospores, basidia, and clamp connections are present in all the three subphyla (although not in all species), suggesting that they have a common origin. Non-molecular characters that have been used to recognize major groups within the *Basidiomycota*, include the form of the basidia (shape and septation), ultrastructure of hyphl septa, presence or absence of yeast phase and "spore repetition" (production of spores directly from spores), and cellular carbohydrate composition, for example of the cell wall. Three major clades (2 witiout basidioma and one with basidioma) are designated as three subphyla, *Agaricomycotina, Pucciniomycotina* and *Ustilaginomycotina*. The monophyly of each of the groups has been supported in phylogenetic analyses (Hibbett *et al.,* 2007). Similarities in ultra-structure of septal pores and **spindle pole bodies** (microtubule-organizing center functionally analogous to centrosome of animal cells) suggest that *Ustilaginomycotina* and *Agaricomycotina* are **sister groups.**

REFERENCES

Ainsworth, G.C. 1973. Introduction and keys to higher taxa. In: *The Fungi. An Advanced Treatise* (G.C. Ainsworth, F.K. Sparrow and A.S. Sussman eds.). Vol. IV A. 1–7. Academic Press, New York, London.

Berbee, M.L. and Taylor, J.W. 1999. Fungal Phylogeny. In: *Molecular Fungal Biology* (R.P. Oliver and M. Schweizer eds.). pp. 21–77. Cambridge University Press, Cambridge.

Buller, A.H.R. 1909-1950. *Researches on Fungi*. Vols. I – VII. London.

Corner, E.J.H. 1932a. The fruit body of *Polystictus xanthopus Fr. Ann. Bot.* (London). 46: 71–111.

Corner, E.J.H. 1932b. A *Fomes* with two systems of hyphae. *Trans. Brit. Mycol. Soc.,* 17**:** 51–81.

Giesy, R.M. and Day, P.R. 1965. The septal pores of *Coprinus lagopus* (Fr.) sensu Buller in relation to nuclear migration. *Amer. J. Bot.*, **52:** 287–293.

Hawkswoth, D.L. *et al.,* 1983. *Ainsworth and Bisby's Dictionary of the Fungi.* C.A.B. International, Oxon, U.K.

Hawksworth, D.L. *et al.,* 1995. *Ainsworth and Bisby's Dictionary of the Fungi* C.A.B. International, Oxon, U.K.

Hibbett, D.S. 2006 A phylogenetic overeview of the Agaricomycotina. Mycologia 98: 917-925

Hibbett, et al., 2007 A higher level phylogenetic classification of the Fungi. Mycological Research 111:509-547

Ingold, C.T. 1966. Aspects of spore liberation: violent discharge. In: *The Fungus Spore*. (M.F. Madelin, ed.) pp. 113–132. Butterworths, London.

Ingold, C.T. 1971. The glebal mass in *Sphaerobolus*. *Trans. Brit. Mycol. Soc.,* **56:** 105–113.

Swann, E.C. and Taylor, J.W. 1993. Higher taxa of Basidiomycetes: an 18S rRNA gene perspective. *Mycologia*. **85:** 923–936.

SUMMARY

The *Baidiomycota*, containing 37% of the 'true fungi', are the highest evolved group of fungi. These are characterized by (1) the production of **basidiospores,** (2) the highly developed **dikaryophase**, (3) **clamp connections** (4) **dolipore** septa in some, and (5) a **double-layered cell wall**, that appears lamellate and electron-dense in electron microphotographs. *Basidiomycota* and *Ascomycota* are regarded as **sister groups** with a common ancestor. This is supported by (1) the similar mode of development of basidiospores and ascospores, (2) dikaryophase and (3) similarity in crozier and clamp connection, which are regarded as homologyous structures

The basidium may be unicellular, club-shaped **holobasidium** or septate **phragmobasidium,** formed terminally on dikaryotic hyphawe. In rusts and smuts the basidia develop from special thick-walled resting spores, formed by the dikaryotic hyphae, called **teliospores**. *Basidiomycota* are typically mycelial, but there are also some yeasts, (or have a yeast phase). Three types of mycelia are involved in the life cycle of *Basidiomycota*, called the **primary**, **secondary**, and **tertiary** mycelium. The primary mycelium is formed by germination of the basidiospore and consists of haploid, monokaryotic, septate hyphae, which may produce oidia as asexual spore. It is of short duration and soon gives rise to the dikaryotic secondary mycelium. The secondary mycelium consists of cells which contain two nuclei of opposite mating types, brought together during plasmogamy and maintained by conjugate divisions. The greater part of the life cycle, and in some genera the whole life cycle, is lived in this stage. The secondary mycelium winds up by forming basidia (teliospores, in rusts and smuts). The dikaryotic secondary and tertiary hyphae are physiologically equivalent to diploid cells, but are mentioned as (n + n) and not (2n) to indicate the separate existence of their two nuclei. The tertiary mycelium consists of specialized dikaryotic hyphae which participate in the formation of the **basidioma** (pl. basidiomata).

Clamp Connections are formed during cell division in dikaryotic secondary hyphae having dolipore septa. Clamp connections indicate the dikaryotic nature of the hyphae, although those lacking them cannot be excluded from being dikaryotic hyphae. These are not formed on the tertiary hyphae. The clamp-like "croziers', formed in *Ascomycota* at the base of the asci (where meiosis occurs), is suggested to be homologous to clamp connections.

The dolipore septum is equal to, or even more important, than the basidium, as a characteristic feature of *Basidiomycota*. However, most of the smuts do not possess dolipore septa. The function of this septum is to prevent movement of nuclei from one cell to another, while permitting cytoplasmic continuity. The dolipore septum is, thus, another mechanism to sustain the dikaryotic nature of the hyphae. (Clamp connections also perform the same function).

Asexual reproduction, which is not common, is brought about by budding, fragmentation, conidia, arthrospores and oidia. Budding of basidio-sphores is quite common in smuts.

Distinguishable sex organs are not formed in *Basidiomycota* except in *Puccinia*, where the spermatia and the receptive hyphae represent the male and female sex organs, respectively. However, the disappearance of definite sex organs has not done away with the three fundamental events of sexual reproduction viz. plasmogamy, karyogamy and meiosis. The incompatibility systems regulate out-breeding and in-breeding.

The basidiomata (rusts and smuts, mostly do not form basidioma) are tremendously variable in colour, size, shape and texture. The hyphae that compose the basidiomata are of three basic types: generative, skeletal and binding. Generative hyphae are the main and first-formed hyphae produced directly from secondary hyphae. The other types of hyphae viz. the skeletal and binding hyphae originate as specialized branches of the generative hyphae. When the basidioma consists exclusively of generative hyphae, the **hyphal system** is called monomitic. When in addition to the generative hyphae, there is any one of the other type of hypha, it is called dimitic. When all the three types of hyphae are present, the hyphal system is trimitic. **Hymenophore** is the surface that bears the hymenium. It may be smooth, toothed, poroid, lamellate etc. The **hymenium** is the row of basidia and the sterile companions, the basidioles, cystidia, hyphoides, setae etc.

The basidiospores are discharged violently (ballistospores) or passively (statismospores). Both are equally common. The four basidiospores, as a rule, are discharged one after the other, and not simultaneously.

Basidiomycota is divided into 3 subphyla: *Agaricomycotina, Pucciniomycotina* and *Ustilaginomycotina.* There are 6 classes which are incertae sedis i.e. not assigned a subphylum.

There is strong molecular phylogenetic evidence that the *Basidiomycota* is monophyletic.

Three major clades inferred from molecular data are designated as the three subphyla. The *Ustilaginomycotina* and *Agaricomycotina* are **sister groups**, as evidenced by similarities in ultrastructure of septal pores and spindle pole bodies.

TEST QUESTIONS

1. What are the diagnostic features of *Basidiomycota?* Describe the three types of mycelium.
2. Write short notes on: (a) Clamp connection, (b) Dolipore septum, (c) Similarities between *Asco-* and *Basidiomycota,*(d) Amyloid and dextrinoid reaction with Melzer's reagent.
3. Distinguish between: (a) Mono-, di-, and trimitic hyphal system, (b) Hymenophore and hymenium, (c) Generative, binding and skeletal hyphae. (d) Basidiomycetous and ascomycetous yeasts
4. Explain:
   (a) Why are the secondary and tertiary hyphae designated as (n+n) and not (2n)?
   (b) What are the various components of hymenium?
   (c) Why are the clamp connections and croziers (of *Ascomycota*) homologous?
   (d) Which two events are associated with the name of Buller?
   (e) Which two subphyla of *Basidiomycota* are sister groups, and why?
5. Fill in the blanks
   (a) Clamp connections and the ------------- ---------- sustain the dikaryotic nature of thc hyphac.
   (b) In absence of sex organs, in-breeding and out-breeding is regulated by----------.
   (c) In "di-mon" mating (Buller phenomenon), the nucleus is donated by ----------------------------.
   (d) The hyphal analysis and hyphal system was introduced by --------.
   (e) The *Basidiomycota* contains ------% of the true fungi.
   (f) The *Asco*-, and *Basidiomycota* are jointly designated as -----------.
   (g) Clamp connections and Dolipore septa are associated with ------------- hyphae.
   (h) Basidia are formed on ---------- hyphae, but in rusts these are formed by -----------------

(i) The primary mycelium is multiplied by -----------, which also serve as ---------

(j) The basidiospore is made of two walls, named as ---------- and ---- ----------.

(k) The two types of basidiospores are ------------- and ---------.

(l) The basidiospores are shot off ------------. (simultaneously/ one after other).

(m) The mechanism of explosive discharge of ballistospores is explained by ---------- theory.

15

Subphylum *Agaricomycotina*

(Former Hymenomycetes)
(Mushrooms, Jelly fungi, Yeasts)

Agaricomycotina is the biggest subphylum containing 20,000 species, which is almost 70% of the total *Basidiomycota*. For comparison, *Puccinio-mycotina* has 7,400 speciess, while *Ustilaginomycotina* has 1500 species. The members are saprobes, symbionts (ectomycorrhizal) or parasites. *Agari-comycotina* is characterized by (1) formation of basidiomata (basidiocarps), (2) presence of dolipore septa with parenthosomes, (3) presence or absence of clamp connections, (4) somatogamous plasmogamy, and (5) basidiospores ballistospores or statismospores.

The baidiomata and doloipore septa are the unifying characters of *Agaricomycootina*. The basidimata are remarkable for their size, diversity and complexity. The dolipore septa with parenthesome (parenthesome-dolipore septum) are present only in *Agaricomycotina. Pucciniomycotina* and *Ustilaginomycotina* have septal pores but not the septal pore cap (=parenthesome). Variation in the form of parenthesome is an important phylogenetic character, which is useful in distinguishing the classes of *Agaricomycotina.*

Classification

There are 3 classes, which represent the three phylogenetic clades. These are *Agaricomycetes*, *Dacrymycetes* and *Tremellomycetes*. About 90% of the species of *Agaricomycotina* belong to the class *Agaricomycetes*, which includes mushrooms, bracket fungi, puff balls and others. The *Dacrymycetes* and *Tremellomycetes* include the jelly fungi, which have gelatinous and translucent basidiomata, and many yeast-forming basidiomycetes. (However, a few former 'jelly fungi', such as *Auricularia*, are classified in the *Agaricomycetes*). The 3 classes have different types of parenthesomes. It is

perforate in *Agaricomycetes*, **imperforate** in *Dacrymycetes* (and some *Agaricomycetes*) and **vesiculate** in *Tremellomycetes*. In perforate form, the septal pore cap has openings of various sizes, while in vesiculate form, the septal pore cap is divided into cup-shaped sections. The imperforate pore cap is smooth without pores. The phylogenetic analyses suggest that the *Agaricomycetes* and *Dacrymycetes* form a monophyletic group and form **sister groups** (Matheny *et al.*, 2007). An outline classification of *Agaricomycotina* is as follows.

Class *Agaricomycetes* (2 subclasses)

Agaricomycetidae (3 orders):

Agaricales (33 families),

Atheliales (1 family),

Boletales (16 families)

Phallomycetidae (4 orders)

Geastrales (1 family)

Gomphales (3 families)

Hysterangiales (5 families)

Phallales (2 families)

Incertae sedis (no subclass) ***Auriculariales***, ***Cantharellales***, *Corticiales, Gloeophyllales*, *Hymenochaetales*, ***Polyporales***, ***Russulales***, *Sebacinales, Thelaphorales, Trechisporales*.

Class *Dacrymycetes* (1 order) *Dacrymytales*, (1 family) *Dacrymytaceae* (9 genera)

Class *Tremellomycetes* (3 orders)

Cystofilobasidiales (1 family) *Cystofilobasidiaceae* (9 genera, 20 species)

Filobasidiales (1 family) *Filobasidiaceae*, (1 genus, 4 species)

Tremellales (9 families)

Class *Agaricomycetes*

All members of this class form basidiomata which include mushrooms, bracket fungi, puff balls and others. These range in size from tiny cups a few millimeters in diameter to giant polypores greater than a meter in diameter, weighing up to 130 kg. Nearly all species are terrestrial, except a few aquatic forms. These are saprobic, lignicolous, ectomycorrhizal, rarely parasitic on plants. These are edible, but also highly poisonous and hallucinogenic. The mycelium may be perennial. The mycelium of *Armillaria gallica* has been estimated to extend over 150,000 square meters (37 acres) having an age of 1500 years, obviously oldest living organism (Smith *et al.,* 1992). There are 17 orders, which are placed in 2 subclasses, *Agaricomycetidae* (3 orders) and *Phallomycetidae* (4 orders); and under incertae sedis (10 orders).

Subclass *Agaricomycetidae*

Order *Agaricales*

Agaricales is the largest order of *Agaricomycetes*, containing 8500 species. Earlier only gill-forming fungi were placed in this order, but now non-gill forms, like puff-balls, birds' nest fungi and false-truffles are also included. The non-gill fungi, which form basidia in closed basidiomata (gasterocarps), were earlier placed under 'Gasteromycetres' (now derecognised). Besides these, some other non-gill fungi, forming reduced or cup-like (cyphelloid) basidiomata, or members of fomer Aphyllophorales are also included in *Agaricales*. Since the gross morphology of basidiomata is highly variable, it is a poor phylogenetic indicator, and is no more used in characterization of families. Similarly, the shape and septation of the basidia (holobasidium or phragmobasidium) are also not used in the formal designation of taxa. They can be used in description of the characters. Consequently, no diagnosis is possible for the families (Kirk *et al.,* 2008). A recent study (Matheny *et al.,* 2006, 2007) of nucleic acid sequences of a large number of species from different genera of *Agaricales* showed that the families could be grouped into 6 clades (informal). These are Agaricoid clade (14 families), Tricholomatoid clade (8 families), Marasmioid clade (7 families), Hygrophoroid clade (3 families), Pluteoid clade (4 families) and Plicaturopsidoid clade (2 families). The order *Agaricales* has 33 families; we shall examine only *Agaricaceae, Amanitaceae, Clavariaceae, Pleurotaceae, Pluteaceae, Schizophyllaceae, Strophariaceae* and *Physalacriaceae.*

The legendary 'mushrooms' and 'toadstools', which first attracted the attention of man towards fungi, belong here. In fact the word mycology means study of mushrooms (mykes = mushrooms, logus = study). The edible species were called 'mushrooms' and the poisonous ones, 'toadstools'. The word toadstool is a distortion of the German word "Todestuhl", which means death chair (Bessey 1950). For the other and more common belief of the origin of the word toadstool see Fig. 15.1. The basidiomata are typically fleshy, umbrella-shaped with a stipe (stem or stalk), a pileus (cap), and gills (lamellae) present on the lower surface of the pileus, where the basidiospores are produced.

Fig. 15.1 A toadstool!

In 'poroid' genera, detachable tubes bearing internal hymenium which open outside through pores, serve as the hymenophore. The basidiomata are usually umbrella-shaped and have a gymnocarpous or hemiangiocarpous mode of development. In gymnocarpous mode of development, the hymenium is external from the beginning, while in hemiangiocarpic development, the hymenium originates in a closed cavity but is exposed much before maturity. The gymnocarpous basidioma starts in form of a columnar mass of hyphae, whose apical end spreads laterally to initiate a pileus. Gills develop on the lower surface of the pileus and bear the hymenium. In hemiangiocarpic basidiomata, the gills develop inside a circular chamber formed in the pileus. In some species, the margin of the developing pileus is connected to the stem by a membrane called the partial veil (Fig. 15.2). When the pileus expands the partial veil ruptures leaving behind a circular skirt-like tissue around the stem, the annulus. The torn fragments of the partial veil attached at the margin of the pileus are called cortina.

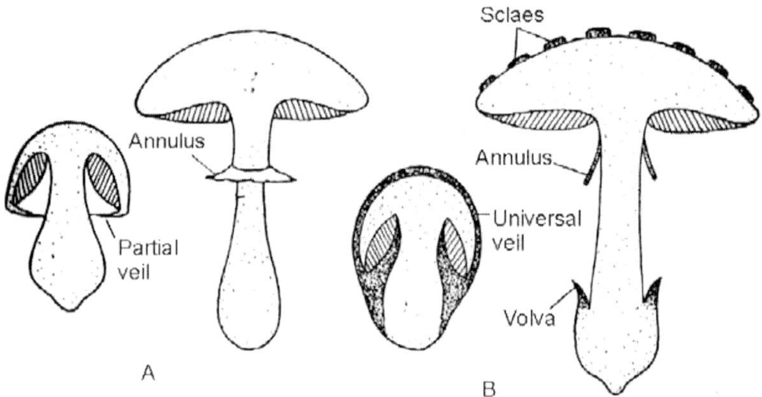

Fig. 15.2. Showing two types of hemiangiocarpic development of basidoma exemplified by A, *Agaricus* and B, *Amanita*.

In another type of development, represented by *Amanita*, the entire basidioma is formed inside a universal veil, which is torn when the stipitate basidioma emerges out. The torn universal veil around the base of the stipe forms a cup-shaped structure called volva. Some fragments persist on the upper surface of the pileus in form of scales. The annulus, cortina, volva and the scales are characteristics used for the identification of agaric genera and species.

The Gills (= lamellae)

Gills, which are usually wedge-shaped and only rarely parallel-sided, hang downwards from the undersurface of the pileus. These start from the periphery of the pileus and converge towards the stalk, like the spokes of a wheel. However, all the gills do not run the entire length. There are half-length, quarter-length and full-length gills. The gills also differ in their extent of attachment with the stalk. The relation of the gills, with respect to the stalk, is an important taxonomic feature and is described by the following technical terms (Fig. 15.3). The gills may be free from the stalk; adnexed just reach near the stalk but don't fuse; adnate, united with the stipe by their whole width, or decurrent, run down the stem for some distance. Sometimes, the gill has an upward curve near the point of attachment with the stem. Such a gill is called sinuate. It is emarginate when the curve becomes a deep notch.

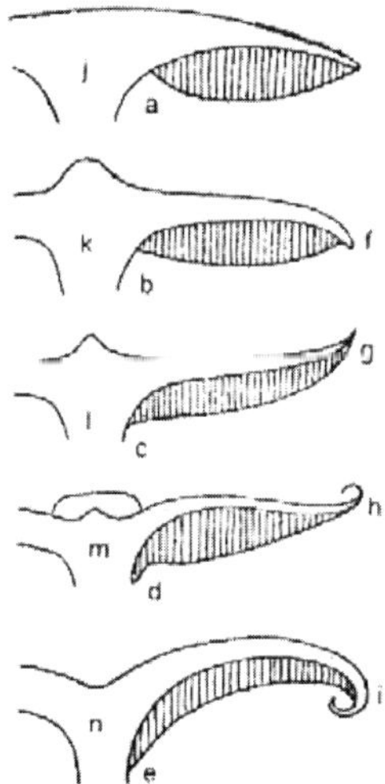

Fig. 15.3. Showing: modes of attachment of gills (a. free, b. adnexed, c. adnate, d. sinuate, e. decurrent) and margin of pileus (f. inflexed, g. reflexed, h. revolute, i. involute); and pileus (j. plano-convex, k. bullate, l. umbonate, m. umbricate, n. depressed)

Internally, the gills are composed of pseudoparenchymatous tissue, called trama, which is made of thin-walled, dikaryotic generative hyphae. It is homoiomerous i.e. composed of a single type of hyphae, the generative hyphae. However, in the family *Russulaceae* the trama of pileus and stipe are heteromerous and contain large, globose or oval cells, the sphaerocysts, alongwith the generative hyphae (Fig. 15.4). In *Lactarius* and some other genera of *Russulaceae*, the flesh contains lactiferous hyphae, which are filled with latex.

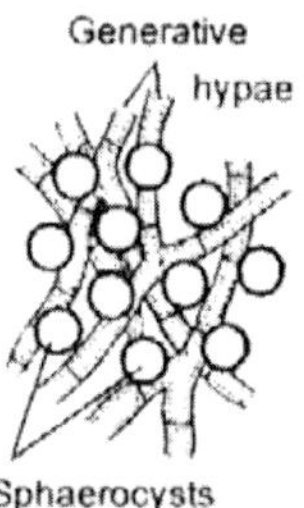

Fig. 15.4. Trama with sphaerocysts

The gills are called aequi-hymeniferous or inaequi hymeniferous, on the basis of their shape and the simultaneous maturation of the hymenium. The aequi-hymeniferous gills are wedge-shaped and the hymenium development is simultaneous all over the surface. The inaequi-hymeniferous gills, exemplified by *Coprinus*, are parallel-sided, and the basidia in the lower region of the gills mature earlier than the rest.

Family *Agaricaceae*

The family includes 85 genera including *Agaricus*, *Calvatia*, *Coprinus*, *Cyathus*, *Lycoperdon* and several other important genera.

Genus *Agaricus* (= *Psalliota*)

Agricus has long been taught in mycology as the typical example of "gill fungi". The 200 species of *Agaricus* grow saprobically on humus-rich soil, decaying leaves, wood, manure-heaps etc. Occasionally, the umbrella-shaped basidiomata appear in circles, especially in grassy lands, forming the so-called fairy ring (see Fig. 15.5). But, like all good things, the beautiful spectacle is short-lived. The basidiomata are ephemeral and survive only for a few days. However, the fungus is not lost. It exists in the soil and continues growing, like a fungus colony, growing in diameter in a Petri dish. A new crop of basidiomata is formed at the fruiting time at the periphery of the fungus colony. This continues for indefinite number of years (Fig. 15.6); the diameter of the, 'fairy ring' extends every fruiting season.

Fig. 15.5. 'Fairy ring' formed by *Agaricus*

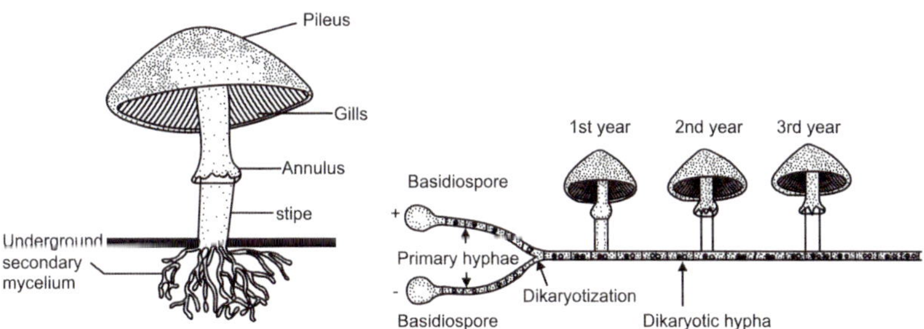

Fig. 15.6. Showing perennial nature of *Agaricus*: the basidiomata continue being formed year after year endlessly unless destroyed.

The basidioma is umbrella-shaped, consisting of a stem (stipe) and a pileus, all made of dikaryotic, tertiary mycelium (Fig. 15.7). The stipe is negatively geotropic and holds the cap high above the ground to enable the vertically-falling spores to be easily drifted away by wind. Anatomically, the stipe consists of an outer cortex, made up of pseudoparenchymatous tissue, and a prosenchymatous central medulla, consisting of less compact, longitudinal hyphae, through which food materials are transported to the pileus.

Fig. 15.7. *Agaricus campestris*

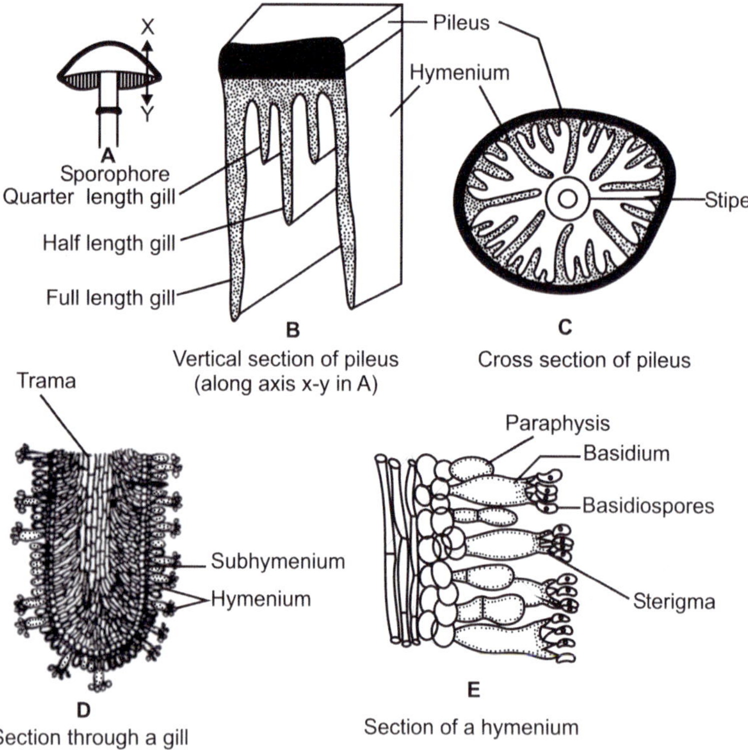

Fig. 15.8. *Agaricus campestris*, anatomy of the pileus

The pileus is a convex, fleshy structure, which later becomes flat. Its upper surface is white or coloured: red, yellow, purple, orange, pink, green or brown. The lower surface bears vertically-hanging gills, which radiate from the periphery and converge towards the stipe. The gills are pink in the beginning, but with age become dark-brown. The gills are of different lengths (Fig. 15.8). There are the full-length gills, which stretch the whole length from periphery of the pileus to the stipe, between two such gills lie the half-length gills, and between each full-length and half-length gill lies a quarter-length gill. Gills are separated from each other by a distance of about 0.5 mm. This distance far exceeds the maximum distance (0.1 mm) to which the basidiospores are shot horizontally in the interlamellar space. This allows the smooth downward fall of the spores.

A vertical tangential section of the pileus shows wedge-shaped gills covered with hymenium. The hymenium consists of elongated, club-shaped basidia of different lengths, arranged at right angle to the surface of the pileus. Intermingled with the basidia are basidioles, which are young, aborted or sterile basidia (earlier called paraphyses). The gills are positively-geotropic. This ensures their downward vertical orientation. The gills are made up of elongated cells in the center, called trama, from which short branches run obliquely towards the surface to form a compact mass of hyphae, called subhymenium. The hymenium arises from the subhymenium.

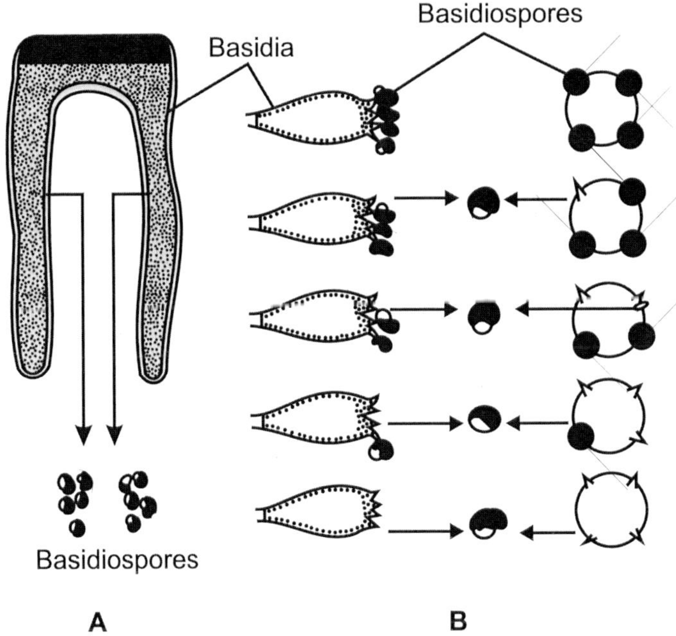

Fig. 15.9. *A. campestris*. A, path (trajectory) of discharged spores; B, showing sequential discharge of the basidiospores.

The basidiospores, which are black or dark-brown, are violently discharged by the 'water-bubble method'. The trajectory (path) of the released spores is a sporobola (Fig. 15.9). Spores are shot horizontally into the inter-lamellar space, from where they fall vertically downwards, and carried away by wind. The distance between the gills far exceeds the distance to which the basidiospores are shot. Even a little tilt of 5° of the gills from the exact vertical position may obstruct the free downward fall of the spores, but this is not allowed to happen. The gills are positively geotropic and any deviation from the vertical position is immediately corrected by growth movements at the place of attachment of the gills with the pileus.

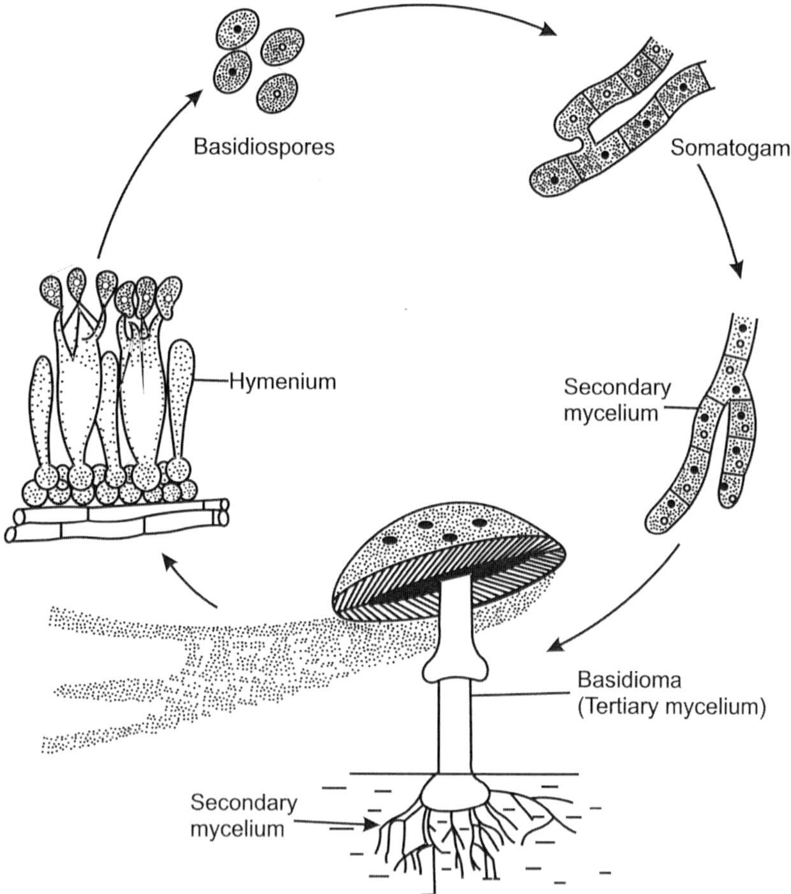

Fig. 15.10. *Agaricus campestris*, the life cycle.

The basidiospores produced in astronomical numbers virtually 'rain down' at a rate of half a million spores per minute, during the 2 or 3 days of the existence of the basidioma. (Fig. 15.10). If the earth is not covered by mushrooms, it is because the chances of failure of spore germination are equally great. The basidiospores may be of 2 or 4 mating types, depending upon the type of heterothallism (bipolar or tetrapolar) operating in the fungus. On germination, they give rise to primary (monokaryotic) hyphae of different mating types. When the hyphae of opposite mating types come together and anastomose, somatogamy occurs resulting in the formation of a dikaryotic cell. A dikaryotic hypha originates from this cell, which gives rise to secondary mycelium. On reaching a particular stage, the umbrella-shaped basidiomata are formed. (Dikaryotic hyphae which form the basidioma are called the tertiary mycelium). As said above, the basidioma lives a short but vigorous life of 2-3 days. After it has fallen and degenerated, the fungus survives through the mycelium, which spread underground extensively, but of course, does not meet the eye.

Mushroom Cultivation

Man as a mycophagist has been interested in mushrooms since ancient times. But the demarcation of the edible from poisonous species was difficult and not fool-proof. This led to mushroom cultivation of reliable and delicious forms. Though it was started in the early 17th century by French horticulturists, mushroom cultivation became a thriving industry only by 1850 in Paris. Until recently it was a gamble growing mushrooms. The crop frequently failed and basidiomata did not appear. But now the knowledge derived from studies of their requirements for fruiting (nutrition, temperature, humidity, etc.) has enabled their cultivation a sure and profitable industry.

There are many species of edible fungi, notably those belonging to genera *Agaricus, Boletus, Lepiota* and *Morchella* (an ascomycete), but *Agaricus campestris* (= *A. bisporus*) is more common. Some other species are now increasingly grown e.g. *Volvariella volvacea* (the 'paddy straw' mushroom) in India and Far East. *A. campestris*, when artificially cultivated shows alterations in its morphology and physiology, and therefore, is designated by the name *A. bisporus* in mushroom industry. It is usually grown on a mixture of horse dung and straw, which must undergo some decomposition before the mushrooms can grow well on it. The mixture is allowed to 'ferment' in compost heaps of 1-2 metres height under controlled conditions of humidity, aeration and temperature. The well composted manure mixture is filled in trays and mixed with mycelial pieces of the size of groundnuts obtained from pure cultures of the fungus. This is called spawning. After 2-3 weeks, the compost beds are covered with a 2.5 cm thin layer of soil or peat and vermiculite mixture. This is called casing. Without this the basidiomata are not formed, but the nature of stimulation or the chemical factors involved are not known. After a month, the basidiomata start emerging. Higher temperatures of the bed (15–21°C) favour mycelial growth while lower temperatures (13–15°C) favour fruiting. So, the temperature is kept below 15°C after the initiation of fruiting.

Genus *Calvatia* ('puff-ball mushroms')

The genus of 'puff-ball mushrooms' has 40 cosmopolitan species It was formerly classified in order Lycoperdales (now invalid) but now the molecular phylogeny has placed it in *Agaricaceae*. The genus came in lime-light on account of the anti-tumor chemical **calvacin**, present in its basidioma. In fact, North-American Indians have long been eating these fungi with the belief (or experience) that it prevents stomach tumors.

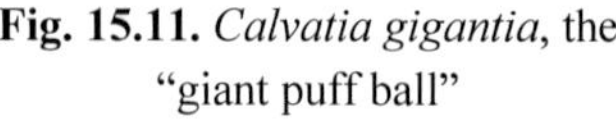

Fig. 15.11. *Calvatia gigantia*, the "giant puff ball"

Buller with a "giant puff ball" (*Calvatia gigantia*)

C. gigantia (Fig. 15.11), the giant puff-ball, is the best-known species, which grows on humus-rich soil near ditches under trees. The basidiomata are creamy-white, globose and large, usually 30-40 cm in diameter, but, occasionally, grows really giant, 1 meter or more in diameter. Even a novice in mushroom hunting cannot miss them. The surface is smooth and leathery. A root-like strand of mycelium remains attached at the base of the basidioma. There is no special exit for the spores. The peridial walls, which are thin and fragile, rupture at maturity, and fall off in flakes. The powdery gleba gets exposed and every gust of wind takes away its share of spores. The fungus has a record in spore production. It has been calculated that a basidioma measuring 40 × 30 × 25 cm would produce 7,000,000,000,000 (7 trillion) spores. What happens to them? Their spore germination is a rare event, but, even then, nature could be so wasteful is not believable!

Genus *Coprinus* (the 'ink-caps')

No student of mycology should go through his course without reading one of Buller's accounts of a species of Coprinus. C.T. Ingold (1974)

The genus is unmistakably distinct in the mushroom world by the property of melting (auto-digestion) of the mature lamellae into an inky fluid, which leaves indelible stain. The whole cap is soon transformed into a black liquid, and hence the name 'ink-caps'. The gills are inaequi-hymeniferous i.e. parallel-sided, and the basidia mature in succession from below upwards (not simultaneously).

There are 10 species of *Coprinus* which grow on humus-rich soil, compost heaps and herbivore dung. The word *Coprinus* means 'living on dung'. The basidiomata vary much in size. *C. comatus* is edible ranking among the best edible fungi. **Coprine,** a metabolite of *C. atramentarius*, gives a reaction in humans similar to **antabuse**. After ingestion it reacts with alcohol to give unpleasant symptoms. Thus, coprine is used in the treatment of chronic alcoholism.

Fig. 15.12 *Coprinus comatus*, the 'shaggy ink caps'

C. comatus, the 'shaggy ink-cap' (Fig. 15.12), initially has an elongated barrel-shaped cap, with white scales all over the surface (hence the name 'shaggy'), but at maturity, it becomes bell-shaped and turns pinkish and finally black in colour. With the rupturing of the **cutis***, the striate nature of

* The thin layer covering the pileus or stipe is called **cutis**. It is made of an **epicutis** and a **subcutis**

the cap becomes evident. The stalk is 1.5 cm, or more, in thickness, white, smooth and shows a distinct annulus. The gills are white when young, turn pink with age, and finally become black at maturity.

As the basidiospores start ripening from below upwards, the gills also undergo digestion in the same direction. The liquified mass, containing basidiospores, drips as a black liquid. The shaggy caps are eaten only in young stage, before the autodigestion sets in.

Coprinus was a large genus until 2001, including 200 species that showed autodigestion of the lamellae. But molecular phylogeny indicated that most of these species were only distantly related to the type species *C. comatus*. These unrelated species were segregated into three genera in the family *Psathyrellaceae* (*Agaricales*).

Genus *Cyathus* ('birds' nest fungus')

Cyathus is the most beautiful among the 'birds' nest fungi'. It has 45 species, which are more prevalent in the tropics. *C. striatus* (Fig. 15.13) is the best known species, which grows on rotting wood, sometimes in large numbers.

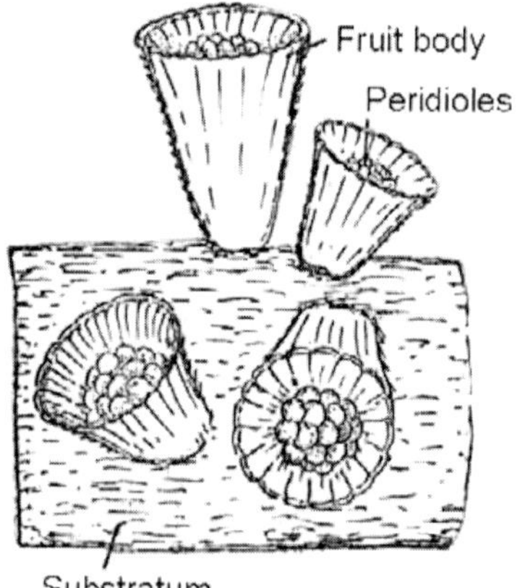

Fig. 15.13. *Cyathus striatus*, the 'bird's nest fungus', showing peridioles nestled inside the basidioma

The basidioma (Fig. 15.14) is initiated at the tip of a mycelial strand, as a knot of hyphae. It is initially globose, but later becomes funnel-, or vase-shaped, due to rupture of the papery **epiphragm,** the membranous covering. The mature basidioma is 1.5 cm high and about 1 cm in diameter at the mouth. It is reddish-brown and hairy outside; the inner surface is shining-gray and marked with longitudinal striations. The species is named striatus on this count.

About a dozen lentil-shaped, white **peridioles**, 2-3 mm in diameter lie neatly nestled inside and attached to the inner peridial wall by means of a short mycelial strand, called **funiculus** (Peridioles are seed-like hard bodies formed by division of the gleba). The wall of the peridiole is many-layered, one of which is very hard. Inside the peridiole is a cavity which, in the young stage, is lined with basidia, but later, when the basidia are digested, the basidiospores lie free in the cavity.

The short funiculus is made of 3 portions: sheath, middle piece and purse. The sheath, which is the lower portion of the funiculus, is a tubular structure and lies attached to the inner wall of the basidioma. The middle piece is a sheaf of hyphae, which connects the sheath to the sac-like purse. The rim of the purse is attached to a short depression on the peridiole. The purse is closed below and packed with a highly-coiled elastic, thread-like extension of the peridiole, the funicular cord, whose lower free terminal end, called hapteron, is sticky.

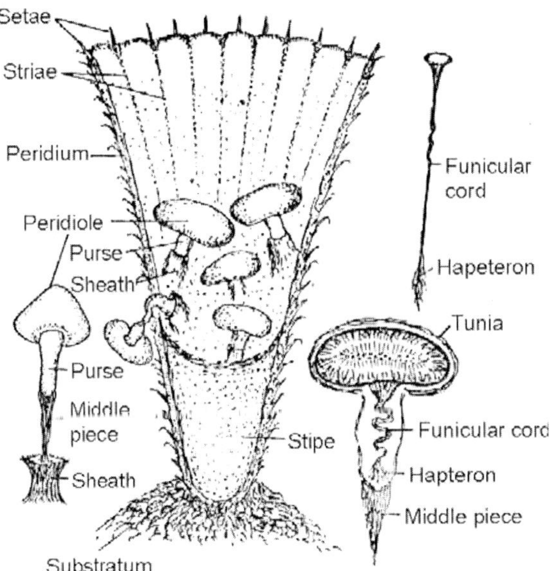

Fig. 15.14. *C. striatus*, semidiagramatic view of a longitudinal section of mature basidioma

The dissemination of the peridioles is brought about by raindrops. The basidioma acts like a 'splash-cup'. When large rain drops (about 5 mm in size) strike the inner wall of the cup, they are reflected back as smaller droplets, which eject some of the peridioles upwards in the air to a distance of 3-4 metres. The peridioles get detached from the bursted purse and fly off, along with the funicular cord behind them. The sticky hapteron gets stuck to some object, like a tree branch or leaf, and the peridiole hangs down by the stretched funicular cord.

The peridioles have long viability period of several years. They germinate on rotten wood and directly give rise to dikaryotic hyphae. Basidiospores, thus, play no role in the life cycle.

Some species, e.g. *C. stercoreus*, are coprophilous, growing on cow–, or horse dung or manure heaps. The germination of the peridioles might be stimulated during their passage through alimentary canal of herbivore animals.

Genus *Lycoperdon* (the 'puff ball')

Lycoperdon (Fig. 15.15), the typical puff-ball, has 50 species, growing usually in pastures on the ground or on tree stumps. The basidiomata are pear-shaped or globose (2-5 cm in diameter), depending on the length of the stalk. These are developed from a mycelial knot at the tip of the mycelial cord, which remains hidden in the substratum, soil or wood and serves as feeding hyphae. It consists of dikaryotic hyphae. Thus, a root like mycelium can be seen at the base of the puff ball. The peridium is 2-layered. The exoperidium is fragile and covered with cone-like spines, which are also fragile and sloughed off on maturity. The exoperidium withers exposing the endoperidium, on which an apical pore is developed.

Fig. 15.15 *Lycoperdon* sp. The 'puff ball'

The mature gleba (fertile portion) consists of numerous irregular chambers, separated by tramal tissue. The chambers are lined with basidia, which bear slender sterigmata of unequal lengths, each capped by a spherical basidiospore. Autodigestion of the gleba, results in break-down of basidia and the tramal tissue, except the hyphae destined to form the capillitium. The mature gleba consists of dry, springy, capillitial hyphae, saturated with dry spores, like a face-powder – 'puff', loaded with powder. In *L. perlatum* (Fig. 15.16), one group of capillitial hyphae form a basal tuft in the cavity, while the remainder converges inwards from the peridial wall.

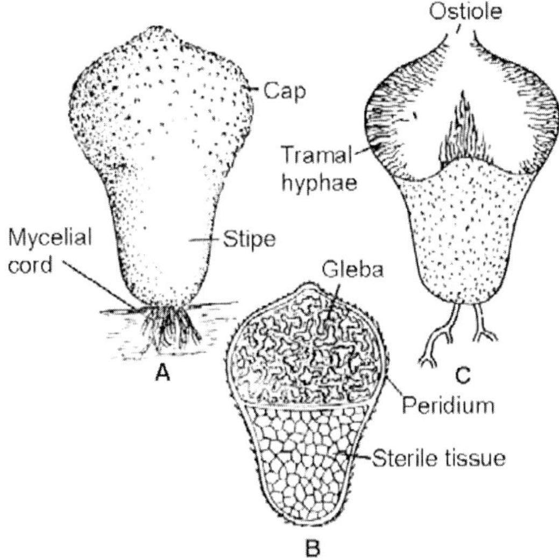

Fig. 15.16. *Lycoperdon perlatum*. A, the mature basidioma, the cap is covered by fragile spines; B, same in longitudinal section showing chambers lined with hymonium; C. capillitium threads in dry condition

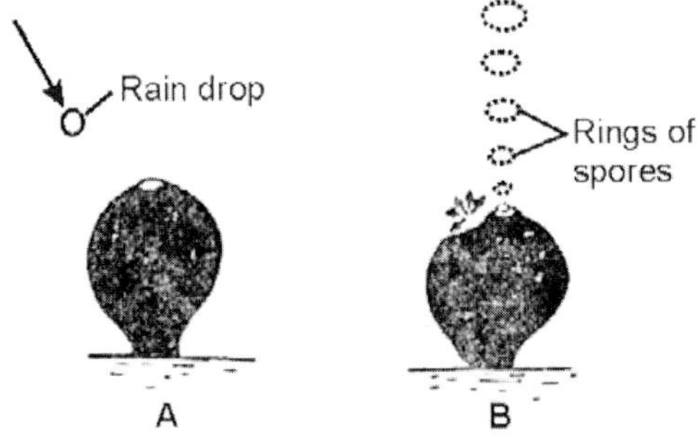

Fig. 15.17. Shows how a puff ball acts like a bellows during spore release. A, before a rain drop hits the puff ball; B, after impact

The endoperidium is unwettable, thin and papery. The 'puff ball' acts like bellows (Fig. 15.17) during spore dispersal. It blows out visible rings of spores (resembling smoke rings blown out by a seasoned smoker). Whenever, any object, like water-drops dripping from tree, strikes it, the endoperidium is depressed expelling a mass of spores in spore rings through the apical pore. The endoperidium soon regains its shape, and is ready for a second exercise. The bellowing continues until all the spores are expelled.

Family *Amanitaceae*

Genus *Amanita*

Amanita has 500 species, and many are ectomycorrhizal. It shows the highest degree of development among the gill fungi. It is also one of the first fungi known to man. On account of their highly poisonous nature, the species were given names which are really horrible! *A. phalloides* (Fig. 15.18, 15.19) by experience was called the 'death-chair' or 'death cap'. *A. verna* was given the name, 'destroying angel' and *A. pantherina*, the 'panther cap! *A. muscaria* (Fig. 15.20), for its use in killing flies, was called, the 'fly agaric'. Most of the species grow underground in mycorrhizal associations, forming external basidiomata in clusters around the base of the trees.

Amanita is easily distinguished from *Agaricus* by (1) the presence of volva, (2) presence of scales on the pileus surface, (3) white basidiospores, (4) free gills, and (5) apically narrow stalk, which is easily separable from the stalk. However, the life cycle of *Amanita* is similar to that of *Agaricus*.

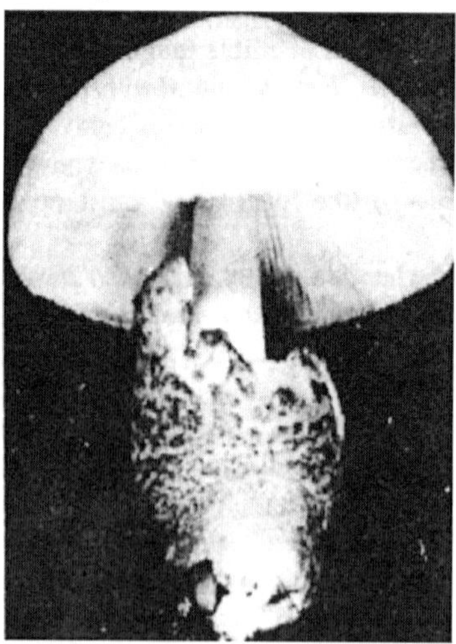

Fig. 15.18. *Amanita phalloides*, the basidioma

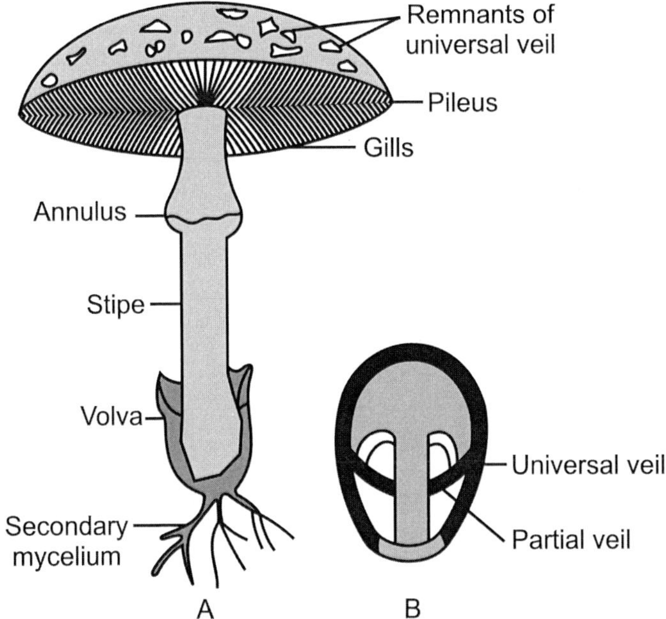

Fig. 15.19. *Amanita phalloides.* A, structure of basidioma; B, showing positions of universal and partial veils.

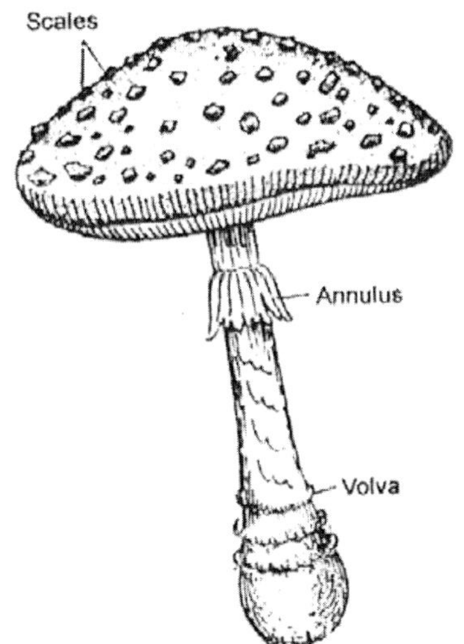

Fig.15.20 *Amanita muscaria*, the 'fly agaric'

A. phalloides is extremely poisonous. A tea-spoonful of the mushroom is enough to cause most painful and lingering death, against which there is no antidote. The cap is about 10 cm in diameter, usually yellowish-green or olive, but, sometimes, it is brownish or even white. *A. verna*, (the destroying angel), which is pure white, with a larger cap (up to 25 cm), is the most deadly mushroom.

A. muscaria was once used as a fly-poison. The cutis of the pileus is sticky and poisonous, and if this is removed, the mushroom becomes innocuous and eatable. The fungus is used in the preparation of intoxicating beverages. In addition to this, the basidioma is favourite of craftsmen, who use it for decorative purposes. The stem is long (25 cm or more), swollen at the base and bear the cap, which is upto 20 cm in diameter.

A. citrina (false death cap) is harmless but unpleasant and unacceptable to mushroom eaters.

Family *Clavariaceae* (club or coral fungi)

'Clavarias' are the most beautiful fungi. They form conspicuous, brightly-coloured, club–, or coral-like, simple or branched basidiomata. The hymenophore is smooth, spread over the entire surface or restricted at the apex. The basidioma is negatively geotropic, but the hymenophore is independent of gravity. There are 7 genera and 120 species, most of which are lignicolous, growing on dead wood, while a few are pathogenic on crop plants. Some are edible e.g., *C. vermicularius*. Thind (1961) in his monograph *Clavariaceae of India* described the Indian collections, some of which now belong to other families.

Genus *Clavaria*

The genus was unknown in India until Thind (1961) reported 28 species. The flesh of *Clavaria* consists of thin-walled hyphae, without clamp connections. It is found in grass fields, growing in clusters on the ground and, sometimes, also on wood. *C. fusiformis* forms spindle-shaped, bright-yellow basidiomata, 5–10 cm high, which are often wavy or irregular, and somewhat flattened or grooved, adhering closely at the base. The tips turn brown with age. The spores are white, but appear yellow in mass.

In *C. vermicularius*, the "wax tapers" or "fairy fingers' (Fig. 15.21) the basidiomata are brittle, pure white and cylindrical with pointed tips, looking like white wax tapers. With age, they become hollow. *C. cineria* is the most common, branched species, forming greyish, highly branched tufts, 2–7 cm high and arising from a short, thick stem. The tips of the branches are blunt, sometimes flattened and toothed. The basidiospores are white. *C. indica* is an interesting species reported from Himalayas, which forms dark-brown to black, club-shaped basidiomata.

Fig. 15.21 *Clavaria vermicularius*, wax tapers or fairy fingers

Family *Pleurotaceae*

Genus *Pleurotus*

There are 20 species which are large and fleshy and grow on tree trunks or stumps in shelf-like layers. It is a 'gilled' mushroom, which includes one of the most widely eaten mushrooms, *P. ostreatus* (the oyster mushroom). The stipe is lateral, eccentric or even absent. The term "pleurotoid" is used for mushrooms having this shape. The pileus is oyster or shell-shaped (Fig. 15.22) and hence the name. The young cap is deep bluish-gray or black, but soon turns pale and brownish. The surface is smooth and moist with the edge bent downwards. The other cultivated species are *P. flabellatus, P. sajor-caju, P. sapidus* and *P. citrinopileatus. P. alearius* is poisonous. *Pleurotus* is now one of the most commonly cultivated mushrooms in the world. Because of its flexible temperature and environmental requirements, *Pleurotus* has more cultivated species than any other mushroom. Present production of this mushroom is over 900,000 tons per year. *Pleurotus* species are used in mycoremediation of pollutants such a petroleum and polycyclic aromatic compounds.

Fig. 15.22. *Pleurotus ostreatus*, the "oyster mushroom"

Family *Plutaceae*

Genus *Volvariella* ('paddy straw mushrooms')

The genus has 50 cosmopolitan species. *V. volvacea* (Fig. 15.23) and *V. diplasia* are called "paddy-straw mushrooms". These are fast growing mushrooms with a slender, fleshy stipe, 3-8 cm in length and a dark-gray cap, 6–12 cm in diameter. It is cultured in rice straw (hence the name) and contributes 16% of the total cultivation of mushrooms in the world. Unfortunately, it closely resembles the highly poisonous *A. phalloides*, and has been the cause of lethal mushroom poisoning. *Amanita* produces white spores while *Volvariella* produces pink spores. But the two can not be distinguished in the "button stage" when these are collected. In DNA sequences the genus is closely related to *Schizophyllum commune*.

Family *Schizophyllaceae*

The family with 2 genera and 7 species is characterized by **split-gills**. The basidioma, which is initially cupulate and later discoid, lies attached to logs or bark of trees, by a narrow base. Longitudinally-split gills are borne on the lower surface. The hyphal construction is monomitic, or rarely feebly dimitic. Basidiospores are hyaline, not amyloid. The species cause wood-rot and are pathogenic for human.

Fig. 15.23. *Volvariella volvacea*

Genus *Schizophyllum*

The important distinguishing feature of *Schizophyllum* (6 spp) is the **marginal proliferation** of the basidioma, and **split gills**. *S. commune* is an important cosmopolitan fungus, used extensively in cytogenetical studies. It grows on decaying logs and branches of beech (*Fagus*), forming white, fan-shaped basidioma, 1-3 cm in diameter (Fig. 15.24).

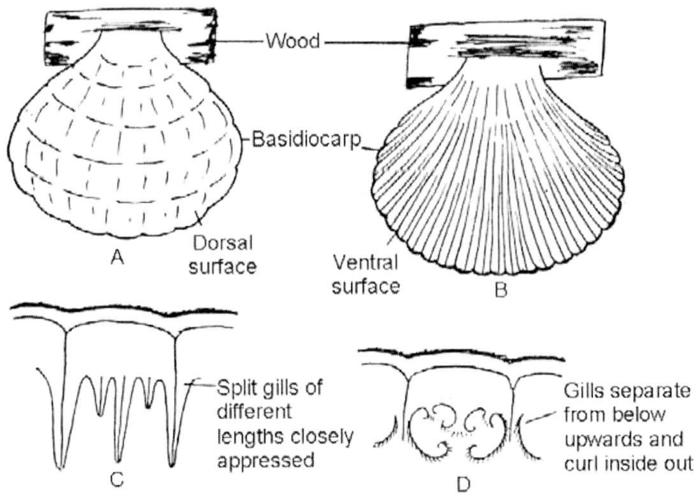

Fig. 15.24. *Schizophyllum commune* basidioma. A, dorsal view; B, ventral view; C, vertical section of gills in humid weather; D, same in dry weather

The basidioma is initially cupulate, later becomes fan-shaped (due to faster growth on one side) and remains attached to woody substratum. It has hairs on the upper surface and gills on the lower. The gills, which are of different lengths and depths, radiate from the point of attachment at the lower surface. These are longitudinally split (hence the name *Schizophyllum* (schizo =split, *phyllum* = leaf or gill). The two halves of the gills, bearing hymenia on their external surface, remain closely appressed, but when the weather becomes dry, they curl outwards to protect the hymenia from drying. The outward curling is brought about by the collapse of a group of thin-walled cells at the base of the gills. The basidiospores are hyaline and binucleate. The dry basidiomata can survive for several years and return to active life as soon as they get moistened. **Buller** (1909) left a large number of air-dried basidiomata of *S. commune*, sealed in evacuated glass tubes, to be opened for testing their viability by later generation of mycologists. When the basidiomata were taken out after 50 years and moistened, the basidiomata came to active life and discharged spores (see Ainswoth, 1962). Love for life is not wanting even in small and lowly organisms (?) like a tiny fungus.

Family *Strophariaceae*

The epicutis* of the pileus is composed of appressed, narrow, tubular, often gelatinous hyphae. The basidiospores are brown and possess an apical germ pore or a minute apical discontinuity. The stipe and pileus are confluent i.e. their flesh are not separable. There are 18 genera and 1316 species. We shall study the genus *Psilocybe*.

Genus *Psilocybe* (the 'sacred fungi')

The genus has 300 species, many of which are hallucinogenic. The Mexican Indians consumed them in religious magical ceremonies and, hence, the name 'sacred fungi'. *Psilocybe semilanceata* (Fig. 15.25) is commonly known as 'liberty cap', on account of the conical and pointed cap of its basidioma, which resembles the 'cap of liberty', the symbol of French Republic. The conical pileus, which is 1.5 – 2 cm in diameter, is pale-buff when dry, but becomes darker when wet. The stipe is long, slender and often wavy. It can be collected from fields, road sides and city lawns.

---

* **Cutis,** the thin layer of hyphae that covers the pileus or stipe, is made of **epicutis** and a **subcutis**.

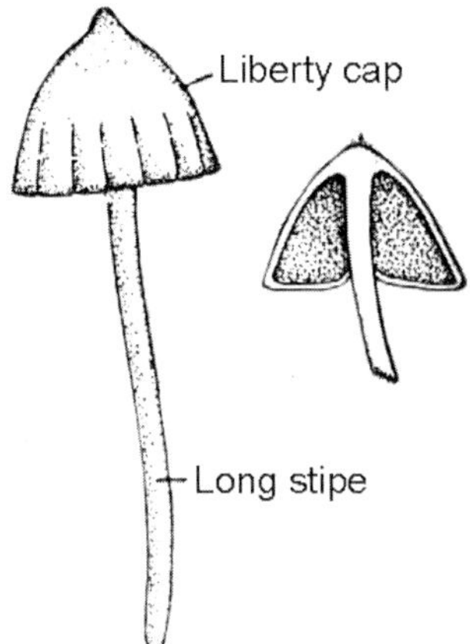

Fig. 15.25. *Psilocybe semilanceata*, the "liberty cap", the halluctinogenic "sacred mushroom".

The hallucinogenic property has been pinned down on two chemicals, named **psilocybin** and **psilocin**, both indole derivatives with similar effects. These simulate the effect of LSD (d-lysergic acid diethyl amide), a product of the 'ergot', (*Claviceps purpurea*) viz. loss of the sense of time, space, and a feeling of extraordinary lightness or hovering. The eater, as they say, "takes a trip" to a colourful dream land! The species, *P. rugoso-annulata* ("strophaire") is edible.

Family *Physalacriaceae*

This is a family segregated from *Tricholomataceae* in 2006. It contains 11 genera, including *Armillaria* and *Strobilurus*. The stipe (if present) is confluent with the pileus. The colour of the spores is variable (white, yellowish, buff, pink or brown) and the lamellae show all possible forms of attachment with the stipe; rarely free. The hyphae lack clamp connections.

Genus *Armillaria* ('Honey fungus' or 'Shoe-string fungus')

The genus has 35 widespread species that caue serious root diseases in woody plants. Some form orchid mycorhizas. The genus is characterized by (1) yellow brown pileus (2) white basidiospores, (3) adnate or slightly decurrent gills, and (4) presence of annulus. *A. mellea* (Fig. 15.26), the most important species, is a parasite causing root-rot diseases, especially of forest trees. It is popularly known as the 'honey fungus' (on account of the tawny colour of the basidioma) and also as the 'shoe-string' fungus (on account of the black rhizomorphs* that it forms). The mycelium of *Armillaria gallica* has been estimated to extend over 150,000 square meters (37 acres) having an age of 1500 years, obviously oldest living organism (Smith *et al.*, 1992).

Fig. 15.26. *Armillaria mellea*

The fungus spreads to long distances in the soil and infects roots of trees with the help of shoe-string-like rhizomorphs. The role of basidiospores in infection and spread of the fungus is negligible. They fail even in artificial inoculations to infect living roots. However, they can colonize dead roots. Previously-infected and dead host trees serve as a food base, from which the rhizomorphs arise and spread in all directions, up to long distances, through the soil to attack living trees (Fig. 15.27).

What a single basidiospore, or an individual hypha, fails to do, is done by the joint action of the aggregated hyphae in the rhizomorph. This is called **inoculum potential** in plant pathology. The rhizomorphs enter the root

---

***Rhizomorphs** are rigid structures formed by aggregation of unbranched secondary hyphae, which act as one unit and grow by an apical meristem. The central hyphae are loosely woven and serve for conduction of food materials from the food base (the tree) to the growing tip. Surrounding the central hyphae is a compact dark rind of hyphae. Branches of the rhizomorph are produced endogenously in a manner similar to the branches of a plant root. A remarkable feature of the rhizomorph is that it grows about five times faster than an unorganized mycelium. The rapid growth is probably due to the fact that the constituent hyphae, being unbranched, grow in one direction only. *A. mellea* forms rhizomophs freely on nutrient media in laboratory.

tissues, encircle the basal portion of the trunk and kill the tissues lying external to the vessels. If the bark of dead trees is removed, one can see the anastomosing rhizomorphs. Basidiomata develop in clusters around the base of the dead tree. *A. mellea* also shows the phenomenon of **bioluminescence**; the luminescent part being the hyphae. Infested wood pieces glow in the dark.

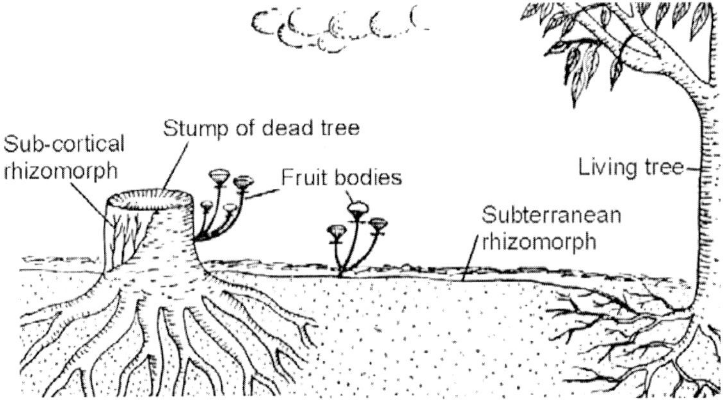

Fig. 15.27. *A. mellea*. Diagram showing mode of attack of living beech tree by rhizomorph originating from a dead stump

Order *Boletales*

The order includes poroid, lamellar and gasteroid fungi. Earlier only poroid fungi were included in this order. There are 16 families, which can not be characterized on morphological characters. These are terrestrial, mostly ectomycorrhizal or lignicolous (causing brown rots); sometimes parasitic on ectomycorrhizal members of *Boletales*. Some are edible while others are fatally poisonous. We shall examine 2 families: *Boletaceae*, which includes mostly poroid forms but some are gilled fungi, and *Sclerodermataceae*, comprising gasteroid fungi.

Family *Boletaceae* ('boletes' or 'pore mushrooms')

'Boletes', as the members of the family are called, are poroid (i.e. having tubular hymenophore) although some, like *Phylloporus*, are lamellate. There are 35 genera; most of them are ectomcorrhizal on conifers and other woody plants. *Boletus* is the most important genus.

Genus *Boletus*

There are 300 species which are ectomycorrhizal. Some are edible. *B. edulis*, the 'king boletes' (Fig. 15.28) is the most popular edible species. It is sold in pulverized form and used for flavouring curries. It has a nutty taste and very favourite of connoisseurs. The brown, rounded pileus (6 – 20 cm) resembles a 'bun'. The lower surface of the pileus is like a sponge with numerous minute pores, which represent the openings of the tubes. The stem is white or pale, 5-10 cm long, with distinct reticulate markings. The tubes and flesh are also white. For cooking, the tubes are removed and the flesh is sliced and cooked immediately, or dried and stored for future use. *B. satanas* is poisonous; *B. parasiticus* grows parasitically on *Scleroderma cirtinum* (*Sclerodermataceae*).

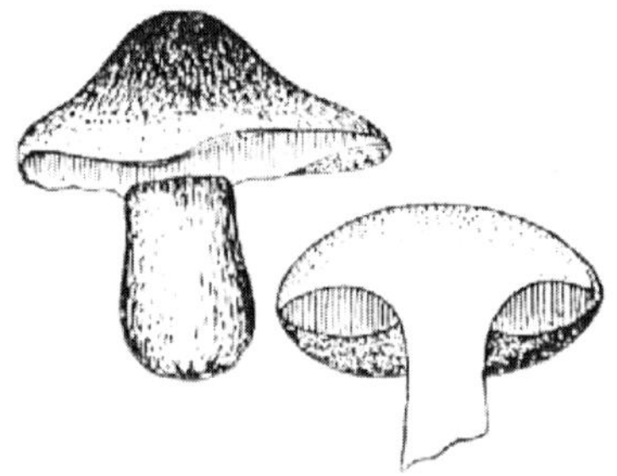

Fig. 15.28. *Boletus edulis* (brown, rounded pileus resembles a bun)

Family *Sclerodermataceae* ('hard-skinned puffballs', 'earthballs', 'prettymouths')

There are 6 genera and 39 species, commonly known as 'hard-skinned puffballs', 'earthballs', 'prettymouths' etc. The best known members are *Scleroderma* ('earthballs'), *Pisolithus* ('dye fungus') and *Calostoma* ('pretty-mouths'), which are mostly ectomycorrhizal with woody plants or grow on rotten woods, or bare soil.

Genus *SCLERODERMA* ('earth balls')

There are 30 species of *Scleroderma* (Fig. 15.29) which grow most often on bare soils, in woods and waste lands. The basidioma is epigeous, sessile, round and flat, about 5-8 cm in diameter and attached to soil by mycelial cords. It has outer peridium surrounding the inner fertile portion, the gleba. The peridium is thick, hard, rough, and yellow in colour. The gleba is of lacunar type containing several cavities lined with clavate basidia.The cavities are separatd by sterile tissue, called trama. The mature gleba crumbles to purple-black powder of spores and disintegrating mass of tissue. The peridium ruptures irregularly exposing the spores for wind dissemination by wind. The spores are over 10 μm in diameter and show reticulate or spiny ornamentation. *S. citrinum* is parasitized by *Boletus parasiticus.*

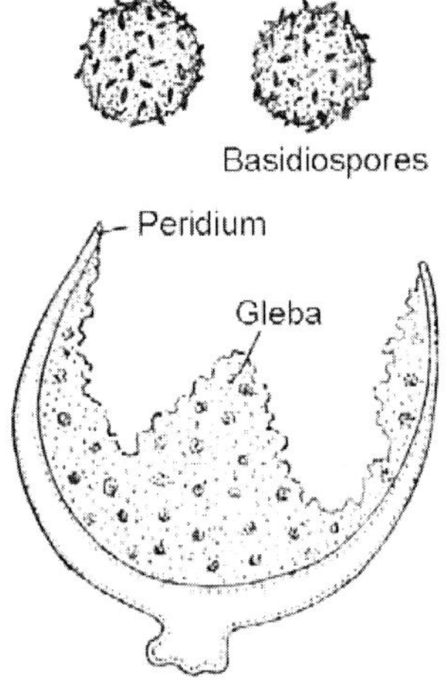

Fig. 15.29. *Scleroderma citrinum.* Longitudinal section of a mature open basidioma showing the spore mass (gleba); magnified view of spores

Subclass *Phallomycetidae* : Order *Geastrales*

Family *Geastraceae*

In this family of 'earth-stars', the peridium is 3-layered, which at maturity, splits from top downwards, into radial lobes, which turn outward and like rays of a star. The exposed endoperidium develops an apical pore. The basidia are capitate; sterigmata are of equal lengths; capillitium is made of unseptate, unbranched capillitial hyphae, which aggregate to form a pseudocolumella. Capillitial hyphae radiate from the pseudocolumella, and also from the undersurface of the endoperidium, into the cavity. There are 7 genera, including *Geastrum* and *Sphaerobolus*.

Genus *Geastrum* (The earth star)

There are 50 species of *Geastrum*, but they are not common. They show much structural variations among themselves. In *G. triplex* (Fig. 15.30), the young, onion-shaped basidioma, lies buried in the leaf-litter of beech woods, with its pointed end projecting above the surface. The 3-layered exoperidium (the innermost of which is fleshy) ruptures along radial lines from apex downward, and the lobes bend outward, like rays. The lobes bend with such a force that the basidioma is lifted up and detached from the mycelial cord at its base (Fig. 15.31). The endoperidium persists and forms an apical pore, through which the spores are expelled in installments by bellowing action.

Fig. 15.30 *Geastrum triplex*, the 'earth star'.

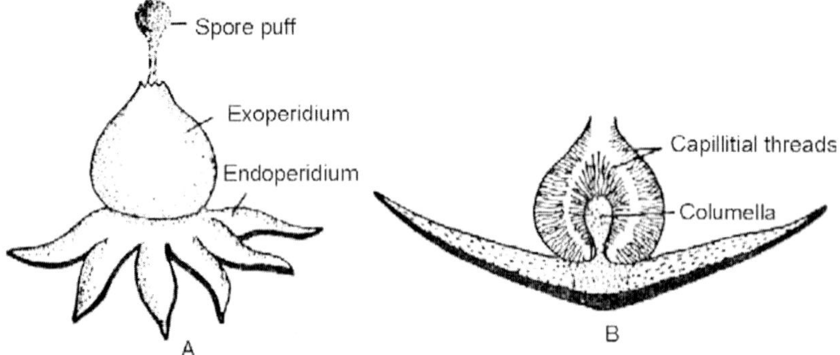

Fig. 15.31. *Geastrum* sp. A, expanded basidioma looks like a star; B, vertical section through the basidioma showing sheafs of capillitium threads attached to the wall and the columella

Genus *Sphaerobolus*

Sphaerobolus is by far the most powerful piece of fungal artillery.

C.T. Ingold

The genus *Sphaerobolus* (Fig. 15.32) has 2 species, which are quite common, growing freely in old, rotten wood and old cow-dung. *S. stellatus* grows well and forms basidioma on suitable culture media, like oat-meal agar. It requires light and a temperature below 25°C. The basidiomata are spherical, minute (2 mm across), with stellate (star-like) peridium. The endoperidium everts to eject the single globose peridiole, which contains the gleba and acts as a unit for dispersal.

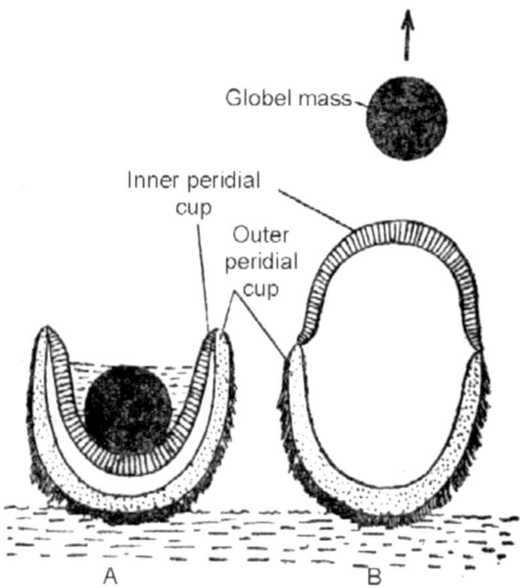

Fig. 15.32. *Sphaerobolus stellatus*. A, L.S. of an open sporophore before discharge of spores; B, same at the moment of discharge of the gelatious mass.

The basidiomata grow crowded in large numbers. Longitudinal section shows a spore-bearing gleba and 6-layered peridium. Finally, when the peridium splits, it separates into 2 cups with toothed margins, one fitting inside the other, but in contact only at the margins. The outer cup is derived from the 3 outer layers of the peridium, and the inner cup from the subsequent two layers. The sixth layer is digested to form a fluid, so that, the glebal mass is loosely placed inside the inner cup. The inner cup, which is made of highly turgid tissues, absorbs water and turns inside out with great speed, hurlıng the glebal mass, containing the spores to long distance (upto 7 metres). *Sphaerobolus*, according to Ingold (1967), "Is by far the most powerful piece of fungal artillery". The basidioma is positively-phototropic and the spores are thrown towards lighted areas. The spores are sticky. All these are features characteristic of coprophilous fungi.

Incertae sedis Orders of *Gasteromycetes*

Order *Auiculariales* (1 family)

Family *Auriculariaceae*

The family is characterized by (1) formation of septate; 1-4 celled basidia (phragmobasidia), (2) gelatinous or waxy basidiomata that are hemiangiocarpic and sessile, and (3) repetitive basidiospores. There are 7 genera and 112 species which are saprobic; some are parasitic on plants and fungi.

Genus *Auricularia*

This is the most common genus with 15 species. The basidioma of *A. polytricha* is edible and highly priced in China and Hong Kong, where it is grown on Quercus poles. *A. auricula-judae,* the 'jew's ear' (Fig. 15.33) is the most common species (The basidioma resembles an ear and hence the name). It grows throughout the year on dead branches of elder (*Sambucus*) trees. The fleshy basidioma is gelatinous, limp, tough, partially translucent and red-brown or maroon in colour. It is velvety on the upper surface and veined on the lower. When water becomes deficient, the basidioma dries up and becomes horny. On wetting, it quickly regains its shape and within a few hours starts discharging basidiospores.

Fig. 15.33. *Auricularia auricula-judae*, the 'Jew's ear'

The basidioma is made up of dikaryotic hyphae having clamp connections. A section of the flesh shows a hairy upper surface, a central gelatinous mass of hyphae and the hymenium at the lower surface (Fig. 15.34). During formation of basidia, a large number of hyphal ends become differentiated into cylindrical probasidia, separated by a basal septum. Karyogamy takes place in the probasidium. The probasidium becomes a metabasidium (=hypobasidium) with the occurrence of meiosis, and the haploid nuclei become separated by transverse septa. An epibasidium arises from each cell, which grows through the gelatinous mass and emerges on the lower surface.

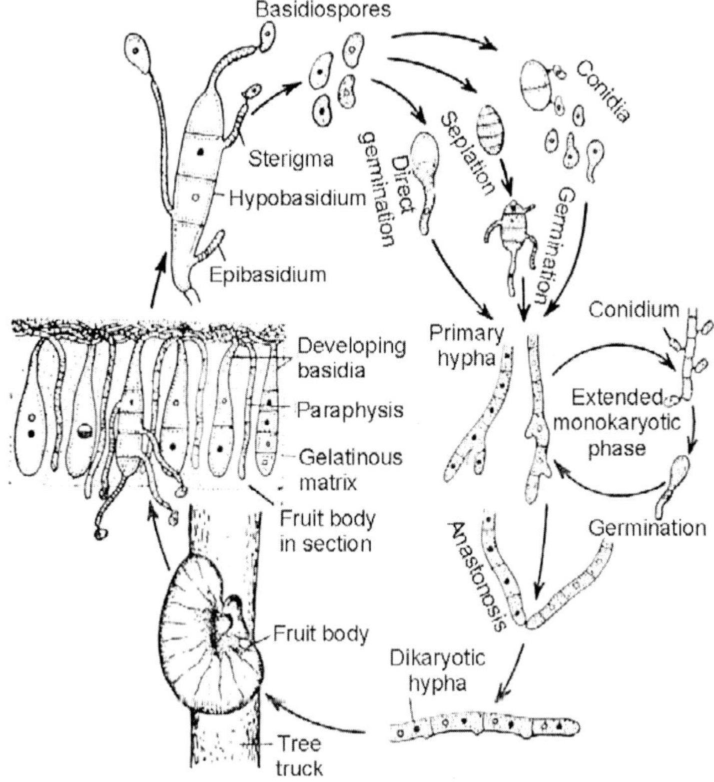

Fig. 15.34. Life cycle of *Auricularia auricula - judae*

A curved basidiospore develops on a pointed sterigma at the apex of each epibasidium, and is violently discharged.

The basidiospores germinate in three ways (1) directly through a germ tube, (2) by forming conida on short sterigmata, formed on the basidiospore after septation, and (3) septation of the basidiospore into 3-4 cells and formation of germ tubes from each of these cells. The primary mycelium established as a result of basidiospore germination, reproduces asexually by means of sickle-shaped conidia. These conidia may germinate directly or form secondary conidia by budding, which then form primary hyphae after germination. Eventually, plasmogamy takes place, when primary hyphae of compatible strains meet and anastomose. This results in the formation of a dikaryotic cell, from which the dikaryotic mycelium develops and forms the ear-shaped basidioma.

Order *Cantharellales*

Cantharellales is an early diverging clade of *Agaricomycetes*. Specifically, members of *Cantharellus*, *Craterellus* and *Tulasnella* have high divergent nuclear rRNA sequences that do not match entirely to other *Agaricomycetres* (Hibbett *et al.,* 2007; Matheny *et al.,* 2007). The order has 7 families and 38 genera, which are mostly ectomycorrhizal or saprobic. Some are plant pathogenic, while some others are highly valued as delicious mushrooms. We shall examine the following 3 families and genera: *Cantharellaceae* (*Cantharellus* and *Craterellus*), *Ceratobasidiaceae* (*Ceratobasidium*), and *Hydnaceae* (*Hydnum*).

Family *Cantharellaceae*

There are five genera and 92 species having pileate basidiomata (i.e. having a pilus). The hymenophore has radial folds.

Fig. 15.35. *Cantharellus cibarius*, the 'European chanterlle'

Genus *Cantharellus*

The basidioma is funnel-shaped with a short, black stem (Fig. 15.35). The hymenophore comprises thick folds, resembling mushroom gills. The folds are irregularly branched and united to form a net-like structure, which runs downwards towards the stem. Several species are edible. *C. cibarius,* the famous 'European chanterelle', has the pride of place, which sells costliest in French markets. Its basidiomata are egg-yellow and emit pleasant smell of apricots. *C. tubiformis* is the next-best relished species.

Genus *Craterellus*

The genus has 20 species which form fleshy, funnel-shaped basidiomata with a short black stem (Fig. 15.36). The hymenophore, which lies on the outer surface, is smooth or occasionally wrinkled but not veined. *C. cornucopioides* is edible and is found in large numbers (called 'horn of plenty') in deciduous forests during autumn. The basidiomata lie on the ground hidden under fallen leaves and, therefore, are not easily visible. The fungus is cooked and eaten fresh.

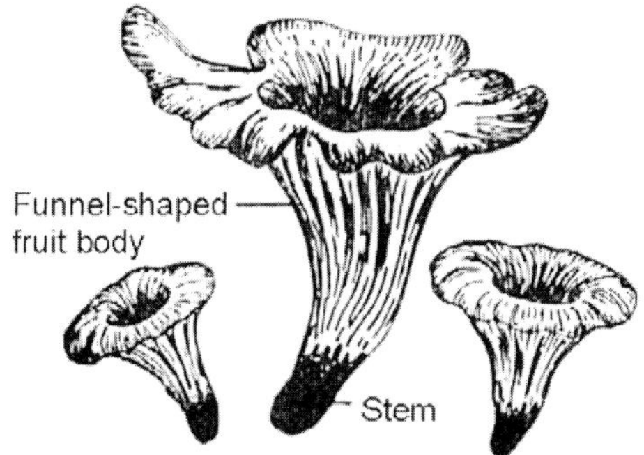

Fig. 15.36. *Craterellus cornucopioides*, "horn of plenty"

Family *Ceratobasidiaceae*

The family *Ceratobasidiaceae* includes 8 genera and 43 species. Many species of *Ceratobasidium* and *Thanetophorus* are the teleomorphs of *Rhizoctonia* which is a destructive plant pathogen.

Genus *Ceratobasidium*

The genus has an inflated, aseptate metabasidium, 2-3 times broader than the supporting hyphae, and abruptly narrowed at the pedicels (Fig. 15.37). Four long epibasidia (as long as the hypobasidium) bear basidiospores apically. Occasionally, the epibasidia are forked, and frequently septate. The basidiospores are repetitive. There are 16 species, which are saprobic or parasitic. Species that form sclerotia represent the teleomorph of *Rhizoctonia.*

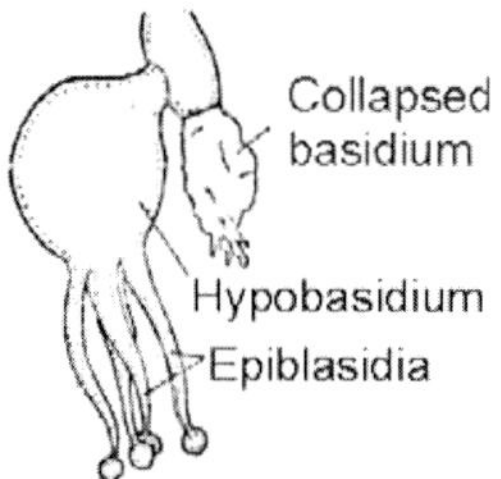

Fig. 15.37 *Ceratobasidium cornigerum*, one collapsed and one mature basidium

C. *anceps* is a parasite on bracken fern (*Pteridium*), forming 'web-blight' on aerial parts. It produces brown sclerotia. *C. cornigerum* (formerly *Pellicularia*) is mycorrhizal. *C. calosporum, C. obscurum* and *C. pseudocornigerum* are saprobic on rotten wood, soil or plant parts that lie in contact with the soil.

Family *Hydnaceae* ('hedgehog' fungi)

The hydnums, as the members of the family are called, bear toothed or poroid hymenium. The basidiomata take many years for maturation, and that is why these fungi are not frequently found. They may be terricolous or lignicolous; the former species occur usually in coniferous or decidous woods under *Pinus, Abies, Picea*, Oaks etc., and grow best in cool humid weather. The lignicolous species are more widespread. The basidiomata are pileate, centrally-stipitate like the agarics, but are easily identified by the absence of the gills. There are 8 genera and 43 species.

Genus *Hydnum*

It is a terricolous genus with 120 species, growing on humus in woodlands. *H. repandum* (Fig. 15.38), the lone species reported from India, is edible. It is common in mixed forests in North-West Himalayas. The basidiomata are pileate and mushroom-shaped. The cap is fleshy in texture, smooth or slightly scaly, pinkish-buff in colour and 5-8 cm in diameter. The hyphal construction of the flesh is monomitic. Positively geotropic spines, of different sizes and bearing the hymenium, hang down from the lower surface of the pileus. The flesh is brittle and tends to crack.

Order *Polyporales* (Former Aphyllophorales, Ganodermatales)

The order includes polypores (but not all), many 'corticoid' (resembling *Corticium, Corticales*) and 'agaricoid' fungi (in the genus *Lentinus*). There are 13 families and 216 genera. We shall study *Ganodermataceae* and *Polyporaceae*.

Fig. 15.38 *Hydnum repandum*, mushroom-like basidioma but toothed, not lamellate

Family *Ganodermataceae* (4 genera)

Genus *Ganoderma*

('Shelf' or 'bracket' mushrooms)

Ganoderma (Fig. 15.39) has 80 species, which are world wide in distribution. These form large, woody, shining, bracket-shaped basidiomata, also called 'conks'. (*Gano* = shining, *derma* = skin). These cause 'white rot' of wood. *G. lucidum*, the type species, is common in India, causing destructive root diseases of *Dalbergia, Albizia, Acacia* and some other trees. It forms annual, bracket-shaped, stalked, brightly-shining basidiomata on the tree runk. *G. applanatum*, the best known species, is a wound parasite. It forms woody, perennial, bracket-shaped basidiomata, 50-75 cm across, broadly and horizontally attached to the tree trunks by means of a stalk. It is called artist's conk. Tube layers are formed every year, for 5-10 years. The annual tube layers can be seen in a vertical section of the basidioma. *Ganoderma* differs from other bracket fungi in having a much longer span of spore release, extending up to 5 months. The basidiospores are 2-layered and truncated apically.

Ganoderma, especially *G. lucidum,* has been used in traditional medicine in Asia for over 4000 years and has recently been rediscovered by the western world. It is called **the magic mushroom** or **miraculous king of herbs**. It is called **lingzhi** in Chinese medicine, **Reishi** in Japan and **Youngchi** in Korea. It promotes general good health and has hormonising effect on body balance. It has therapeutic effects against cancer and has immunoregulatory effects. It strengthens several body functions like blood cholesterol, hypoglycemia, radiation damage and several others.

Fig 15.39. *Ganoderma applanatum*

Family *Polyporaceae*

Genus *Fomes*

The genus has 3 species, which form bracket-shaped basidiomata on tree trunks (Fig. 15.40). The basidiomata are perennial and characterized by corky or woody texture. The pore surface is pure white, smooth and flat, with extremely fine pores. It is also called artist's conk. A new tube layer is added each year beneath the emptied and collapsed layer. This results in stratification of the zone of tube layers (Fig. 15.41). The successive tube layers are separated by a layer of sterile tissue. By continued growth, the basidiomata attain a huge size; sometimes 1 meter or more in diameter. The flesh is white or pale yellow, and hard, like wood.

Fig.15.40 *Fomes fomentarius*

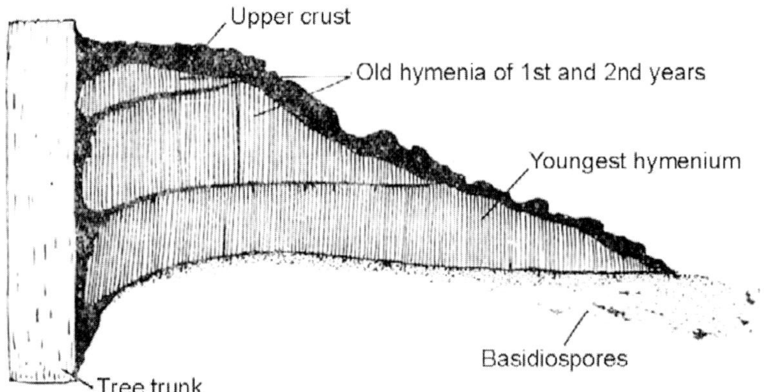

Fig. 15.41. Vertical section through a 3 years old basidioma of *Fomes* showing three zones of tube layers

F. fomentarius, the type species, is very common in North-Western Himalayas causing heart-rot (decay of inner wood) of conifers and hard woods (Thind, 1973). The fruitbody looks like horse's hoof (also called 'hoof fungus'). The fungus has been earlier called as *Boletus fomentarius* (Linnaeus, 1753), *Agaricus fomentarius* (Lamarck, 1783), and *Polyporus fomentarius* (Meyer, 1818, Fries, 1881).

Genus *Polyporus*

It is a large cosmopolitan genus with 26 species which are highly destructive parasites of trees. *P. sulphureus* causes wood rot of oaks; *P. squamosus* (Fig. 15.42) causes heart-rot of elms and sycamore, and *P. schweintzii* causes butt-end rot of several trees. *Polyporus* is now used for production of single-cell proteins.

The basidioma is stipitate and annual. When the single layer of tubes is exhausted, the basidioma degenerates. However, the hyphae survive inside the tree and form the basidioma again next year. In *P. tuberaster* (the type species of the genus), the fruit body is centrally-stalked, but in majority of the species, the stalk is lateral and inconspicuous. The basidiomata are bracket-, or fan-shaped.

P. tuberaster forms large underground, sclerotium-like structure, which, when moistened and placed in a warm place, gives rise to several centrally-stalked, funnel-shaped basidiomata. The fleshy basidioma is edible and very popular in Italy. People store the sclerotia, which they call **pietra fungaia**, and grow the basidiomata from them at will by moistening. *P. umbellatus* (centrally-stalked) is the other edible species growing on ground above buried wood, or around the base of old trees and stumps. The basidioma consists of a highly-branched stalk bearing caps.

Fig. 15.42. *Polyporus squamosus*

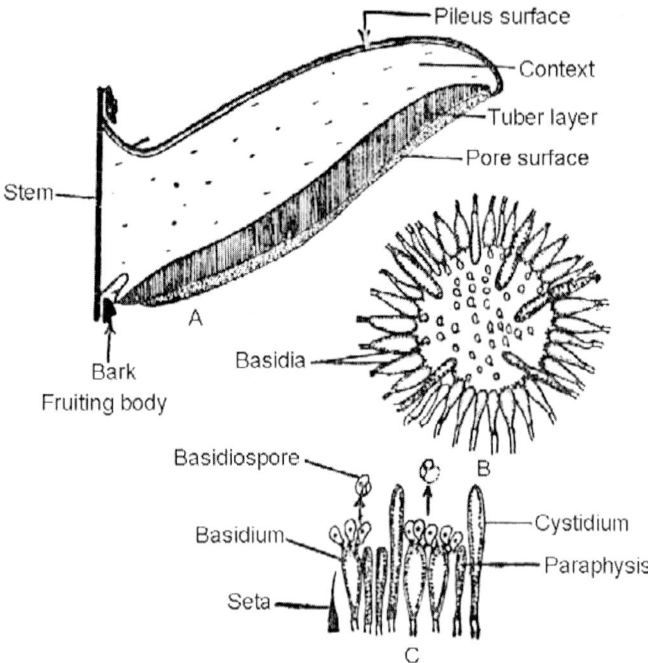

Fig. 15.43. Vertical section of *Polyporus* basidioma

Septate dikaryotic hyphae ramify within the tree, causing discolouration and decay of the wood. A vertical section of the fruit body (Fig. 15.43) shows the following four zones: (1) the pileus surface, (2) context, (3) tube layer, and (4) the pore surface. The species vary in size, colour and mode of arrangement of the basidiomata on the tree trunk. The pileus surface may be smooth or incrusted. The tube layer consists of enormous number of vertical tubes, which are lined internally by the hymenium. Each tube opens to the exterior on the lower surface through a pore. The hymenium comprises clavate, 4-spored basidia, basidioles and cystidia, all arranged at right angles to the long axis of the tubes. The basidiospores are shot into the lumen of the tubes from where they passively fall down and escape through the pore. The spore output is enormous. A single basidioma may release a billion basidiospores. The basidiomata are short-lived, surviving for only a few months, but the hyphae continue growing inside the host tissues, and form the massive basidiomata year after year. The basidiospores germinate and, being a wound parasite, enter the host trees through ruptured surface.

Genus *Trametes*

Trametes has 50 cosmopolitan species, which cause white-rot of wood. *T. versicolor* is the most common of all polypore fungi ('versicolor' means 'of several colours). It is also called 'Turkey tail', becaue of its resemblance to the tail of wild turkey. It grows on logs, old stumps and branches of trees all the year round, forming semicircular or kidney-shaped basidiomata, arranged in tiers (Fig.15.44).

Fig. 15.44 *Trametes versiclor*

These are thin, velvety, 2-5 cm in diameter and marked on the surface by characteristic zonations of different colours; hence the name 'versicolor'. Sometimes, closely-growing basidiomata fuse to form a broad sheet. The flesh is thin, without differentiation into context and tube layers. It is used in China and Japan as an immunoadjuvant therapy for cancer.

Order *Russulales*

The order includes lamellate, poroid, hydnoid (resembling *Hydnum*, with teeth-like hymenium) and gastroid representatives. The basidiomata are epigeous or partly hypogeus. The members are ectomycorrhizal, saprobic or pathogenic on trees. There are 12 families, including *Stereaceae* which we shall study.

Family *Stereaceae* (22 genera)

Genus *Stereum*

The genus has 27 species which are saprobic or parasitic. The infected leaves turn silvery-gray. The fungus is a weak, 'wound' parasite, requiring abrasions on the host surface for infection. When the infection succeeds, the fungus colonizes the xylem vessels, as well as, the cortical tissues. Sometimes, *Stereum* is included among the 'wilt fungi', but it is more appropriate to treat it as a wood-rotting fungus. Small, semi-circular and pretty basidiomata appear in several tiers on the ruptured surface of the dead wood (Fig. 15.45). The upper surface of the basidioma is greyish or brownish and covered with soft hairs. The hymenium is yellow, which later fades to buff colour.

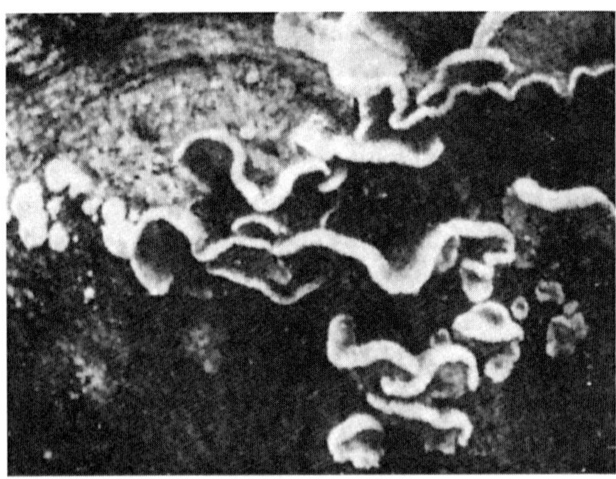

Fig. 15.45. *Stereum hirsutum* on a beech stump

S. hirsutum causes heart-rot of oak and forms bracket-shaped, yellow or orange-coloured basidiomata. It is, in turn, parasitized by *Tremella aurantia*.

S. murraii causes canker and white-rot of maples and birches. Dull-white, resupinate (i.e. crust-like) basidiomata appear on the cankers. *S. complicatum,* one of the first fungi to infect logs through wounds, causes white rot. The basidiomata are thin, with a smooth hymenophore.

Class *Tremellomycetes* (3 orders)

It is a class of dimorphic fungi. Some species have gelatinous basidioma. There are 3 orders.

Order *Cystofilobasidiales* includes teleomorph genera of basidiomycetous yeasts. There is 1 family, *Cystofilobasidiaceae* which has 9 genera; *Cystofilobasidium* is the type genus.

Order *Filobasidiales* too has 1 family, *Filobasidiaceae* and 1 genus *Filobasidium* which is teleomorph of 4 species of *Cryptococcus.* There is no basidioma.

Order *Tremellales* includes both teleomorphic and anamorphic species, placed in 9 families.

Order *Tremellales*

It contains both teleomorphic and anamorphic species, most of the latter being yeasts. All teleomorphic species are mycoparasites (parasite on other fungi). The yeast state is widespread and not restricted to hosts. Basidiomata when produced are gelatinous. These dry up during dry weather and survive for long periods. As soon as wet weather returns, they absorb water and regain the gelatinous texture and return to active life. Many are brightly-coloured. The hymenium consists of basidia, basidioles and cystidia. The basidia are typically cruciate-septate, with long sterigmal apparatus (epibasium and spicule) to ensure formation of basidiospores external to the gelified hymenium. There are 9 families (including *Sirobasidiaceae* and *Tremellaceae*).

Family *Sirobasidiaceae* (2 genera)

Genus *Sirobasidium*

The genus *Sirobasidium* (Fig. 15.46) has 8 species that grow on wood and bark. *S. sanguineum* and *S. albidum* are the common species. The basidia are catenulate (i.e. in chains), formed by successive transformation of cells of the basidiogenous hyphae beginning at the apex. The basidia are 2-4 celled, the septa being oblique in the latter case.

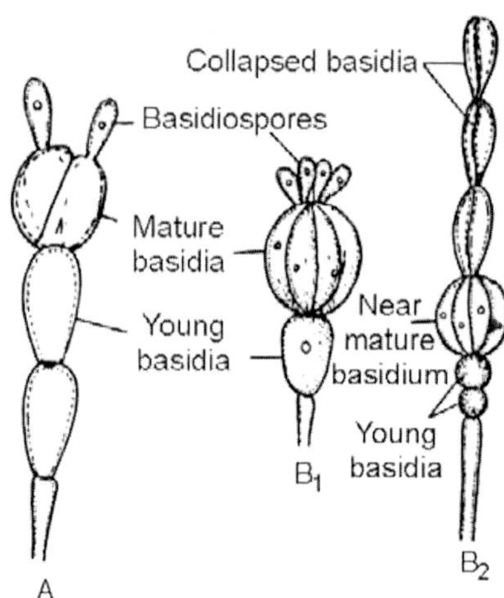

Fig. 15.46. *Sirobasidium sanguineum,* A catenate bicelled basidia (obliquely septate), the terminal is mature; B1 *S. albidum* — two basidia, terminal being mature; B2, chain of basidia, upper three collapsed, the fourth approaching maturity.

Family *Tremellaceae*

There are 17 genera; we shall study *Filobasidiella* and *Tremalla.*

Genus *Filobasidiella*

It has 3 species which are parasitic on other fungi and do not produce distinct basidioma. These are the teleomorphs of three species of *Cryptococcus*, 2 of which *C. neoformans* and *C. gattii*, are human pathogens causing cryptococcosis.

The hyphae of *Filobasidiella* bear haustorial cells with filaments that attach with the hyphae of the host fungus. Basidia are elongate, which form basidiospores from 4 loci at the apex. The basidiospores are released one by one which remain attached to the basidia in chains. The baidiospores germinate to form the yeast cells (*Cryptococcus*). The yeast-foorming species of *Filobasidiella* are: *F. neoformans* and *F. gattii* (named after *C. neoformans* and *C. gattii*).

Genus *Tremella*

There are 90 species which are parasitic on other fungi and produce anamorphic yeast states. Basidiomata when produced are gelatinous. *T. aurantialba* and *T. fuciformis* ('silver ear') are edible and commercially cultivated. Basidiomata vary in size from small (under 5 mm across) and pustular or cushion-shaped, to big (up to several cms across) and lobed wih folds and ridges or foliose (leaf-like). Many species are hymenial parasites, producing spores within basidiomata of host fungus. *T. aurantia* parasitizes *Stereum hirsutum*, while *T. mesenterica* is parasitic on *Peniopora*.

Fig 15.47 *Tremella mesenterica*

T. mesenterica (Fig 15.47), also called 'golden jelly fungus', 'yellow brain' or 'witches' butter', is a common wood fungus and also parasitic on the genus *Peniophora*. The gelatinous orange-yellow fruit body, which can grow up to 7.5 cm in diameter, has a convoluted or lobed surface that is greasy. It grows in crevices in the bark appearing during rainy weather. Within a few days after the rains, it dries into a thin film which can revive when moisture is again available. It is edible but has no flavour. The basidia are spherical, 2-3 septate, forming vertical compartments. The basidiospores give rise to yeast like cells by budding. These haploid yeast-like cells by mating between compatible cells, give rise to dikaryotic cells, which in turn, form dikaryotic hypha. The mating involves pheromones (sex hormones) produced by the mating cells. The dikaryotic hyphae form asexual conidia on conidiophores in abundant quantity. The basidiomata arise from a primordium beneath the wood bark.

Some species of *Tremella* produce polysaccharides useful in preventing cancer and for building immunity.

REFERENCES

Ainsworth, G.C. 1962. Longevity of *Schizophyllum commune.* Nature, London, 195: 1120-1121.

Bessey, E.A. 1950. Morphology and Taxonomy of Fungi. Hafner.

Buller, A.H.R. 1909. *Researches on Fungi*. Vol. I. Longmans, London.

Hibbett, D.S. *et al.,* 2007 A higher level phylogenetic classification of the fungi. Mycological Research. 111:509-547.

Ingold, C.T. 1967. *The Biology of Fungi*. 2nd ed. Hutchinson, London.

Ingold, C.T. 1974. Centenary tribute to A.H.R. Buller. Kavaka, 2: 1-2.

Matheny P.B. *et al.,* 2007 Contributions of rpb2 and tefl to the phylogeny of mushrooms and allies (Baidiomycota, fungi). Molecular Phylogenetics and Evolution (In Press).

Pegler, D.N. 1973. Aphyllophorales IV. Poroid families. In: *The Fungi, An Advanced Treatise* (G.C. Ainsworth, F.K. Sparrow and A.S. Sussman eds.). Vol. IV B. pp. 397–420. Academic Press, N.Y. and London.

Thind, K.S. 1961. *The Clavariaceae of India*. Indian Council of Agric. Res. New Delhi.

Thind, K.S. 1973. The Aphyllophorales in India. Indian Phytopath., 26: 1–22.

Smith M *et al.* 1992 The fungus *Armillaria bulbosai* among the larges and oldest living organisms. Nature 356: 428-431.

SUMMARY

Agaricomycotina is the biggest subphylum containing 20,000 species, which is almost 70% of the total *Basidiomycota*. It is characterized by (1) formation of basidiomata, (2) presence of dolipore septa with septal pore cap (=parenthosomes), (3) presence or absence of clamp connections, (4) somatogamous plasmogamy, and (5) basidiospores may be ballistospores or statismospores. The baidiomata and doloipore septa are the unifying characters of *Agaricomycootina*. The dolipore septa with parenthesome are present only in *Agaricomycotina*. *Pucciniomycotina* and *Ustilaginomycotina* have septal pores but not the septal pore cap. Variation in the form of parenthesome is an important phylogenetic character, which is useful in distinguishing the 3 classes of *Agaricomycotina*.

The 3 classes, which represent the three phylogenetic clades, are *Agaricomycetes*, *Dacrymycetes* and *Tremellomycetes*. About 90% of the species of *Agaricomycotina* belong to the class *Agaricomycetes*, which includes mushrooms, bracket fungi, puff balls and others. The *Dacrymycetes* and *Tremellomycetes* include the jelly fungi (However, a few former 'jelly fungi', such as *Auricularia*, are classified in the *Agaricomycetes*). The parenthesomes are perforate in *Agaricomycetes*, imperforate in *Dacrymycetes* (and some *Agaricomycetes*) and vesiculate in *Tremellomycetes*. The phylogenetic analyses suggest that the *Agaricomycetes* and *Dacrymycetes* form a monophyletic group and are sister groups.

Class *Agaricomycetes* has 2 subclasses: *Agaricomycetidae* (3 orders), and *Phallomycetidae* (4 orders).

There are 10 orders which have not been put in any of the above 2 subclasses. Class *Dacrymycetes* has 1 order *Dacrymytales*, while *Tremellomycetes* has 3 orders.

The orders and families whose representative genera we studied are the following.

Agaricomycetes

Agaricales: *Agaricaceae* (*Agaricus, Calvatia, Coprinus, Cyathus and Lycoperdon*), *Amanitaceae* (*Amanita*), *Clavariaceae* (*Clavaria*), *Pleurotaceae* (*Pleurotus*), *Pluteaceae* (*Volvariella*), *Schizophyllaceae* (*Schizophyllum*), *Strophariaceae* (*Psilocybe*) and *Physalacriaceae* (*Armillaria*).

Boletales: *Boletaceae* (*Boletus*), *Sclerodermataceae* (*Scleroderma*).

Geastrales: *Geastraceae* (*Geastrum*, *Sphaerobolus*).

Auriculariales: *Auriculariaceae* (*Auricularia*).

Cantharellales: *Cantharellaceae* (*Cantharellus, Craterellus*), *Ceratobasidiaceae* (*Ceratobasidium*), *Hydnaceae* (*Hydnum*)

Polyporales: *Ganodermataceae* (*Ganoderma*), *Polyporaceae* (*Fomes, Polyporus* and *Trametes*).

Russulales: *Stereaceae* (*Stereum*).

Tremellomycetes

Cystofilobasidiales: *Cystofilobasidiaceae* (*Cystofilobasidium*).

Filobasidiales: *Filobasidiaceae* (*Filobasidium*).

Tremellales: *Sirobasidiaceae* (*Sirobasidium*), *Tremellaceae* (*Filobasidiella* and *Tremella*).

Many of these fungi are edible, highly nutritious and promote health and also used in medicine. The highly rated edile mushrooms are: *Agaricus campestris, Volvariella volvacea, Pleurotus ostreatus, Coprinus comatus, Amanita muscaria* (after removal of cutis), *Boletus edulis, Auricularia polytricha, Cantharellus cebarius, Craterellus coronucopioides*, and *Polyporus squamosus*. *Amanita phalloides* and several of its species are highly poisonous. Some mushrooms like *Ganoderma,* are known since ages to be food for health and longevity. Some are hallucinogenic (*Psilocybe*), while some others are used in medicine (*Calvatia gigantia*, *Coprinus atramentarius, Trametes versicolor*). *Filobasidiella* species cause cryptococcosis in humans. The lignicolous species of *Boletus, Hydnum, Tremetes* and *Stereum* cause wood rot (brown rot, white rot and heart rot), and cause damage to forest trees. *Boletales* and *Sclerodermatales* are ectomycorrhizal, while *Armillaria mellea* forms orchid mycorrhizas. The corpophilous species belong to *Coprinus* and *Hydnum.*

TEST QUESTIONS

1. What are the diagnostic features of *Agaricomycotina*? Describe its classification up to orders.
2. Distinguish between **any three**: (a) Parenthesomes of the three classes of *Agaricomycotina*, (b) Basidioma of *Amanita and Agaricus*, (c) *Amanita* and *Volvariella*, (d) Autodigestion in *Coprinus* and *Lycoperdon*, (e) aequi hymeniferous and inaequi hymenoherous gills.
3. Write notes on **any three**: (a) rhizomorph of *Armillaria mellea*, (b) Unifying characters of *Agaricomycotina*, (c) *Ganoderma* for health, (d) 'Pietra fungaia', and (e) Telleomorphic genera of *Cryptococcus*.
4. Descibe **any two** the following genera: *Tremella*, *Filobasidiella*, *Polyporus*, *Armillaria*, and *Lycoperdon.*
5. Answer in brief:
   (a) Name three poisonous species of *Amanita.*
   (b) Which are the two subclasses of *Agaricomycetes*?
   (c) Name any four incertae sedis order of *Agaricomycetes.*
   (d) How does *Armillaria mellea* travel long distances from the food base?
   (e) Name any two 'poroid' and two 'gasteroid' genera.
   (f) What is peridiole?
   (g) Why is *Sphaerobolus* called "most powerful piece of artillery?
   (h) Which dried fungus was put by Buller in evacuated tubes to test its longevity?
   (i) Why is *Coprinus* called 'ink cap'?
6. Fill in the blanks:
   (a) *Agaricomycetes* and --------------- are sister groups.
   (b) Calvacin, an antitumor chemical, is obtained from the bsidioma of -------------.
   (c) *Agaricus campestris* is also called as *Agaricus* ---------.
   (d) Coprine, derived from ------------------, is used in the treatment of chronic alcoholism.
   (e) *Boletus parasiticus* parasitises basidioma of --------------.

(f) *Ceratobasidium* and --------------- are the teleomorphs of ---------------------.

(g) *Stereum hirsutum* is parasitised by *Tremellla* ---------------.

(h) The anamorphs of *Tremellomycetes* are mostly --------------.

(i) *Dacrymycetes* and *Tremellomycetes* are called -------------- fungi.

(j) *Tremella mesenterica* is parasitic on -----------------.

(k) Basidiospores of *Tremella* form ---------------- by budding.

(l) The 'puff ball' acts like a ------------- during spore dispersal.

(m) *Amanita muscaria* is called --------.

(n) *Volvariella volvacea* is also called the ------------ mushroom.

(o) Split gills are the characteric property of the family ----------------.

(p) *Psilocybe semilanceata* produces hallucinogenic chemicals,-------- and -----------.

(q) Te mycelium of ----------- ------------- has been found to have the age of 1500 years.

(r) The mycelium of ------- -------- shows the property of bioluminescence.

16

Subphylum *Ustilaginomycotina*
(Smuts and related fungi, former Ustilaginomycetes, Ustomycetes)

The taxonomic status of 'smut fungi' has always been uncertain. **Ainsworth** (1973) placed them under order Ustilaginales (class Teliomycetes). **Hawksworth** *et al.* gave them the rank of a class, Ustilaginomycetes (1983), and Ustomycetes (1995). Now it is subphylum *Ustilaginomycotina* representing one of the 3 **clades** that are discerned in *Basidiomycota*, having a **sister group** relationship with *Agaricomycotina* (i.e. the two clades are formed by splitting of a single lineage).

Characteristics

The *Ustilaginomycotina* are characterized by (1) lack of parenthesomes, most members have membrane caps at both sides of the pore, (2) formation of yeast-phase in culture, (3) presence of **zones of host-parasite interaction** with fungal deposits resulting from special fungal vesicles, and (4) dominance of glucose and absence of xylose in the cell wall. The last 2 characters confirm the monophyly of *Ustilaginomycotina*, which is supported by phylogenetic analyses of large subunit rDNA sequences.

Zones of host-parasite interaction or **interaction zones** are sites of fungal deposits released by specific fungal vesicles (primary interactive vesicles). The contents of the vesicles are transferred to the host plasmamembrane. Two major types of **interaction zones** are recognized, **local** and **enlarged**. In. local interaction zones, found in *Entorrhizomycetes* and *Exobasidiomycetes*, there is a short term production of the vesicles per interaction site, resulting in local interaction zones. In enlarged interaction zones, found in *Ustilaginomycetes*, there is a continuous production of the vesicles and deposition of fungal material at the whole contact area with the host cell.

The subphylum comprises of 1500 species, belonging to 80 genera, which, with few exceptions (*Malassezia* spp.), are parasitic on angiosperms, mainly on members of *Poaceae* (57%) and *Cyperaceae* (12%). *Ustilaginomycotina* not only comprises smut fungi but also non-telelosporic plant parasites such as *Graphiola, Exobasidium, Microstroma* and a few smaller genera. Ultrastructural and molecular investigations suggest that these are closely related to smut fungi although they do not form teliospores. On the other hand, smut fungi of *Microbotryales* forming teliospores and having life cycles typical of smut fungi, are now shifted to *Pucciniomycotina*, again on the basis of ultrastructural and molecular characteristics. There are even a group of human pathogens, *Malassezia*, which are placed within *Ustilaginomycotina*. The number of 'true smut fungus species'(possessing teliospores) is about 1450 (Kirk *et al.*, 2008).

Life Cycle

In contrast to *Agaricomycotina* and *Pucciniomycotina*, members of *Ustilaginomycotina* have essentially similar life cycle with a **saprobic, yeast-like haploid phase** and a **parasitic dikaryophase** (Fig. 16.1).

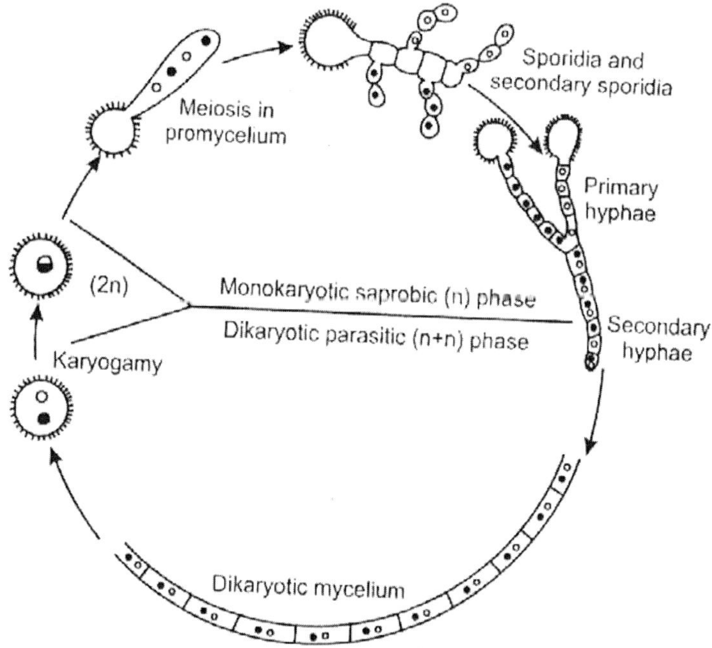

Fig. 16.1 Life cycle of smut fungi

The haploid phase usually starts with the formation of basidiospores after meiosis of the diploid nucleus in the basidium and ends with the conjugation of the compatible haploid cells to produce dikaryotic, parasitic mycelia. It has been shown that mating and dikaryotization are essential for infection of host plants. The dikaryotic phase ends with the production of probasidia (teliospores), which in majority of *Ustilaginomycotina* are thick-walled and serve as the dispersal agent after separation from the sorus. Depending upon the fungal species, the sori appear in or on different organs of the host (root, stem, leaf, inflorescence, flowers, anthers, ovaries, seeds etc.). The powdery, dark-brown or black teliospores are the most conspicuous stage in the life cycle giving the name "smut" to these fungi.

Classification (Begerow, Stoll and Bauer 2006, Kirk *et al.,* 2008)

There are three lines of *Ustilaginomycotina*, which are designated as the 3 classes. These are *Entorrhizomycetes, Ustilaginomycetes* and *Exobasidiomycetes.* The basal dichotomy (Fig. 16.2) is between the *Entorrhizomycetes* and the branch uniting the *Ustilaginomycetes* and *Exobasidiomycetes*, as shown below.

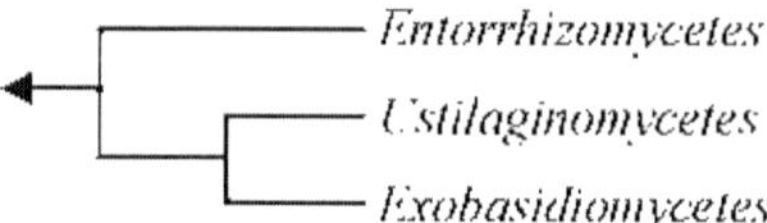

Fig. 16.2. Basal dichotomy in *Ustilaginomycotina*

Entorrhizomycetes is represented by *Entorrhizales*, a small group of **unusual teleosporic root parasites** of *Juncaceae* and *Cyperaceae*. Lack of membrane bands or caps at the pores, and presence of local interaction zones without interaction apparatus characterize the class. ***Ustilaginomycetes*** comprises majority of the *Ustilaginomycotina.* Presence of enlarged interaction zones is characteristic feature of the class. It is teleosporic and gastroid group, consisting of ***Urocystidales*** (holobasidiate) and ***Ustilaginales*** (phragmobasidiate). ***Exobasidiomycetes*** is predominantly holobasidiate and consists of teleosporic and non-teleosporic genera (detailed below). The presence of pore caps (absent in *Entorrhizomycetes*) and local interaction zones (against enlarged interaction zones of *Ustilaginomycetes*) are characteristic feature of the class.

The structures of **cellular interactions** and those of **septal pores** are used for distinguishing the orders, while morphological characters of basidia and sori are important in the differentiation of families of some orders.

Class *Entorrhizomycetes* (1 order)

Order *Entorrhizales*: 1 family *Entorrhizaceae*, 2 genera

Class *Ustilaginomycetes* (2 orders)

Order *Urocystidales*: 4 families, *Urocystidiaceae*, 7 genera

Order *Ustilaginales*: 8 families, *Ustilaginaceae*, 17 genera

Class *Exobasidiomycetes*: (8 orders) teleosporic, non-teleosporic and anamorphic

Teleosporic: 4 orders, including *Tilletiales*

Non-teleosporic: 3 orders, including *Exobasidiales*, and

Anamorphic order *Malasseziales*, no family recognized,

1 genus *Malassezia*, causing pityriasis versicolor (tinea versicolor), a superficial skin disease of humans.

Class *Entorrhizomycetes*

It is a small group of teleosporic root parasites that form galls on plants of family *Juncaceae* (rush) and *Cyperaceae* (sedge). There are 2 genera *Entorrhiza* and *Talbotiomyces.*

Class *Ustilaginomycetes*

Presence of enlarged interaction zones characterizes the *Ustilaginomycetes*. These are highly host-specific and exist in nature as obligate parasites, except *Ustilago maydis* the corn smut fungus, which has good saprobic growth in soil as well. Since smuts can be cultured in laboratory, they are regarded as facultative saprobes.

The hyphae are usually intercellular and form haustoria. In many cases hyphae are perennial giving no evidence of their presence until spore formation sets in. The teliopores are formed in sori, which consist of host tissues, of the spore mass and in several genera also of modified fungal cells or tissues. The spores may be free, in pairs or aggregated in more or less persistent spore balls. The spore balls may be composed entirely of fertile spores or a combination of fertile spores, sterile cells and/or hyphae. The teliospores are formed from intercalary as well as terminal cells of the dikaryotic secondary mycelium. These arise like chlamydospores, by rounding up of the cells and formation of a thick dark wall. The disease manifests in form of smut sori that may be formed on any part of the plant.

Teliospore germination

The teliospores perennate the fungus and germinate at the return of favourable conditions. There are two types of teliospore germination, called ***Ustilago* type** and ***Tilletia* type** (Fig. 16.3). In *Ustilago* type, the teliospore forms a **septate epibasidium** (promycelium), also called **phragmobasidium**, bearing 4 basidiospores, terminally and laterally. In the *Tilletia* type, the epibasidium is hollow (**holobasidium**) bearing variable number of basidiospores only terminally.

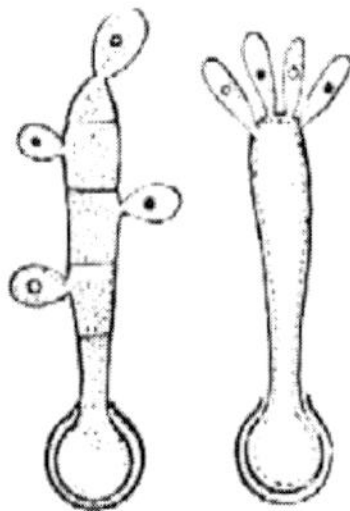

Fig. 16.3 Showing the two modes of teliospore germination in smut fungi

The epibasidia give rise to haploid sessile basidiospores (sporidia) or more rarely to haploid hyphae. These and sometimes, compatible basidial cells, fuse in pairs to restore the parasitic dikarophase. Frequently the basidiospores multiply by budding on non-living substrata as yeast-like cells or also as hyphal anamorphs developed from basidiospores forming forcibly-discharged ballistoconidia.

On germination, the basidiospores give rise to monokaryotic primary hyphae which, except in *U. maydis,* exist only for a short period. It has been observed that even though the monokaryotic hyphae may initiate infection, dikaryotic hyphae are essential for parasitic colonization of host tissue. Heterothallism (bipolar and tetrapolar) is common in smut fungi. If dikaryotiza tion is not achieved, infection fails to progress. This requirement of different types of nuclei for parasitic activity is an example of the phenomenon of **complementation**. Sex organs are absent. The dikaryotization (plasmogamy) may occur by fusion between the various structures formed by teliospores after germination: (1) between two promycelia of compatible mating types, which may or may not have produced basidiospores, (2) between basidiospores on the promycelium itself or after detachment, (3) between primary haphae inside host tissue during early stages of infection, or (4) by any combination of above structures.

Infection types

There are four chief types of infection

(1) Seedling infection: Systemic, initiated at seedling stage by teliospores on the seed, hyphae mostly intercellular, sporulation in host ovary.

(2) **Embryo infection.** Systemic, initiated through infected embryo as a result of teliospore germination on stigma at flowering time, intercellular mycelium remains dormant in seed until infected seed germinates.

(3) **Shoot infection.** Systemic, initiated by infection of shoots or young buds by basidiospores from promycelia among decaying plant material. May result in lack of floral development or aborted inflorescence.

(4) **Local infection.** Mycelium and sporulation restricted to region of infection. Fungus is not systemic.

The dikaryotic hyphae, which grow intercellularly in the host tissue, may be **localized** or **systemic**, and feed on the host cells through haustoria. The infected plants usually do not show much morphological changes, until the telia formation, which may occur as soon as some hyphae have accumulated or, as happens in the grain smuts, only at the time of the 'ear' formation. Depending on the site of formation of sori, the **smut diseases** are designated as: **grain smuts**, **anther smuts**, **leaf** (or **flag**) **smuts**, **inflorescence smuts**, etc. Occasionally, the sori appear on the **stems** or on the underground parts like the **scale leaves** of onion bulbs, (onion smut) and even on **roots** (on *Brassica* sp.). In cereal smuts, the sori resemble the grains. In 'boil smut' of corn, galls, as big as a tennis ball, full of teliospores, are formed.

The nuclear phases in the life cycle comprise of (1) a short **haplophase**, represented by the primary mycelium, (2) a long **dikaryophase**, and (3) a **diplophase**, represented by the diploid nucleus, formed in the teliospore at the time of germination.

Classification

The class *Ustilaginomycetes* is divided into two orders, ***Urocystidales*** (=*Urocystales*) and ***Ustilaginales*** based on teliospore germination. The *Urocystidales* are holobasidiate while *Ustilaginales* are phragmobasidiate. The identification of the genera and the species is based on, (1) teliospore characteristics, (2) type and site of sorus development, and (3) host range. The teliospores may be free or aggregated into hard spore-balls. About half of the smut genera form spore balls. The ornamentation of the teliospore (smooth, reticulate, echinulate, verrucose or tuberculate) is a stable character, highly useful in species delimitation.

The site of sorus development, is another stable character. Soral characters used in the delimitation of genera are: (1) dusty or compact nature of the sorus, (2) presence or absence of a peridium (of fungus origin), (3) enclosed or exposed nature of the sorus, (4) presence of a columella (of host origin) inside the sorus, and (5) erumpent (galls) or nonerumpent nature.

Order *Urocystidales* (=Urocystales)

There are 4 families but we shall study only the family *Urocystidaceae*, which has 7 genera including the genus *Urocystis*.

Genus *Urocystis*

The genus with 170 species forms sori mostly in leaves, stems, forming streaks, swellings or galls.

The spore balls consist of 1-4 dark **fertile central cells** surrounded by a peripheral layer of hyaline **sterile cells**. The teliospores germinate *in situ*. The two most important diseases caused by *Urocystis* are "flag smut" of wheat (*U. agropyri*) and onion smut (*U. cepule*).

Flag smut of wheat

It is one of the three major smut and bunt diseases of wheat (The other two are karnal bunt and dwarf bunt). Symptoms may appear anytime, from seedling stage to the 'ear' formation. Sori, containing spore-balls, form black stripes on the stems, culms and leaves (Fig. 16.4). Infected plants are stunted in growth and bear twisted leaves, which droop down (flagging symptom), dry up and finally fall off. The grains are either not formed, or if formed, are shriveled.

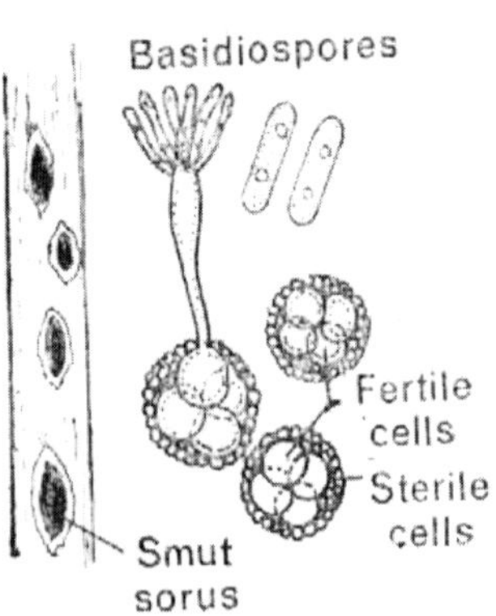

Fig. 16.4 *Urocystis agropyri*

The spore balls, perennating in the soil, germinate at the time of emergence of wheat seedlings. The fertile cells germinate *in situ* and produce a hollow promycelium (*Tilletia* type germination), bearing 2–6 cylindrical basidiospores terminally. These may cause infection through 'infection hyphae', while still attached to the promycelium. The infection hyphae may dikaryotize and give rise to a dikaryotic hypha, before or after penetration of the seedling. The dikaryotic mycelium becomes systemic in the host and forms black stripes of smut sori on leaves and other parts, which become exposed after rupture of the overlying epidermis. The teliospores perennate in the soil.

Order *Ustilaginales*

There are 8 families and 49 genera. We shall examine two: *Tolyposporium* (family *Anthracoideaceae*) and *Ustilago* (family *Ustilaginaceae*).

Genus *Tolyposporium*

There are two species that grow on *Cyperaceae* and *Juncaceae*. Sori, though formed in various parts of the host, are more common in the ovaries. These consist of spore balls, covered by a membrane of host origin. The spore balls may consist of only one type of spores, or differentiated into an outer layer of coloured spores and an inner mass of hyaline spores. A characteristic feature of *Tolyposporium* is the presence of outgrowths on the spores, which interlock the spores into spore balls. The balls break with difficulty and the spores germinate *in situ*.

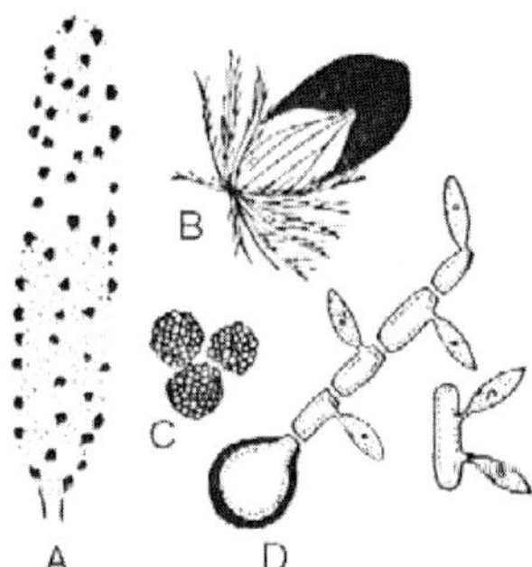

Fig 16. 5 *Tolyposporium penicillariae*. A, smutted ear, B, enlarged view of smutted grain, C soore balls and teliospore germination. (Note breaking of promycelium and budding of basidiospores).

T. penicillariae causes smut of 'bajra' (*Pennisetum typhoides*). Smutted grains, which are pear-shaped and double the size of normal grains, lie scattered or in patches on the cob (Fig. 16.5). Plants are infected at the flowering stage. Germination of teliospores (*Ustilago* type) in the spore balls, which perennate in the soil, coincides with the flowering of the host. The septate promycelium, bearing terminal and lateral basidiospores, has a tendency to break into individual cells. These basidiospores form secondary basidiospores (= secondary sporidia) by budding, which, blown by wind, land on the flowers and infect the ovaries. The infection is localized in the ovaries, which get stuffed with dikaryotic hyphae and transform into smut sori. Secondary infections are caused by successive generations of teliospores formed on early-infected ovaries. The last generation of teliospore balls perennates in soil, waiting for the next crop to come up and flower.

Genus *Ustilago*

There are about 200 species that parasitize members of *Poaceae*. The teliosorus is without a peridium; the black dusty teliospores are covered by a membrane of host origin. The spores are small, less than 20 μm in diameter. Some important species and the diseases they cause are: *U. segetum* var. *tritici* (loose smut of wheat), *U. nuda* (loose smut of barley), *U. maydis* (smut of maize), *U. scitaminea* (whip smut of sugarcane), and *U. occidentalis* (smut of *Cyanodon*).

Loose smut of wheat
***Ustilago tritici* (= *Ustilago segetum* var. *tritici*)**

Plants, produced by internally-infected seeds, contain the hyphae in every nook and corner of their body. But marked symptoms appear only when the 'ears' come out. Sometimes, the infected plants are stunted in growth. The ears, instead of containing grains, contain black mass of spores (Fig. 16.6). The sori in the early stages are covered by a thin membrane, derived from host tissue, but later, the membrane ruptures and the spores become a loose mass; hence the name 'loose' smut.

Fig. 16.6. Loose smut of wheat

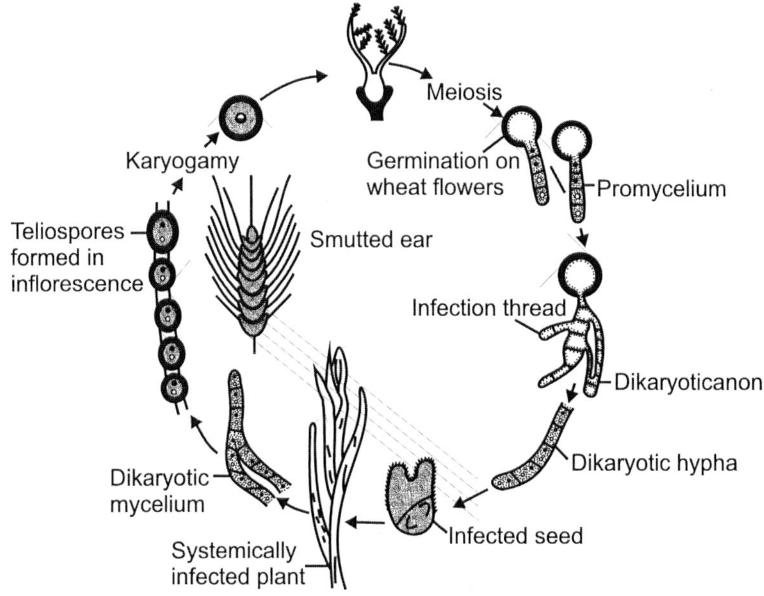

Fig. 16.7 Life cycle of *Ustilago tritici* causing loose smut of wheat

Life cycle (Fig. 16.7) The disease is internally seed-borne i.e., the fungus lives in dormant stage in the embryo of the seed. Such seeds look perfectly normal and there is no method to detect the infection unless they are grown and watched till the 'ear' formation. The seedling, and later the whole plant, is infected by the extensive dikaryotic mycelium, which grows intercellularly and derives nutrition through haustoria. The plants are said to be systemically infected. The hyphae collect in the ovaries and the hyphal cells round off, develop thick walls and become teliospores. The teliospores, after the rupture of the covering membrane, are blown by wind. Only the rachis is left behind. The teliospores land on the feathery style or ovary of wheat flowers and germinate to form septae promycelium. Karyogamy, followed by meiosis, results in the formation of 4 haploid nuclei. Septa are laid down in the promycelium, so as to form four cells, each containing a single haploid nucleus. From each cell a thin hypha-like outgrowth emerges. These are called the **infection threads**. The typical basidiospores are not formed in this species, and their function is performed by the infection threads. The infection threads individually are unable to grow and parasitize the ovary unless two infection threads of opposite mating types fuse and form a dikaryotic hypha. The dikaryotic hypha grows through the ovary wall and reaches the ovule and ultimately infects the embryo. The mycelium becomes established in the seed of the infected ovule. The hyphae lie dormant in the seed and spring to life when the seed is sown and seedling is formed. Such internally-infected seeds give rise to systemically-infected plants. The fungus lives its somatic life, and at the end of the wheat season, forms sori transforming the entire inflorescence, except the rachis, into a black mass of spores.

Smut of maize :*Ustilago maydis* (= *U. zeae*)

This fungus, besides being a destructive pathogen, is now a model organism, like *Saccharomyces cerevisiae*, for studying cell biological processes (Steinberg and Prez-Martin, 2007). It also has the distinction of being the first basidiomycetous plant parasite whose genome has been sequenced. Genome-wide analysis demonstrates that the fungus is more closely related to humans than to budding yeast, and numerous proteins are shared only by *U. maydis* and *Homo sapiens*.

The pathogen is also drastically different from other smut fungi. The important differences are the following:

1. In addition to parasitic life, it grows saprobically in soil.
2. It grows intracellularly and does not form haustoria.
3. Does not grow systemically in the host, and remains localized in several independently-infected parts.
4. Smut sori are formed on almost all the above-ground parts of the plant, the cob, the leaf and the stem.
5. The fungus induces tumors and outgrowths, which are filled with teliospores.
6. The primary mycelium, which in other species is incapable of growth in the host tissue, grows sufficiently well in the host tissue. However, the dikaryotic secondary mycelium grows more vigorously than the primary mycelium.

Ordinarily 2-5% of the plants in a corn field are infected by *U. maydis*, and if conditions are favourable for the pathogen this can go up to 80%. Sori (galls) containing teliospores are formed on all above-ground parts. Galls of different sizes appear, the biggest ones are formed on the cobs. Galls of *U. maydis* are considered a delicacy in Mexico. In U.S.A. it is called "maize mushroom" and "Mexican truffles"

Life cycle (Fig. 16.8) The fungus lives saprobically in soil. When the maize crop is available, the basidiospores, blown by wind, reach the plants and germinate to form a monokaryotic primary mycelium. The primary mycelium does make some growth, but vigorous growth starts only after dikaryotic secondary mycelium is established by anastomosis and fusion between hyphae of opposite mating types. Dikaryotization is necessary for gall formation. The fungus does not spread in the host tissue, but remains localized to the vicinity of the place of infection and forms a sorus of teliospores. Big sori appear in the form of galls or tumors. Each gall is the product of a separate infection. The teliospores are formed by rounding up of the binucleate cells. In the beginning, the telia are covered by the host tissue, but later the covering membrane ruptures and the teliospores are disseminated by wind. They germinate to form a club-shaped promycelium in which karyogamy and meiosis take place, resulting in the formation of four haploid nuclei. Segregation of sex occurs during meiosis and out of the four nuclei formed, two are of one mating type and two of the other. Four septa are laid down which divide the promycelium into four cells, each containing one haploid nucleus. A bud develops on each cell into which the nucleus migrates. The bud later develops into a basidiospore. The basidiospores are passively discharged, and blown by wind to cause more infections. These may form secondary sporidia by budding.

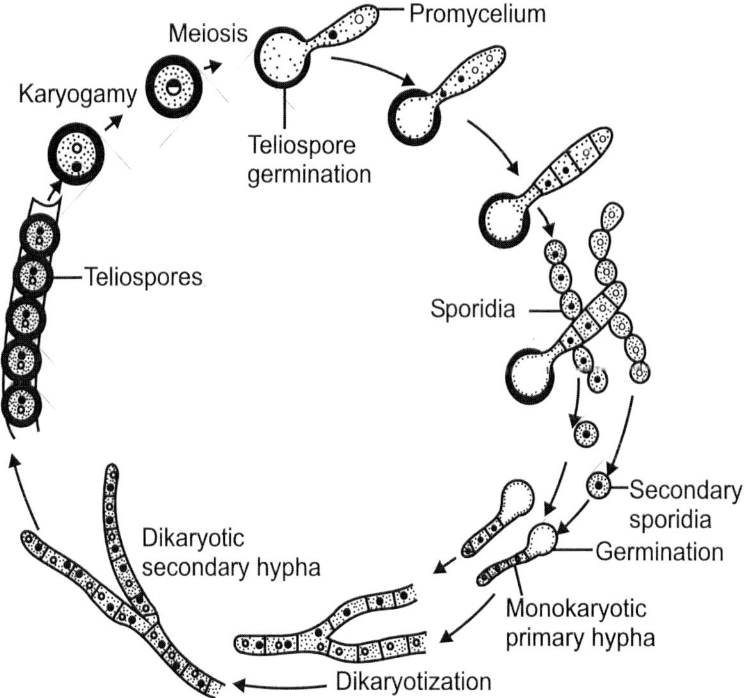

Fig. 16.8. Life cycle of *Ustilago maydis,* which has both saprobic and parasitic mode of life

Class *Exobasidiomycetes*

Order *Tilletiales* (teleosporic order)

The order has one family, *Tilletiaceae* and 6 genera, including *Tilletia* and *Neovossia*. The teliospores of these genera, on germination, form hollow promycelia (called *Tilletia*- type germination), bearing usually elongated or filiform basidiospores (primary conidia or sporidia), which usually fuse and dikaryotize *in situ*, to form secondary conidia (secondary sporidia).

Genus *Tilletia*

The diseases caused by *Tilletia* are called **bunt**. On germination, the teliospores form a holobaidium (promycelium) bearing 8 filiform terminal sporidia (Fig. 16.9). The sporidia form characteristic H-shaped structures, as a result of fusion of compatible sporidia *in situ*. Secondary dikaryotic sporidia arise from one of the fused sporidia. There are 175 species, which parasitize members of the family *Poaceae*. The anamorph is *Tilletiella.*

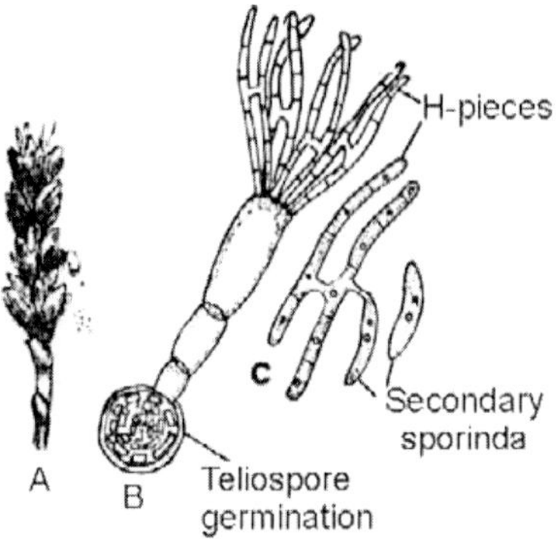

Fig. 16. 9 *Tilletia laevis* (*T. foetida*). A, infected wheat ear; B, teliospore germination; C, H-pieces and budding of basidiospores

Bunt of wheat

There are three bunt diseases of wheat: (1) common bunt or stinking bunt caused by two species, *T. caries* (syn. *T. tritici*) and *T. laevis* (*T. foetida*) (2) Karnal bunt (partial bunt), caused by *Neovossia indica* (mentioned as *T. indica* in western literature), and (3) dwarf bunt, caused by *T. controversa*.

Common bunt or stinking bunt or **old bunt** is caused by two species. *T. caries* (syn. *T. tritici*), which forms rough spores and *T. laevis* (syn. *T. foetida*)`, which forms smooth spores. Both have similar life cycles and may occur in the same 'ear' and grain. Infected plants, which are shorter and produce smutted grains, emit a fishy stink, hence the name 'stinking smut' or 'bunt'. The foul smell is due to a volatile compound, trimethylamine, formed by the fungus. The smut sori have a covering of host origin and the teliospores are released only at the time of thrashing of the grains. The spores adhere to the seeds. The teliospores also infest the soil.

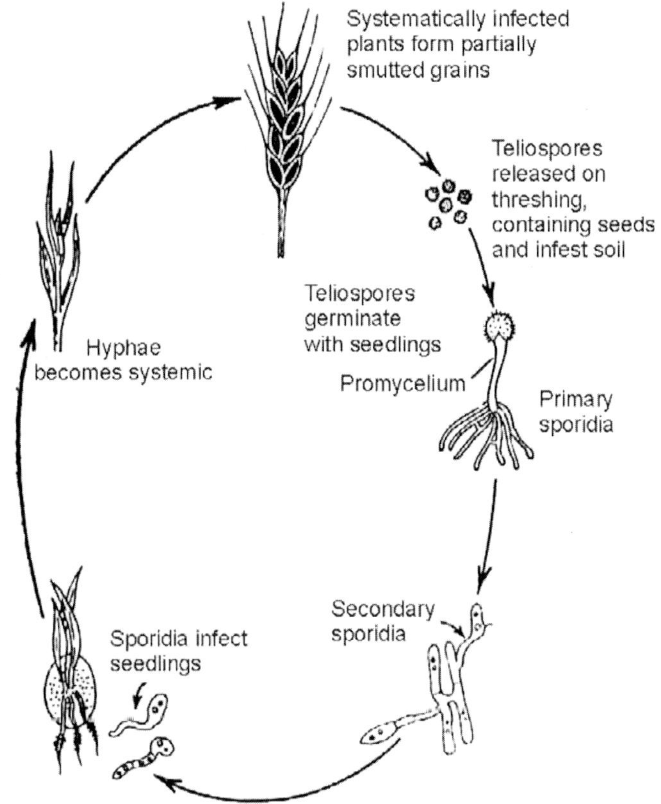

Fig. 16.10. Life cycle of *Tilletia laevis* (*T. foetida*)

Life cycle (Fig. 16.10) Infection takes place at the seedling stage when teliospores, present in the soil or on the seed surface, germinate and produce 8 - 16 cylindrical, terminal basidiospores (primary sporidia) on a hollow, or occasionally septate, promycelium. The basidiospores of opposite mating types fuse in pairs *in situ*, forming characteristic H-shaped structures. The contents of one spore migrate into the other through a bridge bringing about plasmogamy and dikaryotization. A short dikaryotic hypha originates from the binucleate cell and forms a crescent-shaped, binucleate conidium, also called "secondary sporidium". These secondary sporidia are borne a symmetrically on a short sterigma and discharged violently like ballistospores of *Agaricomycotina* and *Pucciniomycotina*. The conidia germinate and produce a dikaryotic hypha which infects the host. The short dikaryotic hypha forming the conidium, can itself initiate infection. The dikaryotic mycelium grows and becomes systemic in the host. Ultimately, the hyphae aggregate in the ovaries and transform into teliospores, which in turn, join to form spore balls. The sorus is invisible under the ovary wall and the spores are released only during thrashing. "Smut clouds" arise from thrashing machines and infest the soils up to long distances.

Genus *Neovossia*

The genus has 1 species *N. indica*, parasitizing members of family *Poaceae*. The promycelium bears numerous (32-130) filiform basidiospores (=primary sporidia) but H-shaped structures are not formed. In addition, the teliospores bear a long pedicel-like appendage, called ***apiculus***. The teliospores are formed singly from the end cells of sporogenous hyphae, whose remnants persist as an investing sheath on the spores.

Karnal bunt (partial bunt) of wheat

N. indica is the causal agent of 'karnal bunt' of wheat. Mitra (1931), who discovered the disease in Karnal district of present Haryana state, named the pathogen as *Tilletia indica*. Later, Mundkur found the pathogen to be actually *Neovossia* and renamed it as *N. indica. i.e. Neovossia indica* (Mitra) Mundukar. The pathogen, however, continues to be designated as *T. indica* in western literature. Karnal bunt now occurs endemically in India, Pakistan, Nepal, Mexico and U.S.A. The disease is included among "new and emerging diseases" at international level (Malik and Mathre, 1998). Only few ears, and in that only few grains are smutted and that too partially (Fig. 16.11). Hence the name "partial bunt".

Life cycle The teliospores germinate (*Tilletia* type germination) and form filiform basidiospores (primary sporidia), which germinate terminally and laterally to produce hyphae or sterigmata on which crescent-shaped secondary sporidia are formed that are discharged forcibly, like ballistospores. The secondary sporidia cause infection of the ovaries by forming dikaryotic mycelium and remain localized there. A fishy stink pervades the infected fields. The foul smell is due to a volatile compound, trimethylamine. When seeds are formed they are partially-smutted and the loss is not much.

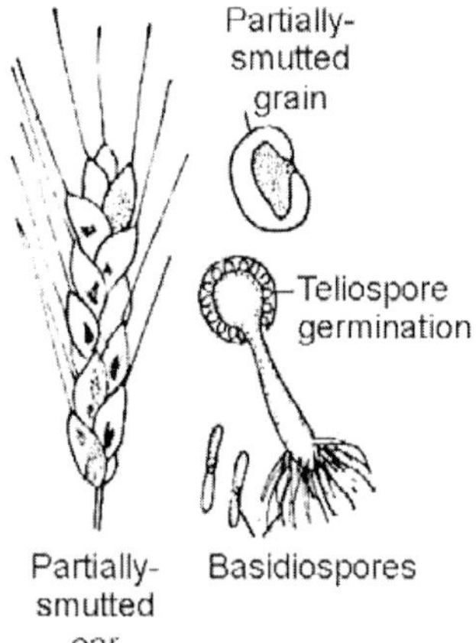

Fig. 16.11 *Neovossia indica.* Only few grains are infected, that too partially. Note filiform basidiospores

Order *Exobasidiales* (Non-teleosporic order)

The order contains plant parasites that form galls on leaves, mostly of the family *Ericaceae* and *Commelinaceae.* The members are characterized by formation of basidia in a layer uncovered on the leaf surface, hence the name *Exobasidiales*. There are 4 families: *Brachybasidiaceae*, *Cryptobasidiaceae*, *Exobasidiaceae*, and *Graphiolaceae*.

Family *Exobasidiaceae*

The family has 5 genera, which are stromatic below epidermis. The basidiospores are ballistospores, i.e., discharged forcibly.

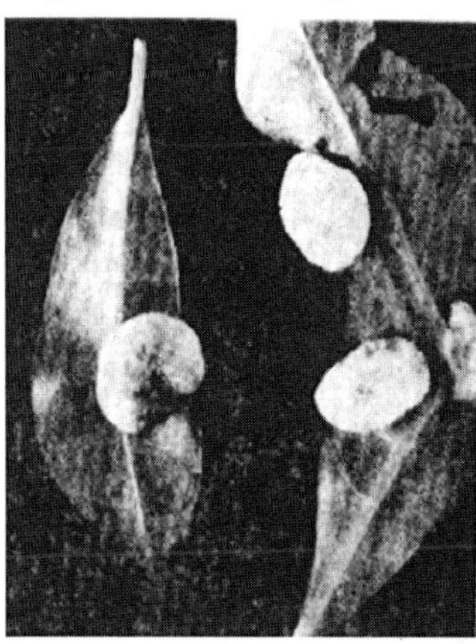

Fig. 16.12 *Exobasidium vaccinii*-induced galls on leaves of *Vaccinium*

Genus *Exobasidium*

Exobasidium is the largest and most common genus of the family. There are about 50 species, which parasitize plants of the family *Ericaceae. E. vaccinii* forms galls on leaves of *Vaccinium* (Fig. 16.12). Many species have been grown in pure culture. *E. vexans* causes the most destructive 'blister blight' disease of tea leaves

Blister blight of tea

E. vexans is an obligate parasite causing the most destructive 'blister blight' disease of tea. It occurs frequently in India, Sri Lanka and Indonesia. Pale green, pale yellow or pinkish, translucent spots appear on leaves, which look like blisters, sunken on the upper and erumpent on the lower surface (Fig. 16.13). More than 50 discrete blisters may occur on a single leaf, which sometimes coalesce to form a big patch. Infected leaves curl and get deformed. The infection often results in the formation of galls. Blisters appear after three days of infection and basidia after about 10 days. The raised lower portion of the blister bears a layer of basidia. Blisters appear also on the stems. The dikaryotic mycelium grows intercellularly in the leaves, deriving nutrition through haustoria. The hyphae lack clamp connections. Hyphae emerging from between lower epidermal cells form a layer of basidia, bearing 4-8 basidiospores on the lower surface. According to Venkata Ram (1975), about one million spores are produced in 24 hours. A relative humidity of 80% is required for sporulation and infection.

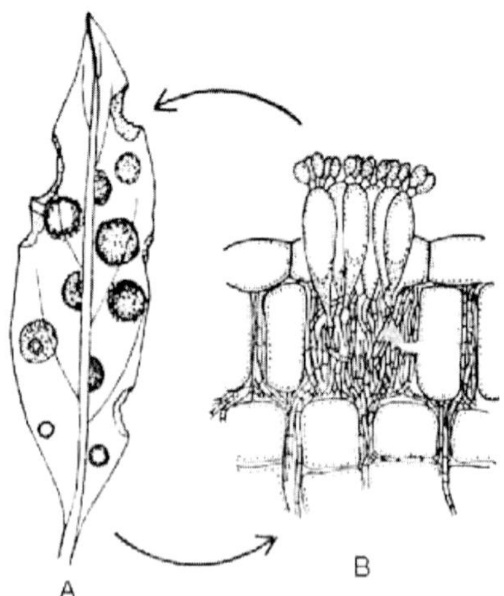

Fig. 16.13. *Exobasidium vexans*. A:'blister blight' on lower surface of tea leaf; B: section through a blister showing emergence of basidia through spaces between epidermal cells.

The basidiospores are disseminated by wind and also by insects, but infection is caused only by the wind-blown basidiospores. The basidiospores germinate on young leaves (old leaves are resistant), and form an appresorium at the tip of the germ tube. An infection hypha arises from the appresorium and penetrates through the leaf epidermis. Occasionally, the basidiospores form blastospores, which also germinate and infect the leaves. The primary hypae, after a short period of growth, give rise to secondary hyphae by homothallic somatogamy. Since the tea crop is grown throughout the year, the inoculum persists endemically in the plantation. The fungus completes the life cycle in 10 to 28 days.

Order *Malasseziales* (no family recognized)

It is an anamorphic order with the lone genus *Malassezia* (named after French medical mycologist Charles Malasssez). It, causes pityriasis versicolor (tinea versicolor), a superficial skin disease of humans. It has 10 species which inhabit the skin surface of animals including humans. About 90% of human adults carry these organisms without noticing any harm. In occasional 'opportunistic infections' some species (*M. furfur, M. globosa* and *M. restricta*) suppress the body's immune response and proliferate to cause skin infections, which include hypopigmentation (whitish patches), dandruff, skin rashes etc. As the fungus requires fat to grow it is common in

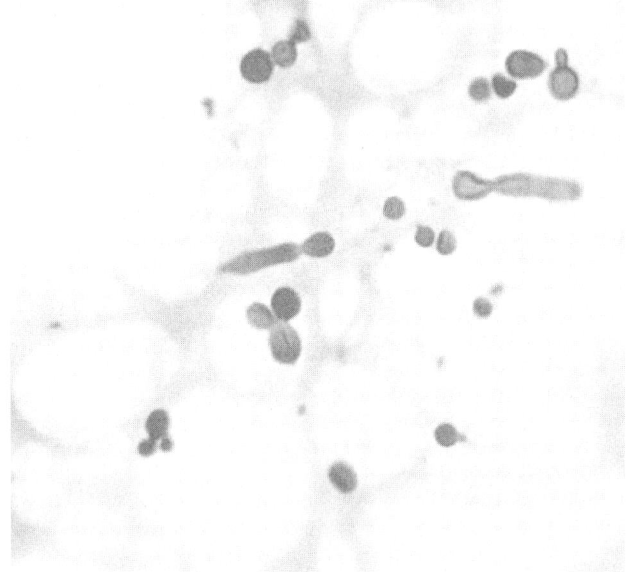

Fig. **16.14** *Malassezia furfur* from pityriasis versicolor-infected skin

areas with many sebaceous glands: on the scalp, face and upper part of the body. The fungus produces 8 different types of lipases and 3 phospholipases that enable it to digest the fat. When the fungus grows too rapidly, the natural renewal of cells is disturbed and dandruff appears with itching. Any of the above 11 enzyme proteins would be a suitable target for dandruff medication.

Diagnosis of *Malassezia* infections is made by microscopic observation of skin scrapings in KOH (Fig. 16.14). Clusters of yeast cells and hyphae are seen in the preparations. These can also be seen directly on the skin in UV light as the cells fluoresce due to production of a fluorescent chemical, pityrialactone. The species are difficult to grow in lab, so scrapings may be reported a "culture negative". Recently, identification of *Malassezia* on skin has been aides by DNA-based techniques.

A project in 2007 has sequenced the genome of dandruff causing *M. globosa* and found it to have 4285 genes. It has been predicted to reproduce sexually but this has not been confirmed.

REFERENCES

Ainsworth, G.C. 1973. Introduction and keys to higher taxa. In: *The Fungi. An Advanced Treatise.* (G.C. Ainsworth, F.K. Sparrow and A.S. Sussman eds.). Vol. IV A. 1–7. Academic Press, New York, London

Begerow D, Stoll, M and Bauer R 2006 A phylogenetic hypothesis of Ustilaginomycotina based on multiple gene analyses and morphological data. *Mycologia* 98 (6): 906-916.

Hawkswoth, D.L. *et al.,* 1983. *Ainsworth and Bisby's Dictionary of the Fungi.* C.A.B. International, Oxan, U.K.

Hawksworth, D.L. *et al.,* 1995. *Ainsworth and Bisby's Dictionary of the Fungi* (8th ed.) C.A.B. International, Oxon, U.K.

Kirk PM *et al.,* 2008 *Ainsworth and Bisby's Dictionary of the Fungi* C.A.B. International, Oxon, U.K.

Steinberg G and Prez-Martin J. 2007 *Ustilago maydis*, a new model for cell biology. Trends in Cell Biology. 18(2): 61-67

Venkata Ram, C.S. 1975. Blister blight of tea, In: *Advances in Mycology and Plant Pathology.*

SUMMARY

The subphylum comprises of 1500 species, which with few exceptions (*Malassezia* spp, an opportunistic human pathogen), are parasitic on angiosperms, mainly on members of *Poaceae* and *Cyperacese* causing mostly 'smut' diseases. These fungi represent one of the 3 **clades** of *Basidiomycota,* and have a **sister group** relationship with *Agaricomycotina*. *Ustilaginomycotina* are characterized by: (1) lack of parenthesomes; most members have membrane caps at both sides of the pore, (2) formation of yeast-phase in culture, (3) presence of **zones of host-parasite interaction** with fungal deposits resulting from special fungal vesicles, and (4) dominance of glucose and absence of xylose in the cell wall. The last 2 characters confirm the monophyly of *Ustilaginomycotina*, which is supported by phylogenetic analyses of large subunit rDNA sequences.

Ustilaginomycotina not only comprises 'smut fungi' but also non-telelosporic plant parasites such as *Graphiola, Exobasidium, Microstroma* and a few smaller genera. On the other hand, smut fungi of *Microbotryales* forming teliospores and having life cycles typical of smut fungi, are now shifted to *Pucciniomycotina* on the basis of ultrastructural and molecular characteristics. There is even a group of human pathogens, *Malassezia* in *Ustilaginomycotina*.

In contrast to *Agaricomycotina* and *Pucciniomycotina*, members of *Ustilaginomycotina* have essentially similar life cycle with a saprobic, yeast-like haploid phase and a parasitic dikaryophase. The haploid phase starts with the basidiospores and ends with the formation of dikaryotic, parasitic mycelia. The dikaryotic phase ends with the production of teliospores in sori that can appear in or on any part of the host

There are 3 classes: *Entorrhizomycetes, Ustilaginomycetes* and *Exobasidiomycetes. Entorrhizomycetes* is represented by *Entorrhizales*, a small group of unusual teleosporic root parasites of *Juncaceae* and *Cyperaceae*. Lack of membrane bands or caps at the pores, and presence of local interaction zones without interaction apparatus characterize the class.

Ustilaginomycetes comprise majority of the *Ustilaginomycotina*. Presence of enlarged interaction zones is characteristic feature of the class. It is teleosporic consisting of *Urocystidales* (holobasidiate) and *Ustilaginales* (phragmobasidiate).

Exobasidiomycetes is predominantly holobasidiate and consists of teleosporic (e.g. *Tilletia*) and non-teleosporic (e.g. *Exobasidium*) genera. The presence of pore caps (absent in *Entorrhizomycetes*) and local interaction zones (against enlarged interaction zones of *Ustilaginomycetes*) are characteristic feature of the class. Anamorphic order *Malasseziales* contains the lone genus *Malassezia*, causing skin diseases in humans.

Ustilaginomycetes are host-specific and exist in nature as obligate parasites, except *Ustilago maydis* the corn smut fungus. The hyphae are usually intercellular and form haustoria. The teliospores are formed from intercalary as well as terminal cells of the dikaryotic secondary mycelium.

The teliospores perennate as well as disseminate the fungus. They germinate at the return of favourable conditions. There are two types of teliospore germination, called *Ustilago* type and *Tilletia* type, also called phragmobasidiate and holobasidiate type, respectively. The basidiospores germinate and form primary hyphae that by fusion with another compatible primary hyphae, form dikaryotic hyphae, which establish the infection. There are four chief types of infections: seedling infection, embryo infection, shoot infection and local infection.

The dikaryotic hyphae, which grow intercellularly in the host tissue, may be local or systemic, The infected plants usually do not show much morphological changes, until the telia are formed. Depending on the site of formation of sori, the smut diseases are designated as: grain smuts, anther smuts, leaf (or flag) smuts, inflorescence smuts, etc.

Exobasidiomycetes form local interaction zones (like *Entorrhizomycetes*) and have membrane caps at the pores (like *Ustilaginomycetes*). It has 'teleosporic' (*Tilletiales*), non-teliosporic (*Exobasidiales*) and an anamorphic order (*Malasseziales*).

We have studied the following genera and diseases caused by them.

Urocystis (*Urocystidales, Ustilaginomycetes*) causing flag smut of wheat.

Tolyposporium penicillariae (*Ustilaginales*), causing smut of 'bajra'.

Ustilago tritici (*Ustilaginales*), causing loose smut of wheat.

Ustilago maydis, causing smut of maize.

Tilletia (T. caries and *T. laevis*) (*Tilletiales, Exobasidiomycetes*), causing common or stinking bunt of wheat.

Exobasidium vaccinii (*Exobasidiales*), causing galls on leaves of *Vaccinium*.

Exobasidium vexans (*Exobasidiales*), causing 'blister blight' of tea leaves.

Malassezia (*Malasseziales, Exobasidiomycetes*) is an opportunistic pathogen causing superficial skin diseases in humans.

It is to be noted that *U. maydis* has emerged as an organism of choice for the study of cell biological processes. It has the distinction of being the first basidiomycetous fungus whose genome has been sequenced. The genomic comparison shows that the fungus is more closely related to humans than to budding yeast, and numerous proteins are shared only by *U. maydis* and *Homo sapiens*. *Malassezia* is a unique addition to the group of former smut fungi though it causes human disease. A project in 2007 has sequenced the genome of dandruff-causing *M. globosa* and found it to have

4285 genes. It has been predicted to reproduce sexually but this has not been confirmed.

TEST QUESTIONS

1. What features characterize *Ustilaginomycotina*? Distinguish between its three classes.
2. Describe the following in *Ustilaginomycetes*: (i) teliospore types, (ii) teliospore germination, (iii) types of infection, and (iv) nomenclature of diseases based on location of sori.
3. In which features *Exobasidiomycetes* show similarities/dissimilarities with *Entorrhizomycetes* and *Ustilaginomycetes*. Describe life cycle of *Exobasidium vexans*. Why is it called *Exobasidium*?
4. Answer in brief:
   (a) Which smut fungi have been transferred to *Pucciniomycotina*?
   (b) Which non-teliosporic genera are now placed in *Ustilaginomycotina*?
   (c) Which human pathogen is now placed in *Ustilaginomycotina*?
   (d) Name the three major smut /bunt diseases wheat.
   (e) Name the pathogens of three bunt diseases of wheat.
   (f) Genome of which two genera of *Ustilaginomycotina* have been sequenced?
   (g) Which smut fungus shows genetic relatedness to humans?
   (h) Which is the fungus that causes hypopigmentation and dandruff in humans?
   (i) How is *Malassezia* identified from skin infections?
   (j) How does *Malassezia* cause hypopigmentation and dandruff?
   (k) Name three genera whose basidiospores (sporidia) are discharged like ballisstospores
   (l) Why is Karnal bunt called 'partial' bunt?
   (m) Which two smut/bunt genera produce fishy smell?
   (n) What is the phenomenon which requires dikaryotic hyphae for infection called?
5. Write brief notes on:
   (a) Types of teliospore germination in smut fungi,
   (b) Methods of formation of dikaryotic hyphae,
   (c) *Ustilago maydis*

(d) Basal dichotomy in *Ustilaginomycotina*.

6. Write names of the causal organism and their classification (class and order only) of:

   (a) Flag smut of wheat.

   (b) Smut of 'bajra'.

   (c) Common or stinking bunt of wheat.

   (d) Karnal bunt of wheat.

17

Subphylum *Pucciniomycotina*
(formerly Urediniomycetes)
(Rusts and related Fungi)

Subphylum *Pucciniomycotina* includes the well known rusts, which are obligate parasites on a wide range of crop plants. Other 'non-rust' plant pathogens like *Helicobasidium*, are important but at much smaller scale. *Microbotryales* the typical smut fungi, are now placed here on the basis of molecular sequence data. The *Pucciniomycotina* also includes fungi that parasitize scale insects and form fungal gardens, which represent common place wonders of nature. The group contains 8016 species distributed among 190 genera (Kirk *et al.,* 2008). More than 90% of the species are placed in the order *Pucciniales* (*Pucciniomycetes*). The next largest orders, *Septobasidiales* (*Pucciniomycetes*) and *Microbotryales* (*Microbotryomycetes*), collectively constitute approximately 5% of the species. Nearly 20% of the rust genera and 60% of the non-rust genera are monotypic i.e. have one species. The rust fungi and some of the yeasts have been extensively studied.

The group is characterized by (1) karyogamy through spermatia, (2) absence of basidioma (except in *Septobasidiales*), (3) obligatory parasitic nature on plants or insects, (4) explosive discharge of basidiospores (ballistospores), (5) presence of dolipore septa (often with plug but without parenthosomes) (6) absence of clamp connections, and (7) formation of teliospores.

The mycelium is intercellular with haustoria and limited to parts of leaves or other aerial pars, causing **local infection**. However, if root stock is infected, the infection becomes **systemic.** Teliospores are formed on secondary (dikaryotic) mycelium at the end of the dikaryotic phase, when conditions become unfavourable for growth. Usually teliospores undergo a period of dormancy before germination (hence called 'resting' spores). Teliospores germinating immediately without a resting period, are called **leptospores**. Sexual reproduction, (karyogamy and meiosis) occurs during germination of teliospores, and thus, they represent the perfect stage. While karyogamy occurs in the teliospore, a little before its germination, meiosis occurs in the germ tube, which develops into a short cylindrical hypha, called **promycelium**. The four recombinant haploid nuclei migrate into four basidiospores, formed externally on the promycelium. Promycelium is septate (usually four-celled) bearing one basidiospore on a pointed stalk called **sterigma**. The basidiospores are discharged violently. Since teliospores are considered homologous to the basidia of "jelly fungi", they are designated by the same terminology*. Before karyogamy the telispore is called a **probasidium**, after karyogamy, it is designated as **hypobasidium** (=metabasidium). The promycelium, which bears the basidiospores is the **epibasidium** (Fig. 17.1). In a typical club-shaped basidium, both karyogamy and meiosis occur in the same cell.

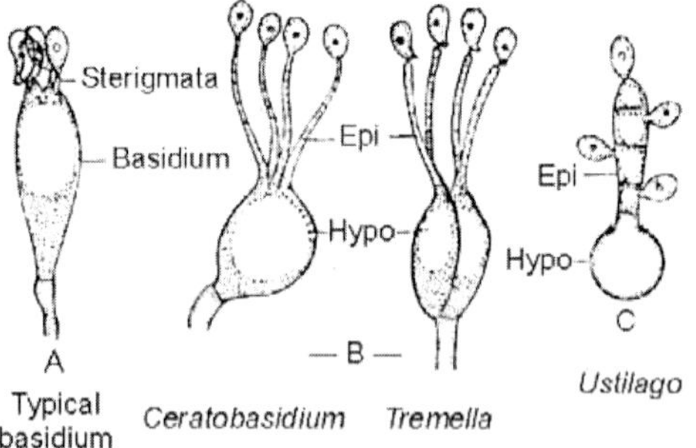

Fig. 17.1 Showing a typical basidium (A) and its modifications in jelly fungi (B), and teliospore (C). Epi = epibasidium, Hypo = hypobasidium.

Classification

There are 8 classes, which represent the 8 **clades** discovered in the *Pucciniomycotina* based on rDNA sequences (Aime *et al.*, 2006). We shall study class *Pucciniomycetes* and *Microbotryomycetes*.

Class *Pucciniomycetes*

There are 5 orders in this class including ***Pucciniales*** and ***Septobasidiales***.

---

* The binucleate cell which develops into the basidium is called **probasidium**. It becomes **hypobasidium** after karyogamy and meiosis have occurred in it and four elongated structures, called **epibasidia** (sing. epibasidium) have developed that bear the basidiospores. Thus, the terms probasidium and hypobasidium represent early and late stages of the same cell. The **basidium** in these fungi, thus, consist of two portions, the basal septate or unseptate, thick-walled hypobasidium and the upper epibasidium that bears the basidiospores.

Order *Pucciniales* (rust fungi)

The order which includes the rust fungi contains 95% of the species of *Pucciniomycotina*. The name 'rust' is used for fungi included in this order, as well as, the plant diseases they cause. These are obligate parasites and show extreme host-specialization. Rust diseases have been known since ancient times. They infect ferns, gymnospersms and angiosperms, but the cereals suffer most. "Rust is the greatest enemy of cereals' observed **Pliny** (29 – 79 A.D.). Romans held a spring festival, the **Rubigalia**, to save their crops from the wrath of god Rubigo.

Rusts, the diseases as well as the fungi, have been at the focus in their respective realms of study, plant pathology and mycology. The vexed biology of the rust fungi and their artificial culture have baffled biologists for long, but axenic culture is now possible from all the states of the life cycle. Several rust fungi have been taken in axenic (pure) culture. These are *Gymnosporangium juniperi–virginianae*, *Puccinia graminis* f.sp. *tritici*, *Melampsora lini* and *Uromyces dianthi* and some others. But in nature, they always live as **obligate parasites**. These fungi grow intercellulary in their hosts and derive nutrition through haustoria. But it is a "peaceful coexistence". The damage is slow, subtle and prolonged. Rusts are considered elite among plant pathogens, whose presence in the host is known not by "symptoms" but by the "signs" of their own sori, as is true also with other obligate parasites like "downy" and "powdery" mildews.

Intercellular growth and formation of haustoria

The intercellular hyphae form haustoria by penetration of the hypha into the cell and making invagination of the host plasma-membrane (Fig. 17.2). Haustoria are the major sites of nutrient uptake from the host cell. But their function is not limited to that. The haustorium is the site where the avirulence (*Avr*) genes, coding for **elicitors** (Avr proteins), are expressed. These proteins, (signal peptides) are secreted by the haustorium into the host cell, which, in incompatible combination, are recognized by the resistance (*R*) gene products (R proteins) of the plant. This initiates hypersensitive response, leading to resistance. In plants lacking *R* gene (and R proteins), the elicitor proteins go unrecognized, resulting in susceptibility and infection. Catanzariti *et al.* (2006) have identified a number of avirulence elicitors (Avr proteins), encoded by genes expressed in haustoria of flax rust *Melampsora lini*. These proteins are called **haustorially-expressed secreted proteins (HESPs)**. The identified proteins contain secretory pathway signal peptides. It suggests the existence of a transport mechanism in the host plasma-membrane that transfers the Avr proteins into the host cell.

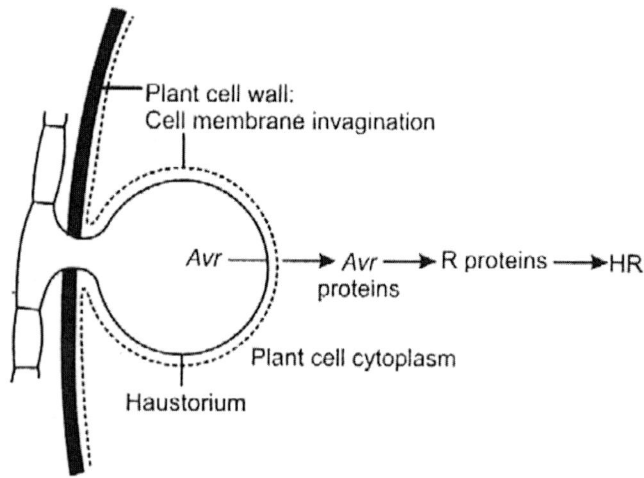

Fig. 17.2 Diagram showing haustorium in flax rust at the site of molecular interaction, determining susceptibility or resistance of the host. *Avr* = avirulence gene of pathogen, R = resistance of host, HR = hypersensitive response.

Complex life cycle

The rust fungi have most complex life cycle in kingdom *Fungi*. There are five **spore types**, numbered 0-IV, in Roman numerals, and each is also given morphological name to avoid confusion. These are as follows: 0 = spermatia; I = aeciospores; II urediniospores; III teliospores and IV basidiospores. Depending on the number of spore states, there are three **types of life cycles**, macrocyclic, demicyclic and microcyclic, as depicted in Table 17.1. A rust fungus may be **autoecious** or **heteroecious.** If the life cycle is completed on one host, the fungus is autoeocious; if two hosts are involved, it is heteroecious.

Table 17.1 Three types of life cycles in rusts.

Life Cycle	Spermatia	Aeciospores	Urediniospores	Teliospores	Basidiospores	Example
Macrocyclic	+ (-)	+	+	+	+	*P. graminis tritici*
Demicyclic	+ (-)	+	-	+	+	*G. juniperi-virginianae*
Microcyclic	+(-)	-	-	+	+	*Endocronartium harknessii*

The order *Pucciniales* is divided into 14 families (Kirk *et al.*, 2008) but phylogenetic studies of two nuclear rDNA genes (18S and 28S) suggest only eight to be genuine families (Aime, 2006). The eight families are: *Coleosporiaceae*, ***Melampsoraceae***, *Mikronegeriaceae*, *Phakopsoraceae*, *Phragmidiaceae*, *Pileolariaceae*, ***Pucciniaceae***, ***Raveneliaceae***. The genus *Hemileia* is "incertae sedis' (not assigned any family). The rust genera form characteristic teliospores by which they are identified (Fig. 17.3).

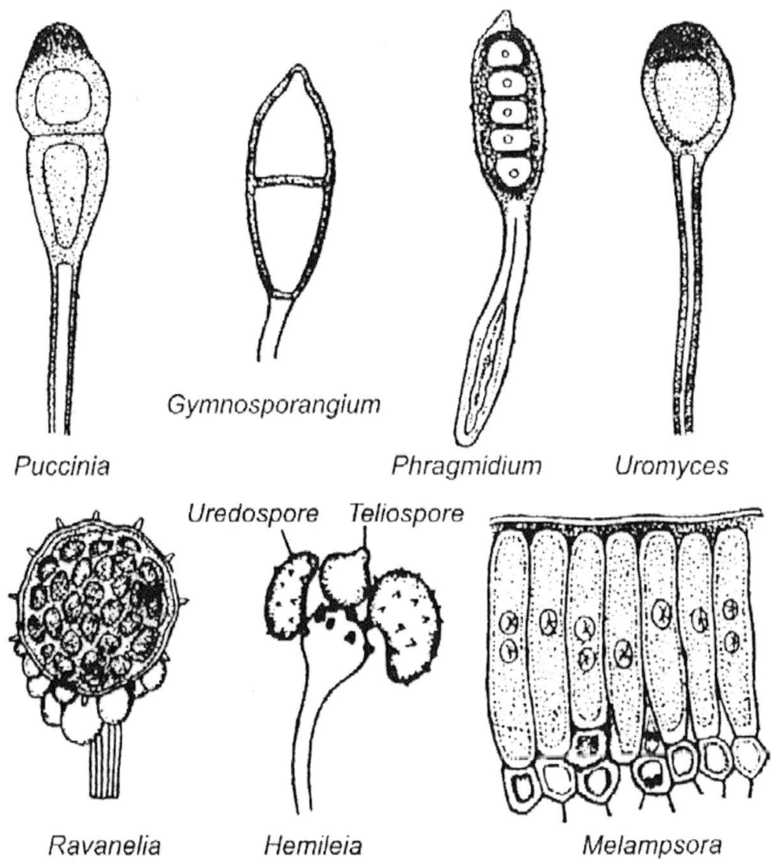

Fig. 17.3. Teliospores of some rust genera.

Family *Pucciniaceae*

The characteristic features of this family are erumpent telia, pedicillate, bi–celled teliospores, flat or discoid, sub-epidermal pycnia, and catenulate (rarely pedicillate) aeciospores. There are 17 genera including ***Puccinia***, ***Gymnosporangium*** and ***Uromyces***, which we shall examine.

Genus *Puccinia*

This is the largest genus with 4938 species, parasitizing angiosperms throughout the world. The important species and the diseases they cause are : *P. graminis* f.sp. *tritici* (black stem rust of wheat), *P. recondita* (orange, or brown rust of wheat), *P. striiformis* (yellow or stripe rust of wheat), *P. coronata* (oat crown rust), *P. purpurea* (sorghum rust), *P. helianthi* (sunflower rust), *P. carthami* (safflower rust), *P. aspargi* (asparagus rust), *P. malvacearum* (holyhock rust), *P. sorghi* and *P. polysora* (maize rust), *P. allii* (on onions), *P. menthae* (on mint), *P. hordei* (barley leaf rust).

Puccinia graminis tritici (Black stem rust of wheat)

It is a macrocylic, hereroecious rust producing all the five types of spores and needs two hosts – wheat and barberry, to complete the life cycle. Uredinio –, and teliospores are produced on wheat, while spermatia and aeciospores are produced on barberry. Basidiospores are produced on promycelium, formed by teliospores on germination.

Life Cycle

On wheat, the first infections of the year are caused by urediniospores (Fig. 17.4) blown from far off places, where rust develops earlier (from hills to the plains of India) or by aeciospores produced on barberry. Both the spores are binucleate and infect wheat leaf. The dikaryotic mycelium grows intercellularly and produces haustoria. After sometime, the hyphae accumulate at several places under the epidermis and form uredinia.

The uredinia appear as rust-red pustules, hence, the name "red rust", and urediniospores develop in them. The urediniospores are single-celled, stalked, thick-walled oval spores, which contain enough reserve food to sustain them during their long-distance dissemination. These are capable of immediate germination on wheat leaves and cause several successive infections and spread the fungus and the disease from field to field. At the close of the wheat season, the uredinia start producing decreasing number of urediniospores and increasing number of teliospores. Finally, the uredinia are turned into telia, producing only teliospores.

Telia are black, and elongated pustules, occurring on the stem and leaf bases. Teliosporess are stalked, bi-celled, spindle-shaped, dark-brown structures with a pointed apex and thick, dark, smooth wall. Each cell is binucleate and has a germ pore. The upper cell has the germ pore at the apex, while the lower cell has it at the side. It is a resting spore, which represents the sexual apparatus of the fungus, in which the karyogamy and meiosis occur. The two nuclei fuse and a diploid nucleus is formed in each cell. It undergoes a resting period in this condition attached to wheat straw or in soil.

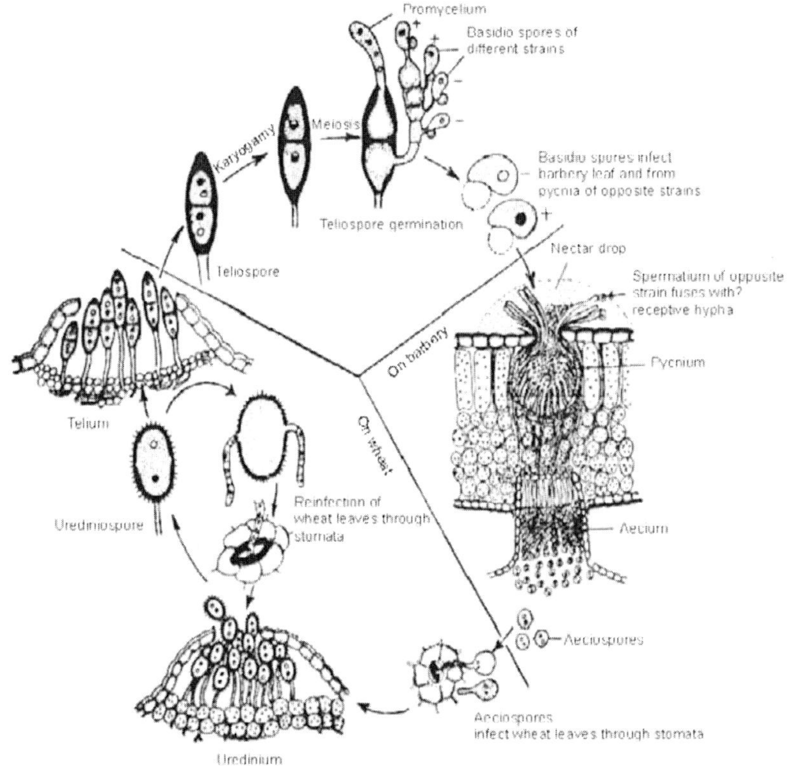

Fig. 17.4. Life Cycle of *Puccinia graminis tricici*

On the return of favourable weather in spring, the teliospores germinate. Each cell forms a curved promycelium. The diploid nucleus undergoes meiosis in the promycelium and forms four haploid nuclei of two mating types. Septa are laid down to separate them into four cells in the promycelium. Each cell produces a single basidiospore, borne asymmetrically on a fine sterigma (pl. sterigmata). Two of the basidiospores are of one strain and the rest two of the other strain. The basidiospores are thin-walled spores violently discharged by the "water bubble method".

The basidiospores infect only barberry and not wheat. Wind blown basidiospores, when reach a barberry leaf, germinate; the germ tube enters the leaf by mechanical penetration of the epidermis. Inside, an intercellular, monokaryotic mycelium (+ or – mating strain, depending on the basidiospore) is established. Several basidiospores may infect the same leaf, each establishing a primary mycelium. In due course, each diseased spot develops a flask-shaped pycnium (Fig. 17.5), which opens on the upper surface of the leaf by an ostiole. The pycnium wall is lined internally by a layer of spermatiophores at the base. These cut minute single-celled, uninucleate spermatia which ooze out through the ostiole in a sugary fragrant solution. The ostiole is surrounded at the fringe, by stiff hairs, called **periphyses** (sing. **periphysis**), which support the sugary droplet containing the spermatia. In addition to spermatiophores, long flexuous hyphae, called **receptive hyphae**, arise from the lateral walls of the pycnium, and grow out through the ostiole. The receptive hyphae represent the female sex organ, while the spermatia are self-sterile male gametes. Several pycnia of opposite mating types (+ or -) are generally formed on the same leaf.

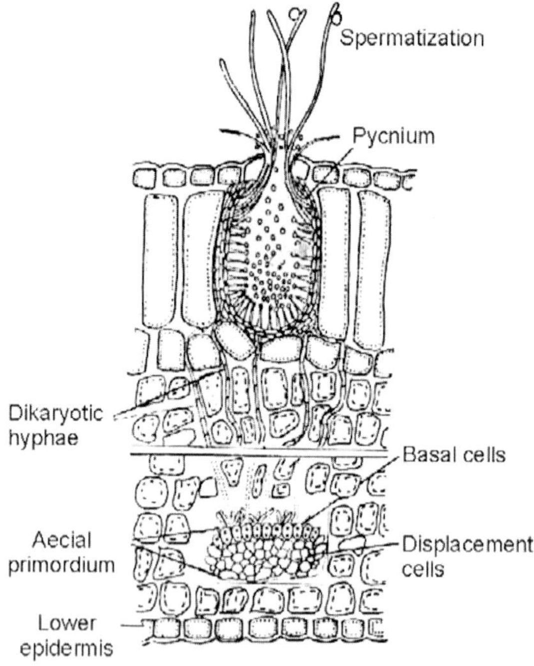

Fig. 17.5 showing pycnium and development of aecium

Alongwith the pycnial development, some of the hyphae of each mating type reach near the lower surface of the leaf and form a knot of hyphae, called the **protoaecium**. It consists of a row of long **basal cells** and a surrounding mass of **displacement cells**. Spermatization takes place through the agency of insects, which are attracted by the fragrance of the nectar. Fusion of adjacent drops also aids in spermatization. When the spermatium of one strain comes in contact with receptive hypha of the opposite strain, the wall dissolves at the point of contact and the male nucleus migrates into the female receptive hypha. Thus, one cell of the receptive hypha is dikaryotized. The male nucleus, by mitotic division, forms a second male nucleus, which moves to the next cell. The male nuclei produced by successive mitotic divisions, pass down and all the cells of the primary mycelium are dikaryotized. Eventually, the palisade-like "basal cells" of the protoaecium are dikaryotized. The protoaecium develops into the aecium. The cells, other than the basal cells, degenerate. Each binucleate basal cell cuts a chain of binucleate cells, which occupy the cavity formed by dissolution of the displacement cells. The binucleate cells of the chain are alternately large and small. The large cells develop into **aeciospores** while the smaller ones, called **disjunctor cells**, gelatinize, liberating the aeciospores. The mass of aeciospores is covered by a membrane called **peridium**, which is derived from the peripheral basal cells. The whole structure is called **aecium**. The chains of aeciospores push through the peridium and emerge out of the epidermis. The broken peridium hangs down the lower epidermis. The aecial stage is often called **cluster-cup** stage because of the formation of cup-like aecia in clusters.

The aeciospores, by crowding become hexagonal, but tend to regain the turgid globular shape. These contain enough reserve food. The aeciospores are unable to germinate on barberry, and germinate on wheat leaf. The germ tube enters through the stomata and establishes a dikaryotic mycelium which develops into a uredium.

Genus *Gymnosporangium*

The teliospores are bi-celled (occasionally single or many-celled) and possess pedicels which gelatinize when moist. The genus has 57 species, which are restricted to temperate regions. Most of the species are heteroecious and demicyclic (i.e. lack uredinia); the telial stage occurs on gymnosperms and aecia on dicotyledons (usually members of *Rosaceae*).

G. juniperi virginianae (Fig. 17.6) causes considerable damage to apple on which it forms the pycnial and aecial stages; the telial stage is formed on juniper. The aeciospores, formed on the apple leaves, infect the junipers in summer but the infection remains latent until next spring when round, reddish galls, bearing telia, appear on the twigs. The telia are initially sub-epidermal, but later become conspicuously erumpent, forming cushions, crests or 'horns', which gelatinize when moist. The gelatinized pedicels intertwine and emerge out carrying the teliospores. The teliospores (leptospores) germinate *in situ* without any rest period and produce four basidiospores. They are blown away by wind for several miles to reach the apple leaves and infect them. The basidiospores constitute the single source of primary inoculum for apple. There are no secondary infections, as urediniospores are not formed. Pycnia develop on the upper surface of apple leaves followed by aecia on the lower surface. The peridium splits at several points and rolls back giving the aecium a star-like (*Roestelia* type) appearance. The aeciospores infect the juniper in summer.

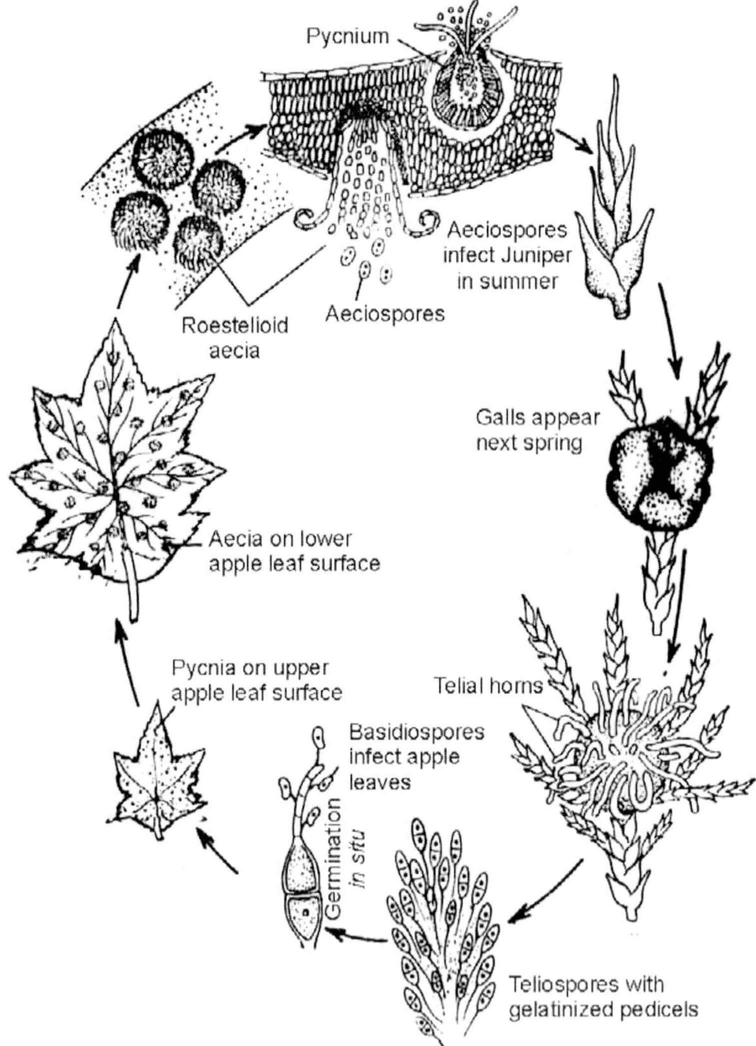

Fig. 17.6. Life cycle of *Gymnosporangium juniperi virginianae*

The fungus alternates between apple and juniper, as only two infectious spore forms are produced, which infect one specific host. The fact that these are repeating spores on either host, has been utilized to break the life cycle in disease control by eliminating either host. But since both hosts are economically important, neither can be destroyed. The disease, however, can be checked by keeping them apart. Thus, 'Juniper-free' zones are established around apple orchards.

Genus *Uromyces*

Uromyces, the second largest rust genus with 800 species, is characterized by unicellular teliospores thickened at the apex. If any bi-celled teliospores are produced in the telia, the fungus is assigned to *Puccinia* by the "gentleman's agreement". This occurs in *Puccinia heterospora*, which produces unicellular teliospores mixed with a few bi-celled spores. One who is not aware of the "gentleman's agreement" will surely identify the fungus as *Uromyces*.

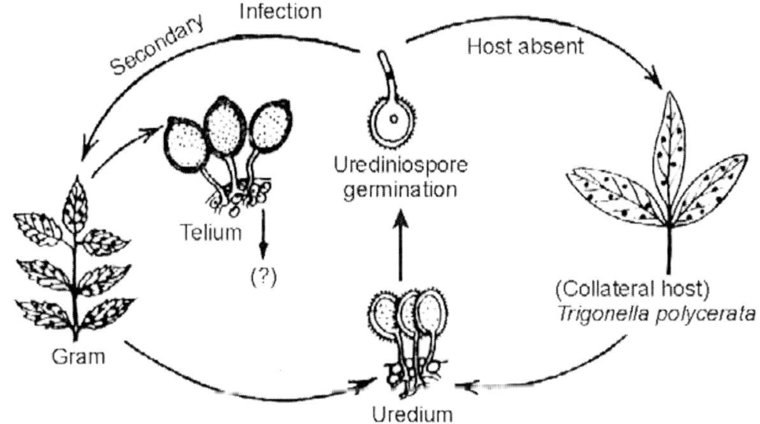

Fig. 17.7. Life cycle of gram rust, *Uromyces ciceris*

Species of *Uromyces* cause important diseases, especially on *Leguminosae*, e.g. *U. appendiculatus* or, *U. phaseoli* (bean rust); *U. ciceris* (gram rust); *U. pisi* and *U. fabae* (pea rust) *U. dianthi* (carnation rust) and several others. The life cycle may be macro-, or microcyclic, and heteroecious or autoecious.

Gram Rust is caused by *U. ciceris* (Fig. 17.7). The pycnial and aecial stages are not known. The uredinia and telia usually appear on leaves, but in severe infections these may be formed on stems and the petioles. The role of teliospores is not known; the annual recurrence is through urediniospores from collateral host, *Trigonella polycerata*, a leguminous weed.

Family *Melampsoraceae*

The family, includes one genus, *Melampsora*, which has 90 species forming sessile teliospores.

Genus *Melampsora*

The teliospores are single-celled, sessile and formed in a palisade layer. The telia are erumpent or crustose, covered by the host epidermis. Pycnia are flat and diffuse; aecia caeomoid i.e. lack peridium. The uredinia are erumpent and have capitate paraphyses, interspersed among urediniospores. The species may be autoecious or heteroecious. *M. lini*, causing rust of flax (*Linum utissimum,* Fig. 17.8), is the best known. The 'gene-for-gene' hypothesis of host-parasite interactions was postulated by Flor (1956) with *M. lini.*

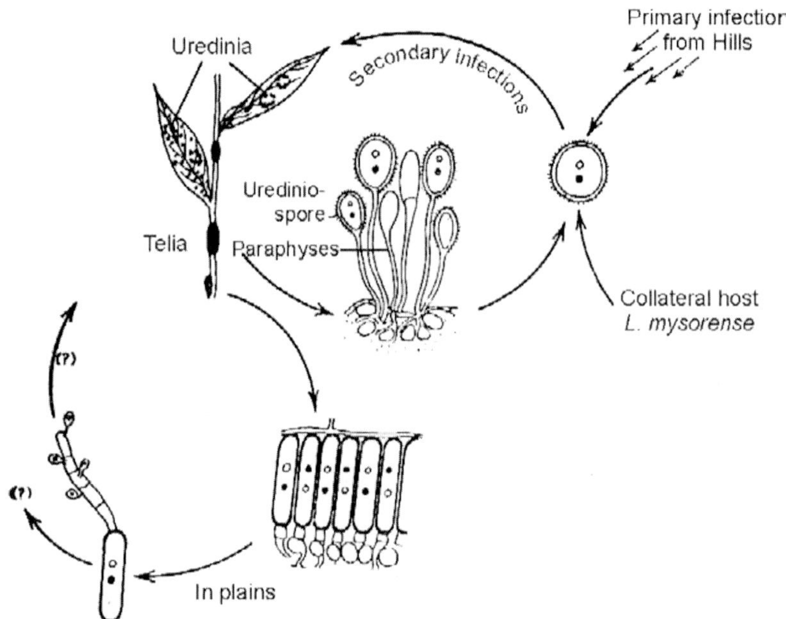

Fig. 17.8. Life cycle (disease cycle) of *Melampsora lini*

Life Cycle (Fig. 17.8)

The fungus, which infects several species of *Linum*, is an autoecious rust. The pycnial and aecial stages are not recorded in nature. The urediniospores, growing locally on collateral hosts, or blown down from hills, initiate infection every year. Bright reddish-yellow spots of uredinia appear in December or January on leaves, and occasionally on stems and petioles. These are formed in circles on the leaves and as elongate patches on the stems. Ovate, stalked urediniospores are formed intermingled with capitate paraphyses. These spores serve to spread the fungus and the disease, until the end of the host season. The infected leaves die and fall off prematurely. Black, crust like telia appear on the stems at the end of the season. These consist of a palisade layer of brown to black, sessile cylindrical teliospores, which are released only after the disintegration of the overlying host tissues. The teliospores germinate (when exposed to freezing temperature) and produce four basidiospores on a septate epibasidium, but they are unable to infect the host. The teliospores fail to survive the intervening hot summer. *Linum mysorens* is reported to act as the collateral host.

Family *Phragmidiaceae* (14 genera)

Genus *Phragmidium*

The genus is identified by the stalked, transversely septate, many celled teliospores (see Fig. 17.3). Each cell has 2-3 germ pores. Pycnia are subcuticular, confluent and indeterminate; aecia caemoid: uredinia subepidermal, bearing the urediniospores, surrounded by a peripheral ring of paraphyses. About 110 species are known; all autoecious, macro- or demicyclic on *Rosaceae*. The most common hosts are *Rosa*, *Rubus* and *Potentilla*. *P. mucronatum* and *P. tuberculatum* are important parasites of ornamental roses, while *P. rubidaei* attacks raspberries.

A group of 10 brachycylic species of *Phragmidium* (lacking aecia or producing uredinoid aecia) are segregated as genus *Phragmotelium* which is incertae sedis. i.e. has not been assigned to any family. These species produce uredinoid aecia, and leptospores (smooth-walled teliospores that germinate without a resting period), which are borne on bi-celled, non-hygroscopic pedicels.

Family *Raveneliaceae* (26 genera)

Genus *Ravenelia*

It is a large genus of about 200 autoecious species, parasitizing legumes in tropical countries including India. The genus is characterized by teliospore 'heads', formed by fusion of several teliospores, each with a separate stalk (see Fig. 17.3). Each fused teliospore is subtended by a sterile cyst cell. The telial heads may be smooth or ornamented with several small projections.The pycnia are sub-cuticular and indeterminate. Aecia bear uredinoid (i.e. pedecillate) aeciospores, but occasionally cupulate aecia are also formed. Uredinia have paraphyses at the periphery. *R. emblicae* is the most common species found on fruits of *Phylanthus emblica*.

Incertae sedis rust genera (*Hemileia* and *Masseeella*)

Genus *Hemileia*

Pycnia and aecia are not known, uredinia are formed characteristically in sub-stomatal chambers, and the urediniospores are typically reniform or hump-backed. Many species are known only in uredinial stage, which are easily identifiable. Telia are formed like the uredinia. *H. vastatrix* forms a mixed sorus containing both uredinio-, and teliospores. The teliospores, which are unicellular, thin-walled, and pale, germinate *in situ* without any rest period and therefore, are termed leptospores. About 40 species are known which mostly parasitize angiosperms (especially members of family *Rubiaceae*) in tropical countries of Asia and Africa.

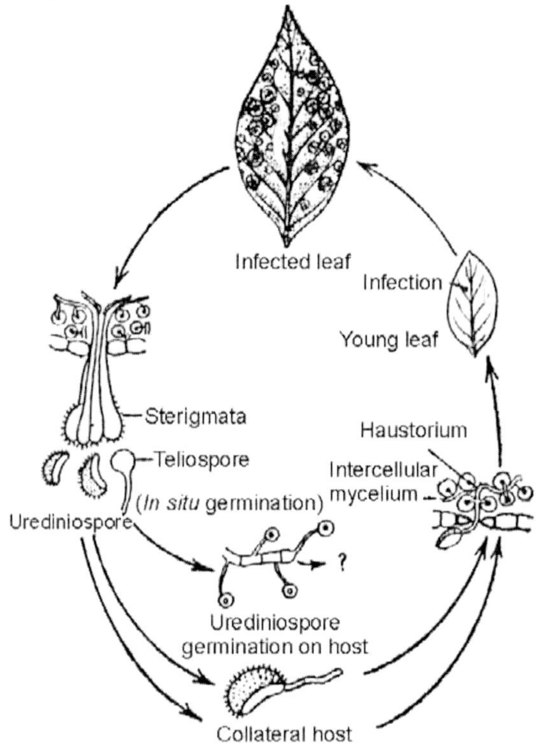

Fig. 17.9 Life cycle of *Hemileia vastatrix,* causing rust of coffee

Hemileia vastatrix (Coffee Rust)

It is an autoecious, microcyclic rust; pycnia and aecia are lacking (Fig. 17.9). The uredinio- and teliospores are formed in orange to brown mixed sori formed in stomatal chambers, mainly on the lower surface of the leaves, and rarely on berries and young shoots. The spores are borne in a typical manner. Hyphae accumulate in the sub-stomatal chambers, from which vertical hyphae arise that emerge through the stomata and become swollen at the tip. Sterigmata are formed all over the swollen apex which bear the uredinio-, and teliospores. The urediniospores are hump-backed, resembling an orange segment, with echinulate upper and smooth lower surface. The teliospores (leptospores), which are formed in small numbers, are turnip-shaped and germinate *in situ*, forming four basidiospores on a septate slender promycelium (=epibasidium). Such teliospores which germinate immediately, without a rest period are called **leptospores**.

The basidiospores fail to infect the host. No alternate host is known. The successive generations of the urediniospores infect coffee plants and keeps the life cycle going single-handed.

Genus *Masseeella*

Masseeella forms characteristic "telial hairs" (Fig. 17.10), projecting out of the leaves. These are gelatinous masses containing single-celled sessile teliospores (Fig. 17.10). A peridium surrounds the base of the telial hairs. To begin with, the telia are sub-epidermal, but later become detached and extruded. The genus has 5 macrocyclic, autoecious species, which grow on several members of *Euphorbiaceae*. The pycnia are diffuse, aecia cupulate, and the uredinia erumpent and paraphysate.

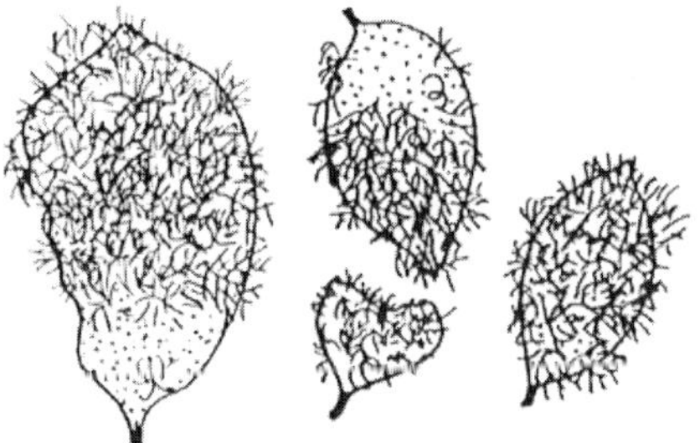

Fig. 17.10 *Masseeella*. Telial hairs emerged out of infected leaf

Order *Septobasidiales*

The order has a single family, ***Septobasidiaceae***, comprising of 7 genera which form unique symbiotic–parasitic relationship with scale insects* that feed on xylem sap of living plants through probosis (sucking

* **Scale insects** are small, barely visible, sedentary creatures that remain attached to one spot on the bark, leaves or fruits of plants. They push in a slender, sucking tube

tube) (Fig. 13.2). The basidioma is dry, crustaceous, or spongy (non-gelatinous) and the basidia are 'auricularioid' i.e., like members of *Auriculariales* (class *Agaricomycetes*), where *Septobasidiales* were once placed. However, because of the presence of the teliospores, dolipore septa, lack of clamp connections and similar DNA sequences, they are now placed along with rusts under *Pucciniomycetes*.

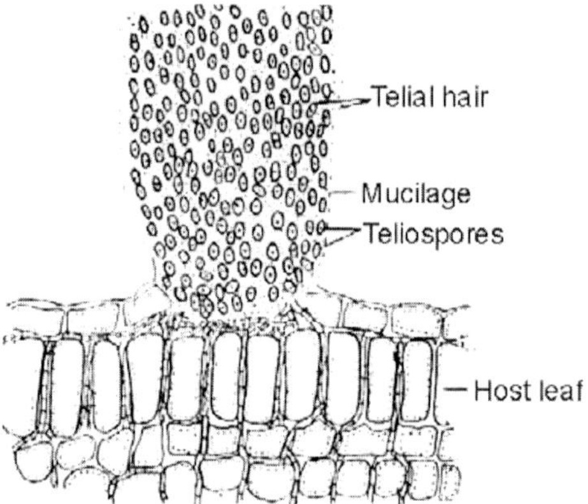

Fig. 17.11 Section of leaf showing emergence of telial hairs

Genus *Septobasidium*

Though the genus was first reported by **Patouillard** in 1892, its interesting biological relationship with scale insects was discovered by **von Hohnel** and **Litschauer** in 1907. Later, **Couch** (1938) contributed much of what we know today on the taxonomy and biology of this fungus. The genus has *Johncouchia* as the anamorph.

Septobasidium is reported from several tropical, sub-tropical and warm temperate regions. There are 170 species some of which are directly parasitic on trees causing hypertrophy, but majority of them enter a symbiotic association with scale insects. The relationship is more than a dual partnership. Besides the fungus and the insect, the host tree is also a partner in the association (Fig. 17.12). The fungus forms a mat of dikaryotic hyphae over the parasitized insects which later develops into an intricate structure (basidioma) called a **fungal garden**. The ever-growing fungal garden derives

---

into the vessels and suck he sap. For protection, they secrete a chitinous scaly covering around them; hence he name scale insects

its nutrition from the parasitized insects, which in turn, grow on the sap of the xylem vessels of the tree through their probosis.

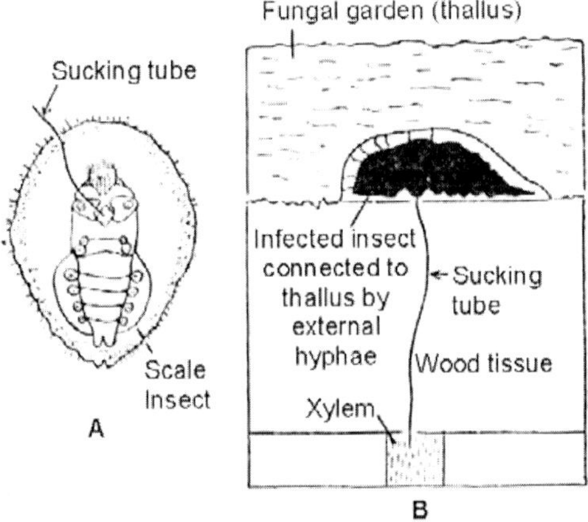

Fig. 17.12 A, scale insect, ventral view showing sucking type of mouth parts; B, showing schematically the interrelationship between fungus, single infected insect and tree.

The basidioma of *Septobasidium* is one of several examples of commonplace wonders of nature, which humbles even the most sophisticated modern architects. It consists of numerous vaulted chambers, arranged in several storeys, supported by pillars. Each chamber opens independently to the exterior (through a door, closed by a flap of hyphae hinged only at one point), and also opens into tunnels formed inside the mycelial structure which lead to the exterior. Healthy or previously-infected scale insects, in search of shelter, enter the chambers through the openings and become a part of the fungal garden. However, several healthy entrants remain uninfected for long. The infected insects are rendered sterile, which is the only adverse effect of infection. The uninfected female insects reproduce in the mating season, and, for this, they protrude their posteriors through the door of the chamber for fertilization by male insects passing by. The fertilized females produce young insects, which moult and move out through the tunnel. After getting out, they crawl over the surface of the fungal mat i.e. the hymenium, where they contact the basidiospores and get infected.

The release of the first brood of young insects coincides with the time of basidiospore formation, which is usually during rains. The infected insects settle under the bark to initiate a new fungal home or, alternatively, they enter a previously existing one, and add some more chambers.

The fungus gives shelter to the scale insects from rough weather and also protects them from a deadly wasp which lays eggs in its body by inserting the ovipositor into the body cavity. The larvae that develop devour the scale insects. The insects resting in chambers of the "fungal garden" are beyond reach of the wasp. Thus, the fungus pays off well for the food it derives from the insects; a good example of mutualism.

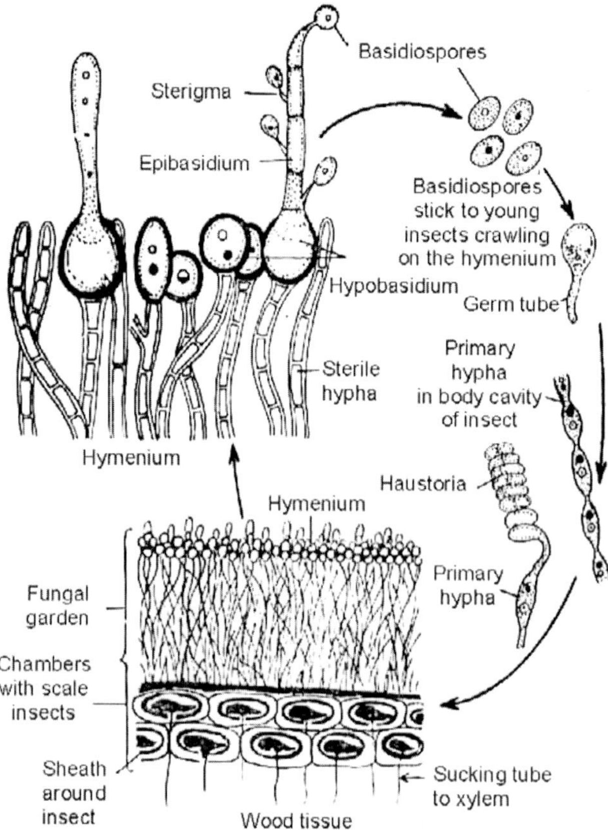

Fig. 17.13 Life cycle of *Septobasidium burtii*

Septobasidium burtii (Fig. 17.13) is the best known species. Basidiospores are formed during the moist weather, which is also the time for release of young scale insects from the fungus garden, through the tunnels. The hymenium, which is formed at the upper surface of the basidioma, consists of basidia and sterile hyphae. The teliospores (probasidia) are thick-walled, binucleate, ovate, sub-globose or pyriform. After karyogamy, it is designated as hypobasidium. Cylindrical epibasidia arise from the hypobasidium, in which the diploid nucleus migrates and undergoes meiosis. The four haploid nuclei are separated into four cells, formed by septation of the epibasidium. Each cell later bears one basidiospore on a pointed sterigma.

When a healthy scale insect crawls over the hymenium, basidiospores stick to its body. These spores germinate immediately or form secondary basidiospores. The germ tubes penetrate the integument, enter the body cavity (haemocoel) and establish primary hyphae. Strings of spindle-shaped cells develop, from which coilded haustoria arise after a few weeks, that penetrate the host tissues and derive nutrition. The infected insect, after wadding for some distance, settles under the bark of the tree and pushes its long sucking tube into the xylem vessels. As it does this, dikaryotic hyphae (formed by anastomosis and fusion of primary hyphae of the opposite mating types), protrude out of the body and form a mat of hyphae on the surface of the sedentary insect, which in turn, feeds on the xylem sap of the tree. The mycelial mat develops into the fungal garden, an intricate structure made of roofed rooms and tunnels, which receive more scale insects. Instead of initiating a new independent basidiocarp, the infected insect may enter a previously-formed one and occupy a cell to become an integeral part of it.

Class *Microbotryomycetes* ("pucciniomycetous smuts")

It is a small class created in 2006. It has 4 orders, including *Microbotryales* and *Sporidiobolales*.

Order *Microbotryales*

The members of this order have poreless septa at maturity. There are 2 families: *Microbotryaceae* and *Ustilentylomataceae*. We shall study the genera *Microbotryorum* and *Sphacelotheca*, both belonging to *Microbotryaceae*. These two genera show typical smut-like life cycle and were earlier placed along with other smuts in *Ustilaginomycotina*. These are better called "pucciniomycetous smuts".

Genus *Microbotryorum*

This genus, long considered to be a 'smut' and placed under genus *Ustilago*, is now a confirmed member of *Pucciniomycotina*. It is characterized by a dimorphic life cycle with a haploid yeast-like stage and a dikaryotic plant-parasitic stage, which bears the teliospores as dispersal agent. *M. violaceum* which causes 'anther smut' is now regarded as a model organism for the study of evolution of host-parasite interactions, fungal genetics, host race differentiation and 'host shifts'. It has 87 species that parasitize dicotyledonous angiosperms, especially members of family *Caryophyllaceae*. After transfer of these fungi, *Ustilago* is restricted to monocot hosts.

M. violaceum, (former *Ustilago violaceum)* is known as "anther smut" fungus. The fungus produces teliospores in the place of pollen (Fig.17.14). In dioecious plants, the infected female plants develop spore-containing anthers and ovaries are aborted. The fungus sterilizes but does not kill the host. The teliospores which are the dispersing spores are transported to healthy plants by insects that usually serve as pollinators. The diploid nucleus of the teliospores undergoes meiosis and forms four haploid nuclei from which four haploid cells are formed. These bud off yeast-like sporidia locally on the plant. Conjugation between the yeast-like cells of opposite mating types gives rise to dikaryotic cell which is infectious. Dikaryotic hyphae formed by the dikaryotic cells infect the host and grow endophytically. Thus, it is a sexually- transmitted disease. If the fungus reaches the root stock before winter, the plant gets systemically infected and produces diseased flowers, with smutted anthers. Recovery from the disease is rare.

Fig. 17.14. *M. violaceum* (anther smut)

Genus *Sphacelotheca*

There are 7 species. *S. reiliana* causes head smut of maize and sorghum. The 'ear' emerging from the sheath is practically a mass of teliospores covered with a whitish membrane. The vascular bundles are exposed and shredded (Fig. 17.15).

Order *Sporidiobolales*

The order includes yeast-like saprobes, which generally form teliospores in the mycelial phase. The dikaryotic hyphae have simple septa and form clamp connections. There are 7 genera placed in a single family *Sporidiobolaceae*. The important genera are *Rhodosporidium* (anamorph

Rhodotorula), *Aessosporon* (anamorph *Bullera*, *Sporobolomyces*) and *Sporidiobolus* (anamorphs *Rhodomyces, Sporobolomyces*).

Fig. 17.15 Head smut of maize, caused by *S. reiliana*.

Genus *Rhodosporidium*

The genus was discovered by **Banno** (1967) as the perfect stage of *Rhodotorula*. The life cycle of *Rhodosporidium sphaerocarpum* (Fig. 17.16) comprises an asexual yeast phase, *Rhodotorula*, and a mycelial phase. The cells in the yeast phase are red to orange in colour, uninucleated and haploid. These multiply by budding, which is **enterogenous**, i.e. the wall of the bud cell is formed afresh and is not a continuation of the parent cell.

The sexual phase starts with conjugation between two compatible haploid cells, leading to formation of a dikaryotic mycelium. The mycelium forms intercalary or terminal teliospores, in which karyogamy takes place. At the time of teliospore germination, the diploid nucleus undergoes meiosis forming four haploid nuclei. A germ tube is formed by the teliospore, which grows to form a septate promycelium bearing basidiospores (sporidia). The basidiospores bud and restart the haploid yeast phase. The life cycle resembles that of a smut.

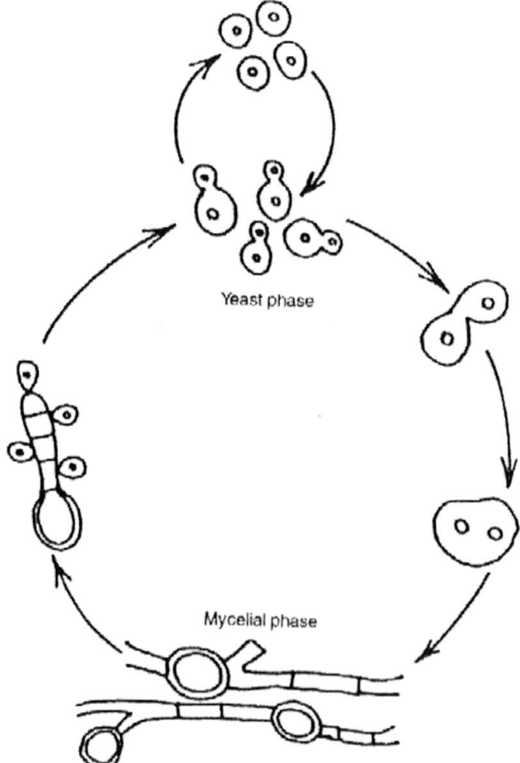

Fig. 17.16. Life cycle of *Rhodosporidium sphaerocarpum*: mycelial and yeast phases

REFERENCES

Aime MC *et al.,* 2006. An overview of the higher levlel classification of *Pucciniomycotina* based on combined analyses of nuclear large and small subunit rDNA sequences. *Mycologia* 98(6): 896-905

Aime MC *et al.,* 2006. Toward resolving family level relationships in rust fungi (Uredinales). Mycoscience 47:112-122

Banno, I. 1967. Studies on the sexuality of *Rhodotorula.* J. Gen. Microbiol. 13: 167-196

Catanzariti AM *et al.* 2006. Haustorially expressed secreted proteins from flax rust are highly enriched for avirulence elicitors. The Plant Cell. 18:243-256

Couch, JN 1938. *The genus Septobasidium.* Univ. of North Carolina Press Chapel Hill.

Flor, HH 1956. The complementary genetic systems in flax and flax rust. *Adv. Genet.*, **8:** 29–54.

Kirk PM *et al.,* 2008 *Ainsworth and Bisby's Dictionary of the Fungi* C.A.B. International, Oxon, U.K.

SUMMARY

Pucciniomycotina includes rusts and "pucciniomycetous smuts" (*Microbotryales*) as well as fungi that parasitize scale insects and form fungal garden. More than 90% of the species are placed in the order *Pucciniales* (*Pucciniomycetes*). The next largest orders, *Septobasidiales* (*Pucciniomycetes*) and *Microbotryales* (*Microbotryomycetes*), collectively constitute approximately 5% of the species. The group is characterized by (1) karyogamy through spermatia, (2) absence of basidioma (except in *Septobasidiales*), (3) obligatory parasitic nature on plants or insects, (4) explosive discharge of basidiospores (ballistospores), (5) presence of dolipore septa (often with plug but without parenthosomes) (6) absence of clamp connections, and (7) formation of teliospores. There are **8 classes**, which represent the **8 clades** of *Pucciniomycotina* based on rDNA sequences.

Class ***Pucciniomycetes*** has 5 orders including ***Pucciniales*** and ***Septobasidiales***. The order *Pucciniales* includes the rust fungi which comprise 95% of the *Pucciniomycotina*. These are obligate parasites and show extreme host-specialization. The mycelium is intercellular with **haustoria** and cause **local infections**. Haustoria are the major sites of nutrient uptake from the host cell. Haustorium is also the site where the avirulence (*Avr*) genes, coding for **elicitors** (Avr proteins), are expressed. These proteins, (signal peptides) are, in incompatible combination, recognized by the resistance (*R*) gene products (R proteins) of the plant. This initiates hypersensitive response, leading to resistance. In plants lacking *R* gene (and R proteins), the elicitor proteins go unrecognized, resulting in susceptibility and infection. These proteins are called **haustorially-expressed secreted proteins (HESPs).**

The rust fungi have a complex life cycle involving five **spore types**, numbered 0-IV, in Roman numerals, and each is also given morphological name to avoid confusion. Family *Pucciniaceae* is represented by *Puccinia*, *Gymnosporangium* and *Uromyces*. *Puccinia* is the largest genus with *Uromyces* at the second position. *Puccinia* and *Gymnosporangium* have bi-celled teliospores, while in *Uromyces* it is single-celled. The genus *Malampsora* (family *Melampsoraceae*) is the organism for which the 'gene-for-gene' hypothesis of host-parasite interaction was postulated by Flor in 1956. *Hemileia vastatrix,* the coffee rust causing organism, produces leptospores and *Masseeella* is distinct by its telial hairs (and also by its spelling).

The *Septobasidiales* are distinct in *Pucciniomycotina* by forming basidioma and by the unique symbiotic–parasitic relationship with scale insects that feed on xylem sap of living trees. The fungus forms a mat of dikaryotic hyphae over the parasitized insects, which later develops into an intricate basidioma called **fungal garden**. The ever-growing fungal garden derives its nutrition from the parasitized insects, which in turn, grow on the sap of the xylem vessels of the tree through their probosis. Healthy or previously-infected scale insects that enter the chambers become a part of the fungal garden. The uninfected female insects reproduce in the mating season and produce young insects, which crawl over the surface of the fungal mat i.e. the hymenium, where they contact the basidiospores and get infected. The infected insects settle under the bark to initiate a new fungal home or enter a previously-existing one, and add some more chambers to the garden.

Class *Microbotryomycetes* was erected in 2006 to accommodate "pucciniomycetous smuts", *Microbotryorum* and *Sphacelotheca*. Genus *Microbotryorum* has a dimorphic life cycle with a haploid yeast-like stage and a dikaryotic plant-parasitic stage, which bears the teliospores as dispersal agent. *M. violaceum* ('anther smut') is now a model organism for the study of evolution of host-parasite interactions, fungal genetics, host race differentiation and 'host shifts'. *Sphacelotheca reiliana* causes the 'head smut' of maize.

The order *Sporidiobolales* includes yeast-like saprobes, which generally form teliospores in the mycelial phase. The dikaryotic hyphae have simple septa and form clamp connections. The lone family *Sporidiobolaceae* contains *Rhodosporidium* (anamorph *Rhodotorula)*, *Aessosporon* (anamorph *Bullera*, *Sporobolomyces*) and *Sporidiobolus* (anamorphs *Rhodomyces, Sporobolomyces*). *Rhodosporidium sphaerocarpum* comprises an asexual yeast phase, (*Rhodotorula*) and a mycelial phase. The mycelium forms teliospores, which produce four basidiospores (sporidia). The basidiospores bud and restart the haploid yeast phase. The life cycle resembles that of a smut.

TEST QUESTIONS

1. What are the characteristic features of *Pucciniomycotina*? Give its classification up to orders and families studied by you.
2. Describe (a) the role of haustoria in disease resistance, and (b) names of 'spore types' in rusts.
3. Distinguish between: (a) teliospores and leptospores, (b) demicyclic and microcyclic rusts, (c) autoecious and heteroecious rusts with two examples, (d) probasidium and hypobasudium.
4. Write notes on: (a) *Microbotryorum*, (b) *Rhodosporidium*, (c) The anamorphs of *Asessospora* and *Sporidiobolus*.

5. Describe the life cycle of *Melampsora* or *Rhodosporidium* and give its classification.

6. Answer briefly:
   (a) Which rust fungi have been taken in axenic culture?
   (b) Which is the biggest class and which is the biggest order in *Pucciniomycotina*?
   (c) Which fungi, other than rusts are placed in *Pucciniomycotina*?
   (d) How many 'clades' are there in *Pucciniomycotina*?
   (e) Why rusts are called 'elite' among plant pathogens?
   (f) Which are the three participants in the formation of 'fungal garden'?
   (g) What is "gentleman's agreement" in *Uromyces* identification?
   (h) Which rust was studied by Flor (1956) for his gene-for-gene hypothesis?
   (i) Name two incertae sedis genera of *Pucciniales*.
   (j) Which fungus is called "anther smut"?
   (k) What is the benefit of 'fungal garden' to scale insects?
   (l) What are leptospores? Name 2 genera which form leptospores.

7. Fill in the blanks:
   (a) Romans held a spring festival called ------------ to save their crops from rust.
   (b) Rust is the greatest enemy of -----------.
   (c) Presence of rust disease is known by -----------, not by symptoms.
   (d) *Gymnosporangium juniperi virginianae* is a ---------------- rust.
   (e) 'Telial hairs' are formed by --------------, while 'telial horns' are formed by -----------.
   (f) *Johncouchia* is the anamorph of -------------.
   (g) The basidioma of *Septobasidium* is called ---------------.
   (h) *Puccinia graminis* f.sp. *tritici* and --------------- are macrocyclic rusts.
   (i) *Microbotryum violaceum* was formerly called as ---------- *violaceum*.
   (j) The teliospores of *Microbotryorum* is transported to healthy plants through -----------.
   (k) ------------------- is now a model organism for the study of host-paarasite interactions and fungal genetics.

18

Kingdom *Chromista (Straminopila)*
Phylum *Hyphochytriomycota*
(Uniflagellate chromistan fungi)

Chromista* is a eukaryotic kingdom created by Cavalier-Smith in 1981. The chromistans, as the members of this kingdom are called, were earlier placed under former kingdom Protista (now derecognized and replaced by two kingdoms *Chromista* and *Protozoa*). The members are characterized by (1) unicellular, filamentous or colonial nature, (2) presence of tinsel type flagellum, (3) photosynthetic or heterotrophic nutrition, (4) cell wall made of a mix of cellulose and glucans; absence of chitin, (5) mitochondria with tubular cristae, (6) presence of Golgi bodies and peroxisomes (peroxisomes are indispensable for metabolism of several compounds in eukaryotic microorganisms), (7) storage product mycolaminarin, and (8) diploid nucleli. The photosynthetic members include golden and brown algae, diatoms chrysophytes, and cryptomonads (but not chlorophyte or red algae, which belong to *Plantae*). The heterotrophic members are the fungi. There are 10 phyla in the kingdom.

The difference between the chromistan fungi and true fungi (kingdom *Fungi*) lies in the nature of the cell wall and the flagella. Also, these organisms are not genetically close relatives of true fungi. In 18S rDNA phylogenetic studies these fungi show closer genetic relatedness to brown algae and diatoms than to true fungi. Nevertheless, these are similar to true fungi in the form of their thalli, nutrition and ecological roles (parasitism and decomposition). In the evolutionary sequence, these fungi are basal to the diatoms, brown and golden-brown algae (Fig. 18.1), which diverged before these

---

* *Chromista* means 'coloured', while *Straminopila* means flagellar-hair (stramen = flagellum, pilos = hair). Thus the term *Chromista* emphasizes photosynthesis and *Straminopila* the 'hairy' tinsel-type flagellum

algae acquired the photosynthetic machinery by endosymbiosis (**Burns** *et al.,* 1991).

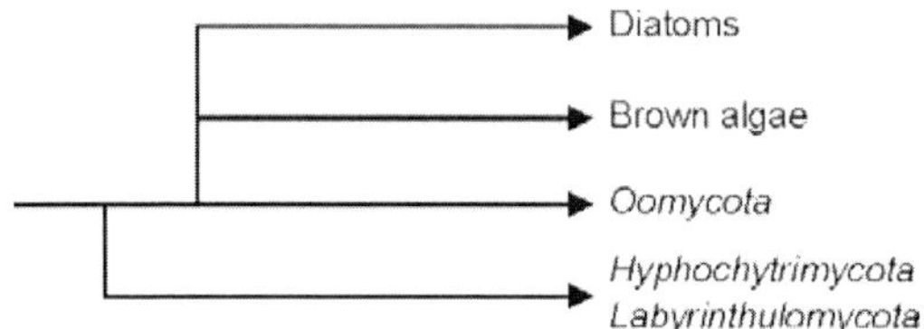

Fig. 18.1. Evolutionary sequence in *Chromista* (after **Burns** *et al.,* 1991)

The classification of the 'chromostan' fungi within the kingdom is not settled. Corliss (1994) put them ("the organisms studied by mycologists") in 2 phyla, while Cavalier-Smith put them in 2 (differently named) phyla. Dick (2001) puts all the fungal groups under 1 phylum *Heterokaonta.* In *The Dictionary of the Fungi* (Kirk *et al.,* 2008), these fungi are placed in 3 phyla: *Hyphochytriomycota, Labyrinthulomycota* and *Oomycota.*

Phylum *Hyphochytriomycota* is characterized by their anteriorly uniflagellate swarmers – zoospores and gametes, having tinsel type flagellum, and cellulose in the cell wall. The fungi show great resemblance to *Chytridiales* in thallus organization and life cycle pattern. Once, most of these fungi were included under *Chytridiales* (*Chytridiomycota*).

These are endo-parasites of fresh water and marine algae, aquatic fungi or saprobes, growing on a variety of organic remains of plant and animal origin. The thallus is microscopic, eucarpic, monocentric or polycentric. These may be epibiotic and monocentric, having a single reproductive organ, anchored to the substratum by means of rhizoids, or endobiotic and polycentric i.e. having many reproductive organs, connected by branched hyphae with occasional septa.

The zoosporangia are inoperculate, and the zoospores are released through discharge tubes. Zoospore elaboration may occur inside, or outside the zoosporangia, at the mouth of a discharge tube. Sexual reproduction is not reported. Resting spores have been found in a few genera.

There is one class *Hyphochytriomycetes,* one order *Hyphochytriales*, 2 families, *Hyphochytriaceae* and *Rhizidiomycetaceae,* which can be distinguished by the following 'key'.

Thallus polycentric, endobiotic, intramatrical, lacking rhizoids : ***Hyphochytriaceae***: *(Hyphochytrium, Cystochytrium, Canteriomyces)*

Thallus monocentric, epibiotic with rhizoids, which may be apophysate : ***Rhizidiomycetaceae*** : *(Rhizidiomyces, Reessia, Latrostium)*

Family *Hyphochytriaceae*

Genus *Hyphochytrium*

Hyphochytrium species (6) occur as parasites of algae and fungi or as saprobes on plant debris. *H. hydrodictyii* grows on the green alga *Hydrodictyon*; *H. infestans* is parasitic on apothecium of *Helotium*. *H. peniliae* parasitizes a marine copepod *Erytemora*, (a minute crustacean animal), which occurs as zooplankton, and is important food for fishes. *H. catenoides* is saprobic on plant debris.

H. catenoides (Fig. 18.2) grows inside the substratum e.g. decaying maize, forming a branched tubular thallus, which enlarges here and there to form sporangia. The zoospores are only incompletely formed inside the sporangium. A discharge tube emerges, which develops a pore at the tip. The uninucleate cytoplasmic segments (incipient zoospores) move into the discharge tube and develop into zoospores. These are finally released through the pore of the discharge tube. Sexual reproduction is not reported.

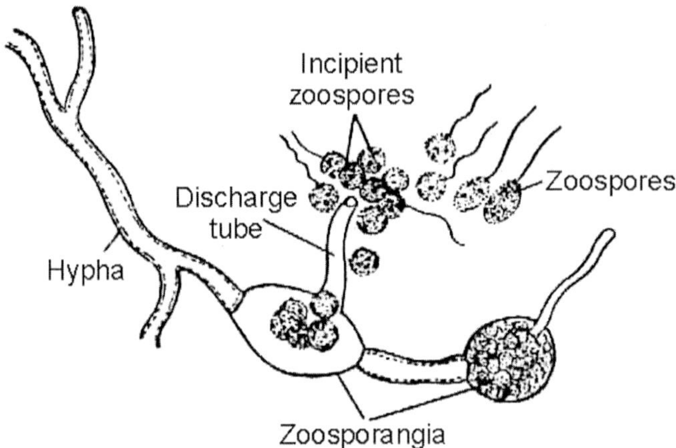

Fig. 18.2. *Hyphochytrium catenoides*

H. hydrodictyii (Fig. 18.3) differs from other species by its resting spores. The zoospores settle on the alga *Hydrodictyon,* develop a wall, and push in a germ tube, which grows into a swollen spindle-shaped structure, called the **primary swelling**. Hyphae originate from this and form **secondary swellings**. These swellings develop into sporangia and form zoospores, which are released through discharge tubes.

Spindle-shaped resting spores, whose origin and function are not known, are frequently formed.

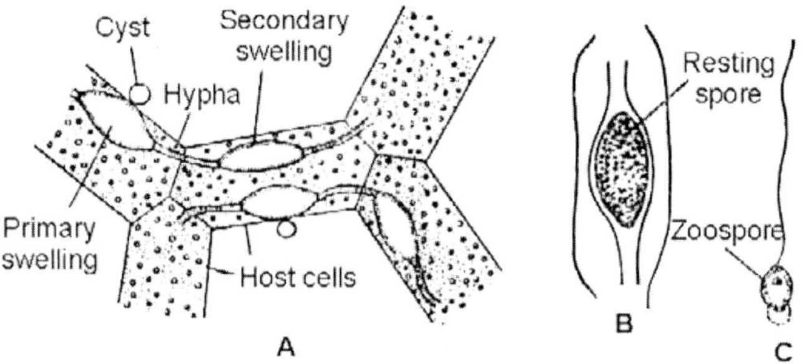

Fig. 18.3. *Hyphochytrium hydrodictyii*

Family *Rhizidiomycetaceae*

Genus *Rhizidiomyces*

Rhizidiomyces, with 13 species, is the biggest genus of this family. *R. apophysatus,* which is the best-studied member, is a parasite, growing on the oogonia and oospores of water molds, like *Saprolegnia*, and on the alga *Vaucheria*. It is also reported growing saprobically in soil. Flies serve as good 'bait' for isolation of the fungus from the soil.

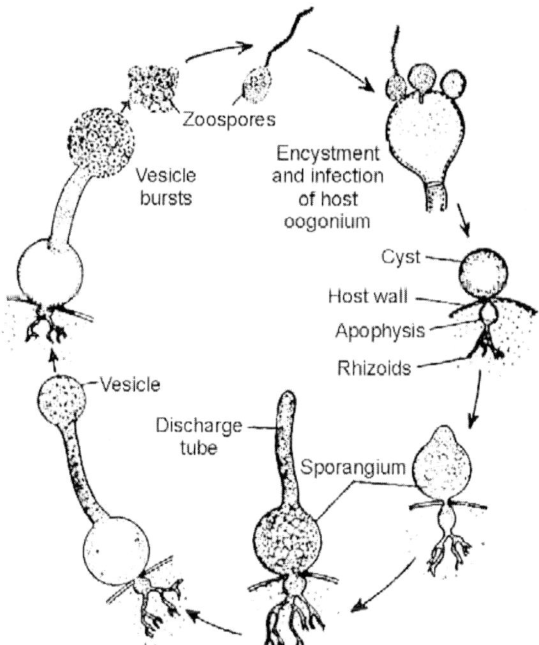

Fig. 18.4. Life Cycle of *Rhizidiomyces apophysatus*

The anteriorly uniflagellate zoospores swim for sometime and settle down on the host or the substratum (Fig. 18.4). The flagellum is absorbed, and the body becomes globular and encysted. A germ tube is formed, which penetrates the host cell wall and gives rise to branched rhizoids. The germ tube, which lies at the top of the rhizoids, swells and forms the **apophysis**. The rhizoids originate from this swelling and, hence, the name, apophysis (*apo*-from, *physis* = growth). The cyst, which lies external to the host, grows and develops into a zoosporangium. A discharge tube emerges from the sporangium. The protoplast, which by this time becomes multinucleate, flows out through the discharge tube and comes to lie in a vesicle, formed at the tip of the discharge tube. The protoplast is cleaved into uninucleate segments in the vesicle, which metamorphose into anteriorly uniflagellate zoospores. The mature zoospores are freed by dissolution of the vesicle. They swim away and repeat the life cycle. Sexual reproduction is not known.

REFERENCES

Burns, TD *et al.,* 1991. Fungal Molecular Systematics. *Annu. Rev. Ecol. Syst.* **22:** 525–564.

Kirk PM *et al.,* 2008 *Ainsworth and Bisby's Dictionary of the Fungi* C.A.B. International, Oxon, U.K.

SUMMARY

The kingdom *Chromista* is characterized primarily by the photosynthesis and tinsel-type flagellum. Other characters are related to cell wall composition, comprising cellulose and glucans, tubular cristae of the mitichondia, presence of Golgi bodies and peroxisomes, mycolaminarin as the storage product and diploid nature of the thalli. The chromistam fungi are more related to photosynthetic members of the *Chromista* than to true fungi. This is evidenced by the 18S rDNA sequences of the two groups. Nevertheless, these are similar to true fungi in the form of their thalli, nutrition and ecological roles. In the evolutionary sequence, these fungi are basal to the diatoms, brown and golden-brown algae, which diverged before these algae acquired the photosynthetic machinery by endosymbiosis. These fungi are placed in three phyla: *Hyphochytriomycota, Labyrinthulomycota* and *Oomycota*.

Phylum *Hyphochytriomycota* is characterized by their anteriorly uniflagellate zoospores and gametes, having tinsel type flagellum, and cellulose in the cell wall. It is a small group comprising endo-parasites of fresh water and marine algae, aquatic fungi or saprobes, growing on a variety of organic remains. There is one class *Hyphochytriomycetes,* one order *Hyphochyt-*

riales, and 2 families, *Hyphochytriaceae* and *Rhizidiomycetace*. *Hyphochytrium* and *Rhizidiomyces* are the representative species.

TEST QUESTIONS

1. Give the characteristic features of the kingdom *Chromista*. How are the fungi placed here similar and dissimilar to true fungi?
2. Answer briefly:
   - (a) Which are the parasitic and saprobic species of *Hyphochytrium*.
   - (b) Which are the three fungal phyla of *Chromista*?
   - (c) Who created the kingdom *Chromista*?
   - (d) Which two kingdoms have replaced the former kingdom Protista?
   - (e) Which are the basal groups in the evolution of *Chromista*?
   - (f) Who put the chromistan fungi under one phylum *Heterokonta*?
   - (g) What do the terms *Chromista* and *Straminopila* mean?
   - (h) What is the storage product in *Chromista*?
   - (i) What are peroxisomes?

3 Write notes on any **two**:
   - (a) *Hyphochytrium hydrodictyii*
   - (b) *Rhizidiomyces apophysatus*
   - (c) *Hyphochytrium catenoides*

19

Phylum *Labyrinthulomycota* (*=Labyrinthista*)

(Labyrinthids and Thraustochytrids)

The most important feature of this phylum is the formation of a wall-less, ectoplasmic network, in which the spindle-shaped or spherical thalli live and move by gliding. The cytoplasmic network is formed by special invaginated cell surface organelles, called **bothrasomes**. The other important characteristic, is the cell wall formed of scales produced by Golgi bodies. The cell wall scales (as reported for thraustochytrids) are made up of L-galactose and sulphated galactan. Zoospores, where produced, are laterally biflagellate, with an anterior tinsel-type flagellum and a posterior whiplash flagellum. The tinsel flagellum is longer than the whiplash flagellum. In sexual reproduction, which is known only in labyrinthids, biflagellate zoospores are formed. These possess an **eye-spot** which is a very rare feature, not found in any other fungal group. The eye-spot helps in positive phototactic movement. The zoospores lose the eye-spot, along with the flagella, at the time of transformation into spindle-shaped thallus.

The members of this phylum are mainly inhabitants of estuaries and near-shore waters, and found on plant leaves and algae. Some are reported to be endobiotic in marine algae (**Raghukumar** *et al.*, 1992). Saprobic species grow on organic debris.

Classification. There is one class *Labyrinthulomycetes* (*=Labyrinthulia)* and 2 orders: *Labyrinthulales* and *Thraustochytriales*, each having one family viz. *Laybyrinthulaceae* and *Thraustochytriaceae*. The former contains a single genus *Labyrinthula*, while the later contains 10 genera.

Order *Labyrinthulales*

Family *Labyrinthulaceae*

The family is characterized by spindle-shaped somatic cells, extensive ectoplasmic network enveloping the somatic cells, and occurrence of sexual reproduction, in which biflagellate zoospores with an eye-spot are formed.

Genus *Labyrinthula*

The genus has 10 species, which are mostly found in marine habitats, including terrestrial saline soils. The most important species is *L. macrocystis* (= *L. zosterae*), which causes a serious disease of 'eel-grass' (*Zostera marina*), called 'eel grass wasting disease". The spindle-shaped somatic cells glide inside the ectoplasmic network (Fig. 19.1). Asexual reproduction is achieved by transverse division of the somatic cells. This results in an increase in the number of the somatic cells inside a single network. Sexual reproduction is described in *L. vitellina*. The spindle-shaped cells, which are diploid, become globular and increase in size. The meiotic division results in 4 haploid nuclei (n = 9), which are separated by cytokinesis. An additional mitotic division results in 8 cells which transform into biflagellate zoospores with an eye-spot. The zoospores show positive phototaxis and, eventually, give rise to somatic cells, accompanied by loss of flagella and the eye-spot. The site of plasmogamy and karyogamy, leading to diploid thallus, is not known.

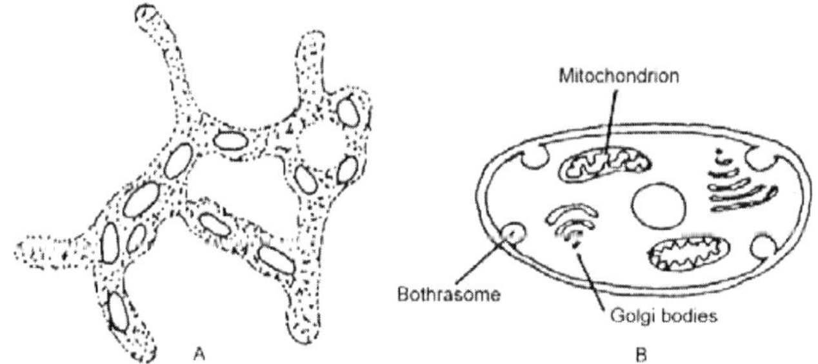

Fig. 19.1. *Labyrinthula.* A, cells in net; B, one cell showing organelles

Order *Thraustochytriales*

Family *Thraustochytriaceae*

The family has 10 genera, which are unicellular, live in marine and estuarine habitats, as saprobes on decomposing marine plants and algae. Some are parasites on shell-less molluscs, echinoderms and sponges. The important genera are *Althornia*, *Aplanochytrium*, *Japonochytrium*, *Schizochytrium*, and *Thraustochytrium*. **Raghukumar** (1992) has reported feeding on bacteria (bacteriovory) for some *Thraustochytrids*, which is unique and unknown in fungi.

The somatic cells are globular (not spindle-shaped), the ectoplasmic network is very limited (not enveloping the cells) and sexual reproduction is not reported. Asexual reproduction is brought about by biflagellate zoospores (which differ from sexually formed zoospores of *Labyrinthula* in lacking eye-sport).

Genus *Thraustochytrium*

The genus, which is best known among thraustochytrids, has 16 species. It closely resembles *Rhizophydium* (*Chytridiomycota*) and was once placed in *Saprolegniales* (*Oomycota*). The thallus is unicellular, globose, surrounded by cell wall. The thallus is epibiotic (Fig. 19.2), attached to the substratum (a living host or organic debris) by ectoplasmic filaments (not rhizoids, as believed earlier), formed by **bothrasomes**, as described above for *Labyrinthula* Fig. 19.3. Thus, the ectoplasmic network is highly restricted and does not envelop the cell. A sexual reproduction is achieved by biflagellate zoospores.

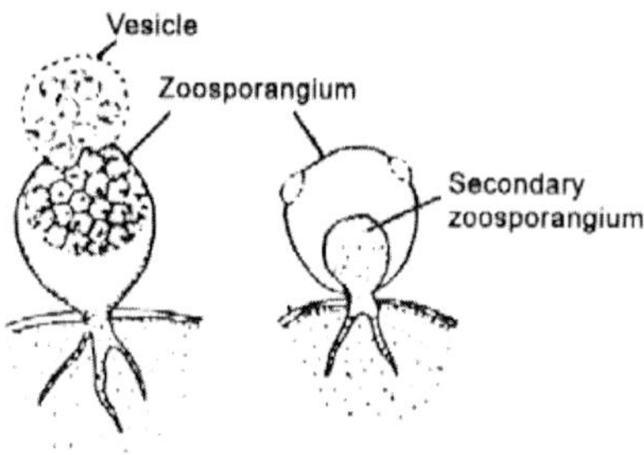

Fig. 19.2. A. *Thrastochytrium roseum*; B. *T. aureum*

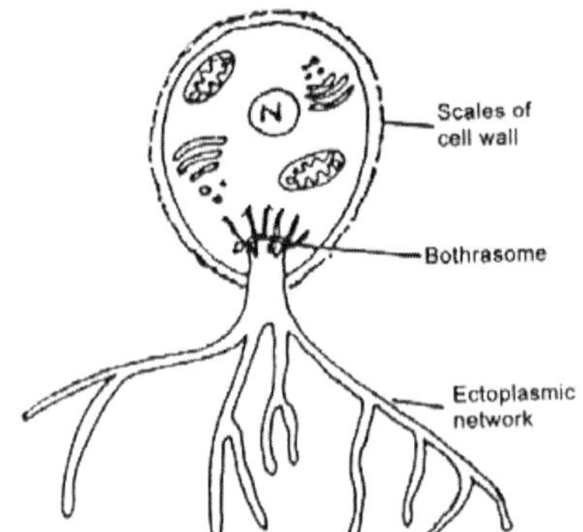

Fig. 19.3. Cellular structure of *Thraustochytrium,* showing rhizoid - like ectoplasmic network formed by a basal bothrasome

The globose thallus grows and transforms into a zoosporangium. The zoospores are released by gelatinization of the cell wall. The laterally biflagellate zoospores swim for sometime and finally settle on the substratum. They lose the flagellum and develop into a somatic globose thallus, attached by ectoplasmic filaments.

REFERENCES

Raghukumar, S. 1992. Bacterivory: a novel dual role for thraustochytrids in the sea. *Marine Biology*. **113:** 165–169.

SUMMARY

The phylum ***Labyrinthulomycota*** (= *Labyrinthista*) is characterized by (i) wall-less, ectoplasmic network, formed by **bothrasomes**, in which the spindle-shaped or spherical thalli live and move by gliding, (ii) cell wall made of 'scales' (L-galactose and sulphated galactan) produced by Golgi bodies, (iii) laterally biflagellate zoospores, with anterior tinsel-type and posterior whiplash flagellum, (iv) in sexual reproduction (known only in labyrinthids) biflagellate zoospores are formed (v) presence of an **eye-spot** (a rare feature, not found in any other fungal group) which helps in positive phototactic movement.

There is one class *Labyrinthulomycetes* (= *Labyrinthulia)* and 2 orders: *Labyrinthulales* and *Thraustochytriales*, each having one family viz. *Laybyrinthulaceae* and *Thraustochytriaceae*. The former contains a single genus *Labyrinthula*, while the later contains 10 genera (including *Thaustrochytrium*. The 2 orders differ in (i) nature of somatic cells, (ii) extent and role of ectoplasmic network, (iii) mode of asexual reproduction, and (iv) function of the zoospores.

TEXT QUESTIONS

1. What are the characteristic features of phylum *Labyrinthulomycota*? Describe the differences between the two orders, as shown by *Labyrinthula* and *Thraustochytrium*.
2. Describe the life cycle of *Labyrinthula* or *Thraustochytrium*.
3. Write notes on: (a) bothrasomes, (b) cell wall composition, (c) eye-spot.

20

Phylum *Oomycota*
(Pseudofungi)

The characteristic features of *Oomycota* are: (1) diploid somatic phase (like plants), (2) laterally biflagellate zoospores with one tinsel type flagellum, (3) mycolaminarin as storage product (4) protoplasmic or nucleus-associated microtubules in the zoospores, (5) cell wall made up of glucan-cellulose (rarely with minute amount of chitin), (6) thick-walled oospores, and (7) differences in path way of lysine biosynthesis, and nature of sterol.

The **tinsel-type flagellum,** and **mycolaminarin** (which is similar to storage compound laminarin of red algae) are the key characters of these fungi that place them in *Chromista*. The two flagella, which are laterally attached, are **heterokont** and **anisokont** i.e. dissimilar in structure, and unequal in size. One is 'tinsel' and other of 'whiplash' type. During movement, the tinsel flagellum, which is longer, takes anterior position, while the whiplash flagellum, which is shorter, lies in the posterior position.

The **cell wall** is made up of β 1-3 -, and β 1-6 glucans with cellulose (4–20% of dry wt.). Though small in amount, the presence of cellulose is an important feature. However, chitin is reported in small amounts in the cell walls of *Apodachlya*, *Achlya*, and *Saprolegnia*. It is also present as chitinous cellulin plugs in *Leptomitus*. The presence of hydroxyproline is another important characteristic of oomycete cell wall.

The *Oomycota* radically differ from 'true fungi' in **lysine biosynthetic pathway**. It is synthesized by diaminopimelic acid pathway (like plants), in contrast to alpha aminoadipic acid pathway in true fungi. The **sterol** in *Oomycota* is **fucosterol** and not **ergosterol** found in true fungi. (*Peronosporales* do not have sterols). The reserve product is mycolaminarin in *Oomycota* and glycogen in true fungi. Also, the **sugar alcohols**, found in true fungi, are absent in *Oomycota*.

The name *Oomycota* means "egg fungi" and refers to oogamous type of sexual reproduction. It involves large non-motile gamete called **egg** or

oosphere and smaller, motile male gamete called **sperm**. The oosphere or egg is contained in globular **oogonia**, while the male gametes are formed in club-shaped **antheridia**. In some primitive oomycetes, the gametangia are not differentiated into male and female. The holocarpic thalli of different sizes act as gametangia and fuse to form a zygote. Asexual reproduction is mainly by zoospores, which are of two types: the **primary** and **secondary** (or **principal**) zoospores. The life cycle is diploid (only the gametes represent the haploid phase).

The *Oomycota* vary from unicellular parasites to forms having well-developed mycelium. These are found in both, aquatic and terrestrial environments. Most of the aquatic forms grow in fresh water as saprobes, and play an important role in biodegradation of plant and animal remains. Some are parasites of algae and fishes. The terrestrial forms are mostly important plant parasites, causing serious diseases.

Classification

The phylum *Oomycota* has 1 class, *Oomycetes* (= *Peronomycetes*) and 8 orders (Kirk *et al.,* 2008). These are (in alphabetical sequence) *Albuginales*, *Leptomitales, Myzocytiopsidales, Olpidiopsidales, Peronosporales, Pythialels, Rhipidiales,* and *Saprolegniales.*

Dick (2001) segregated the orders into 3 subclasses, *Saprolegniomycetidae, Rhipidiomycetidae* and *Peronosporomycetidae*, which is supported by 18S rDNA data. Earlier, **Sparrow** (1973) had divided the *Oomycetes* into two groups, containing **saprolegnian** and **peronosporacean** oomycetes. Sparrow's two groups, based on characteristics of zoospores and oospores, are now confirmed by molecular data. The saprolegnian fungi, or *Saprolegniomycetidae*, include basically the **water molds**, while the peronosporacean fungi (*Peronosporomycetidae*) include the terrestrial genera, which are mostly plant parasites. The *Peronosporomycetidae* are derived, or evolved, from the *Saprolegniomycetidae* in the following characters: (1) a change from saprobic mode of nutrition to parasitic and obligately parasitic nature, (2) loss of requirement for sterol, (3) loss of primary zoospores, (4) germination of sporangia like conidia, (5) development of differentiated sporangiophores, (6) decrease in number of oospheres in the oogonium and development of periplasm around the egg, and (7) shift from homothallism to heterothallism.

The order *Rhipidiales* (*Rhipidiomycetidae*) is basal to *Leptomitales* which, in turn, is basal to *Saprolegniales* (*Saprolegniomycetidae*). In the *Peronosporomycetidae*, *Pythiales* is basal to *Peronosporales*. The classification is summarized below. We shall study representative genera of some of the orders, shown in bold type.

Subclass *Rhipidiomycetidae* : *Rhipidiales.*

Subclass *Saprolegniomycetidae*: *Leptomitales, Saprolegniales, Olpidiopsidales,* and *Myzocytiopsidales*.

Subclass *Peronosporomycetidae*: *Pythiales, Peronosporales, Albuginales.*

Subclass *Saprolegniomycetidae*

Order *Leptomitales*

The order comprises aquatic saprobic fungi, which have a characteristically constricted and segmented thallus. The presence of chitin in the cell wall has been reported in *Apodachlya* and some other genera. Two fungi belonging to family *Leptomitaceae*, *Leptomitus lacteus* and *Apodachlya* spp. are peculiar in their nutrition, in not being able to utilize glucose, fructose, galactose, sucrose or any other sugar. They derive their carbon nutrition from fatty acids. The order comprises 4 families. We shall study family *Leptomitaceae*.

Family *Leptomitaceae*

There are 3 genera including *Leptomitus* and *Apodachlya*. These are all aquatic saprobes, growing either on plant debris or as free-living organisms in polluted waters, rich in organic content.

Genus *Leptomitus* (the sewage fungus)

It is a monotypic genus i.e. has a single species, viz. *L. lacteus*. It is called the **sewage fungus**, because it is found in polluted waters of high organic content, like sewage and drainage water of sugar factories and breweries. The fungus frequently causes trouble by blocking filters of drainage pipes. The hyphae are extensive and constricted at regular intervals (Fig. 20.1). Internally, the constrictions are plugged with chitin-glucan granules, called **cellulin**.

Zoosporangia are formed in basipetal succession i.e. first the terminal segment and then the succeeding segments transform into sporangia. The sporangia are slightly swollen, as compared to the hyphal segments. Each zoosporangium forms numerous apically biflagellate, primary zoospores, which swim and finally settle down, encyst and germinate to give rise to a kidney-shaped, laterally biflagellate secondary (= principal) zoospore. After completing the swimming period, it encysts and germinates to form 1–3 buds, which give rise to germ tubes that develop into the hyphae. Sexual reproduction is unknown in this genus.

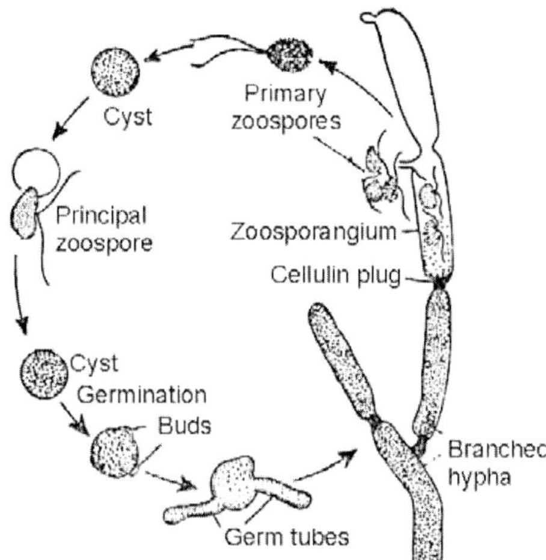

Fig. 20.1. Life cycle of *Leptomitus lacteus*

Genus *Apodachlya*

The genus has **chitin** in the cell wall. It forms ovoid sporangia, distinct from the hyphal segments. The thallus (Fig. 20.2) consists of highly-branched and constricted hyphae. Like *Leptomitus lacteus*, it also fails to utilize sugars, and derives its carbon requirements from fatty acids.

The zoosporangia, which are formed terminally are ovoid tapering at the base to form a short pedicel. These produce apically biflagellate, primary zoospores, which, on germination give rise to the thallus through buds, as in *Leptomitus*. The oogonia may be terminal or lateral. The antheridium may lie immediately below the oogonium or on a stalk, borne by the oogonium. The oospore is plerotic (i.e. completely fills the oogonium) and two-layered. The endospore is thick. The oospore germinates and forms a hypha.

Order *Saprolegniales*

Saprolegniales are the typical **water molds**. There are 2 families *Saprolegniaceae and Leptolegniaceae* that contain 24 genera. Most of these fungi are aquatic saprobes but some are soil-inhabitants. *Saprolegnia parasitica* and *Aphanomyces invadans* cause diseases of fishes and their eggs, frequently resulting in epizooics in fish hatcheries. Species of *Aphanomyces* also cause root diseases of several plants. The thallus is mycelial. The members of *Saprolegniaceae* have the stoutest hyphae among all fungi, exceeding 40 μm in diameter. **Zoosporangial proliferation** and **dimorphism** of zoospores (involving encystment–emergence sequence of the zoospores) are frequently encountered in *Saprolegniales* (Fig. 20.3).

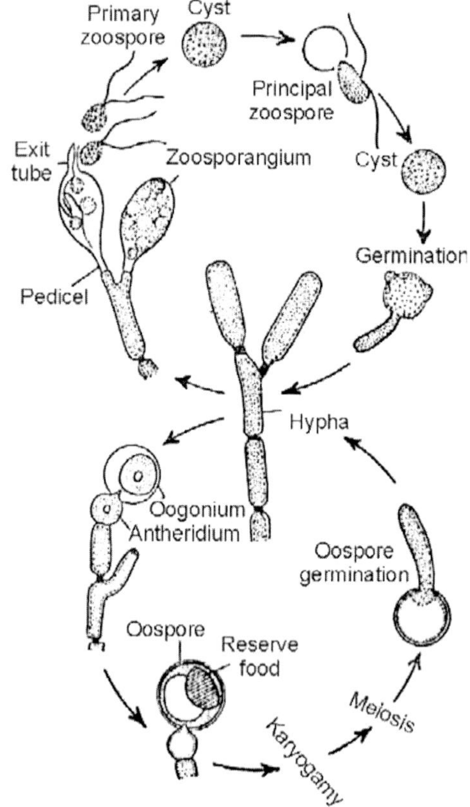

Fig. 20.2. Life cycle of *Apodachlya* sp.

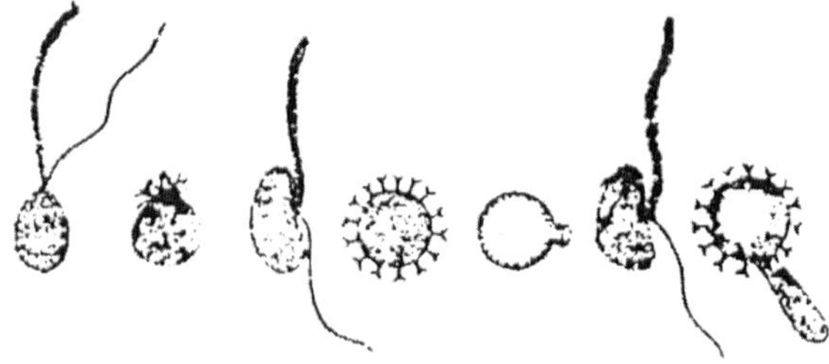

Fig. 20.3. Dimorphism of zoospores and zoospore cysts and diplanatism of the secondary zoospore

Family *Saprolegniaceae*

There are 19 genera including *Saprolegnia*, *Achlya*, *Isoachlya*, *Aplanes*, *Dictyuchus*, *Geolegnia*, *Thraustotheca*, *Aphanomyces* and *Pythiopsis*.

Generic distinctions are based on the bahaviour of the zoospore. The species are delimited on the basis of host for parasitic forms, and zoospore diameter. Both primary as well as secondary zoospores are formed by most genera. The encystment-emergence sequences of zoospores of some genera are depicted in Fig. 20.4.

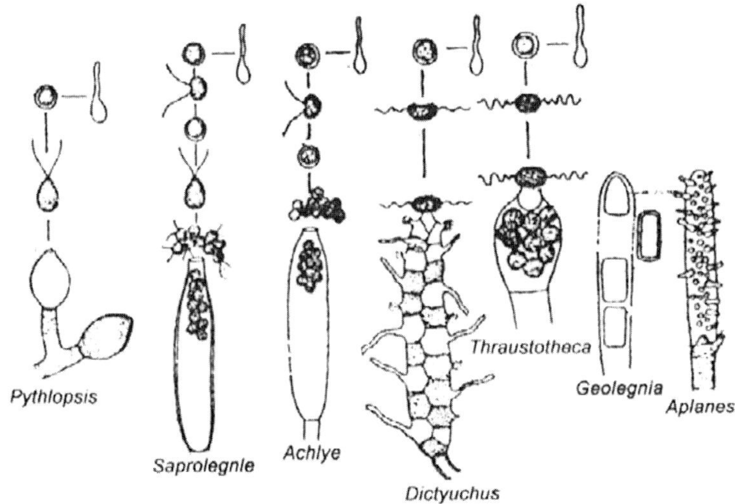

Fig. 20.4. Different modes of discharge and development of zoospores

Isolation and Laboratory Culture

These are isolated from pond water or soil, by **baiting** technique, using seeds of poppy, onion, mustard, corn etc. The seeds are sterilized by boiling, split and placed in pond water or the soil suspension. In a few days, hyphae appear on the seeds, which may be removed and grown in pure culture on liquid media. Autoclaved tap water alone, or mixed with pond water, is a highly-suitable medium for growth. Plain distilled water is toxic.

Genus *Saprolegnia*

Species of *Saprolegnia*, which are 15 in number, are common in soil and well-aerated, fresh water ponds. Most of them colonize dead plant and animal remains. A few, like *S. ferax* and *S. parasitica*, parasitize fishes and, frequently, cause their death, sometimes in epizooic proportions. Thus, these fungi occasionally become problematic in fish hatcheries. When inoculated artificially in wounded fishes, death occurs within 24 hours. These are **primary sugar fungi**, and fail to utilize polysaccharides like cellulose.

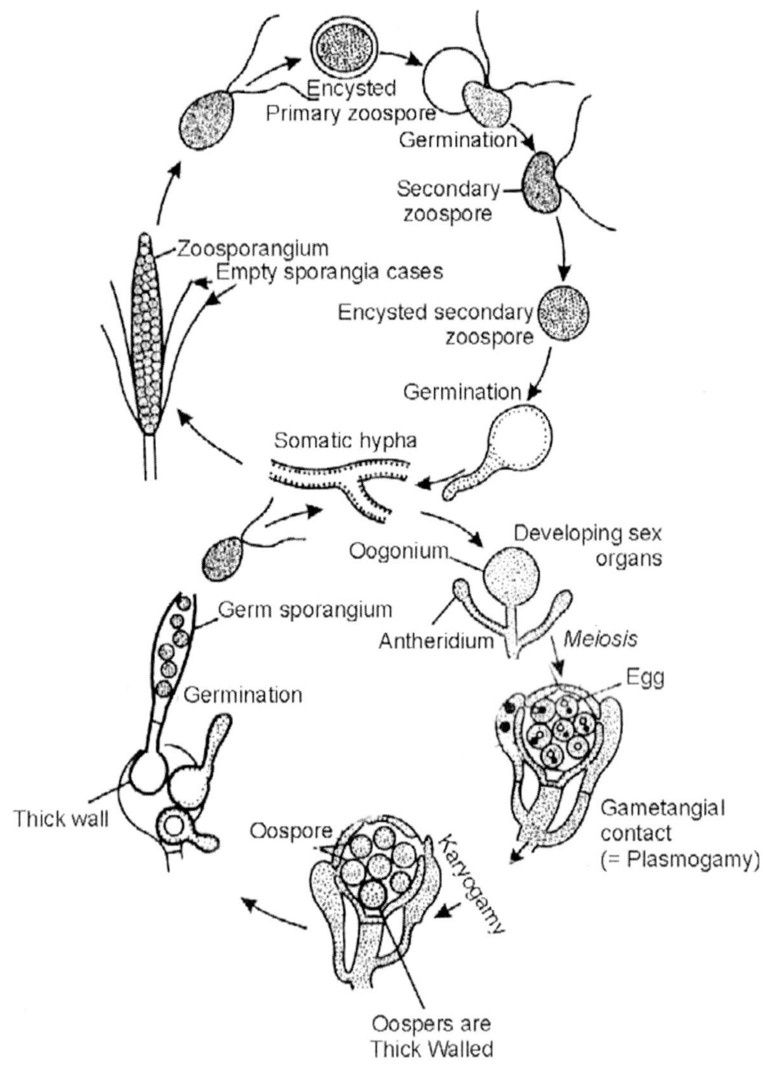

Fig. 20.5. Life cycle of *Saprolegnia*

Zoospores are formed in cylindrical zoosporangia (Fig. 20.5) at the tips of undifferentiated hyphae. Mature biflagellate zoospores can be seen jostling one another within the zoosporangium. The apex of the zoosporangium suddenly breaks and the zoospores rush out as if discharged by pressure. But afterwards, the remaining zoospores escape slowly. A cine film of the emptying sporangium has shown that the zoospores escape with the blunt end forward, but once they are expelled, the direction of movement reverses and they swim with the pointed end forward. Dimorphism and zoosporangial proliferation are typical characteristics of *Saprolegnia*.

The zoospores that are liberated are primary zoospores. These swim for sometime, encyst and after a non-motile, rest period of a few hours, release a secondary zoospore, through a papilla formed on the cyst. The empty cyst is left behind. The secondary zoospore, after a swimming period, also encysts and later germinates to form a hypha.

Zoosporangial proliferation (Fig. 20.6). Successive sporangia are formed within emptied ones; the remnants of the previous sporangia can easily be seen. The new sporangium is formed by bulging out of the septum, at the base of the empty sporangium. A new septum separates it from the hypha. Chlamydospores too are formed.

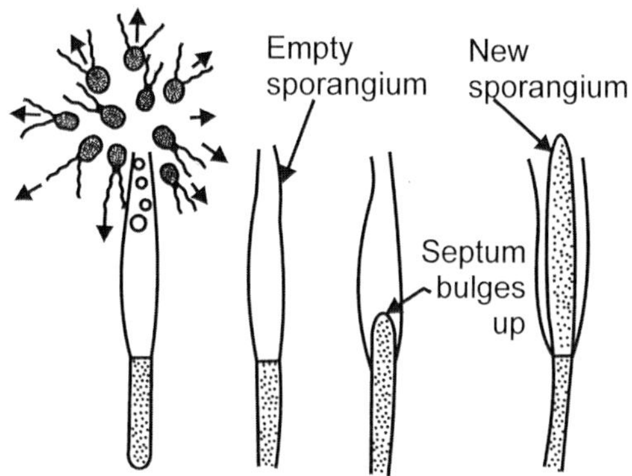

Fig. 20.6. Showing sporangial proliferation in *Saprolegnia*

The sex organs, oogonia and antheridia, begin to appear almost simultaneously with the zoosporangia. They are formed on the same hypha. The oogonium is globular, bigger and has a comparatively thicker wall than the antheridium, which is club-shaped. At maturity, the oogonium has one to several oospheres (eggs). The mature antheridium is multinucleate. Plasmogamy involves gametangial contact. Several antheridia may surround the oogonium. Fertilization tubes originate from the antheridia, which penetrate the oogonia and release the male nuclei. Occasionally, the fertilization tube forms branches inside the oogonium and all oospheres may be fertilized by a single antheridium. Each fertilized oosphere secretes a thick, smooth wall and develops into an oospore, inside the oogonium, and is never released. It germinates *in situ* after completion of dormancy. The germ tube breaks through the oogonial wall and forms a hypha, which develops an apical 'germ-sporangium'. Apically-biflagellate zoospores, formed in the germ sporangium, encyst and give rise to a new thallus.

Genus *Achlya*

This is the largest genus of the family. There are 50 species, which like *Saprolegnia*, are common in soil and fresh-water habitats, growing saprobically or parasitically on fishes. These can be isolated using hemp seeds and dead flies as 'baits'. Sporulation does not occur on ordinary media. On hemp seeds in water, sporangia and zoospores appear within 48 hours. The sex organs can be seen 3-4 days later.

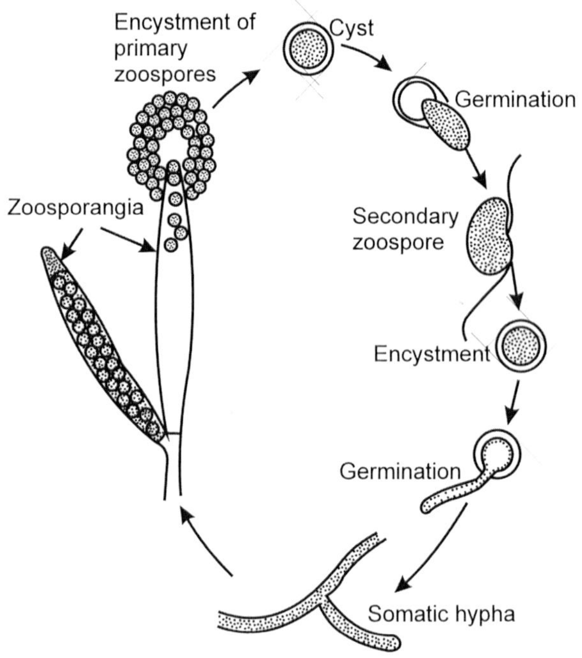

Fig. 20.7. Asexual reproduction in *Achlya*

Zoosporangial development is similar to that of *Saprolegnia*. Cylindrical zoosporangia appear terminally on undifferentiated hyphae (Fig. 20.7). The difference from *Saprolegnia* lies in the ejection of the zoospores. All primary zoospores are ejected simultaneously which encyst at the mouth of the zoosporangium, without any swimming period. The cysts form a hollow sphere. After a period of rest, each cyst germinates and releases a secondary zoospore, which, unlike the primary zoospore, has a swimming period. These also encyst and, on germination, give rise to a hypha. *Achlya* lacks dimorphism of zoospores and sporangial proliferation. However, a new sporangium is formed beside the old one.

Sexual reproduction is similar to that of *Saprolegnia*, except that it is heterothallic. Sex organs, formed on different (compatible) thalli, undergo plasmogamy. A very detailed study of the sex hormones of *Achlya* (*A. bisexualis*) has been made (*see* Chapter 31 for details).

Subclass *Peronosporomycetidae*

Order *Pythiales*

The order has 2 families, *Pythiaceae* and *Pythiogetonaceae*, and 10 genera.

Family *Pythiaceae*

The family contains 7 genera: *Diasporangium*, *Lagenidium*, *Lucidium, Peronophythora, Pythium*, *Myzocytium*, and *Trachyshaera* (*Phytophthora* has been assigned to *Peronosporaceae*). These are aquatic, amphibious or terrestrial, many of which are important plant parasites. The sporangiophores are indistinguishable from hyphae and are also of indeterminate growth.

Genus *Pythium*

With 150 species, this is the largest genus of the family. Majority of them are soil-inhabitants, while mainly cause root-rot (particularly of *Poaceae*) and 'damping off' diseases of seedlings. A few are mycorrhizal, while some others are aquatic. Some attack fruits (prominent being the cucurbit fruits, lying usually on the ground).

Some plant pathogenic species and the diseases they cause are *P. debaryanum* (damping off of tobacco and chillies), *P. aphanidermatum* (soft rot of papaya and damping off of potato), *P. graminicolum* (rhizome rot of turmeric) and *P. myriotylum* (root-rot of ginger). *P. thalassium* is marine, growing on pea-crab (*Pinnotheres*).

Isolation and Culture

Pythium can be obtained by germinating mustard seeds in dish containing soil. Some seedlings topple down due to damping off and the fungus can be easily isolated from the rotted, basal portion of stem on PDA (potato dextrose agar) medium. *Pythium* grows very rapidly. Some species form sporangia only when grown on natural substrates like sterilized grass, oat-meal, or by supplying certain sterols in the medium.

Life Cycle

The fungus grows saprobically in soil, as well as, parasitically in host tissues. The hyphae grow inter-, or intracellularly in the host tissues. Haustoria are not produced. Globose or oval sporangia arise apically on unspecialized hyphae; the sporangium is pushed aside and the hyphal tip gives rise to the next sporangium.

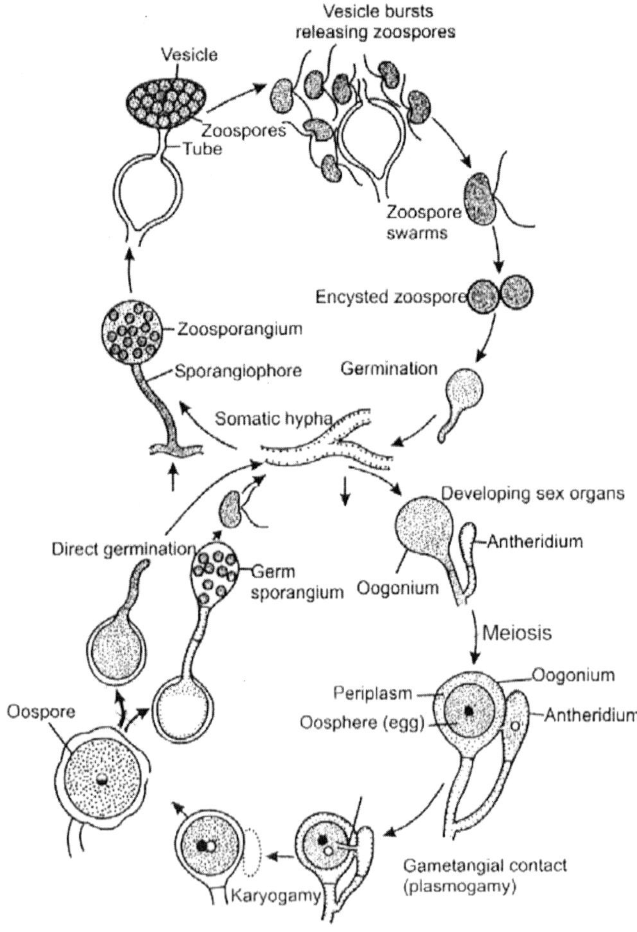

Fig. 20.8. Life cycle of *Pythium debaryanum*

At the time of zoospore formation (Fig. 20.8), a discharge tube, with an apical vesicle, appears on the sporangium, into which the contents of the latter migrate. The zoospores show a rocking motion and bounce on the delicate wall, causing its rupture and consequent release of the zoospores. The discharge tube often persists after the dehiscence of the vesicle. The zoospores (reniform, laterally biflagellate), after the swimming period, encyst and germinate by forming a germ tube, which grows into a hypha. In dry weather, the zoosporangia behave as conidia and germinate directly by forming a germ tube.

The gametangia may be formed on same or different hyphae, The oogonia are globose, periplasmatic, with a single oosphere. Antheridia, which are club-shaped and smaller, may be hypo-, or paragynous in position. At maturity, the gametangia are uninucleate. The plasmogamy is by gametangial contact. A fertilization tube takes the male nucleus to the female nucleus in the oosphere. Karyogamy may take place immediately or a little before oospore germination. The oospore is usually colourless, unornamented and often plerotic. (In *P. debaryanum* it is aplerotic). It undergoes a period of rest before germination. The mode of germination of oospore is determined by temperature. At higher temperatures (28°C), oospore germinates directly by forming a germ tube, while at lower temperatures (10–17°C) it germinates by forming a 'germ sporangium' at the tip of a stout hypha. Zoospores or the germ tube, formed by the oospore, give rise to new thalli.

Order *Peronosporales*

The genera belonging to *Peronosporales* show morphological features, which are more suited to the terrestrial habitat. Sporangia in dry weather, instead of forming zoospores, germinate directly by germ tube, thus, behaving as conidia. The intercellular hyphae form haustoria, which enter host cells and absorb nutrients. The sporangiophores are distinct in shape and determinate in growth. The oogonium shows differentiation into a central **ooplasm** and peripheral **periplasm**. The ooplasm (oosphere) is the female gamete.

The order has a single family *Peronosporaceae* with 20 genera and 365 species. Some important genera are *Peronospora, Phytophthora, Plasmopara, Pseudoperonospora, Basidiophora, Bremia, Sclerophthora* and *Sclerospora*.

Genus *Phytophthora*

The genus is of great historical, as well as, of economic importance. Studies of the large subunit (LSU) rDNA indicated *Phytophthora* to be nearer to *Peronospora* than to *Pythium*, and it was transferred from *Pythiales* to *Peronosporales* (Petersen and Rosendahl, 2000).

Except some apparently saprobic species (which may be only casually so), the members are parasitic, a few have been brought in culture. Many, which have been isolated from soil, attack underground plant parts, but there is little evidence that they can grow and reproduce in soil. Many species attack aerial parts of plants and cause fruit rot, canker and leaf-blight.

The species are segregated into two groups: one group with little or no apical thickening of sporangium (non-papillate), proliferating through the empty sporangium (Fig. 20.9) and non-deciduous. In the other group, the sporangia have conspicuous apical thickenings, are papillate, non-proliferating and deciduous. The species of the first group are aquatic and cause root and tuber-rot, in contrast to species of the other group, which are common on aerial plant parts. A computer analysis, using 32 diagnostic morphological characters, has supported the segregation of the species into two groups.

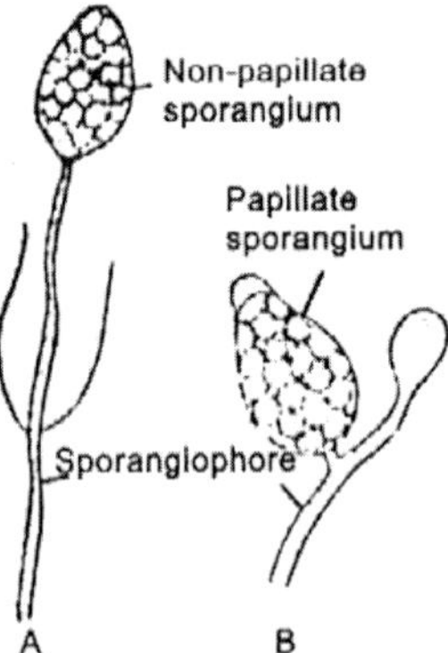

Fig. 20.9. A, non-papillate sporangium *of P. cinnamomi,* B, *papillate* sporangium of *P. nicotianae*

There are 90 species, some of which cause important plant diseases e.g. *P. infestans* (late blight of potato), *P. colocasiae* (late blight of *Colocasia*), *P. parasitica var. sesami* (*Sesamum* leaf-blight), *P. palmivora* (bud-rot of toddy palm, coconut palm and cocoa), and *P. nicotianae* (foot and buck-eye root of tomato).

Isolation and Culture

Phytophthora can be isolated from the soil by using apple as a bait. Take an unripe apple. Stab one side of it with needle and press it into the soil. After a few days, take it out. A brown rot appears on the fruit. Wash the fruit thoroughly and place a portion of the rotted tissue in dish containing tap water. A fluffy mycelial growth appears in 24-28 hours. A pure culture of *Phytophthora* is obtained if these hyphae are transferred to PDA tubes.

The fungus survives in potato tubers as dormant mycelium. When such tubers are used as 'seeds', they give rise to systemically-infected plants. The hyphae grow intercellularly and form haustoria to derive nutrition from adjacent cells. The optimum temperature for growth is 20°C; above 26°C the hyphae die.

Asexual reproduction is brought about by zoospores. Sporangial production needs a low temperature (18–20°C) and high relative humidity (91–100%). Sporangiophores emerge from the lower surface of leaves through stomata (Fig. 20.10) and bear terminal, lemon shaped sporangia with distinct papilla. The renewed growth of the sporangiophore gives it a zigzag, sympodial shape. The sporangiophores also arise in large numbers from the cut surfaces of infected potatoes.

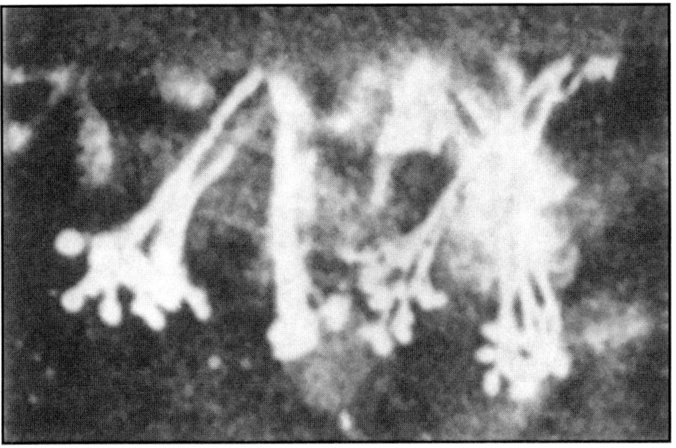

Fig. 20.10. Sporangiophores and sporangia of *Phytophthora infestans* hanging down from the lower surface of potato leaf

The sporangia are deciduous and get widely disseminated by wind or soil-water (especially irrigation water). Depending on the availability of water, the sporangia germinate directly (behaving as conidia) or form zoospores. The wind-disseminated conidia germinate directly, but the germination of water-borne sporangia is governed by temperature. A low temperature of 12°C or so, favours zoospore production, while higher temperatures (24°C) induce direct germination.

The reniform, biflagellate zoospores, as usual, swim for sometime, then encyst and finally germinate to form a new hypha. The tip of the germ tube, on host leaf surface, forms an **appressorium** (a bulbous apical swelling), from which an infection hypha enters the host and establishes an intercellular mycelium. Dead, blighted areas appear on the leaves and indicate infection. Zoospores present in the soil infect the tubers.

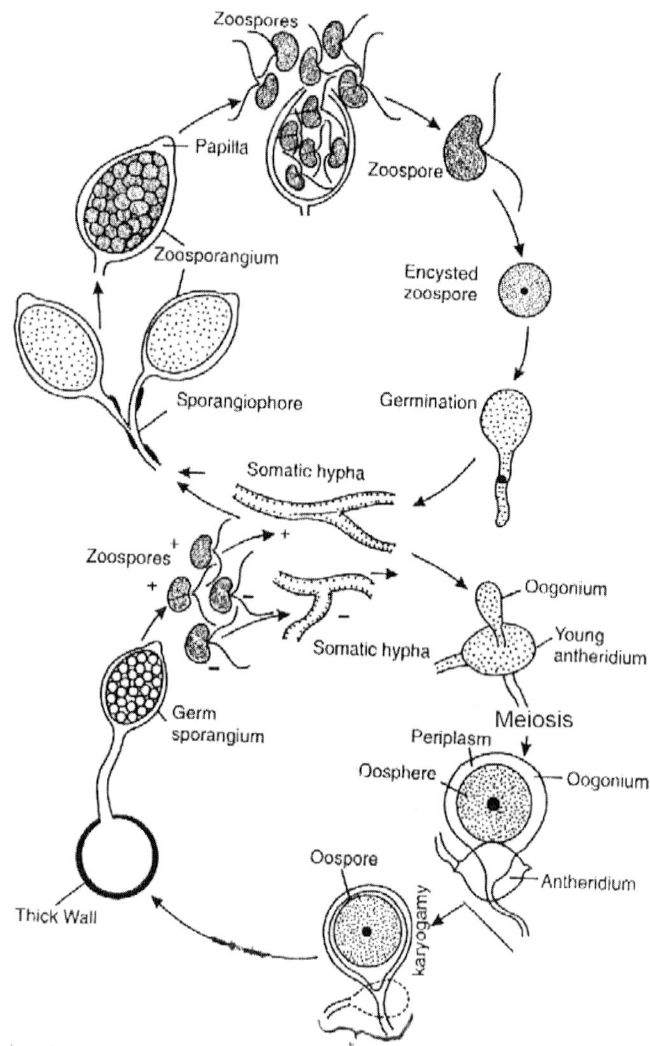

Fig. 20.11. Life cycle of *Phytophthora infestans*

Sexual reproduction is infrequent in *P. infestans*, for the reason that it is heterothallic and the + and – strains are not reported from the same place. But it has been observed in other species. On the basis of the position of sex organs, the species are designated as **paragynous** or **amphigynous**. In the paragynous type (*P. cactora*), the sex organs lie laterally, while in the amphigynous type (*P. infestans* and *P. colocasiae*), the antheridium forms a funnel or collar-shaped structure at the base of the oogonium. The oogonium, which is initially multinucleate, at maturity contains a single nucleus. The protoplasm, at the center of the oogonium is called **ooplasm**, which contains the female nucleus and serves as the female gamete (egg or oosphere). The cytoplasm surrounding the ooplasm is called **periplasm**. The club-shaped antheridium, which too is initially multinucleate, finally contains a single male nucleus. The male nucleus reaches the ooplasm through a fertilization tube formed by the antheridium. Plasmogamy, followed by karyogamy, results in a diploid zygote nucleus. Parthenogenetic development of unfertilized ooplasm into oospore is also known. The oospore fills the cavity of the oogonium (plerotic type of oospore). After a rest period, the oospore germinates and forms a zoosporangium at the tip of the germ tube. Zoospores of the two mating types are formed in equal numbers, which give rise to thalli of opposite mating types.

The Downy Mildew Genera

In **downy mildews** characteristic sporangiophores (or conidiophores), of determinate growth, emerge singly or in tufts through stomata, on the lower surface of leaves. The term downy mildews etymologically means **hairy fungi**. The sporangiophores look like hairs, emerging from the stomata. The genera are based on sporangiophore and sporangial characteristics Since the sex organs are not much different, asexual characters are used for delimitation of genera and species.

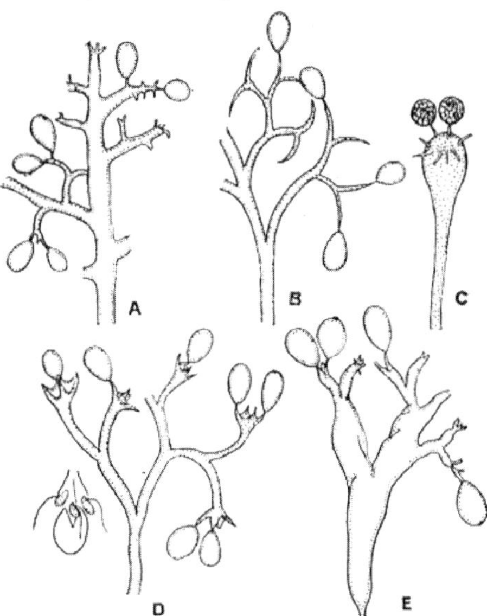

Fig. 20.12. 'Downy mildew' genera. A, *Plasmopara*; B, *Peronospora*; C, *Basidiophora*; D, *Bremia*; E, *Sclerospora*

Some downy mildew fungi viz. *Peronospora parasitica*, *Pseudoperonospora humuli*, *Plasmopara viticola* etc. have been grown in tissue culture. The downy mildew genera are identified by their characteristic sporangiophores (Fig. 20.12). In *Plasmopara,* the branches of the sporangiophores arise at right angles and are irregularly spaced. In *Peronospora*, the dichotomously-branched sporangiophores taper to fine points, bearing conidia. If similar branches bear sporangia, the fungus is designated as *Pseudoperonospora*. *Bremia*, too, has similar branches, but the apices of the branches terminate in disc-like structures, bearing sporangia on sterigmata. *Basidiophora* forms basidium-like, club-shaped sporangiophore, which is unbranched and bears sporangia on sterigmata. The sporangiophore of *Sclerospora* is swollen and bear upright branches bearing sporangia.

Genus *Plasmopara*

P. viticola made history in France in 1930 by devastating the grape vines. It is reported from India from all the grape growing areas. All the succulent parts are affected. Chlorotic patches appear on the upper surface of

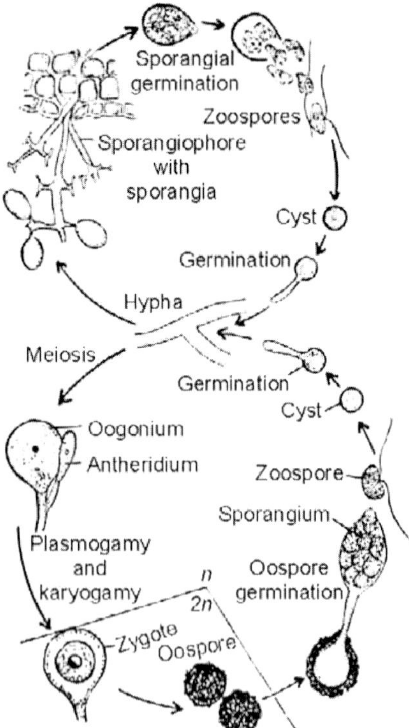

Fig. 20.13. Life cycle of *Plasmopara viticola*

leaves making them brittle, which fall off in shreds. In humid weather, **downy growths** (sporangiophores) emerge on the lower leaf surfaces, tendrils, pedicels, flowers, and fruits. Infected fruits shrivel and mummify.

The life cycle of the fungus starts with the germination of the soil-borne oospores (Fig. 20.13). The germ tube bears an apical sporangium (= germ sporangium), in which biflagellate zoospores are produced. The zoospores infect the host and cause primary infection. The mycelium grows intercellularly and establishes systemic infection. Globular haustoria enter the host cells and draw nutrients. When the weather becomes favourable, the fungus forms sporangiophores and sporangia. Zoospores, produced in the sporangia, infect more plants and spread the fungus, and the disease, extensively.

Sexual reproduction, leading to oospore formation, takes place only at the end of the host season. The oospores find their way into the soil through dead host issues.

Genus *Sclerospora*

Sclerospora contains 3 species, pathogenic on *Poaceae*. The sporangiophore is swollen and bear upright branches terminating in sporangia that form laterally-biflagellate zoospores.

S. graminicola causes the **green ear** disease of 'bajra' or 'pearl millet' (*Pennisetum typhoides*), reported by **E.J. Butler** (1907) from India. The symptoms are most pronounced on the ears, which turn green and leafy; hence the name 'green ear' disease (Fig. 20.14). Bristles, glumes, stamens and pistils become leafy and green. But anatomically, the 'green ear' leaves are much different from the normal leaves. Patel (1978) found that, in addition to differences in orientation, the vascular bundles of the 'green ear' leaves lack the sclerified bundle sheaths and the marginal fibers. The linear arrangement of epidermal cells and the stomata is missing. Symptoms appear also on leaves as chlorotic patches and twisting of the lamina. Tillering too occurs.

The oospores, surviving in the soil (up to five years), germinate when the hosts become available, and cause primary infection by attacking the seedlings. The fungus spreads intercellularly and soon becomes systemic. Conidia, as a rule, are formed in the early hours of morning when the air is cool and humid. Dewy nights are best for sporulation of *Sclerospora* (and all downy mildews). The sporangiophores are characteristically swollen and bear hyaline sporangia at the tips of apical dichotomies. The sporangia germinate (optimum temperature 18-23°C) and release biflagellate zoospores, which cause secondary infections and spread the fungus and the disease. The oospores develop late in the season. They reach the soil, along with host debris, and undergo a long dormant period.

Fig. 20.14 Showing symptoms of "green ear" disease of "bajra"

Genus *Sclerophthora*

The genus, as is evident in the name, combines characters of *Sclerospora* and *Phytophthora*. The sporangiophores are branched, like *Sclerospora*, and are indeterminate in growth and have bigger sporangia, like *Phytophthora*.

Order *Albuginales*

Family *Albuginaceae*

This is the lone family which contains 3 genera, *Albugo* (Cystopus of older literature) being the best-known.

Genus *Albugo*

The 45 species comprising the genus, are obligate parasites, characterized by a unique asexual stage, which consists of sub-epidermal, erect, club-shaped sporangiophores, budding off sporangia. White blisters appear on the leaves. The fungus also causes hypertrophy of floral parts and forms tumors of varying sizes on all parts–stems, leaves, veins, petioles etc.

The important species are: *A. candida* (on crucifers), *Albugo ipomae–panduranae* (on plants *Convolvulaceae*), *A. bliti* (on *Amarantaceae*), *A. tragapogonis* (on *Compositae*), *A. occidentalis* (infects spinach), and *A. portulaceae* (on *Portulaca*). The species are based on host specificity, oospore ornamentation and conidial characteristics.

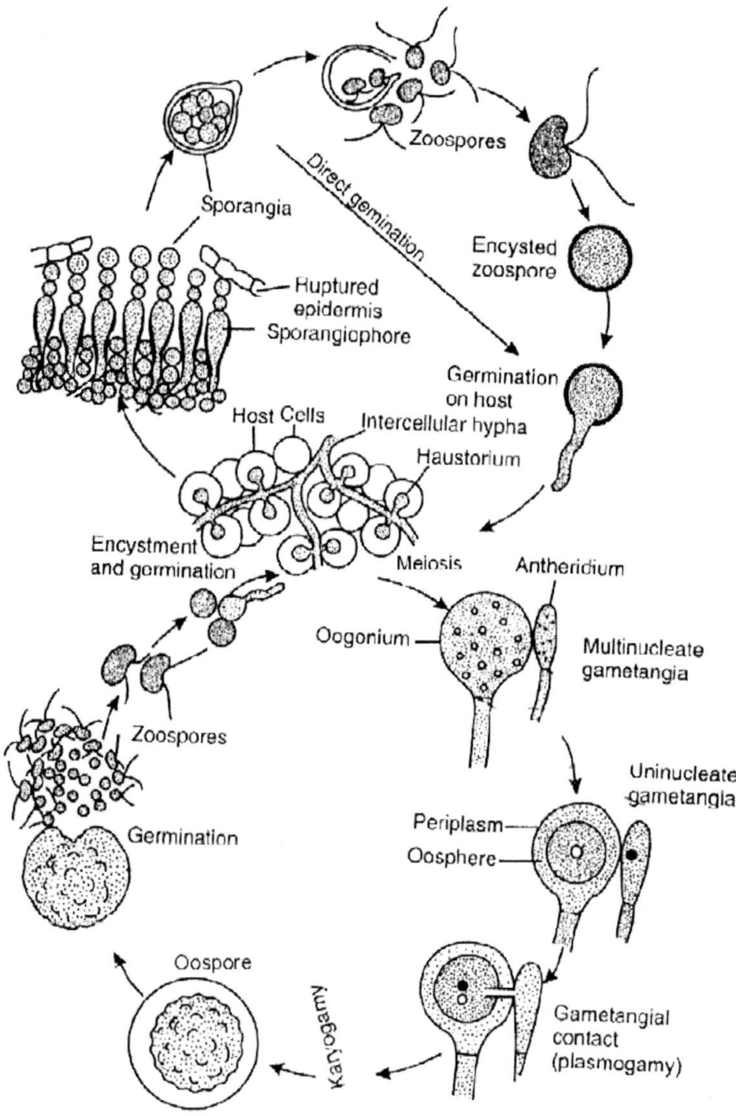

Fig. 20.15. Life cycle of *Albugo candida*

Life Cycle

In the absence of host, the fungus survives through oospores or dormant mycelium in plant debris. On germination, biflagellate, reniform zoospores are formed and held in a vesicle on the oospore, before they are set free. These swim, encyst and germinate on the host leaf. The mycelium grows intercellularly, forming haustoria in host cells.

Asexual reproduction is initiated by aggregation of hyphae at certain places, below the epidermis. Club-shaped sporangiophores, with thick lateral walls and a thin apex, arise in a palisade layer. These cut multinucleate sporangia which remain attached to form a chain at the apex. The oldest sporangia lie at the top and youngest at the base of the chain. The walls between the sporangia fuse to form a gelatinous, disc-like structure called **disjunctor** or **separation disc**. The disjunctors are dissolved by water, and the sporangia are set free. The sporangia push against the epidermis, which bulges out and ultimately ruptures. The creamy white masses of sporangia, appear like pustules or blisters (white rust) on the leaves.

The sporangia are globose or hexagonal, and disseminated by wind. Germination occurs when the sporangia reach a suitable host. The mode of germination depends on the availability of water. If water is available, zoospores are formed; otherwise, the sporangia germinate directly and, thus, behave like conidia. Biflagellate, reniform zoospores, after a swimming period, encyst and germinate by a germ tube. The germ tube enters the host and establishes an intercellular mycelium. The repeated production of sporangia spreads infection to more plants.

Sexual reproduction occurs at the end of the host season. The hyphae penetrate deeper into the tissues and form sex organs. The onset of sexual reproduction is usually indicated by hypertropohy of the infected parts. Plasmogamy is by gametangial contact between the club-shaped antherida and globular oogonia, borne terminally on somatic hyphae. There is a single oosphere (ooplasm) surrounded by periplasm. The fertilization tube, formed by the antheridium, carries the male nucleus into the oosphere, where the two nuclei fuse to form a diploid zygote nucleus. The oosphere develops a warty wall and becomes the oospore. The multinucleate oospores, after a prolonged resting period, germinate and form reniform biflagellate zoospores, which after swimming for sometime, encyst and germinate to form a germ tube, which infects the host plant.

REFERENCES

Butler, E.J. 1907. Some diseases of cereals caused by *Sclerospora graminicola Mem. Dep. Agric. India. Bot. Ser.,* **1:** 1–60.

Dick M.W. *et. al.,* 1999. 18S rDNA for species of Leptolegnia and other Peronosporomycetes: justification for the subclass taxa Saprolegniomycetidae and Peronosporomycetidae and division of the Saprolegniaceae sensu lato into the Leptolegniaceae and Saprolegniaceae. Mycological Research **103:** 1119–1125.

Patel, J.D. 1978. Histopathology of "Green ear" of Bajra. *Phyton* (Austria), **1:** 301.

Petersen, A.B. and Rosendahl, S. 2000. Phylogeny of the Peronosporomycetes (Oomycota) based on partial sequences of the large ribosomal subunit (LSU rDNA). Mycological Research **104:** 1295–1303.

Skalicky, V. 1966. (Title in German) *Prestia*. **38:** 117–129.

Sparrow, F.K. 1973. Lagenidiaceae. In: *The Fungi, An Advanced Treatise.* (G.C. Ainsworth *et. al.,* eds.), **Vol. IV B**, pp 159–163. Academic Press, N.Y. and London.

SUMMARY

Diploid somatic phase, tinsel type flagellum, mycolaminarin, glucan-cellulosic wall, oospore, differences in path way of lysine biosynthesis, and nature of sterol are the characteristic features of phylum *Oomycota.* The cell wall is made up of β 1-3 -, and β 1-6 glucans with cellulose. The presence of cellulose is an important feature. Chitin is reported in small amounts in the cell walls of *Apodachlya*, *Achlya*, and *Saprolegnia*. It is present as chitinous cellulin plugs in *Leptomitus*. The presence of hydroxyproline is another important characteristic of oomycete cell wall.

The *Oomycota* differ from 'true fungi' in lysine biosynthetic pathway (diaminopimelic acid pathway, like plants in contrast to alpha aminoadipic acid pathway in true fungi), sterol (fucosterol and not ergosterol found in true fungi), storage product (mycolaminarin and not glycogen) and absence of sugar alcohols (found in true fungi). The oogamous type of sexual reproduction, typical of these fungi, involves large non-motile egg and smaller motile male gametes. Asexual reproduction is mainly by zoospores, primary and secondary.

The phylum has 1 class, *Oomycetes* (= *Peronomycetes*) and 8 orders that are segregated into 3 subclasses, *Rhipidiomycetidae*, *Saprolegniomycetidae* and *Peronosporomycetidae.* The order *Rhipidiales* (*Rhipidiomycetidae*) is basal to *Leptomitales* which, in turn, is basal to *Saprolegniales* (*Saprolegniomycetidae*). In the *Peronosporomycetidae*, *Pythiales* is basal to *Peronosporales*. The orders are placed in the subclasses as follows.

Subclass *Rhipidiomycetidae*: *Rhipidiales.*

Subclass: *Saprolegniomycetidae*: *Leptomitales, Saprolegniales*, *Olpidiopsidales,* and *Myzocytiopsidales*.

Subclass: *Peronosporomycetidae*: *Pythiales, Peronosporales, Albuginales*

Order *Leptomitales* comprises aquatic saprobic fungi, which have a characteristically constricted and segmented thallus. The presence of chitin in the cell wall has been reported in *Apodachlya* and some other genera. Two fungi belonging to family *Leptomitaceae*, *Leptomitus lacteus* and *Apodachlya* spp. are peculiar in not being able to utilize any sugar and deriving energy from fatty acids.

L. lacteus, the 'sewage fungus', is found in polluted waters like sewage and drainage water of sugar factories and breweries and frequently causes trouble by blocking filters of drainage pipes. Sexual reproduction is unknown in this genus.

Order *Saprolegniales* includes the typical water molds. Most of these fungi are aquatic. *Saprolegnia parasitica* and *Aphanomyces invadans* cause diseases of fishes and their eggs, frequently resulting in epizooics in fish hatcheries. Species of *Aphanomyces* also cause root diseases of several plants. The members of *Saprolegniaceae* have the stoutest hyphae among all fungi, exceeding 40 μm in diameter. 'Zoosporangial proliferation' and 'dimorphism of zoospores' (involving encystment-emergence sequence) are frequent.

Order *Pythiales* include aquatic, amphibious or terrestrial members, many of which are important plant parasites. The sporangiophores are indistinguishable from hyphae and are also of indeterminate growth. *Pythium* species cause 'damping off'' of seedlings and soft rots.

The order *Peronosporales* shows morphological features suited to the terrestrial habitat. Sporangia in dry weather behave as conidia. The intercellular hyphae form haustoria. The sporangiophores are distinct in shape and are determinate in growth. The oogonium shows differentiation into a central ooplasm and peripheral periplasm. The ooplasm (oosphere) is the female gamete. The single family *Peronosporaceae* includes *Peronospora, Phytophthora, Plasmopara, Pseudoperonospora, Basidiophora, Bremia, Sclerophthora,* and *Sclerospora*. Except *Phytophthora*, the genera are called 'downy mildews', causing downy mildew diseases. These genera are identified by their sporangiophore characteristics.

Order *Albuginales* (1 family *Albuginaceae*) includes obligate parasites, characterized by asexual stage, which consists of sub-epidermal, erect, club-shaped sporangiophores, budding off sporangia. White blisters appear on the leaves. The fungus causes hypertrophy of floral parts and forms tumors of varying sizes on all parts : stems, leaves, veins, petioles etc. The species are based on host specificity, oospore ornamentation and conidial characteristics.

TEST QUESTIONS

1. Describe the characteristic features of phylum *Oomycota*. Compare it with 'true fungi'.

2. Write notes on: (a) *Leptomitus lacteus*, (b) *Apodachlya*, (c) sporangial proliferation, (d) diplanetism of zoospores.
3. Describe the diagnostic features of downy mildew genera and the life cycle of *Plasmopara viticola.*
4. Distinguish between: (a) Asexual reproduction in *Saprolegnia* and *Achlya*, (b) *Peronospora* and *Pseudoperonospora*, (c) *Sclerospora* and *Sclerophthora,* and (d) primary and secondary zoospores.
5. Answer briefly:
   (a) What are (i) mycolalminarin, (ii) cellulin , (iii) fucosterol and (iv) ergosterol?
   (b) What is the characteristic feature of asexual reproduction in *Albugo*?
   (c) Which downy mildew genera form conidia and which form sporangia?
   (d) Which orders are included under subclass *Peronosporomycetidae*?
   (e) Which fungus is called "sewage fungus"?
   (f) Which genera in *Saprolegniaceae* are diplanetic?
6. Fill in the blanks:
   (a) *Leptomitus lacteus* and ----------- are unable to utilize simple sugars.
   (b) *Rhipidiales* are 'basal' to ------- , which in turn, is 'basal' to ------.
   (c) *Pythiales* is 'basal' to --------------.
   (d) Sexual reproduction is ------------ in *Leptomitus* and --------- in *Apodachlya.*
   (e) The formation of successive zoosporangia within emptied ones is called --------------------.
   (f) DNA data suggest *Phypphthora* to be nearer to ----------------- than to -------------.
   (g) ------------- ---------------- made history in 1930 in France by devastating the ----------- crop.
   (h) The term 'downy mildew' etymologically means -----------------.
   (i) 'Green ear' disease of bajra, caused by ------------- ----------, was discovered in 1907 by ---------------.
   (j) -------------- causes hypetrophy of floral parts and tumors in all parts of the host plant.

21

Kingdom *Protozoa*
Phylum *Mycetozoa*
(= slime moulds: Protozoan fungi)

(including former Myxomycota, and Dictyosteliomycota)

Kingdom *Protozoa*

Mycologists have been studying the organisms called 'slime moulds' by tradition, and with the full knowledge that they were not fungi. These were placed separately under division **Myxomycota**, while the 'true fungi' were placed under **Eumycota. C.T. Ingold** (1967), called them 'organisms studied by mycologists'. **Anton de Bary** (1887), one of the founders of mycology, who was unable to reconcile himself in accepting these organisms as fungi, called them **Mycetozoa** (Mycet – fungi, zoa – animals). Since then, zoologists have also studied them and placed them under phylum *Protozoa*. These continued to be treated as 'organisms of uncertain affinity'.

The slime moulds are placed under kingdom ***Protozoa***, a new eukaryotic kingdom segregated from the former from kingdom ***Protista***. The kingdom *Protozoa* includes predominantly unicellular, plasmodial or colonial organisms that are wall-less during their somatic, trophic phase. These are phagotrophic on bacteria, yeasts, spores etc. in dung or decaying plant materials. Ciliary hairs are never rigid or tubular. Chloroplasts when present (as in *Euglena*) lack starch or phycobilisomes. The multicellular species show minimal cell differentiation. There are 18 phyla in *Protozoa* (Cavalier-Smith 1993) .

Earlier the slime moulds were placed under 4 phyla (Hawksworth *et al.,* 1995), which later (Kirk *et al.,* 2001) were restricted to three: (1) *Acrasio-mycota*, (2) *Myxomycota*, and (3) *Plasmodiophoromycota*. Recent molecular and ultrastructural evidences suggest that slime moulds are highly heteroge-neous and polyphyletic, and belong to five separate lineages, defined mainly by the shape of their mitochondrial cristae (Kirk *et al.,* 2008). These are:

1. **Ramicristates** (defined by branched mitochondrial cristrae), under which are placed three slime mould classes: ***Protostelea*** (Protosteliomycetes), ***Myxogastrea*** (Myxomycetes), and ***Dictyostelea*** (Dictyosteliomycetres). These are put under phylum ***Mycetozoa.***
2. **Heterolobosea** (defined by flattened mitochondrial cristae) includes the schizopyenoid amoebae and cellular slime moulds which are treated at order level; 1 order ***Acrasida*** (Acrasiales).
3. **Copromyxida** (tubular mitochondrial cristae), *Copromyxa*, formerly placed in Acrasiales, is now separated.
4. **Fonticulida** (plate-like mitochondrial cristae) includes *Fonticula*, also earlier placed in Acrasiales, is now separated.
5. **Plasmodiophorids**, a group defined by unique 'cruciform' chromosomes and nucleolus at mitosis. Includes one class ***Plasmodiophorea*** (Plasmodiophoromycetes).

The lineage Heterolobosea are grouped with **Euglenozoa** and **Excavata** (unranked taxa), the other four lineages are grouped as **Amoebozoa**. ***Mycetozoa*** is the only phylum recognized by Kirk *et al.* (2008) for slime moulds. These best-studied slime moulds belong to the lineage called **Ramicristates.** These are characterized by (1) free-living, unicellular or plasmodial somatic phase, (2) branched (rami = branched) mitochondrial cristae, (3) unicellular or multicellular sporocarps with single to many spores, (4) spore wall cellulosic or chitinous, and (5) spores germinating to form 1 or 2-flagellated cells. These organisms are most common during humid conditions on decaying logs, on bark of living trees, dead leaves etc. in cool shady places. During dry weather they form sporocarp. Though of little economic importance, these are invaluable experimental organisms.

Classification The phylum *Mycetozoa* has three classes which are further classified as follows. (Only selected taxa which we shall study are given).

Class ***Protostelea***: 1 order, ***Protostelida*** (Protoseliales); 4 families, including family *Ceratiomyxaceae*: *Ceratiomyxa*.

Class ***Myxogastrea***: 5 orders and 14 families

Order **Physarida** (Physarales): *Physaraceae: Physarum*.

Class **Dictyostelea:** 1 order, *Dictyostelida* (*Dictyosteliales*); 2 families; *Dictyostelidiaceae: Dictyostelium*.

Class *Protostelea*

Order, *Protostelida* (Protoseliales)

Trophic phase are simple amoebae or plasmodia having no shuttle (forward and backward) movement of the protoplasm.

Family *Ceratiomyxaceae*

Genus *Ceratiomyxa*

Ceratiomyxa is the lone genus in this family. It has three species: *C. fruticulosa*, *C. sphaerosperma* and *C. morchella*. The spores are not formed inside a sporocarp.

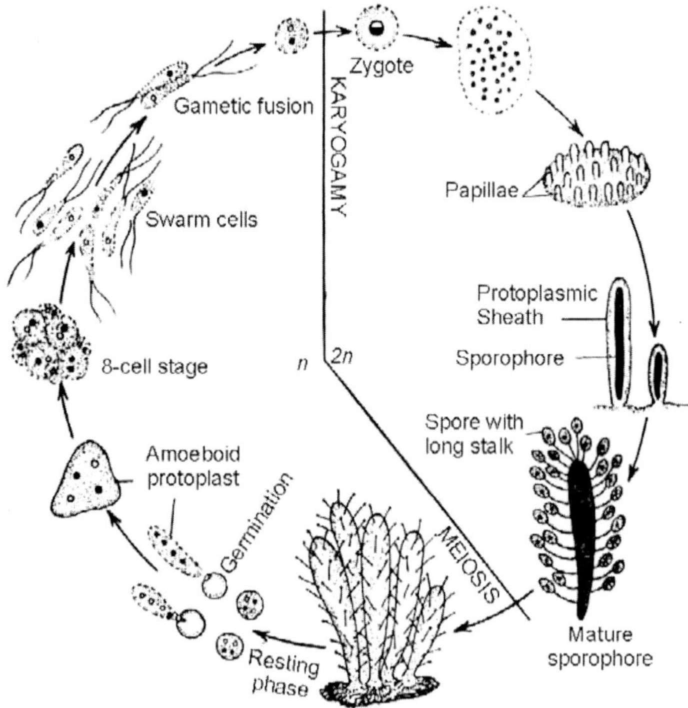

Fig. 21.1. Life cycle of *Ceratiomyxa* sp.

The spores, on germination, release a thread-like, quadrinucleate amoeboid protoplast through a pore (Fig. 21.1). The four haploid nuclei are linearly and equidistantly distributed in the cytoplasm. Soon, the protoplast becomes globose and the four nuclei are tetrahedrally arranged. Each haploid nucleus divides mitotically, around which 8 cells are formed. These cells develop 2 anterior, 'whiplash' flagella and dissociate as swarm cells. The swarm cells are unequal in size and behave as gametes. Two swarm cells of compatible mating types fuse and give rise to a diploid zygote, which enters decaying wood and feeds on bacteria and hyphae of wood-rotting fungi. It grows and gives rise to a multinucleate diploid plasmodium, which represents the somatic phase.

At the time of sporulation, the plasmodium is replaced by several papillate masses. Each papilla differentiates into a central non-living gelatinous portion, called the **sporophore**, and an ensheathing layer of protoplasm. The sporophore increases in length and becomes columnar. Consequently, the protoplast is restricted to the apical portion of the sporophore (Fig. 21.2). The nuclei divide mitotically and the protoplasm is cleaved into uninucleate portions. Each uninucleate mass secretes a cell wall and transforms into a spore. At this stage, the spore is attached to a small projection, formed on the sporophore, with its long axis parallel to the projection. The projection elongates to become a long stalk, bearing the spore terminally. Meiosis occurs in each spore and, at the time of release, each spore has four haploid nuclei. The spores undergo a resting period and germinate when favourable conditions become available.

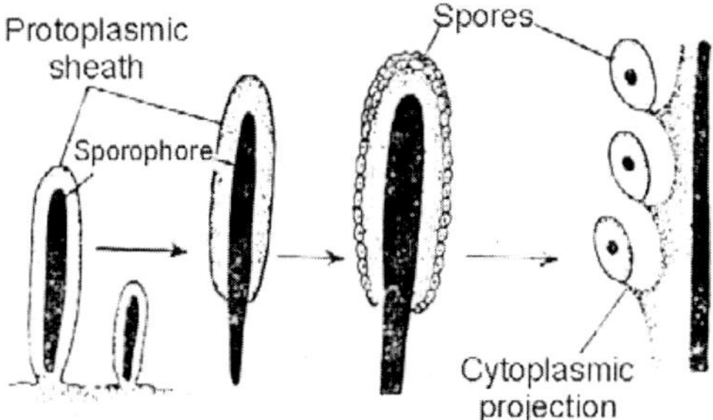

Fig. 21.2. Spore formation in *Ceratiomyxa* sp.

Class *Myxogastrea* (true slime moulds, former Myxomycetes)

These are the **true slime moulds**. The trophic phase is a plasmodium, which is a giant multinucleate amoeboid mass of cytoplasm, lacking cell wall and a fixed form. There is a distinct shuttle movement (forward and backward) of the protoplasm. It moves and feeds on encysted amoebae, bacteria etc. by pseudopodia. The nuclei divide synchronously by mitosis. At the advent of unfavourable conditions, like depletion of food or changes in light-intensity, the plasmodium forms a **sporocarp**, which is essentially like a fungal sporangium, containing uninucleate, walled spores. The spores at the time of germination undergo meiosis and form four haploid nuclei. But, 3 nuclei degenerate and eventually the spore contains a single haploid nucleus. From each spore, a haploid uninucleate **amoeboid cell** extrudes out. It feeds on bacteria and multiplies to form a 'clone' of genetically similar amoeboid cells (**myxamoebae**). If free water is available, the amoebae readily develop into anteriorly uniflagellate **swarm cells**. The flagellum is of 'whip lash' type. When food runs out, the amoeboid cells encyst and turn into dormant **cysts**.

The **myxamoebae**, or the **swarm cells**, act as gametes. When compatible mating types come in contact, they fuse and give rise to a zygote, which grows into a diploid, multinucleate plasmodium. The conversion of amoebae into plasmodium is little understood and poses problems similar to differentiation in higher organisms.

Under certain conditions, the plasmodium instead of forming a sporocarp, forms a thick-walled **sclerotium**. The sclerotium germinates to give rise to a plasmodium, when conditions, like moisture, become favourable.

Three types of plasmodia are known - Protoplasmodium, Aphanoplasmodium and Phaneroplasmodium (Fig. 21.3). **Protoplasmodium** is microscopic and never grows beyond a diameter of 1 mm. **Aphanoplasmodium** (aphano – invisible) is transparent and invisible. **Phaneroplasmodium** (phanero – visible), which is most common, is easily visible due to its granular cytoplasm.

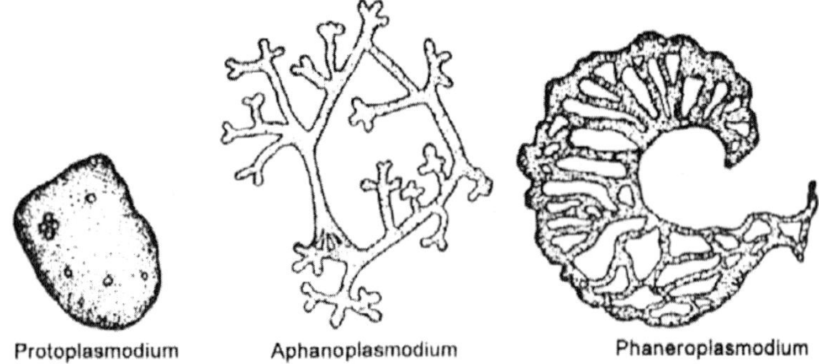

Fig. 21.3 Plasmodial types: Protoplasmodium, Aphanoplasmodium and Phaneroplasmodium

The class *Myxogastrea* has 5 orders. The order *Physarida* is characterized by dark-coloured spore mass and calcareous peridium and capillitium. There are 2 families: *Didymiaceae* and *Physaraceae*. We shall study *Physaraceae*, genus *Physarum*.

Genus *Physarum*

Physarum, with 135 species, is the largest and best-studied genus. Most of our knowledge about *Myxogastrea* is derived from this genus, in general, and *P. polycephalum,* in particular. *P. cinereum* forms large, bluish, unsightly patches on city lawns.

The life cycle (Fig. 21.4) starts with the germination of spores, which are spherical, thick-walled, dark-brown or black, and varying in size. These contain a single diploid nucleus, which undergoes meiosis and forms 4

haploid nuclei, out of which only one survives. The spores germinate when moistened by water, and release haploid cells, either a myxamoeba or a biflagellate swarm cell (depending on the amount of water available). The two stages are interchangeable, as the flagella are readily synthesized or discarded. The myxamoebae, which are less than 10 μm in diameter, move with the help of pseudopodia and feed on bacteria by phagocytosis. These can absorb liquid nutrients also. The swarm cells are sticky at the posterior end. The myxamoebae and the swarm cells divide by fission. Soon, a clone is built up around each parent cell. The myxamoebae or swarm cells, under conditions of shortage of food, develop a wall and turn into **cysts**, which give rise to a myxamoeba, or a swarm cell, as soon as the food becomes available.

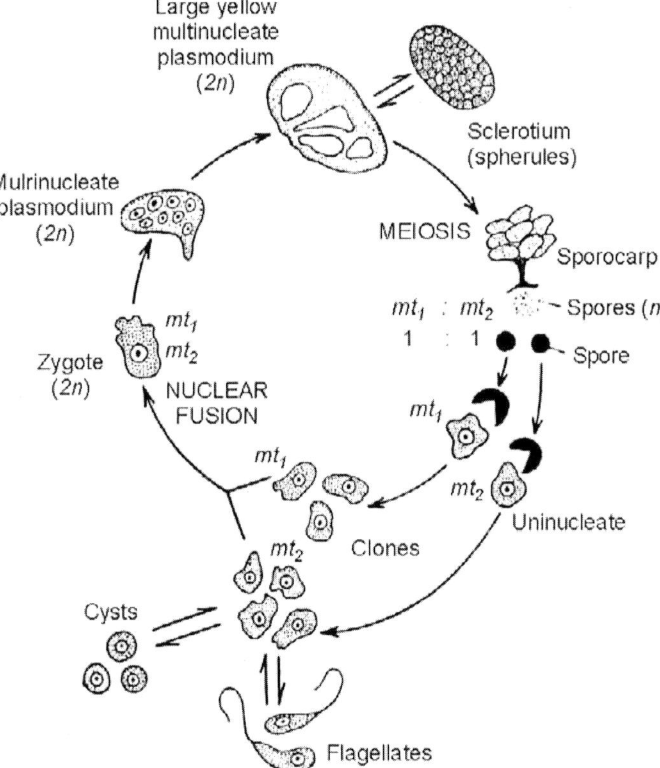

Fig. 21.4 Life cycle of *Physarum polycephalum* (mt_1 and mt_2 represent myxamoebae of different mating types, n and 2n denote haploid and diploid nuclear stages)

The swarm cells act as gametes and fuse by their sticky posteriors and give rise to a diploid cell. It grows, becomes multinucleate (all nuclei are diploid) and develops into a bright yellow, moist and shapeless plasmodium. Series of changes, which distinguish a plasmodium from a myxamoeba, take place. Eventually, it develops into a **phaneroplasmodium** (Fig. 21.5). It looks like a perforated sheet of protoplasm, spreading with a fan-shaped anterior end, made of a network of veins. The cytoplasm in the veins is differentiated into an outer, thick, jelly like, clearer region, the **ectoplasm** that surrounds the more fluid and granular **endoplasm**. The endoplasm shows rhythmic and reversible streaming (**shuttle movement**). The ecto-, and the endoplasm, readily, change into one another and, in this way, the plasmodium resembles a giant amoeba. The plasmodia feed on encysted myxamoebae and move around secreting a slimy track (made of polygalactose), which protects them from dehydration.

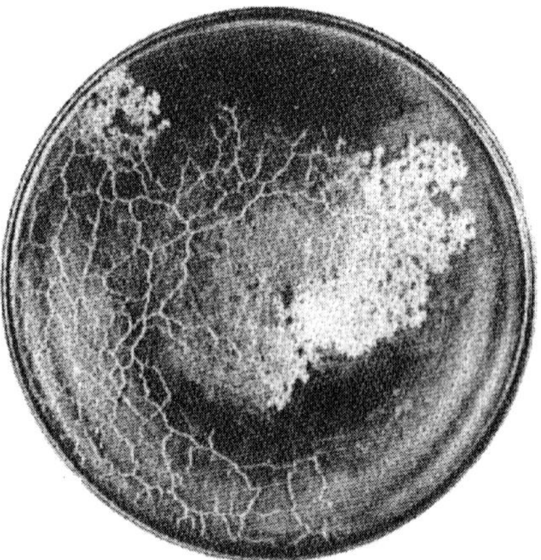

Fig. 21.5. Phootomicroraph of phaneroplasmodium

The slimy track is left behind, as a collapsed slimy streak, indicating the path of movement of the plasmodia. The plasmodia live their life, and, when conditions become unfavourable, they undergo sporulation and transform into the sporocarp. Starvation, a definite age and light, are the factors that initiate sporulation. In laboratory, it is induced by certain chemicals, the most important being *niacin*.

The mature sporocarp consists of large number of stalked sporangia globular in shape, borne on a common cellophane-like basal sheet, called **hypothallus**. The sporangia dehisce and spores are disseminated by wind, which withstand the adverse conditions and germinate, when favourable conditions return.

Occasionally, especially under conditions of drought, the plasmodium transforms into a hard **sclerotium**, rather than developing a sporocarp. The sclerotium contains multinucleate cells, the **spherules**. The plasmodium is regenerated from the sclerotium, when conditions become favourable.

Class *Dictyostelea* (cellular slime moulds, dictyostelids, former Dictyosteliomycota)

Dictyostelids, as the members are called, are abundantly present in soil, feeding on bacteria. They are of no direct importance to us. Their 'niche' in the soil is also not known. Nevertheless, they are enthusiastically studied for their use as a 'tool' in morphogenesis (the science which explains how an organism gets its particular 'pattern' of development). *Dictyostelium discoideum*, has found much use in such studies.

These organisms do not grow on chemically-defined media. For experimentation, they are grown in two-membered cultures, in association with a bacterium, usually *Klebsiella aerogenes*, on which they feed and grow. However, Watts and Ashworth (1970) successfully cultured the *Ax* strain of *Dictyostelium discoideum* in axenic culture i.e. (in the absence of any other organism).

General Characteristics and classification

The somatic phase consists of small, independent, uninucleate haploid amoeoboid cells that feed on bacteria by phagocytosis and multiply by binary fission. As soon as the population of the amoebae attains a 'critical number', they organize themselves into a **slug** (land snail without shell) or bullet-shaped colony, called **grex**. The grex moves and after a definite period, develops into the **sorocarp***. It consists of a stalk, bearing an apical mass of spores, which are not surrounded by a peridium. The spores have a cellulosic cell wall, and are disseminated by wind. On germination, they give rise to amoebae, which restart the life cycle.

The order ***Dictyostelida*** has two families: *Dictyosteliaceae* and *Acytosteliaceae*, which differ in their **sorophores** (stalk of sorocarp). The family ***Dictyosteliaceae*** has 3 genera and 42 species, which form stalked, stout sorocarps, filled with empty cells. We shall study the genus *Dictyostelium*.

Genus *Dictyostelium*

The life cycle (Fig. 21.6) has four distinct phases, growth phase, aggregation phase, migration phase, and sorocarp (or morphogenetic) developm-

---

* The term **sorocarp** is used for the fructification that arises from communities of cells. **Sporocarps,** on the other hand, arise from single protoplasts.

ent phase. The **growth phase** consists of free amoeboid cells, which move with the help of pseudopodia and feed on bacteria by 'phagocytosis'. The amoebae contain large number of lysosomes which fuse with the food vacuoles and digest the bacteria. The amoebae grow and eventually divide by binary fission to form two daughter amoebae. Soon, a population (or clone) of amoebae is built up. During this phase, the amoebae repel one another.

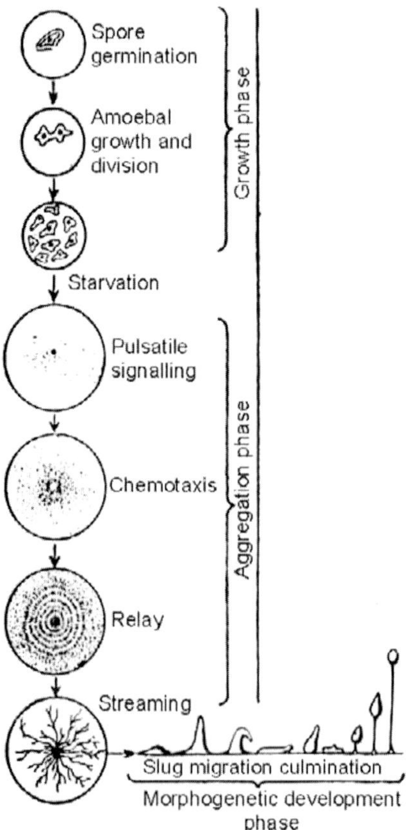

Fig. 21.6 Life cycle of *Dictyostelium discoideum* depicting the growth phase, aggregation phase and morphogenetic development phase. The outer circles show Petridish

Aggregation phase starts when the population of the amoebae attains a critical number, or when the food supply becomes exhausted. A **colonial life** sets in, which ends in the formation of the **sorocarp**. The amoebae start coming closer by chemotaxis (Fig. 21.7) and growth, feeding and multiplication stops. The amoebae start streaming towards a central meeting point called the **aggregation centre**. The streaming amoebae look like tributaries

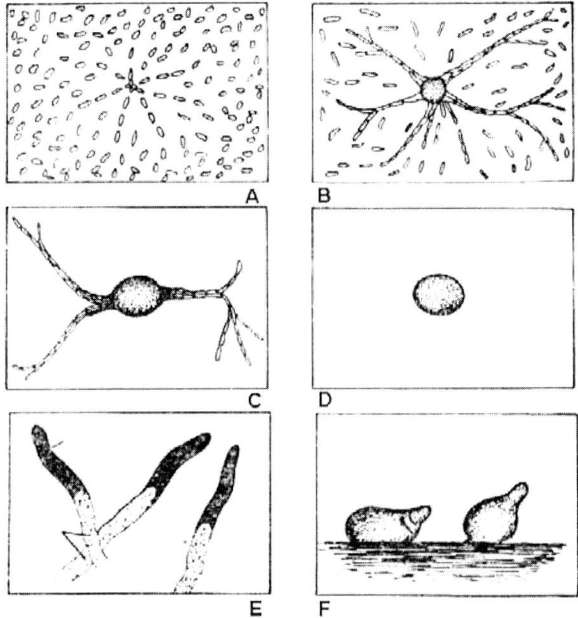

Fig. 21.7 Showing slug formation and its movement. A, beginning of aggregation with the formation of a small centre; B, and C, thickening of aggregation streams and enlargement of centre, D, final cell mass or grex; E, grex moves with a defined apical tip followed by excreted slimy tract; F, grex at the end of the moving phase; ready for sorocarp development

converging to form a river. The nature of the cells that constitute the aggregation centre is not known. They may be (1) **macrocysts** – zygotes, formed by fusion of two amoebae, (2) specialized, large cells, called **initiater cells** or I-cells, or (3) ordinary amoebae. Whatever be their nature, the cells at the aggregation center emit a pulse of **acrasin,** identified as cyclic-AMP (Fig. 21.8), and the near-by amoebae start moving towards it by chemotaxis. The collecting amoebae, in turn, emit their own acrasin pulse, and in this way, an acrasin gradient is established over a long distance. Amoebae, sensing the cyclic–AMP signal, move up the concentration gradient. The amoeboid cells of the aggregation streams develop polarity and later fuse, end-to-end, at newly developed adhesive zones on their body, which are proteinaceous in nature. The fused cells move by their individual pseudopodia in a most coordinated manner. In this way, a large number of cells (a few hundred to 1,00,000) collect at the centre and form a multicellular, 2–4 mm long, bullet-shaped pseudoplasmodium, called **slug** or **grex**. The grex starts moving, i.e. enters the **migration phase**. The moving grex looks like a slug, with a clearly-defined tip at the anterior end. The component amoebae move inside a slimy tubular track, exuded around the colony by the cells themselves. The track is stationary and is left behind as a sheath, which collapses leaving a slimy streak. The grex migrates from the interstices in the soil towards the surface in search of more light, warmer temperature and greater humidity. At the end of the migration phase, which varies from short or no migration to a period of about 2 weeks, the grex enters the **sorocarp development phase** and transforms into a sorocarp. The grex stops, rounds up and develops a delicate stalk, with a spore mass at the apex (Fig. 21.9).

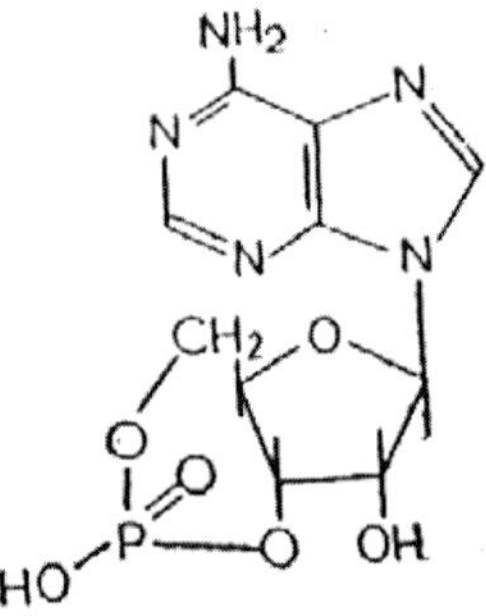

Fig. 21.8 Cyclic AMP, which brings about intercellular communication in *Dictyostelium discoideum*

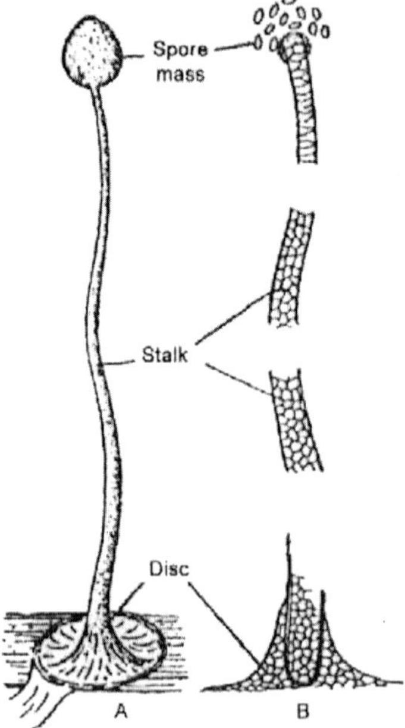

Fig. 21.9 Sorocarp of *D. discoideum* A, morphological zones; B, the constituent amoebae

During sorocarp formation also, the individual amoebae do not lose their identity. Some of the amoebae become large and highly vacuolated. These constitute the **stalk cells**. Some of the amoebae transform into thick, smooth **spores**, with dense cytoplasm. The stalk cells are incased by a cellulose tubular sheath, which gains its rigidity from the turgor pressure of the stalk cells. The undifferentiated amoebae form a disc at the base of the sorocarp. The stalk cells become necrotic and die, but the spores remain viable for years together. These are disseminated by wind and, on germination, release amoebae, which restart the life cycle.

Sexual Reproduction

D. discoideum is heterothallic. When cells of opposite mating types come in contact, they fuse in pairs to form a large diploid cell. The number of chromosomes at the 2n stage is 16. The zygote feeds on smaller amoebae by engulfing them and grows larger until it becomes a **giant cell**. It can feed no more. It secretes a thick wall and transforms into a **macrocyst.** The macroysts possibly constitute the aggregation centre. It is multinucleate; the nuclei are haploid, formed by meiosis and repeated mitotic divisions. When the macrocyst germinates, after the resting phase, it gives rise to several amoebae. All attempts to germinate the macrocysts of *D. discoideum* in laboratory have failed. However, success has been achieved with *D. mucoroides*.

Parasexual genetic recombination has been reported in *D. discoideum* by Sinha and Ashworth (1969), described in Chapter 30 (Heyterokaryosis and Parasexual Cycle).

REFERENCES

Kirk, P.M. *et al.* 2008 *Ainsworth and Bisby's Dictionary of the Fungi.* C.A.B. International, Oxon, U.K.

Sinha, U.K. and Ashworth, J.M. 1969. Evidence for the existence of elements of a parasexual cycle in the cellular slime mould *Dictyostelium discoideum*. *Proc. Roy. Soc.* B, **173:** 531.

SUMMARY

The 'slime moulds' are placed under kingdom ***Protozoa***, which includes predominantly unicellular, plasmodial or colonial organisms that are wall-less during their somatic, trophic phase. These are phagotrophic, possess ciliary hairs, chloroplasts (when present) lack starch or phycobilisomes, and show minimal cell differentiation. There are 18 phyla in *Protozoa*. Molecular and ultrastructural evidence suggest that slime moulds are highly heterogeneous and polyphyletic, and belong to five separate lineages, defined mainly by the shape of their mitochondrial cristae. These are: Ramicristates, Heterolobosea, Copromyxida, Fonticulida, and Plasmodiophorids. The **Ramicristates** include three slime mould classes: ***Protostelea***, ***Myxogastrea***, and ***Dictyostelea***. These are put under phylum ***Mycetozoa.*** **Heterolobosea** is represented only at order level, ***Acrasida***. The lineage **Plasodiophorids** has one class ***Plasmodiophorea.***

Phylum Mycetozoa has 3 classes. We have studied example of each.

Class ***Protostelea***: *Protostelida****:*** *Ceratiomyxaceae: Ceratiomyxa.*

Class ***Myxogastrea***: ***Physarida***: *Physaraceae****:*** *Physarum.*

Class ***Dictyostelea:*** *Dictyostelida Dictyostelidiaceae****:*** *Dictyostelium.*

In *Ceratiomyxa*, the diploid multinucleate plasmodium (the thallus) grows by phagocytosis. It does not form a sporocarp. During sporulation, it forms naked sporophores which bear stalked spores. After a rest period the spores germinate and release haploid amoeboid protoplast which forms flagellated swarm cells. These serve as gametes and by fusion give rise to diploed zygote which develops into the plasmodium.

Physarum polycephalum has been used extensively to study the biology of slime moulds. *P. cinereum* forms ugly blue patches on city lawns. The plasmodium is diploid, multinucleate, bright yellow, moist and shapeless plasmodium (a phaneroplasmodium). It feeds and moves around secreting a slimy track, which protects them from dehydration. The slimy track is left behind. Under unfavourable conditions, it undergoes sporulation and transform into the sporocarp. The sporocarp consists of stalked globular sporangia borne on a basal sheet, called hypothallus. The sporangia dehisce and the spores are disseminated by wind.

The spores germinate when moistened by water, and release haploid cells, either a myxamoeba or a biflagellate swarm cell. The two stages are interchangeable. The myxamoebae and the swarm cells grow and divide by fission forming a clone. The myxamoebae or swarm cells, under conditions of shortage of food, develop into cysts, which give rise to a myxamoeba, or a swarm cell, as soon as the food becomes available. These act as gametes and fuse to form a diploid cell that grows into a diploid plasmodium.

Dictyostelium has been studied as a tool to study morphogenesis. It is abundantly found in soils. The somatic phase consists of small, independent, uninucleate, haploid amoeoboid cells that feed on bacteria by phagocytosis and multiply by binary fission (growth phase). As soon as the population of the amoebae attains a 'critical number', they organize themselves into a slug (land snail without shell) or bullet-shaped colony, called grex (aggregation phase). The grex moves (migration phase) and after a definite period, it develops into the sorocarp (sorocarp development phase). The sorocarp consists of a stalk formed by vacuolated stalk cells, bearing an apical mass of thick-walled spores, which are not surrounded by a peridium, the undifferentiated amoebae form the basal disc. The spores are disseminated by wind. On germination, they give rise to amoebae, which restart the life cycle. A parasexual cycle is reported in *D. discoideum*.

TEST QUESTIONS

1. Name the lineages of slime moulds. Describe the classification of slime moulds based on these lineages.
2. Describe the life cycle of *Ceratiomyxa* sp. **or** *Physarum polycephalum*.
3. Name the various stages in the life cycle of *Dictyostelium discoideum*. Describe any two stages.
4. Answer in brief;
    a. Which are the three types of plasmodia in *Myxogastrea*.?
    b. Which *Physarum* species is used in the study of morphogenesis?
    c. Which are the four phases in the life cycle of *Dictyostelium discoideum*?
    d. Which are the possible cells that constitute the aggregation centre in *Dictyostelium discoideum*?
    e. In which slime mould was parasexual cycle discovered?
5. Distinguish between:
    a. Sorocarp and Sporocarp,
    b. Myxamoebae and Swarm cells,
    c. Proto-, Aphano-, and Phaneroplasmodium,
    d. Trophic phases of *Ceratiomyxa* and *Physarum*.
6. Fill in the blanks.
    a *Physarum* ------------ forms blue ugly patches in grassy lawns.
    b. Members of class -------------------- are called 'true slime moulds'.
    c. The classes of the lineage Ramicristates are placed in the phylum -------------.
    d. The Ramicristates are defined by ---------------- mitochondrial cristae.
    e. The plasmodiophorids are unique in having -------------- type of mitoisis.
    f. Heterolobosea have --------------- mitochondrial cristae.

g. The spores in genus ---------------- are not formed within a sporocarp.

h. Shuttle movement of protoplasm is not found in the plasmodia of ---------------.

i. Under conditions of drought the plasmodium of *Physarum* transforms into a ---------, rather than developing a -------------.

j. The amoebae of *Dictyostelium* migrate towards the aggregation centre under the influence of the chemical ---------------.

7. Write briefly on any three:

a. Formation of slug (grex) in *Dictyostelium*.

b. Phaneroplasmodium.

c. Structure of sorocarp in *Dictyostelium*.

d. Parasexual cycle in *Dictyostelium*.

22

Order *Acrasida* (=Acrasiales)

(*Cellular slime molds, Acrasids,* former Acrasiomycetes)

The slime mould 'lineage' **Heterolobosea,** which is characterized by flattened mitochondrial cristae, includes the **cellular slime moulds** and schizopyrenoid amoebae. 'Order' is the highest taxa in this group. Heterolobosea is grouped with the Euglenozoa (unranked), while other lineages are grouped with another unranked group, Amoebozoa (Kirk *et al.,* 2008). The *Acrasida* have a life cycle similar to *Dictyostelea.*

Order *Acrasida*

The somatic phase comprises small, unicellular, independent, haploid, amoeboid cells, (amoebae). These are typically cylindrical, with a clear demarcation of an inner granular **endoplasm** and outer, non-granular **ectoplasm**. There is a single 'lobed' pseudopodium in the anterior region and several small 'filose' pseudopodia on the posterior. The amoebae are phagotrophic and feed on bacteria, yeasts and spores and multiply by fission. The individual amoebae encyst under unfavourable circumstances, which on germination give rise to amoebae. Under favourable circumstances, the amoebae aggregate to form a **sorocarp**[*] (also called pseudoplasmodium), which looks like a microscopic tree, with stem and branches. It is surrounded by membranous sheath. The stem comprises somatic cells while the cells in the branches are spores. Both give rise to amoebae on germination. Flagellated cells are absent, and sexual reproduction is not reported. The order has 4 families. The family *Acrasiaceae* includes one genus *Acrasis*, which has two species.

---

* The term **sorocarp** is used for the fructification that arises from communities of cells. **Sporocarps**, on the other hand, arise from single protoplasts.

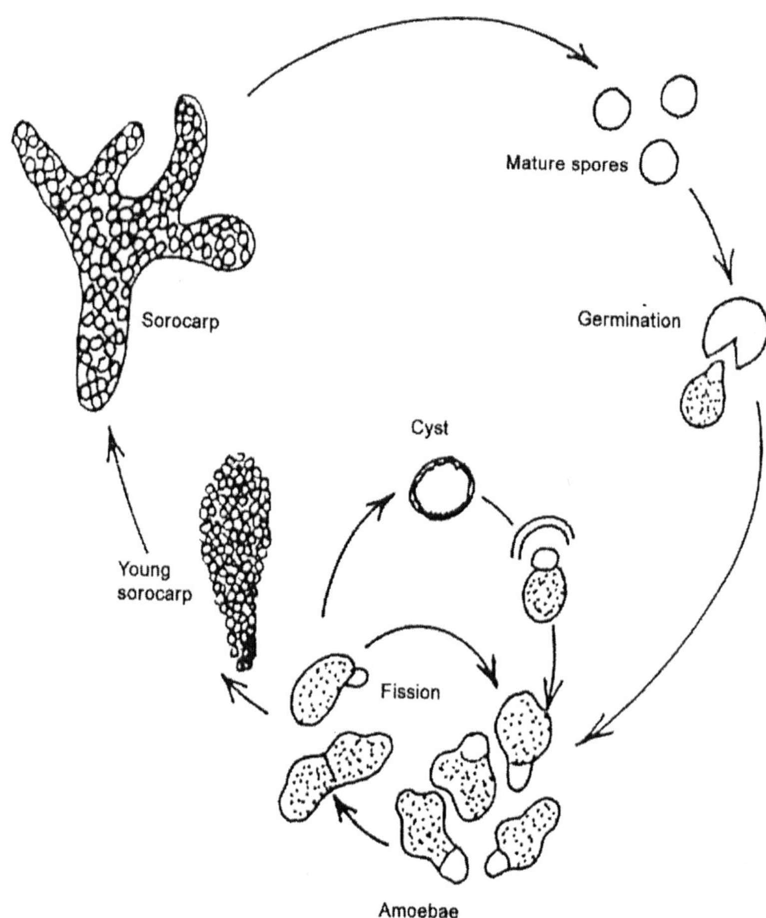

Fig. 22.1. Life cycle of *Acrasis rosea*

A. rosea (Fig. 22.1), which grows on yeast beer, shows the typical life cycle of an acrasid. The spores of the sorocrap germinate and produce amoebae that engulf bacteria, yeasts, spores, etc. and multiply by fission. Under favourable conditions, the amoebae form sorocarp, covered by a fragile membrane. As mentioned above, the sorocarps contain somatic cells and spores. Both give rise to amoebae on germination.

REFERENCES

Kirk *et al.* 2001, 2008 *Ainsworth and Bisby's Dictionary of the Fungi.* C.A.B. International, Oxon, U.K

SUMMARY

Cellular slime moulds represent the slime mould lineage Heterolobosea, characterized by flattened mitochondrial cristae. Order is the highest taxon in this group. There is one order *Acrasida* which have a life cycle similar to *Dictyostelea*. The somatic phase comprises haploid amoebae, which are typically cylindrical, with **endo-,** and **ectoplasm**. The individual amoebae encyst under unfavourable circumstances, which on germination give rise to amoebae. Under favourable circumstances, the amoebae aggregate to form a **sorocarp** (also called pseudoplasmodium). The order has 4 families. The family *Acrasiaceae* includes one genus *Acrasis*, which has two species. *A. rosea* shows the typical life cycle of an acrasid.

TEST QUESTIONS

1. Answer in brief:
   a. What feature places *Acrasida* in Heterolobosea?
   b. What do the amoebae of *Acrasis* form under unfavourable and favourable conditions?
2. Describe the trophic phase and the sorocarp of acrasids.
3. Describe the life cycle of *Acrasis rosea*.

23

Class *Plasmodiophorea*

(former Plasmodiophoromycetes)
(obligate endoparasitic slime moulds; plasmodiophorids)

These organisms belong to the slime mould 'lineage' **Plasmodiophorids**, a group uniquely defined by the 'cruciform' cell division. Class is the highest taxon. The 'plasmodiophorids', as these organisms are called, are obligate endoparasites in plants. Some attack plants and cause economically important diseases, while others attack algae and fungi. One species, *Polymyxa graminis*, is known to serve as a vector of several plant pathogenic viruses e.g. wheat mosaic virus (WMV), barley yellow mosaic virus (BYMV) and oat mosaic virus (OMV).

The thallus is a plasmodium, which develops only in plant cells as obligate parasites. Except the reinfection phase that occurs in water, the life cycle is completed entirely within the host plant. The plasmodium, which may be haploid or diploid, is holocarpic and develops into thin-walled sporangia, aggregated in sori, or into resting spores (cysts). The plasmodium is designated as **zoosporogenous** or **cystogenous**, depending on whether it produces **zoospores** or **cysts (=resting spores)**.

The zoospores are uninucleate, anteriorly biflagellate, anisokont i.e. unequal in size, and, both flagella are of 'whip lash" type. During movement, the flagella take opposite directions. The short flagellum, as indicated in electron microphotographs, lacks the tail piece.

Nuclear division (in family *Plasmodiophoraceae*) is unique, not found in fungi, but known in protozoa. It is called **cruciform** division or **promitosis**. During division, an intranuclear spindle is formed and the chromosomes lie in a circle. The nucleolus lies in the center of the circle of chromosomes and forms a dumbbell-shaped structure. When viewed from a side, the chromosomes and the nucleolus appear like a 'cross' hence the name, "cruciform" division.

The resting spores, or cysts, which are formed by all the genera, are surrounded by smooth or variously ornamented, two-layered wall, made of chitin. These spores can survive in soil, in absence of the host plants, for many years. The resting spores, except in one genus (*Plasmodiophora*), are held together in small or large groups, called **cystosori** (sing. **cystosorus**). On germination, which is initated only in presence of the host or its metabolic products, the resting spores give rise to amoeboid bodies or zoospores. Sexual reproduction has been reported only in few cases, where it occurs by planogametic copulation (isogamy).

Differences from other slime moulds

The plasmodiophorids differ from other slime moulds. These are (1) parasitic (not saprobic), (2) lack of phagptrophy, (3) have chitinous cyst wall (not cellulosic), (4) presence of Golgi bodies in plasmodia, (5) presence of zoospores/resting spores.

Classification

There is a single order *Plasmodiophorida* (=Plasmodiophorales), and two families, *Endemosarcaceae* and *Plasmodiophoraceae*. The former lacks the cruciform nuclear division. Its placement here is disputed, and is regarded as incertae sedis.

Order *Plasmodiophorida* (=Plasmodiophorales)

Family *Plasmodiophoraceae*

The family *Plasmodiophoraceae* includes 13 genera and 44 species that are endoparasitic in flowering plants (*Plasmodiophora*, *Spongospora*), algae (*Woronina* in *Vaucheria*, *Sorodiscus* in *Chara*) and fungi (*Sorodiscus*), and frequently induce hypertrophy in infected cells. These organisms are co-cultured with their host.

The genera (Fig. 23.1) are identified primarily by the morphology of the resting spores formed by them. The zoosporangia are often lacking and, therefore, are not of much use in taxonomy. The resting spores, except in *Plasmodiophora* aggregate to form cystosori, which are characterized by the maximum number of the spores that they contain, or by their shape. Some important genera and their identifying features are as follows:

Genera identified by the maximum number of resting spores in the cystosorus.

1. *Plasmodiophora*: resting spores numerous and free

2. *Tetramyxa* : resting spores 4 (or less), cling together.
3. *Octomyxa* : resting spores 8 (or less). cling together.

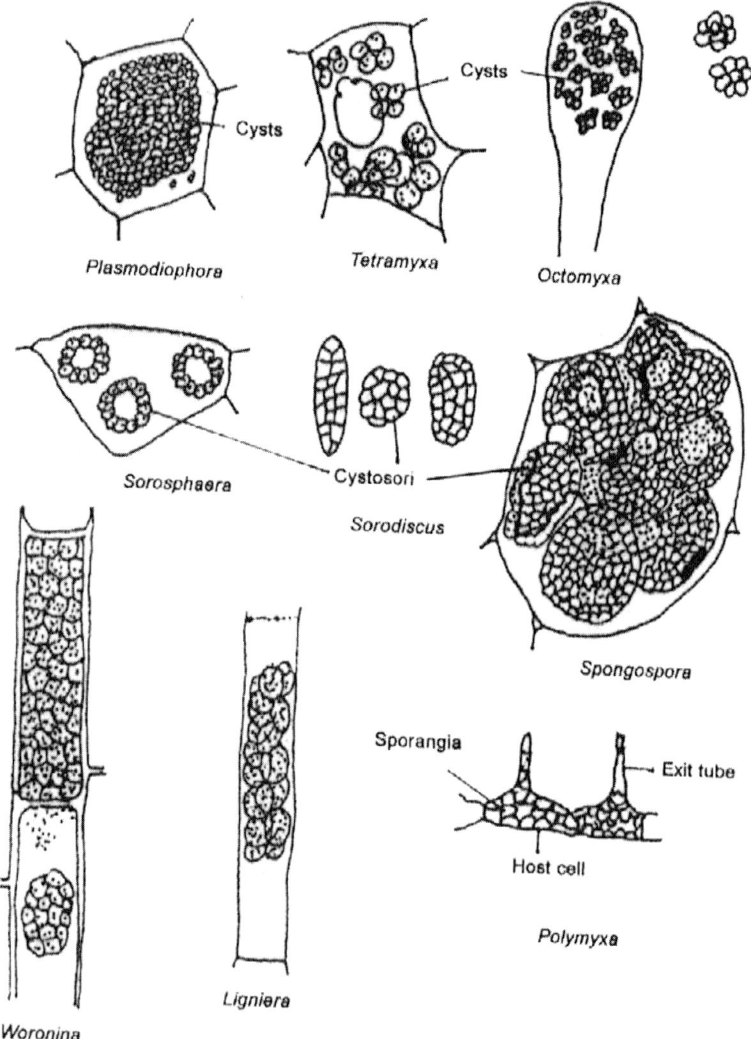

Fig. 23.1. Genera of *Plasmodiophoromycetaceae*

Genera identified by shape of their cystosorus

4. *Sorosphaera* – cystosorus a hollow sphere of resting spores.
5. *Sorodiscus* – cystosorus disc-shaped.

6. *Spongospora* – cystosorus sponge-like, of various shapes.

Genera identified by sporangial form, size, host range and hypertrophy of the host

7. *Ligniera* – little or no hypertrophy of the hosts, (viz. angiosperms) sporangia numerous (2–10 μm).
8. *Woronina* – attack fungi or algae; sporangia large (8–35 μm), one or many.
9. *Polymyxa* – little or no hypertrophy of angiosperm hosts, sporangia elongated, lobed or segmented, exit-tubes long.

The identification of genera, based on number and form of resting spores, has been criticized on the ground that these characteristics, although convenient, are not good generic criteria. Nevertheless, in the absence of knowledge about the life cycles of most of the genera, the present classification, even though artificial, serves the purpose and is currently accepted. It doesn't tell their taxonomic status, but certainly identifies them.

Genus *Plasmodiophora*

The genus has 6 widespread species which are obligate endo-parasites of plants. *P. brassicae* causes the 'club-root' disease of crucifers, especially mustard (Fig. 23.2), and is distributed throughout the world. The roots become hypertrophied and club-shaped.

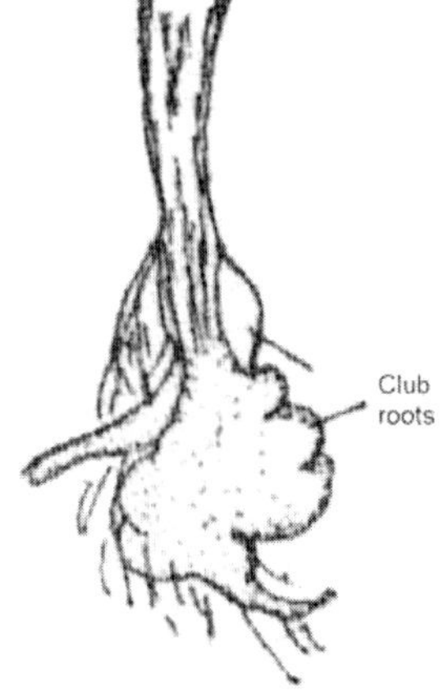

Fig. 23.2. 'Club root' disease of crucifers (musturd)

The life cycle (Fig. 23.3) consists of two phases: the primary phase and the secondary phase. The primary phase occurs in the root hairs and the secondary phase in the cortical cells of the root. The hypertrophy and hyperplasia, which result in the formation of club-like thick roots, are associated with the secondary phase.

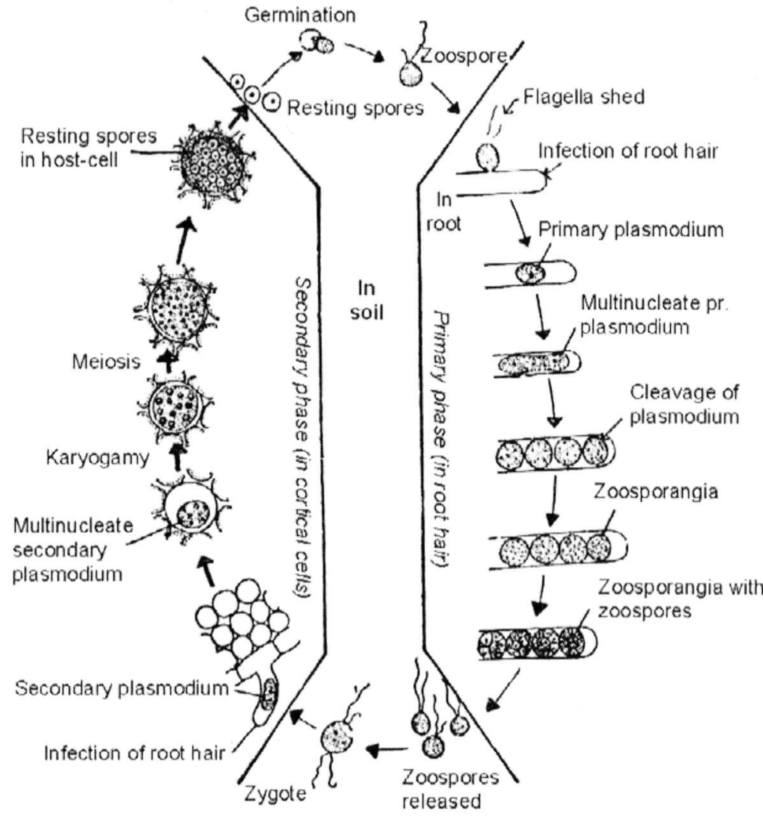

Fig. 22.3. Life cycle of *Plasmodiophora brassicae.*

The Primary Phase Infection of the root hairs takes place during the seedling stage. Restining spores (cysts) lie dormant in the soil and can survive for several years. Germination is stimulated by root exudates of the host. During germination, a pore is dissolved on the wall of the resting spore, through which an apically biflagellate zoospore comes out. The zoospore starts swimming with the longer flagellum backward, and the short flagellum forward. It settles down on root hair and penetrates to give rise to a uninucleate **primary plasmodium** inside the root hair. The plasmodium, by growth and mitotic divisions of its nucleus, becomes large and multinucleate. It acts as a **zoosporogenous plasmodium**. By cleavage, multinucleate segments are delimited, each of which develops into a zoosporangium. Thus, the plasmodium becomes a sorus consisting of several sporangia. Each sporangium forms 4–8 uninucleate, anteriorly-biflagellate zoospores, which are released through exit tubes or through pores. Whether the zoospores reinfect the host, is not known. The zoospores fuse (isogamy) and form a binucleate quadriflagellate motile zygote. Karyogamy does not follow plasmogamy immediately.

The secondary phase. The zygote settles on the root hair, sheds the flagella and enters the cells of the cortex. It forms a multinucleate secondary plasmodium. Karyogamy takes place, and the **secondary plasmodium** becomes a diploid plasmodium. It acts as a **cystogenous plasmodium** forming resting spores. Meiosis follows, and the resulting haploid nuclei multiply by mitotic divisions. By cleavage of the cytoplasm, and wall formation, several uninucleate haploid resting spores are formed. The secondary plasmodium induces hypertrophy of the infected cells. The club-shaped roots develop during the secondary phase. After death and decay of the roots, the resting spores (cyst) enter the soil, where they perennate, until the host is available.

SUMMARY

These organisms belong to the slime mould lineage Plasmodiophorids, characterized by a unique 'cruciform' cell division. These are obligate endoparasites in plants, algae and fungi. *Polymyxa graminis* serves as a vector of several plant pathogenic viruses. The thallus is a plasmodium, which develops only in plant cells as obligate parasites. Except the re-infection phase that occurs in water, the life cycle is completed entirely within the host plant. The plasmodium is designated as zoosporogenous or cystogenous, depending on whether it produces zoospores or cysts (=resting spores). On germination the resting spores give rise to amoeboid bodies or zoospores. The plasmodiophorids differ from other slime moulds in (1) parasitic (not saprobic) nutrition, (2) lack of phagptrophy, (3) chitinous cyst wall (not cellulosic), (4) presence of Golgi bodies in plasmodia, and (5) presence of zoospores/resting spores.

The genera are identified by the (I) morphology of the resting spores, (ii) features of the cystosorus and sporangia, and (iii) range of hypertrophy induced. *P. brassicae c*auses the 'club-root' of crucifers. The life cycle consists of two phases: the primary phase (in root hairs) and the secondary phase (in cortex). The hypertrophy and hyperplasia, which result in the formation of club-like thick roots, are associated with the secondary phase.

TEST QUESTIONS

1. Describe in brief the life cycle of *Plasmodiophora brassicae*.
2. Answer in brief:
   a. Which plasmodiophorid serves as a vector of plant viruses?
   b. On which features are the plasmodiophorid genera identified?
   c. What is the unique feature of the lineage Plasmodiophorids?
   d. How is primary phase in *P. brassicae* started and in what does it end?
   e. How is the secondary phase in *P. brassicae* started, and how it ends?
   f. During which phase is the hypertrophy induced in 'club root' of crucifers?
3. Distinguish between:
   a. Plasmodiophorids and other slime moulds,
   b. Primary and secondary plasmodia,
   c. Resting spores of *Plasmodiophora* and other plasmodiophorids.

24

Symbiotic Associations of Fungi

I. The Mycorrhizae

SYMBIOSIS

De Bary (1887), who coined the term symbiosis, defined it as "Phenomena of living together of dissimilar organisms". The "living together" could vary from the mere side-by-side existence to intimate anatomical connections between the two dissimilar partners, unable to grow independently. De Bary distinguished various forms of symbiosis viz. **mutualism**, in which both partners benefit from the association, and **parasitism**, in which one of the two partners lives at the expense of the other. The concepts of mutualism and parasitism are difficult to be divorced from each other. The existence of pure mutualism, in which both partners would benefit equally, was doubted by De Bary. There is yet another type of "growing together" or symbiosis, called **commensalism**. It is an association in which only one of the partners benefits and there is no disadvantage to the other partner.

Although many biologists have accepted symbiosis as enunciated by De Bary, some restrict it only to strict mutualism. It is in the latter restricted sense, that we shall study symbiosis. A basic criterion of symbiosis (**mutualism**) is that the association is a permanent feature, established during the long process of evolution, and is not a casual association. Permanent association implies physical contact between the two associating organisms, which brings about physiological interplay in form of movement of metabolites (unilateral or bilateral), and increase in the ecological amplitude of the composite organism, the **holobiont**.

MYCORRHIZA (The Fungus Root)

(mykes =fungus, rhiza = root)

Under agricultural field conditions crops do not, strictly speaking, have roots, they have mycorrhiza. S. Wilhelm (1966)

Mycorrhiza (pl. mycorrhizae or mycorrhizas) is defined as a symbiotic, or feebly pathogenic association of the fungus and roots of plants. The mycorrhizal association is found in about 95% of plant species (Morten, 1996). The structure and composition of the mycorrhizae are highly variable. Though numerous fungi live in intimate contact with plant roots, only a few enter into symbiotic association. The mycorrhizae were discovered by the German botanist, Frank, in 1885, but as happens with every new idea, this too had its detractors. McDougal (1922) stated that the mycorrhizae were infrequent and abnormal pathological structures. But, the work of several scientists later proved that mycorrhizae were beneficial to plants and were prevalent throughout the world.

Types of Mycorrhizae

There are six types of mycorrhizae viz. (1) ectomycorrhizae, (2) vesicular-arbuscular mycorrhizae VAM (now called simply arbuscular mycorrhizae AM), (3) ericoid, (4) orchid, (5) arbutoid, and (6) monotropoid mycorrhizae. The six types of mycorrhizae are classified into three broad categories—**ectomycorrhizae**, **endomycorrhizae** and **ectendomycorrhizae** (The suffix trophic, e.g. endotrophic, is now not used). In **ectomycorrhizae**, the fungus forms a **sheath (fungal sheath** or **mantle**) around the rootlets and also enters the root, forming in the outer cortex, an intercellular net of hyphae, called **Hartig net**. The hyphae of the Hartig net do not enter root cells. In the **endomycorrhizae**, which is the predominant type, includes the ericoid, orchid and AM mycorrhizaae. The fungus invades and forms finely divided haustorial branches (arbuscules) or fungal coils (pelotons) in the cells of the root cortex. A portion of the fungus, called external or extraradical hyphae, grows in the soil, but external sheath is not formed. The **ectendomycorrhizae,** which include the arbutoid and monotropoid mycorrhizae, share characteristics of both ecto-, and endomycorrhizae. They form the hyphal sheath and intercellular, as well as, intracellular hyphae.

Mutual Benefits of Mycorrhiza

The fungal partner benefits by the continuous supply of carbohydrates by the photosyntheic plant.

The plant benefits by:

(a) Increased supply of minerals e.g. (i) of **phosphorus** in VA infections, (ii) of **nitrogen** in ericoid infections, and (iii) of both **nitrogen** and **phosphorus** in ecto-, arbutuoid, monotropoid and probably orchid mycorrhizae.

(b) Disease and drought resistance

(c) Colonization of barren soil, and

(d) Resistance to toxicity

The mechanism of increased supply of minerals is both physical and chemical. The mycorrhizal mycelia are much thinner than the finest rootlet and thus can expose a greater volume of soil and providing a larger surface area for absorption. The mycorrhizae can accumulate ions from very dilute solutions. The better growth of mycorrhizal plants is due to the increased absorption of K, P and N. The most conspicuous increase in absorption is for phosphate, which is commonly deficient in natural soils. The mycorrhizae formed between orchid *Goodyera repens* and the basidiomycete *Ceratobasidium cornigerum*) showed 100 times greater rate of uptake of phosphorus than the non-mycorrhizal plant. It is now a common practice in forestry that, whenever trees are introduced into new habitats, they are inoculated with mycorrhizal fungi, to ensure normal good growth without need of excessive fertilizers.

Ericoid mycorrhizae form the dominant flora in climatically and edaphically (soil-related) stressed environments. These plants, characteristically, become dominant when levels of acidity become extremely low and mineralization of nutrients is inhibited. There is evidence that species of *Calluna* and *Vaccinium* have very high tolerance to some stresses.

Mycorrhizal fungi play a protective role for plants in soils contaminated with toxic metals. Pine trees inoculated with *Pisolithus tinctorius* (*Sclerodermataceae*, *Agaricomycotina*) planted in several toxic soils displayed high metal tolerance.

ECTOMYCORRHIZAE (Sheathing mycorrhizae)

The ectomycorrhizae occur only in about 3% of plant species, majority (90%) of which are temperate forest trees, viz. pines, spruces, larches, firs, oaks, beeches, birches, sweet chestnut, eucalyptus etc. There is a general view that ectomycorrhizae are rare in tropics, but this can be an underestimation, as knowledge of the mycorrhizal states of angiosperm trees in the tropics, is extremely limited. The associated fungi belong to *Basidiomycota* and occasionally to some species of *Endogone* (*Zygomycota*: *Mucoromycotina, Endogonales*).

Morphology and Anatomy

The ectomycorrhizae are conspicuous and easily recognizable. The **fungal sheath** gives the roots a swollen and stumpy look (Fig. 24.1). It is a felt-like mass of pseudo-parenchymatous hyphae that virtually acts as a water sponge. It takes the place of root hairs, which are physically prevented from growing out. The entire root, including the apex, is often enclosed by the sheath, so that there is no direct contact between the young roots and the soil. Thus, the sheath insulates the host from the soil and all minerals absorbed by mycorrizal roots, must pass through the sheath.

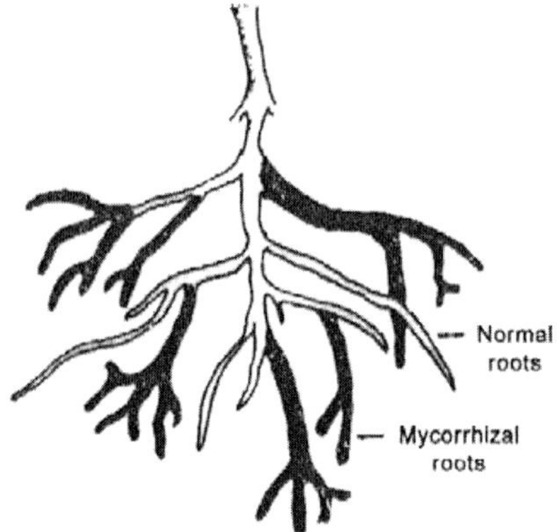

Fig. 24.1. An ectomycorrhizal root

While the uninfected rootlets remain unbranched and short lived, the mycorrhizal roots, by repeated branching, form a crowded cluster of swollen stumps that persist for several years. The new rootlets that emerge from the mycorrhizal roots become infected during their passage through the cortex and the hyphal sheath. The two partners, the fungus and the host root, grow in an organized manner with equal pace. Occasionally during spring season, some of the root apices grow out of the hyphal sheath and extend beyond it, though temporarily. The ways in which the mycorrhizal branches grow and ramify is shown in Fig. 24.2. In *Pinus*, the mycorrhizal rootlets branch dichotomously, while in *Fagus* and all other species, the branching is racemose.

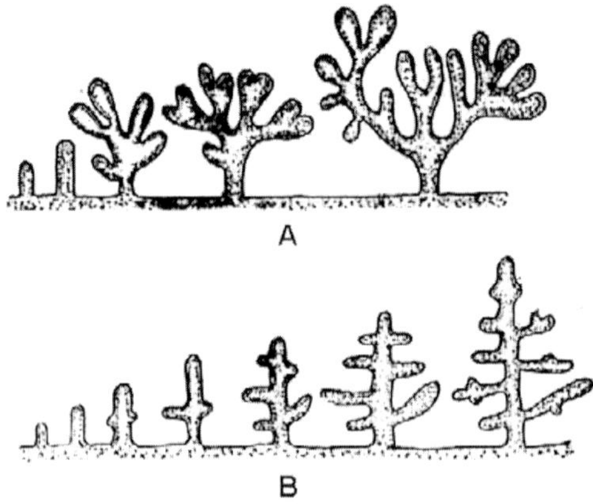

Fig. 24.2. Showing modes of growth and branching of mycorrhizal rootlets. A, dichotomous (*in Pinus*); B, racemose (in *Fagus* and all other species)

Behind the zone of cell division (i.e. apical meristem), hyphae from the inner side of the sheath, penetrate between the root cortical cells, producing a network of hyphae called the **Hartig net** (named after Hartig, who in 1851 first described it without knowing its fungal nature) (Fig. 24.3). The hyphae never enter the cortical cells. The cells of the cortex, which are in close contact with the Hartig net, are larger and more radially elongated than the healthy cells. The mycorrhizal roots have a poorly developed apical meristem. Hyphae pass out from the hyphal sheath and ramify among the soil particles.

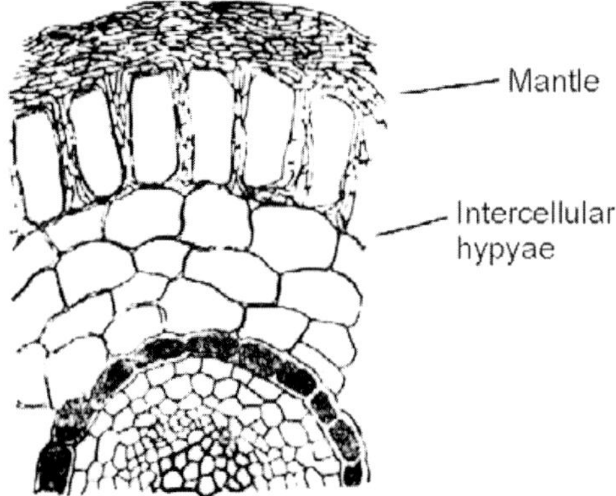

Fig. 24.3. T.S. of an ectomycorrhiza showing 'Hartig net'.

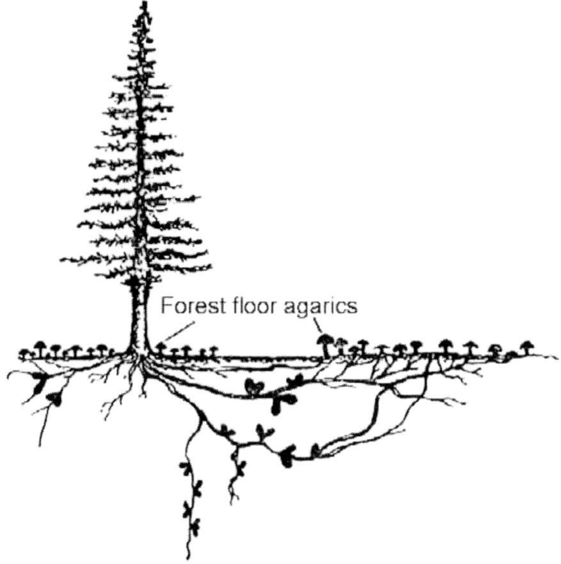

Fig. 24.4. Shows formation of forest-floor agarics

Depending on the fungal species, various secondary structures, in addition to radiating absorptive hyphae, may be found associated with the sheath. These hyphae give rise to the basidiomata above ground, during fruiting seasons (Fig. 24.4). The ectomycorrhizae are differentiated on the basis of their mode of ramification, colour and the structure of the sheath.

The Ectomycorrhizal Fungi

The fungi, associated with the ectomycorrhizae, are mostly *Basidiomycota* (*Agaricomycotina*) and two species of *Endogone* (*Zygomycota, Mucoromycotina*). The common ectomycorrhizal basidiomycetous fungi (belonging to *Agaricomycetes*) are the following:

Amanita, Boletus, Cantharellus, Cortinarius Gamphidius, Entoloma, Hebeloma, Inocybe, Lactarius, Paxillus, Pisolithus, Russula, Rhizopogan, and *Scleroderma.*

While some of these fungi are restricted only to a few plant species, others have a wide host range. Also, many species form mycorrhizae on a single tree (multiple mycorrhiza) e.g. more than 40 fungal species are known to form mycorrhiza with *Pinus sylvestris*.

The persistent association of basidiomata of mushrooms with specific plants gives a clue to their possible symbiotic mycorrhizal association. However, it can be confirmed only by growing the fungi in pure culture, and inoculating them into healthy plants, which are grown aseptically. More than a hundred species of common basidiomycetes have been shown to be mycorrhizal by the method of artificial inoculations.

The mycorrhizal fungi are not easy to grow in pure culture, and the basidiomata are especially difficult to induce. This makes the identification of these fungi difficult. These fungi (1) grow very slowly in culture (2) require exogenous supply of amino acids, especially thiamine, and vitamins, (3) have poor "competitive saprobic ability", compared to other soil fungi, and (4) are susceptible to toxins present in humus and leaf litter. Most of these fungi are unable to utilize complex, carbohydrates like cellulose and lignin, widely present in soils. It was proved by Frank, the discoverer of mycorrhiza himself that although plants with ectotromycorrhiza may grow faster than uninfected ones, mycorrhizal infection is not an absolute necessity, especially if the mineral supply in the soil is quite high. The mycorrhizal fungi, therefore, are useful only in mineral deficient soils.

The minerals that are absorbed by the fungal sheath from the soil are not immediately transported to the tree. As much as 80–90% of the absorbed elements are accumulated in the sheath and only a restricted amount is supplied to the plants, that too in installments. However, when the mineral supply of soil is exhausted, the sheath supplies greater amounts of minerals to the plant. The movement of minerals from the fungus to the plant, is a metabolically controlled process, dependent on adequate supply of oxygen and optimum temperature.

The morphological changes that take place in host roots due to fungal association, are brought about by the action of auxins and other fungal metabolites. Auxins are essential for branching of the mycorrhiza. It also stimulates hydrolysis and translocation of starch from places of storage to site of auxin production (the fungus). Thus, the auxin production by the fungi may account for the high carbohydrate level in the fungal sheath.

Initiation of Ectomycorrhiza

The symbiotic association starts with the germination of the fungal spores under influence of exudates of the host plant, usually at the seedling stage. Melin (1963) observed that some mycorrhizal fungi made only scant growth in absence of the root exudates. Other factors required for mycorrhizal establishment are, (i) mineral-deficiency in soil, and (ii) high carbohydrate level in the host. Mycorrhizae develop only under conditions of high light intensity (Fig. 24.5), which increases the carbohydrate level in the plant. A high carbohydrate level is pre-requisite for exudation of sugars and chemical factors responsible for germination and growth of ectomycorrhizal fungi.

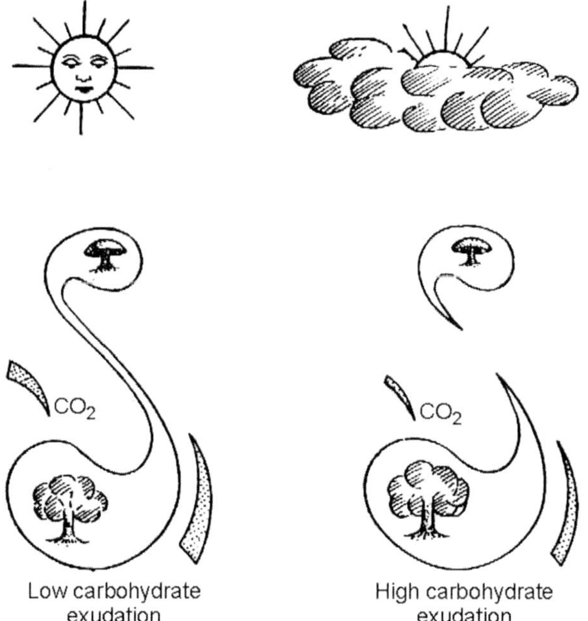

Fig. 24.5. Mycorrhizae develop only under high light intensity when the carbohydrate status of the plant is high.

ENDOMYCORRHIZAE (Ericoid, Orchid and Arbuscular Mycorrhizae)

The fungi in endomycorrhizae invade the cells of the root cortex, with a portion lying externally, as a loose mass of hyphae in the soil. The endomycorrhizae are separated into three types: **ericoid** mycorrhizae, **orchid** mycorrhizae and **arbuscular** mycorrhizae **(AM)**.

Ericoid Mycorrhizae are formed between ascomycetous fungi and plants of the family *Ericaceae* and *Epacridaceae* (order *Ericales*). Members of *Ericaceae*, besides ericoid, also form **arbutoid** and **monotropoid** mycorrhizae. The important plant genera showing ericoid mycorrhizae are *Calluna, Erica, Rhododendron* and *Vaccinium*. The fungus grows loosely over the lateral root hairs and the hyphae penetrate the single layer of cortical cells, often at several points and fill them with hyphal coils, called **pelotons** (sing. peloton). The stele is never penetrated. The fungus can occupy as much as 80% of the total volume of heavily colonized roots.

Ericoid mycorrhizae have coevolved with plants, which form the dominant flora in climatically and edaphically (soil-related) stressed environments. These plants, characteristically, become dominant when levels of acidity leading to decreased mineralization and consequent mineral deficiency in the soil. There is evidence that species of *Calluna* and *Vaccinium* have very high tolerance to some stresses. The formation of ericoid mycorrhizae increases the tolerance and their presence is essential for survival of the plant.

Orchid Mycorrhizae. Orchids must form mycorrhiza. They are **obligately dependent** on mycorrhizal association for seed germination and further development. These mycorrhizae are formed between plants of the *Orchidaceae* and fungi belonging to *Basidiomycota* (including *Rhizoctonia*) and rarely *Ascomycota*. Orchid mycorrhizae lack external sheath, but hyphae penetrate the soil around the infected root. The fungi form intracellular coils of hyphae, pelotons, and usually persist only for a short period.

The plant growth and uptake of phosphorus and nitrogen are enhanced in mycorrhizal orchids (Alexander *et al.,* 1984). The mycorrhizae formed between orchid *Goodyera repens* and the basidiomycete *Ceratobasidium cornigerum*) showed 100 times greater rate of uptake of phosphorus than the non-mycorrhizal plant. There was also a decrease in root:shoot ratio due to greater nutrient availability.

Seeds of orchids are typically minute and are produced in enormous numbers (up to several millions in a single capsule). These have undifferentiated small embryo and with little or no reserve food material. After a few cell divisions the food supply gets exhausted. At this critical point, the fungal symbiont infects the embryo. The infection occurs through specific entrance cells in the suspensor when the imbibed seeds rupture. Only seeds with infected embryos germinate. Growth and cell division is dependent on external source of carbohydrate, which is supplied by the fungus. This supply of carbohydrates continues until the orchids form their own leaves and become autotrophic. The symbiosis, exemplified by the orchids, which after sometime become independent, is called **phase separation symbiosis**. As soon as the orchid becomes an independent photosynthetic unit, the carbohydrate diffusion gradient is reversed (Fig. 24.6) and the orchid starts supplying food partly to the fungus.

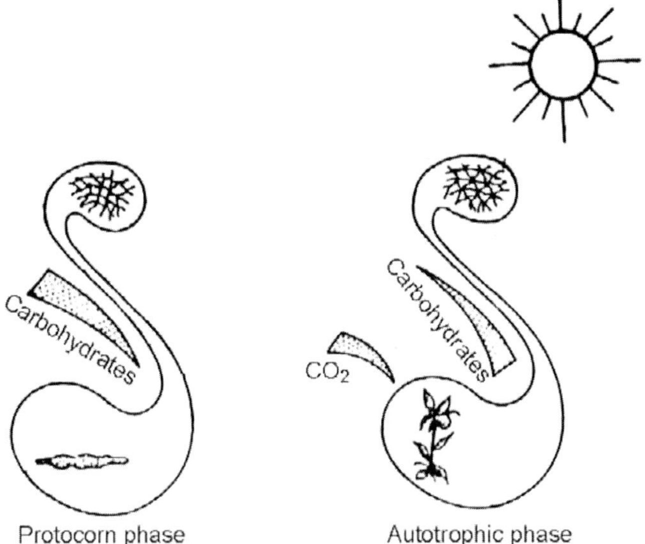

Fig. 24.6. Showing "Phase separation symbiosis" in orchids. In protocorm phase the orchid receives carbohydrates from the fungus; in autotrophic phase it "pays off"

The fungi that form orchid mycorrhizae are mostly basidiomycetous genera e.g. *Armillaria mellea*, species of *Fomes, Xerotinus, Corticium* and *Marasmius*. The *Rhizoctonia*, associated with these mycorrhizae are identified as anamorphs of *Corticium*, *Sebacina* and *Tulasnella.* Unlike the fungi of ectomycorrhizae, these fungi are capable of utilizing complex carbohydrates, like cellulose and lignin in soil. Consequently, these fungi have good "competitive saprobic ability" and are successful "soil inhabitants".

The fungus penetrates cells of the embryo and later of the cortex, forming pelotons. The hyphae and the penetrated host cells, both swell. Later, the hyphae are digested by the host cells. This appears to be an attempt on part of the host to forestall the fungus to become aggressive and parasitic. Gaumann (1950) reported the production of a phenolic substance, called **orchinol** by the resistant cells of the orchid, *Orichis millitaris*, which kills the fungus.

Arbuscular Mycorrhizae (AM) are defined as associations formed by glomeromycotan fungi in plants that form arbuscules and often vesicles. These are ubiquitous and occur in 85% of all plant families, including Bryophytes, Pteridophytes, Gymnosperms (except *Pinaceae*), and Angiosperms. Large numbers of agricultural crops form AM. Fossil evidence and DNA sequence studies suggest that this mutualism appeared 400-460 million years ago when the first plants were colonizing land.

An important feature of these mycorrhizae is the formation by the fungus, of two characteristic organs in the root cortex (Fig. 24.7): (1) **arbuscules** (dichotomously branched, bush-like haustoria) and (2) **vesicles**, formed intra-, and intercellularly. Since vesicles are not formed by some genera (*Gigaspora* and *Scutellospora*), the term AM (arbuscular mycrrhiza) is now more commonly used. Some external hyphae extend into the soil up to a distance of about 1 cm around the roots. Sometimes, these external hyphae grow closely appressed to the root surface, but they never form a mantle. The roots do not show any morphological changes.

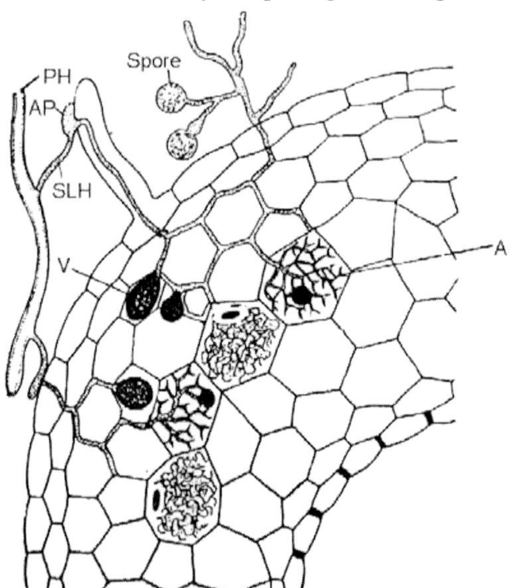

Fig. 24.7. T S. of a vesicular-arbuscular mycorrhiza. (A. arbuscules, AP. appresorium, PH. permanent hyphae, SLH. short-lived hyphae, V. vesicle)

Arbuscules are formed shortly after infection. These are later digested by the host cells. First the tips are eroded and soon the entire arbuscule is degraded and digested. The host nucleus responds to the formation of arbuscules by swelling in size to attain double the normal size. The nuclei revert to normal size with the degradation of the arbuscules.

Vesicles are spherical or oval, thick-walled structures, formed terminally in the intercellular spaces or occasionally in the cortical cells. These contain fat granules and serve as storage organs. In old mycorrhizal roots, the vesicles are produced in enormous numbers, which completely obliterate the cortex cells. The roots, in extreme cases, may also get deformed.

The external hyphae or **extraradical hyphae** are characteristically dimorphic and are composed of (1) 'thick-walled' or 'permanent hyphae' (20–30 μm wide), and (2) 'thin-walled', or 'short-lived hyphae' (2–7 μm wide), which arise as lateral branches of the former. The external hyphae bear various kinds of spores (thick-walled spores and sporocarps), in the soil. The hyphae form appresoria at the site of penetration of the root (i.e. on root hair or epiblema cells). The fungus invades the cortex but never encroaches upon the endodermis, stele or meristem.

The AM Fungi

The AM fungi belong to phylum *Glomeromycota*. There are twelve genera placed in 4 orders and 9 families (see Chapter 8: Coenocytic AM fungi). A few more common are the following:

Glomus, Sclerocystis: Family *Glomeraceae*

Acaulospora, Entrophospora: Family *Acaulosporaceae:*

Gigaspora, Scutellospora: Family *Gigasporaceae:*

While the four genera, belonging to *Glomeraceae* and *Aculosporaceae*, form both arbuscules, and vesicles, the two genera under *Gigasporaceae* (*Gigaspora* and *Scutellospora*) form only arbuscules (and not vesicles). The genera are identified on the basis of formation of chlamydospores, auxillary cells, spores and sporocarps (see fig. in Chapter 8).

ECTENDOMYCORRHIZAE (Arbutoid and Monotropoid Mycorrhiza)

Mycorrhizae, sharing characteristics of both, ecto-and endo-mycorrhizae, are called ectendomycorrhizae. The fungus forms hyphal sheath and the Hartig net, like the ectomycorrhizae, and haustoria or pelotons like the endomycorrhizae. There are two types of ectendomycorrhizae: arbutoid and monotropoid.

Arbutoid Mycorrhizae. These associations are formed between members of plant families *Ericaceae* (*Arbutus*, *Arctostaphylos* and *Arctous*) and *Pyrolaceae* (*Pyrola*), of the order *Ericales*, and fungi belonging to *Basidiomycota*, that are mainly species that form ectomycorrhizae with forest trees. There is a well-developed fungal sheath and a Hartig net, in between the outer cells. Extensive penetration of the outer cortical cells occurs and pelotons fill the cells.

Monotropoid Mycorrhizae. One of the best examples of ectendomycorrhizae, is the mycorrhiza of *Monotropa indica*, the Indian pipe (Fig. 24.8), which grows on forest floors under *Fagus, Pinus, Quercus*, *Carpinus*, *Salex* and other trees. Earlier, *M. indica* was thought to be a root parasite of these trees, but it was disproved when no tissue connections could be found between the roots of the two plants. Now, it is known for sure that *Monotropa* and the associated tree are connected by a common mycorrhizal fungus, *Boletus* (*Agaricomycetes*, *Boletaceae*) (Fig. 24.9). The fungus forms ectendomycorrhiza with *Monotropa* and ectomycorrhiza with the trees.

Fig. 24.8 *Monotropa indica*, the 'Indian pipe', a non-chlorophyllous plant receives nutrients from a green plant through a 'fungal bridge'

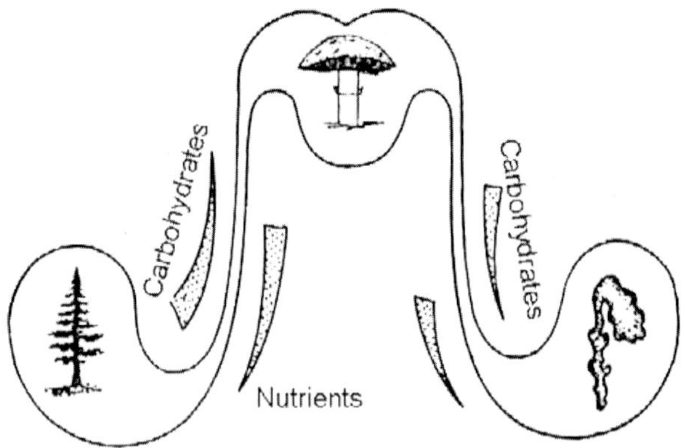

Fig. 24.9. *Monotropa* symbiosis. The fungus serves as a bridge in the conifer-*Boletus*-*Monotropa* association

In *Monotropa*, the fungus forms an external hyphal sheath and the Hartig net, one cell deep in the root cortex. It also forms invaginated structures, called **hyphal pegs**, inside the cortical cells. These resemble

transfer cells formed in a variety of plant species, when there is need for rapid short distance transport of minerals.

Bjorkman (1960), by using tracer carbon technique, proved that the mycorrhizal fungus receives nutrients from the trees and transports it to *Monotropa*. If labeled sugars were introduced into the phloem of the trees, some of them found their way into the fungus, and from there into *Monotropa.* Injection of ^{32}P-phosphate to the tree phloem resulted in similar translocation. The *Monotropa* symbiosis is, thus, a **three-tier (tripartite) system**. The fungus *Boletus* serves as a 'fungal bridge', connecting two plants, one chlorophyllous (a tree) and *Monotropa* a nonchlorophyllous plant.

REFERENCES

Alexander, C. *et al.,* 1984. Phosphate uptake by *Goodyera repens* in relation to mycorrhizal infection. *New Phytol.* 97: 401–411.

Bjorkman, E. 1960. *Monotropa hypopytis* L. an epiparasite on tree roots. *Physiol. Pl.,* 13: 308–329.

Cleland, R.E. 1961. The relation between auxin and metabolism. In: *Handbuch der Pflanzenphsiolgie* (W. Ruhland, ed.), **14:** 754–783. Springer, Berlin.

De Bary, A. 1887. Comparative Morphology and Biology of the Fungi, Mycetozoa and Bacteria. English translation. Clarendon Press, Oxford.

Frank, A.B. 1885. (Title in German) Ber. Deut. Botan. Ges., **3:** 128.

Gaumann, E., Braun, R. and Bazzigher, G. 1950. (In German). Phytopathol. Z., **17:** 36–62.

Gerdemann, J.W. 1968. Vesicular-Arbuscular Mycorrhiza and Plant Growth. Annu. Rev. Phytopathol. **6:** 397–418.

Hartig, T. 1851. (In German). Vlgs. Buchhdlg. A. Forstner, Berlin.

Mc Dougal, W.B. 1922. Mycorrhizas of coniferous trees. Jour. For., **20:** 255–260.

Melin, E. and Nilsson, H. 1957. Transport of C^{14}–labeled photosynthate to fungal associate of pine mycorrhiza. Syensk. Bot. Tidskar. **51:** 166–186.

Morten, J.W. 1996. Mycorrhizae. In: McGraw-Hill 1997—Year Book of Science & Technology. pp. 324–327. McGraw-Hill Co., N.Y.

Wilhelm, S. 1966. Chemical treatments and inoculum potential of soil. Annu. Rev. Phytopathol., **4:** 53–78.

SUMMARY

Mycorrhiza is a symbiotic or feebly pathogenic association of the fungus and roots of plants. It is found in about 95% of plant species, 85% being contributed by the arbuscular mycorrhizae (AM). Fossil records and DNA sequences indicate that mycorrhizae appeared 400-460 million years go when first plant appeared on earth. The mycorrhizae were discovered by the German botanist, Frank, in 1885.

The are six types of mycorrhizae classified into three broad categories — **ectomycorrhizae**, **endomycorrhizae** and **ectendomycorrhizae**. In **ectomycorrhizae**, the fungus forms a **sheath** (or **mantle**) around the rootlets and also enters the root, forming in the outer cortex, an intercellular net of hyphae, called **Hartig net**. The hyphae of the Hartig net do not enter root cells. In the **endomycorrhizae**, which is the predominant type, includes the ericoid, orchid and AM mycorrhizae. The fungus invades and forms finely divided haustorial branches (arbuscules) or fungal coils (pelotons) in the cells of the root cortex. A portion of the fungus lives in the soil, but external sheath is not formed. The **ectendomycorrhizae,** which include the arbutoid and monotropoid mycorrhizae, share characteristics of both. They form the hyphal sheath and intercellular, as well as, intracellular hyphae.

The fungal partner benefits by the continuous supply of carbohydrates by the photosyntheic plant.

The plant benefits by: increased supply of minerals mainly **phosphorus** (in VA infections), **nitrogen** in ericoid infections, and (iii) both **nitrogen** and **phosphorus** in ecto-, arbutuoid, monotropoid and probably orchid mycorrhizae. Mycorrhizal plants also show increased disease and drought resistance, colonization of barren soil, and resistance to toxicity.

The Ectomycorrhizae occur only in about 3% of plant species, majority (90%) of which are temperate forest trees. The associated fungi belong to *Basidiomycota* and occasionally to some species of *Endogone* (*Zygomycota*)

Endomycorrhizae comprise Ericoid, Orochid and Arbuscular mycorrhizae.

Ericoid mycorrhizae are formed between ascomycetous fungi and plants of the family *Ericaceae* and *Epacridaceae* (order *Ericales*). Members of *Ericaceae*, besides ericoid, also form arbutoid and monotropoid mycorrhizae. The formation of ericoid mycorrhizae increases the tolerance and their presence is essential for survival of the plant.

Orchid mycorrhizae are formed between plants of *Orchidaceae* and fungi belonging to *Basidiomycota* (including *Rhizoctonia*) and rarely *Ascomycota*. Seeds of orchids are typically minute and are produced in enormous numbers (up to several millions in a single capsule). Seed germination progresses only if the embryo gets infected when the imbibed seeds rupture. Only seeds with infected embryos germinate. The symbiosis, exemplified by the orchids is called *phase separation symbiosis*. As soon as the orchid becomes an independent, it starts starts supplying food partly to the fungus.

Arbuscular Mycorrhizae (AM) are associations formed by Glomeromycotan fungi in plants that form arbuscules and often vesicles. These are ubiquitous and occur in 85% of all plant families.

The AM fungi belong to phylum *Glomeromycota*. Some genera form both arbuscules and vesicles (e.g. *Acaulospora, Entrophospora, Glomus* and *Sclerocystis*), while some others (like *Gigaspora* and *Scutellospora*) form only arbuscules.

Ectendomycorrhizae comprise Arbutoid and Monotropoid mycorrhiza.

Arbutoid mycorrhizae are formed between members of families *Ericaceae* and *Pyrolaceae* of the order *Ericales* and fungi belonging to *Basidiomycota*, that are mainly species that form ectomycorrhizae with forest trees. There is a well-developed fungal sheath and a Hartig net, in between the outer cells. Extensive penetration of the outer cortical cells occurs and pelotons fill the cells.

Monotropoid mycorrhizae is exemplified best by *Monotropa indica*, the Indian pipe which grows on forest floors. *Monotropa* and the tree under which it grows are connected by a common mycorrhizal fungus, *Boletus* (*Agaricomycetes*, *Boletaceae*). The fungus forms ectendomycorrhiza with *Monotropa* and ectomycorrhiza with the trees. The common mycorrhizal fungus conducts nutrients from forest trees to *Monotropa*. The *Monotropa* symbiosis is, thus, a *three-tier (tripartite) system*, involving chlorophyllous and nonchlorophyllous symbionts, and a 'fungal bridge', connecting the two.

TEST QUESTIONS

1. Define mycorrhiza and give characteristics of the various types of mycorrhizae. What are the benefits of the association to plants?
2. Describe the ectomycorrhizae under the following heads: morphology and anatomy, ectomycorrhizal fungi and initiation of ectomycorrhizae.
3. Describe ericoid and orchid mycorrhizae. Explain the importance of the symbiosis to the plant.
4. Describe the arbuscular mycorrhizae and their importance. Which are the associated fungi?
5. What are ectendomycorrhizae? Describe the arbutoid and monotropoid mycorrhizae mentioning the mycorrhizae formed by the fungus with the two plants.
6. Explain the following terms: external (extraradical) hyphae, sheath (mantle), Hartig net, arbuscules, pelotons, vesicles, and 'hyphal peg'

7. Answer in brief.
   a. Which are the most widespread mycorrhizae?
   b. What is the basic criterion of symbiosis (mutualism)?
   c. Who discovered mycorrhizae?
   d. When did mycorrhizae appear on earth in view of fossil record and DNA sequences?
   e. What makes the identification of ectomycorrhizal fungi so difficult?
   f. What is the role of auxins in ectomycorrhizae?
   g. What are the edaphic and plant factors essential for mycorrhizal initiation?
   h. Which fungi are associated with ericoid mycorrhizae?
   i. Why orchids must form mycorrhiza?
   j. How do the ericoid mycorrhizae represent "phase sepeartion symbiosis"?
   k. Which are the fungal partners in arbutoid mycorrhizae?
   l. How does the fungus serve as a bridge in monotropoid mycorrhizae?
   m. Which fungus serves as a fungal bridge in monotropoid mycorrhizae?
8. Fill in the blanks.
   a. Mycorrhizal association is formed in about 95% of the plants of which ....% is contributed by the AM.
   b. The mycorrhizae enhances absorption of ----,-----------, and-------.
   c. The ectomycorrhizal fungi mostly belong to -----------------.
   d. The mineral absorbed by the fungal sheath is mostly accumulated in ----------.
   e. The fungus *Boletus* forms ectendomycorrhiza with ------------, and ectomycorrhiza with the tree.
   f. The 'hyphal pegs' formed in *Monotropa* mycorrhizae resemble ------------------, found in certain plants for short distance transport of minerals.
   g. In *Monotropa* mycorrhizae, the carbohydrates are translocated from------ to -------- and from there to ----------.

25

Symbiotic Associations of Fungi
II. The Lichens

Lichens are organisms formed by symbiotic association of a fungus (mycobiont) and an alga or a cyanobacterium (photobiont). The lichen thallus shows no resemblance to either partner. Their complete integration eluded scientists for long, who considered them as single autonomous organisms. In 1868, a Swiss botanist **Schwendener,** discovered their dual nature. Lichens are identified on the basis of their morphology. Since the fungi and algae or cyanobacterial partners, are organisms in their own right, and genetically autonomous, they deserve their rightful place in fungal and algal or cyanobacterial taxonomy. Thus, lichens are a biological and not a systematic group.

Schwendener, who discovered the dual nature of lichens

The lichen symbiosis is mutualistic, the fungus derives carbohydrates from the photobiont, which in turn, receive water, minerals, and protection (against desiccation and lethal light radiations) from the fungus. Some, however, believe the lichen association to be case of controlled parasitism. The association is not a casual one. It involves complete integration and formation of a morphologically new thallus. The two components are individually incapable of free existence in nature. That their association is "forced" on them by circumstances, is proved by the fact that in rich nutrient solutions, the two components fall apart. Could they grow alone, as they do in laboratory cultures, they would never form the lichen.

Distribution

Lichens grow in varied environments, including such extremes as hot deserts and chilly mountain tops. One example of their resistance to extreme cold is the report (Scholander *et al.,* 1953) that specimens frozen in liquid oxygen at –183°C for 18 hours, respired after thawing. Similar resistance is reported against extreme heat and prolonged drying. *Cladonia pocillum* could survive exposure to 101°C. Dried specimens survive even higher temperatures. Lichens survive long periods of drought. Interestingly enough, they are not found in and around modern cities! It is because lichens cannot withstand air pollution and, therefore, serve as indicators of air pollution by their absence. Lichens colonize rocks and also grow on fertile soils. Tree trunks on hills are the most common site of lichens. In Greenland, Iceland and Alaska, they grow in abundance and serve as fodder for cattle.

The Photobionts

The algal and cyanobacterial partners of lichens are spread over at least 40 genera (25 algae and 15 cyanobacteria). Among photobionts, 8% are cynobacteria and 90% green algae. The most common cynaobacterial partner is *Nostoc* and the most common green algal partner is *Trebouxia* (Fig. 25.1). In pure culture, these do not show any striking difference from non-symbiotic organisms, but their astonishingly slow growth, even in nutrient rich medium, is a noteworthy feature. Although the photosynthetic efficiency is similar to free-living organisms, the fixed carbon is not directed to growth processes. In nature, the algae (except *Trebouxia*) are total failures, hence the recourse to symbiosis. Their cultural requirements are no different from other algae or cyanobacteria. The behaviour of the isolated photobionts in culture is different from what occurs in the lichen thallus. For example, most trebouxiae form zoospores in culture but never in lichens. Again, the gelatio-nous sheath, formed in cultures (also formed by *Nostoc*), is suppressed in lichen association. The reason for this change in behaviour is not difficult to visualize. In lichen associations, the photobiont has to share the photo-synthetic products with the fungus and therefore, unable to store much for production of sheath, pigments or zoospores. Majority of the photobionts show improved growth in culture, when supplied with sugars and organic nitrogen. Some grow even in absence of light, i.e. heterotrophically, if the supply of sugars and organic nitrogen is maintained.

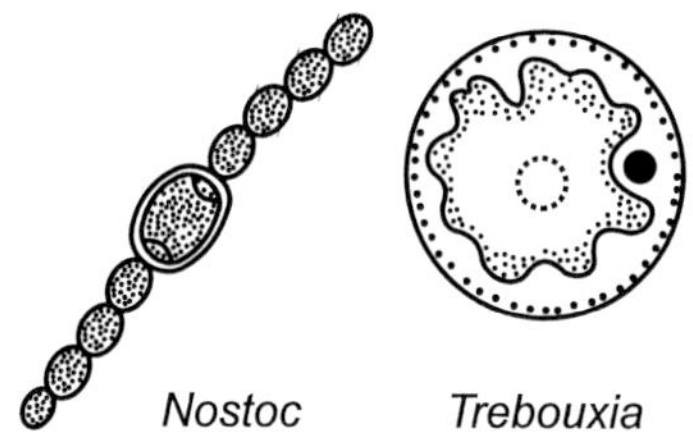

Fig. 25.1. The two common photobionts

The Mycobionts

Mycobionts, or the fungal partners, belong mostly to *Ascomycota*; rarely to *Basidiomycota.* The lichen fungi have never been reported as independently-growing organisms. It is important to note here that 20% of all fungi and 40% of *Ascomycota* are lichenized (lichen-forming) fungi (Kirk *et al.,* 2008). A recent estimate of global diversity suggests that there are 17500-20000 lichen-forming species and 1500 lichenicolous fungi (fungi growing on lichens as parasites, commensals or saprobes). All ascomycetous lichenized fungi belong to subphylum *Pezizomycotina.* Majority of these belong to *Arthoniomycetes, Eurotiomycetes, Lecanoromycetes* and *Lichinomycetes.* Four orders comprise only lichenized fungi. There are many families comprising exclusively the lichenized fungi, while some families and genera have lichenized as well as non-lichenized fungi. Basidiomycetous lichenized fungi belong to *Agaricomycotina,* which are agarics (e.g. *Lichenomphalia*), clavarioid fungi (e.g. *Multiclavula* species) and corticoid fungi (like species of *Dictyonema*).

Some fungi show **optional lichenization** as an ecological adaptation (Sedin *et al.,* 2004). *Conotrema,* a lichenizing fungus, and *Stictis,* a non-lichenizing fungus can live their whole life cycle (including asexual and sexual phases) as lichens or as independent saprobes under different environmental conditions. This optional lichenization allows separate individuals to exploit different ecological niches. Both fungi belong to *Lecanoromycetes, Pezizomycotina*).

Identifying lichen fungi is a difficult task, as they do not produce spores or other characteristic structures. Even if asexual spores are formed occasionally, sexual bodies are never formed. Whether this is due to absence of a compatible mating partner, or presence of rich growth medium, is not known. However, Ahmadjian (1966) successfully induced the perfect stage of *Cladonia cristatella* (a fruticose lichen) in culture.

The lichenized fungi can be grown in axenic culture (with approximately 42% success) from ascospores or from the thallus. The form of the lichen is determined by the genes of the fungus, but its expression requires the presence of the photobiont. In laboratory cultures the fungus, when grown alone, develops into an undifferentiated mass of hyphae. If combined with its photobiont, the characteristic form emerges (Brodo *et al.*, 2001). So morphogenesis (differentiation) is controlled by the photobiont. It is remarkable that the same fungus forms entirely different lichen form in association with different photobiont.

Morphology and Anatomy of the Lichen

The thallus construction may be **homoiomerous** or **heteromerous** (Fig. 25.2). The homoiomerous is primitive type of construction, in which the fungal an algal or cyanobacterial elements lie mixed, inseparable into distinct zone. The more advanced heteromerous construction, resembles a dorsiventral angiosperm leaf.

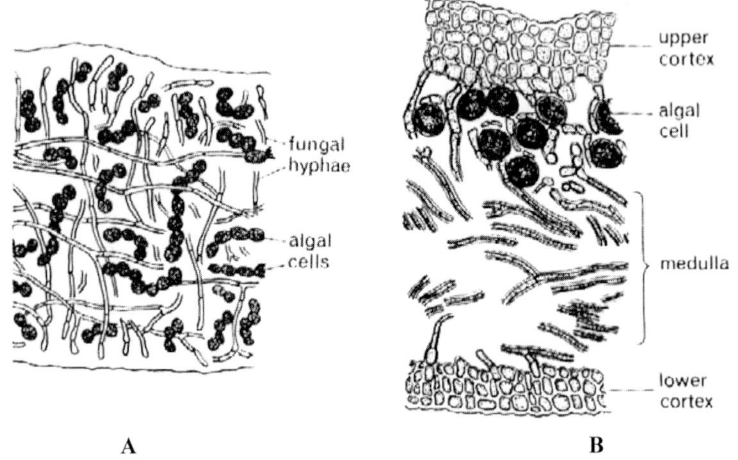

Fig. 25.2. Showing homoiomerous (A) and heteromerous (B) thallus construction

Despite the varied morphology of the thalli, which is mainly controlled by the fungus partner, the lichens are neatly grouped in three broad categories. These are (1) **crustose** - thin, flat, crust-like thalli, lying closely appressed to the rocky substratum, (2) **foliose** - flat thalli, with leaf-like lobes, and (3) **fruticose** – highly branched upright or pendulous thalli.

The **crustose lichens** lie so close to the substratum, that they cannot be separated without damaging the thallus. In some endolithic (under-rock) species of *Lecidea* and *Sarcogyne*, the thallus penetrates the rocks up to several mm. The borders of the thallus are not sharply delimited from the substratum. Anatomically, the crustose lichens are undifferentiated, and the algal or cyanobcterial cells lie mixed among fungal hyphae. *Lecanora, Lecidea, Pertusaria Rhizocarpon* (used in lichenometry) and *Endocarpon* are the important genera.

The **foliose lichens** are dorsiventral, lobed, leafy structures, attached to the substratum by means of **rhizines**. Some of the foliose lichens attain a fairly large size, 30 cm or more in diameter. Anatomically, these lichens show striking similarity with an angiosperm leaf. In a transverse section, starting from above, it shows an upper cortex, an algal or cyanobacterial layer, medullary layer and lower cortex. Except the algal or cyanobacterial layer, all the layers are fungal in nature. The upper and lower cortex are pseudoparenchymatous, while the medullary layer consists of loosely–oriented hyphae. Pores (cyphellae) or hair-like rhizines are found protruding downwards from the lower surface (Fig. 25.3). The **cyphellae** serve for aeration and, to this extent, these are analogous to stomata. But in structure and frequency, they are more akin to lenticels. *Dermatocarpon, Parmelia, Peltigera, Lobaria, Sticta* and *Xanthoria* (used in lichenometry) are important foliose lichens.

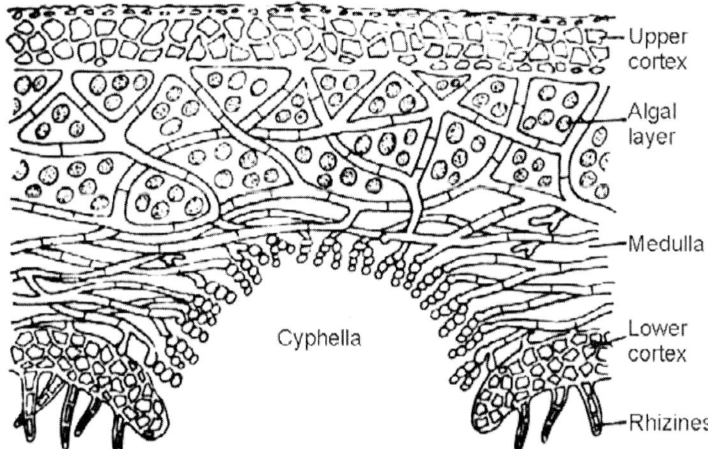

Fig. 25.3. A diagramatic view of vertical section of *Sticta* sp. showing a. cyphella and rhizines

Fruticose lichens, which are shrubby, with finger-like or strap-shaped branches, range from minute, 1-2 mm high species of *Cladonia* (Fig. 25.4), to 5 metres, or more, long *Usnea* (Fig. 25.5). A transverse section reveals a central medulla, formed by lax hyphae, surrounded by more compact mass of pseudocortex, above and below (*see* Fig. 25.2). The algal or cyanobacterial cells lie in patches or in a layer in the pseudoparenchymatous cortex. *Cladonia*, *Cetraria* and *Evernia* are important frutricose lichens.

Fig. 25.4. *Cladonia verticillata,* a fruticose lichen usually 1-2 mm high, with finger-like branches

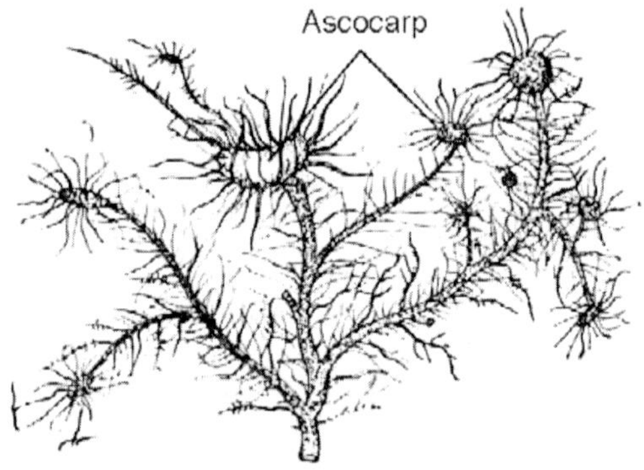

Fig. 25.5. *Usnea barbata.* another fruticose lichen, which is shrubby 5 metres or more in height

Growth of Lichens

The lichens grow at an extremely slow rate, usually less that 1 mm in diameter per year. The slowest to grow are the crustose lichens, whose size and general appearance, according to one study, remained unchanged for 20 years. Species of *Peltigera* are among the fastest growers, with average annual increment of 2-3 cm. The maximum yearly growth of 4.5 cm per year has been recorded for *Peltigera paretextata*, a foliose lichen. The other foliose lichens have annual growth rates of 0.01-1.3 cm.

The slow growth of lichens is ascribed to two reasons (1) the environmental factor, to which they are exposed, permit only brief periods for optimal metabolic activity, and (2) the intrinsic slow growing nature of the component symbionts. However, despite slow growth, lichens form a significant biomass in certain parts of Greenland, Iceland and Alaska. The lichens are fastidious in their growth requirements. They need a low light intensity, cool temperature and moisture, in form of fog. These conditions exist only for a few hours in the morning. Water requirement is the critical factor. The thalli behave like agar gel and absorb moisture from the humid environment in amounts up to 100-300 times more than their dry weight. They also lose the water with the same rapid speed.

Equally surprising is their longevity. The age of some species of *Lecidea* and *Rhizcarpon* has been calculated to range between 1000 to 4500 years; and they are still going strong (Beschel, 1961). Most lichens, however, are not as persistent as the above ones, and are destroyed by the changing microclimates.

Reproduction

Lichens are perpetuated exclusively by asexual reproductive bodies, containing both the components. The fungi do produce conidia and ascospores in pycnidia and ascocarps respectively, but these are of no avail, as their chance of contacting the right photobiont is one in a million.

Asexual Reproduction

(1) **Fragmentation**. Fragmentation of thallus may be natural, due to death and decay of the older portions, or accidental. The fragments give rise to new thalli.

(2) **Soredia** (sing. Soredium). The soredia are minute propagules (25-100 mm in diameter) comprising of a few algal or cyanobacterial cells, wrapped in fungal hyphae (Fig. 25.6). Several soredia adhere to form granular masses on the surface of the thallus, in distinct pustules called **soralium**. Soredia originate inside the thallus and protrude out through pores or cracks, developed in the cortex. Being dry, they are easily blown and disseminated by wind.

(3) **Isidia** (sing. Isidium). These are stalked outgrowths of the thallus, 1-3 mm in diameter and 0.5 to 50 mm in height (see Fig.25.6). These consist

of portions of the upper cortex, enclosing some algal or cyanobacterial cells and fungal hyphae. The isidia, which are formed in great numbers on the upper surface, are separated only by accidental breakage.

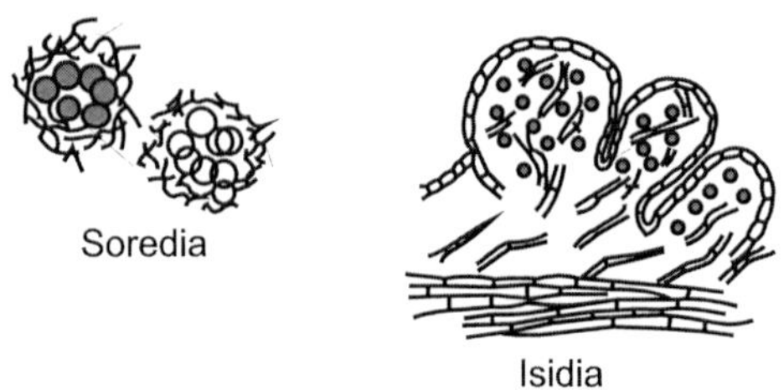

Fig. 25.6. Section through a soralium showing soredia

(4) **Conidia**. Many lichens produce conidia in pycnidia, but, since their getting into contact with the right algal or cyanobacterial partner is a remote possibility, their role in establishing new thalli is negligible.

Sexual Reproduction

Several lichens e.g *Physcia, Parmelia, Usnea* and *Cladonia* form distinct apothecia (Fig. 25.7), which are usually borne on long stalks. The fate of the ascospores, like that of the conidia, is not known. Perhaps their production is an exercise in futility! The photobiont does not show sexual reproduction.

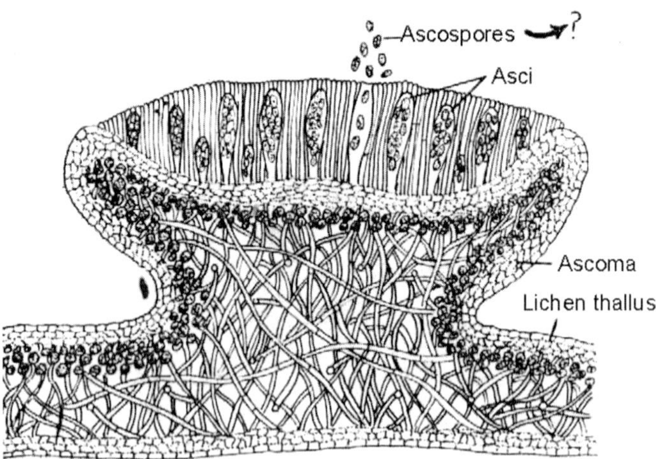

Fig. 25.7. The mycobiont forms large apothecia but their efficacy is doubted

Lichen Acids

Lichen acids are secondary metabolites produced by lichens. The distinctive colour of lichen thalli is due to some of these acids. In sections, coloured lichens show crystals of these acids, deposited on the fungal hyphae, which are easily extractable. Most of the lichen acids belong to chemical groups called **depsides** and **depsidones**, which are condensation products of 2-, or 3 phenyl carboxylic acids, mainly orsellinic or β-orsellinic acids held together by an ester linkage. It is known that these aromatic acids, are formed by the mycobiont and the ester linkage is provided by the photobiont. But *Ramalina crassa*, an isolated mycobiont, brings about both steps in the production of salazinic acid, a lichenic acid.

About 400 lichen acids are reported, of which 230 are found only in lichens, and not elsewhere. Some well-known lichen acids are evernic, gyrophoric, lecanoric, lobaric, physodic, psoromic, stictic, thamnolic and usnic acids. Lichens produce other aromatic compounds like quinones and pulvic acids, xanthone, and dibenzofuran derivatives.

The role, which these excretory substances play in the life of lichens, is not clear. However, some of the known properties are given below:

(1) Many lichen acids possess antibiotic properties, especially against Gram +ve bacteria. Usnic acid, from *Ramalina scopulorum,* is active against *Staphylococcus aureus* and some other pathogenic organisms. It is proposed that lichen acids may provide protection against invasion of lichens by bacteria. But, since most soil bacteria are Gram-ve, the protective role of lichen acids may not be of much significance.

(2) Lichen acids, because of their bitter taste, serve as defense chemicals, against attacks by slugs and other minute animals. But the giant millipeds, in the tropics, are voracious eaters of coloured lichens, such as *Dermatiscum* and *Acarospora.*

(3) Some lichen acids have good chelating (metal binding) properties and, may be, they help in absorption of minerals from rock particles.

(4) Lichen acids are also assigned role in the translocation of substances between the symbiotic parteners.

(5) The lichen acids are known to increase permeability of algal cells, causing increased exudation of carbohydrates from them.

(6) Being coloured, the lichen acids might be functional in stabilizing the amount of light reaching the photobiont layer.

Importance of Lichens

l. As Initiators of Vegetation on Rocks

Lichens, especially the crustose forms, are of primary importance in the colonization of rocks. The acids, produced by some of them, bring about disintegration and erosion of rock surface. The minerals that are released are utilized by the lichens, which then get a foothold on the barren rock. Their death and decay provide nutritive ground for the growth of other lichens and mosses that extend the colonization.

2. Use as Fodder

The lichens**,** *Cladonia rangiferina* (reinder moss) in tundra, *Cetraria islandica* (Iceland moss) in Iceland, and species of *Lecanora* (carpeting the rocky forests and sea beaches) are used as fodder. The bitter taste (due to lichen acids) is removed by soaking the thalli in water, containing some sodium and potassium carbonate Their nutritive value lies in the presence of **lichenin** (= lichen starch), a carbohydrate allied to starch (Starch and cellulose are not found in lichens).

3. As Food

The commercial Iceland moss (*Cetraria islandica*), which is rich in lichenin, is used as food after removing the bitter chemicals by dipping the thalli in solutions of sodium and potassium carbonate. The thalli are usually dried and sold in form of powder, which, on boiling with water or milk, forms a delicious jelly. The powder may also be used in making soups or other dishes. *Dermatocarpon* (=*Emdocarpon*) *miniatum*, the "stone mushroom" is freely sold in vegetable markets in Japan. In India, one species of *Parmelia* (Fig. 25.8) is eaten in curry.

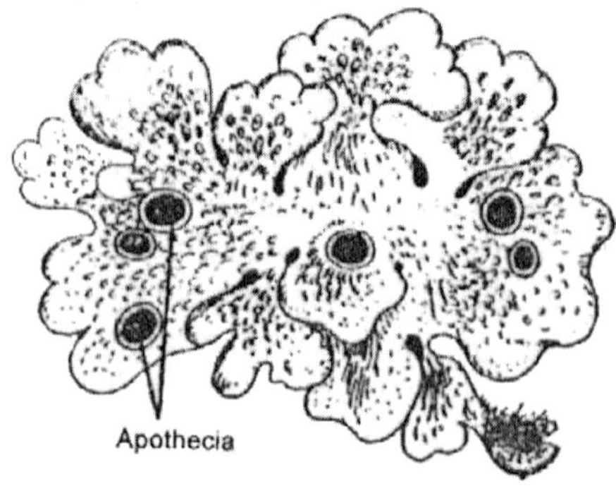

Fig. 25.8. *Parmelia* sp. (a foliose lichen), eaten as curry

4. Medicinal Properites

Some lichen products are antibiotics that are most effective against Gram +ve bacteria. **Usnic acid,** which is used commercially (as Usno, Binan, Usni plant), strongly inhibits *Mycobacterium.* It also inhibits *Neurospora crassa* and this, as well as, lichen extracts inhibit wood-rotting fungi. Sodium asnate is effective against tomato canker (*Corynebacterium michiganese*) while some lichen acids are active against *Trichoderma.*

Lichenin and stringent chemicals, present in the lichens, are of great medicinal value against various ailments. A preparation from *Peltigera canina* (the dog lichen Fig. 25.9) is useful against hydrophobia. *Parmelia saxatilis* (the skull lichen) has a medicinal property, which it is claimed can control epilepsy. *Lobaria pulmonaria* contains chemicals, which are helpful in curing lung infections. Since these medicinal properties have been assigned to lichens after centuries of astute observation by intelligent people all over the world, their efficacy has to be taken seriously.

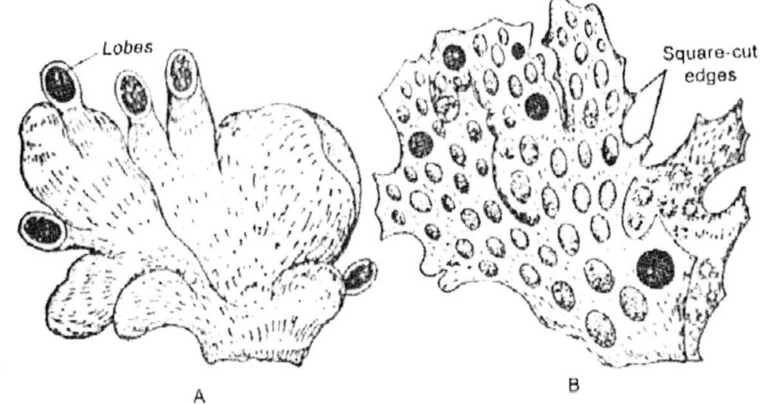

Fig. 25.9. *Peltigera canina* (the dog lichen) and *Lobaria pulmonaria* (tree lung wort), both foliose lichens.

5. Use as Perfume

Several French perfumes, considered best in the world, have ingredients derived from lichens. *L. pulmonaria* is the lichen of choice for this purpose.

6. Preparation of Dyes

The colour (orchill), used in dyeing of woolen and silk fabrics, is extracted from species of *Roccella* and *Lecanora.* The stain **orcein,** used in microscopic preparation, is a purified derivative of orchill. Litmus, the well-known pH indicator, is also extracted from same lichens.

7. Uses in Tanning and Distillaries

The astringent chemicals, extracted from *Cetraria islandica* and *Lobaria pulmonaria*, are used in tanning of leather. *P. lobaria* has also been used in the brewing of beer. *Cladonia rangiferina* and certain species of *Usnea* and *Ramalina*, have been used in the preparation of alcohol.

8. Use in Determining Age of Rocks (Lichenometry)

The exposure age of rocks is determined by the size/diameter (proportional to age) of lichen thalli. The technique though first employed by **Faegri** in 1933, was published in 1950 by **Beschel**. Lichen can be preserved on old rocks for up to 10,000 years, providing the maximum age limit for the technique (with 10% error). But it is most accurate when applied to rocks less than1000 years. The lichens most commonly used in the technique are those of the ascomycetous genera *Rhizocarpon* (*R. geographicum*), a crustose lichen and *Xanthoria elegans* (Fig. 25.10) a foliose lichen.

Fig. 25.10. *Xanthoria* sp.

9. Harmful Effects

Lichens frequently cause fire. During hot months, the dry lichens on tree trunks catch fire quickly, which is soon transmitted to the tree tops. Some species of *Cladonia* and *Amphiloma* parasitize and kill mosses.

Resynthesis of Lichens in the Laboratory

The lichen fungi and algae or cyanobacteria, are not obligate symbionts. Given suitable conditions, they can live as independent organisms. When both components are grown together in nutrient-rich solutions, the first reaction is that of parasitism of the alga or cyanobacterium by the fungus. Thus, when the mycobiont of the lichen *Collema tenax*, was grown in nutrient rich medium with *Nostoc*, or with any other photobiont, the algal/cyanobacterial cells died. The dead cells were found stuffed with hyphae. The symbiotic association never ensued. However, in a nutrient-deficient medium, the two partners are forced into symbiotic association. Ahmadjian, who successfully synthesized *Acarospora fuscata* (1962) and *Cladonia crisystella* (1966) in laboratory, found that the most important conditions for including lichen associations, are nutritionally-poor growth medium and slow drying (see Ahmadjian, 1993). The medium, which failed to support growth of either individual separately, supported joint growth, leading to lichen synthesis.

Nomenclature and Classification

Lichens do not have independent scientific names. The myco- and photobionts, each have separate names, but the lichen is referred by the name of the fungus only. The classification of the lichens has, therefore, has to be integrated with the classification of the fungi. This is because lichens are recognized as a nutritional, rather than a taxonomic group.

REFERENCES

Ahmadjian, V. 1993 *The Lichen Synbiosis*. John Willey. NY.

Beschel, R.E. 1961. Dating rock surfaces by lichen growth and its application to glaciology and physiography (lichenometry). In: *Geology of the Arctic*. (G.O. Raascgm ed.). pp. 1044–1062 of Toranto Press, Toranto.

Brodo, I.M. 2001 The lichens of North America. Yale Univ. Press

Kirk *et al.* 2008 *Ainsworth and Bisby's Dictionary of the Fungi* C.A.B. International, Oxon, U.K.

Scholander, P.E. *et al.*, 1953. Studies on the physiology of frozen plants and animals in the arctic. *J. Cellular Comp. Physiol*. 42: Suppl. 1.1.56

Schwendener, S. 1868. (Title in German). *Bot. Z.*, 26:289–292.

Sedin, M. *et al.* 2004 Saprotrophy and lichenization as options for the same fungal species in different substrata : environmental plasticity and fungal life styles in the *Stictis-Conotrema* complex. New Phytologist. 164: 459-465.

SUMMARY

Lichens were considered as single autonomous organisms until 1868 when a Swiss botanist **Schwendener** discovered their dual nature. Lichens are identified on the basis of their morphology, but they are a biological and not a systematic group. Therefore, the fungi and algal or cyanobacterial partners, which are autonomous organisms, deserve their rightful place in fungal and algal or cyanobacterial taxonomy. The lichen symbiosis is mutualistic; the fungus derives carbohydrates from the photobiont, which in turn, receives water, minerals, and protection (against desiccation and lethal light radiations) from the fungus. Among **photobionts**, 8% are cynobacteria and 90% green algae. The most common cynaobacterial partner is *Nostoc* and the most common green algal partner is *Trebouxia*. **Mycobionts** belong mostly to *Ascomycota*; rarely to *Basidiomycota.* 20% of all fungi and 40% of *Ascomycota* are lichenized (lichen-forming) fungi. The form of the lichen is determined by the genes of the fungus, but its expression requires the presence of the photobiont. Some fungi, like *Conotrema* and *Stictis* show optional lichenization.

Lichens are grouped by their form into three categories: **crustose** - thin, flat, crust-like thalli, lying closely appressed to the rocky substratum, **foliose** - flat thalli, with leaf-like lobes, and **fruticose** – highly branched upright or pendulous thalli. The lichens grow at an extremely slow rate, usually less that 1 mm in diameter per year. The slowest to grow are the crustose lichens. This property has been used in **lichenometry** for determination of age of rocks by age of the lichens. Species of *Peltigera* (foliose lichen) are among the fastest growers. The age of lichens, however, is very long. The age of some species of *Lecidea* and *Rhizocarpon* has been calculated to range between 1000 to 4500 years.

Lichens reproduce asexually by fragmentation, soredia and isidia., containing both the components. The fungi do produce conidia and ascospores but these are of no use as their chance of contacting the right photobiont is one in a million. Most lichens (except the gelatinous ones and some others) produce **lichen acids**, as secondary metabolites. Mostly these provide colour to the thalli. The lichens produce over 400 compounds; 230 of these is not produced by any other organism. Lichen acids have many uses for the lichens. It protects them from bacterial pathogens (antibiotic property), from slugs etc. (bitter taste), absorption of minerals (chelating property), translocation of chemicals between the two partners, release of carbohydrates from photobiont (permeability alteration), as pigments they protect the photobiont from damage by sunlight. The lichens are highly useful and important, as initiators of rock vegetation, fodder, food and medicines.

TEST QUESTIONS

1. Describe the types of lichen thalli with examples.
2. Write on the mycobiont and photobiont components of lichens.
3. What are lichen acids? Write on the synthesis and uses of lichen acids to the lichens.
4. Describe the importance of lichens as initiators of vegetation, as food and feed, and in medicine.
5. Write notes on any two of the following :
   a. Optional lichenization.
   b. Asexual reproduction in lichens.
   c. Growth of lichens.
   d. Resynthesis of lichens in laboratory.
   e. Lichhenometry.
6. Answer in brief.
   a. What are cyphelle and rhizines?
   b. What are soredia and isidia?
   c. Who discovered the dual nature of lichens?
   d. Who first used lichenometry and who first published this technique?
   e. Who induced the perfect state of *Cladonia* in laboratory?
   f. Which lichens have the slowest growth?
   g. Which is the fastest growing lichen?
7. Fill in the blanks.
   a. Lichens are identified by their----------.
   b. The morphology of the lichen is controlled by --------- genes, but its expression is dependent on the presence of -------------.
   c. The most common green algal photobiont is --------, and the most common cyanobacterial photobiont is -------------.
   d. Cyanobacteria are photobionts in ---% lichens.

e. *Trebouxia* forms ------- in culture but not in the lichen.

f. Lichenized fungi form ---% of *Ascomycota* and ---% of all fungi.

g. The lichenized fungi mostly belong to ------------.

h. Foliose lichens are attached to the substratum by ----------.

i. Some lichens have been shown to withstand a low temperature of ----, and a high temperature of -------------.

26

Nutrition of Fungi

Fungi are heterotrophs dependent on external supply of readymade organic food. Unlike animals, which are also heterotrophs, fungal nutrition is **absorptive**. Ingestion is rare and restricted only to the slime moulds. The hyphae lie in direct contact with the nutrients and absorb dissolved smaller molecules, such as simple sugars and amino acids. Large insoluble substances, like polysaccharides, fats, proteins, etc., are first broken into smaller fragments until the soluble monomers are released (see Chapter 3, Fig. 3.1). This is called **digestion** and is achieved by extra-cellular enzymes.

Fungi are extremely versatile in their nutrition. Except certain fluorine-containing plastics and a few detergents, there is nothing, which fungi cannot degrade and derive nutrition. Naturally, the numerous digestive enzymes, needed to cope up with varied organic compounds, cannot be present in the cells all the time. They are produced only when needed and, fall in the category of **adaptive** or **inducible** enzymes. Fungal saprobism and fungal parasitism are subjects, which have been studied in great depth. Saprobic fungi abound in the soil and live on dead plant or animal remains or organic materials, leached out or exuded by plant or animal tissues. Parasitic fungi, on the other hand, invade living tissues. It is the ability to invade living tissues that distinguishes parasites from saprobes. There are gradations, of course, in both modes of life. **Obligate parasites** (rusts, smuts, powdery mildews and downy mildews) live only on their hosts. But there are fungi, which can live both as saprobes and parasites, as the need may be. These are called **facultative parasites.** Physiology of parasitism is an extensive and independent field of study. Studies on fungal saprobism, however, have lagged behind, but lately are receiving due care at the hands of fungal ecologists. The role of fungi in soil ecology and soil fertility has become an interesting field of study, and a wealth of information is now available. Physiology of obligate parasitism has been a challenge to biochemists, who are unceasingly working to grow them in laboratory cultures. This may enable better experimentation with these fungi.

NUTRITIONAL REQUIREMENTS

Essential Elements

For fungi also, like other organisms, some elements are essential for growth. Fungi need about 17 elements, some in extremely small or trace amounts (0.001 to 0.05 ppm), called **micro-**, or **trace elements,** and others in comparatively large amounts and, therefore, called **macro-elements.** If any of these macro- or micro elements is lacking in the medium, the fungus fails to grow, despite the presence of other elements in abundance. A fungus is unable to grow in the absence of a trace element, say iron, as it can not grow in absence of macronutrients, like nitrogen. In a balanced medium, the ratio between iron and carbon is 1: 50,000. But both are indispensable.

The macroelements include the **non-metallic elements** (carbon, hydrogen, oxygen, nitrogen, sulphur, phosphorus) and **metallic elements** (potassium and magnesium). Calcium, which is an essential macroelement for green plants, is needed in traces or not at all by fungi. The non-metallic macroelements, in general, serve as **structural elements** in protoplasm and wall synthesis. The metallic elements are **functional elements,** active in metabolism. However, all elements participate in metabolic events of life processes. Hydrogen is obtained from water, and oxygen from air. Other elements are taken in form of compounds or ions.

The essential microelements of fungi are the metallic elements - iron, zinc, copper, manganese and molybdenum. Boron, an essential microelement of green plants, is not essential for fungi. Yeast extract, which contains all the microelements, plus several growth factors, is usually added to the media to furnish these elements. The microelements act as **cofactors** or **activators** of enzymes, and are also used in the synthesis of vitamins. These are essential for growth and sporulation. We shall concentrate only on some of the macroelements.

CARBON SOURCES

Fungi are **chemoorganotrophs** i.e. derive energy by oxidation of organic compounds. A few fungi do fix atmospheric CO_2 but that does not serve as the sole carbon source. About 50% of the dry weight of fungi is contributed by carbon compounds. This suggests the importance of carbon requirement of fungi. Carbon serves as (1) a source of energy, and (2) as chief structural element. Practically, all the components of the cell wall viz. cellulose, chitin, hemicellulose, and pectic substances, are synthesized from carbon compounds, derived from the medium. Carbohydrates are, all the more, significant, as they serve as almost exclusive source of energy. Fats and proteins are of secondary importance.

A. CARBOHYDRATES AS CARBON SOURCES

Utilization of Monosaccharides

Pentoses. Among pentoses, xylose supports growth of largest number of fungi and, may be good or better than glucose. Next important pentose is arabinose, but it is always inferior to both xylose and glucose.

Hexoses. Glucose is utilized by almost all fungi. Whenever a fungus of unknown nutrition is grown in laboratory, glucose is the first sugar tried. But, fungi excel in variety. There are fungi e.g. *Leptomitus lacteus* and *Apodachlya* sp., which fail to grow on glucose or any other sugar. They grow on acetates and fatty acids in sewage and sewage-polluted waters. There are some more fungi, which do not utilize glucose, these are — *Sphaeronema fimbriatum, Fusarium lini, Sclerotium rolfsii, Ozonium texanum* var. *parasiticum* and *Tieghemiomyces parasiticus.* Glucose prevents utilization of other carbohydrates. This is called **glucose effect** or **catabolite repression**. It occurs by repression of synthesis of degradative enzymes or inhibition of action of other degradative enzymes.

Fructose (fruit sugar), **mannose** and **galactose** are usually inferior to glucose but are utilized frequently by fungi, and, serve as the exclusive carbon source. Their efficiency as carbon source depends on their conversion into phosphorylated derivatives of glucose, which can enter the main respiratory pathways.

Disaccharides

Sucrose, maltose, cellobiose, and lactose are the disaccharides, which the fungi frequently encounter in nature (Fig. 26.1). While sucrose is the main storage product of green plants, maltose and cellobiose are formed during degradation of starch and cellulose, respectively. Fungi do not utilize them as such. The glycosidic bond is first broken by extra cellular hydrolytic enzymes and the sugars released are then absorbed. The ability to utilize the disaccharides depends on the production of appropriate hydrolytic enzymes.

Sucrose. It consists of α - glucose and β-fructose molecules, joined by α 1- 2 β glycosidic linkage. Many fungi utilize sucrose by splitting it into glucose and fructose moieties with the help of enzyme sucrase. The disaccharide is inferior to glucose. Out of the two monomers released, glucose is absorbed faster than fructose. Some members of *Chytridiales, Mucorales* (*Rhizopus oligosporus*) and *Sphaeriales* (*Sordaria fimicola*) are unable to utilize sucrose.

Maltose. This is formed during degradation of starch. It consists of 2 α glucose molecules, linked by α 1–4 glycosidic linkage, which are released after hydrolysis of the bond by the enzyme maltase. Compared to sucrose, maltose is utilized by more fungi. Even fastidious ones, with limited substrate range, grow well on maltose and record better growth than on glucose.

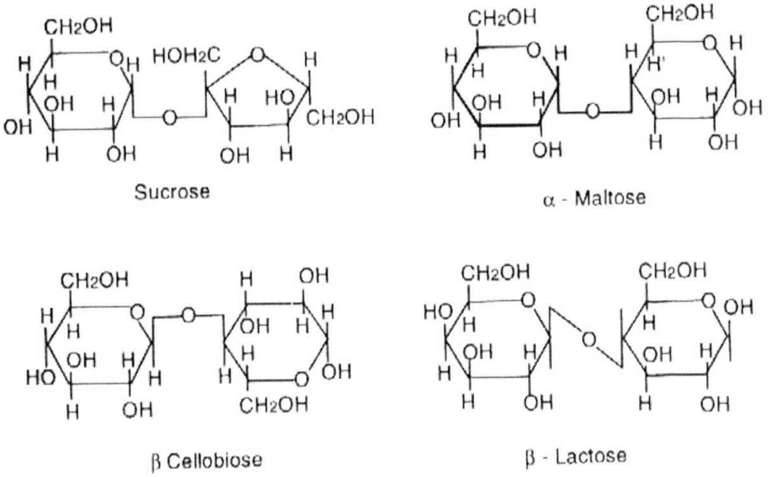

Sucrose α - Maltose

β Cellobiose β - Lactose

Fig. 26.1. Structural formulae of some disaccharides

Cellobiose is a hydrolytic product of cellulose and is made of 2 β glucose units, linked by β 1-4 glycosidic linkage. The glucose moieties are released by hydrolytic action of cellobiase. Saprobic fungi utilize cellobiose with avidity.

Lactose (=milk sugar) is usually not a good source of carbon for fungi. It is made of one α glucose and one β galactose molecules, joined by β 1–4 galactoside linkage.

Polysaccharides

Polysaccharides constitute the bulk of carbohydrates synthesized by plants and animals. On hydrolysis, involving usually more than one enzyme, simple sugars are released, which are then absorbed by the fungi.

Cellulose. This is the most abundant organic substance on earth. Its degradation by fungi for their nutrition is an important event in the maintenance of the carbon cycle in nature. Cellulolytic fungi are indispensable for this reason. Cellulose is the insoluble polysaccharide that forms the structural framework of cell walls. It is a polymer of β glucose units joined by β 1-4 glycosidic linkages and may be considered as made of repeating units of cellobiose (Fig. 26.2). About 280 to 800 chains of cellulose molecules, bound laterally by hydrogen bonds and other physical forces, give rise to micro-fibrils, which aggregate to form the fibrils. The polymer chains at some places in the fibrils show highly compact orientation, while at other places they are relatively loose. Places of compact orientation are called **crystalline areas** (= **crystallites**), while regions with loose chains are known as **amorphous areas**. A single cellulose chain passes through several crystalline and amorphous areas in the fibril. Crystalline areas are resistant to enzymatic attack on account of their compactness, which prevent entry of enzymes.

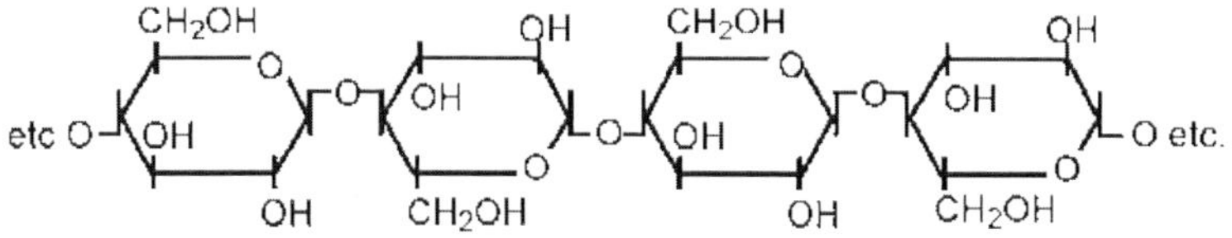

Fig. 26.2. Cellulose, a polymer of β-glucose units

Cellulolytic Enzymes

There are five types of cellulolytic enzymes (= cellulalses) based on type of reaction catalyzed.

1. **Endoglucanase** that breaks the internal bonds to disrupt the crystalline structure of cellulose, and exposes individual chains.
2. **Exocellulase (= cellobiohydrolase CBH)** that cleaves 2-4 units from the ends of the exposed chains produced by endoglucanase, resulting in tetrasaccharides and disaccharides (cellobiose). There are two main types of cellobiohydrolases: CBH I and CBH II. **CBH I** works progressively from the reducing end (i.e. C_1 end), while **CBH II** works progressively from the non-reducing end (C_4 end).
3. **Cellobiase (β glucosidase**) which hydrolysaes the exocellulase-product into individual monosacccharides i.e. β glucose.
4. **Oxidative cellulases** that depolymerize cellulose by radical reactions, as for instance cellobiose dehydrogenase (acceptor).
5. **Cellulose phosphorylases** that depolymerize cellulose using phosphates instead of water.

The three main enzymes (endoglucanase, exoglucanase and cellobiase) form a cellulase complex, which by their sequential, synergistic and cooperative action bring about total breakdown of cellulose to β glucose (Fig. 26.3).

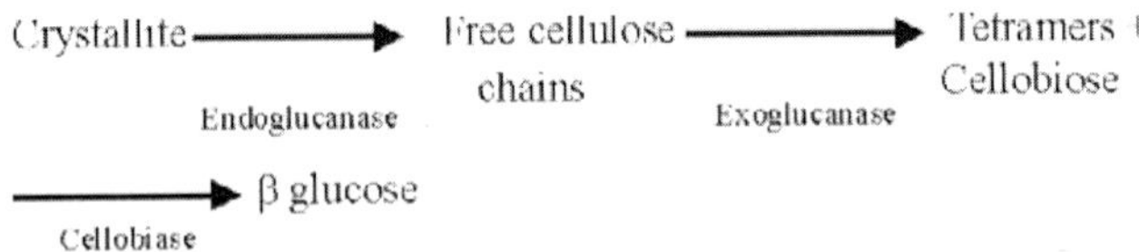

Fig. 26.3. Sequential breakdown of cellulose

In many bacteria cellulases *in vivo* are organized in 'supramolecular complexes' called **cellulosomes**. They contain roughly five different enzymatic subunits, representing the five enzymes mentioned above. But only three viz. endoglucanases, exoglucanases and cellobiases participate in the actual hydrolysis of the β 1-4 linkage. Most fungal cellulases have a two domain structure with one **catalytic domain** and one **cellulose-binding domain** that are connected by a flexible **linker**. This spatial separation of catalytic and binding sites is important for efficient hydrolysis.

Pectic Substances (= Pectins)

Pectic substances form the middle lamella. Besides, they form an integral part of the plant cell wall. These are the most complex macromolecules in nature. A number of polysaccharides together form pectin. The constituent polysaccharides are galacturonans (*homogalacturonans, rhamnogalacturonan II, xylogalacturonan*) rhamnogalacturonan I, arabinogalactan I, arabinan, and arabinogalactan II. They collectively form the pectic polymer structure, as describe below.

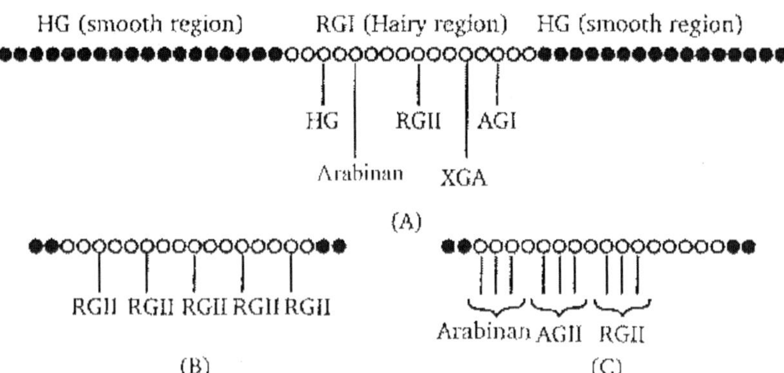

Fig. 26.4 Structure of pectic polysaccharides. (A) Pectic back bone made of HG (smooth) and RG I (hairy regions). (B) Hairy region showing one type of side chains (all RG II). (C) Showing different type of side chains. HG = homogalacturonan; RG I = rhamnogalacturonan I; AG I = arabinogalactan I; XGA = xylogalacturonan, RG II = rhamnogalacturonan II; AG II arabinogalactan II

Macromolecular Structure of Pectic Polysaccharides

In the model (Fig. 26.4) proposed by **Ridley** *et al.* (2001), and **Voragen** *et al.* (2003), the constituents of pectic polymers are covalently linked to each other. There is a **pectic backbone** comprising HG (homogalacturonans) and RG I regions. The RG I region bears several side chains viz. HG, RG II, Arabinan, AG I, AG II, XGA, and is called the **hairy region.** The HG is called the **smooth region**. The types of side chains and their arrangement on the RG I region vary with the pectin sample. There may be one type of side chains, or of different types arranged at random or in clusters (B,C in Fig. 26.4). The pectin macromolecules, as described above, are further cross-linked to form the noncovalently linked pectin net work, in which the cellulose–hemicellulose net work is embedded.

Constituents Pectic Polysaccharides: There are 5 types of constituents; 1. Galacturonans, 2. rhamnogalacturonan I, 3. Arabinogalactan I, 4. Arabinan, and 5. Arabinogalactan II.

1. **Galacturonans** have a backbone made of 1-4 linked α galacturonic acid residues. There are three types of galacturananans: (i) homogalacturonan (HG), (ii) rhamnogalacturonan II (RG II), and (iii) xylogalacturonan.

(i) Homogalacturonans (HGs), which constitute 60 % of the pectic polysaccharides, are made of 1-4 linked α galacturonic acid residues (Fig. 26.5). Some of the carboxyl groups, at C_6 position, are methyl-esterified. Depending on the degree of esterification (DE), the HGs are classified as: **pectin** or **high methoxyl pectin** (DE >50%), **pectinic acid** or **low methoxyl pectin** (DE < 50%), and **pectic acid** (DE 0%), which have no methyl esters.

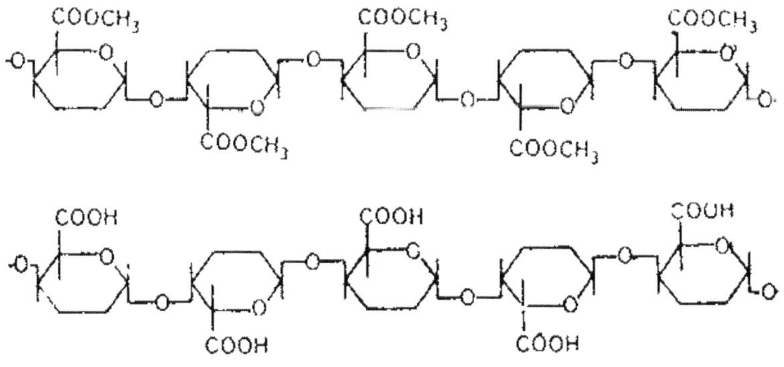

Fig 26.5 Homogalacturonan: pectim (above) and pectic acid (below).

(ii) Rhamnogalacturonan II (RG II) is the most unusual polysaccharide in nature. The name RG II is misleading, because it does not have a rhamnogalacturonan back bone.

(iii) Xylogalacturonan (XGA) has a galacturonan back bone (i.e. made of 1-4 linked α galacturonic acid residues) having side chains of β xylose residues.

2. Rhamnogalacturonan I (RG I) has a backbone comprising of alternating rhamnose and galacturonic acid residues. The rhamnosyl residues carry galactose as side chains.

3. Arabinogalactan I (AG I) is made of 1-4 linked β galactose backbone, which has α arabinose residues attached to the galactose residue.

4. Arabinans have 1-5 linked α arabinose backbone with one or two arabinose residues as side chains.

5. Arabinogalactan II (AG II) is composed of 1-3 linked β galactose backbone, with short side chains of arabinose and galactose residues

Pectolytic Enzymes

These enzymes degrade the homogalacturanan i.e. the "smooth region" of the pectic macromolecule. The enzymes are classified broadly into two broad categories, the **chain modifying (PME)** and **chain splitting** or **depolymerisig** enzymes (glycosidases and lyases).

Pectin methyl esterase (PME) hydrolytically removes the methyl group (Fig. 26.6) from pectin or pectinic acid, to give methyl alcohol, pectinic acids of lower methoxyl content, and ultimately pectic acid. The chain length remains unaffected, yet removal of the methoxyl groups gives new properties, like formation of salts with polyvalent cations. It also predisposes pectin to cleavage by chain-splitting enzymes, which, otherwise, would not degrade pectin. PME is widely distributed in plants, but to a much lesser extent in microorganisms. Enzymes of microbial origin have lower pH optima, as compared to those of plants.

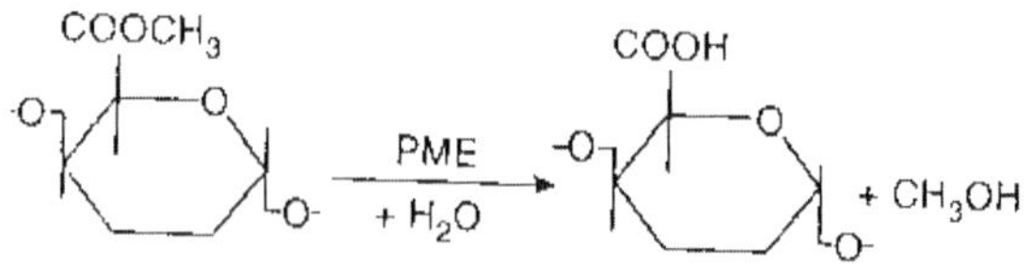

Fig. 26.6. Shows PME action

Chain Splitting Enzymes break the α 1-4 glycosidic bonds, forming smaller fragments, and ultimately releasing galacturonic acid residues. There are two types of chain-splitting enzymes that are classified on the basis of **hydrolytic** or ***trans-eliminative*** mode of split of the glycosidic bond, hence called **glycosidases** (= **hydrolases**) and **transeliminases** (= **lyases**).

Glycosidases (=Polygalacturonase, PG, and Polymethyl galacturonase, PMG)

These enzymes break the glycoside bonds hydrolytically (Fig. 26.7). Their nomenclature is based on the preference of the substrate (pectin or pectic acid) and site of split (random or terminal). If the enzyme shows greater activity on pectic acid, the enzyme is called PG, and if the activity is greater on pectin, the enzyme is called as PMG. The two enzymes are further characterized as endo-, or exo-enzymes. Those attacking the linkages randomly are called endoenzymes (endo-PG or endo-PMG), while those attacking terminal bonds are called exo-PG, or exo-PMG. The endo-PG or endo-PMG split the polymers into smaller fragments, and are assayed by measuring the loss in viscosity of the substrate. The exo-enzymes release reducing sugars from the terminal reducing, C4 ends, and are assayed by measuring the increase in reducing groups in the reaction mixture.

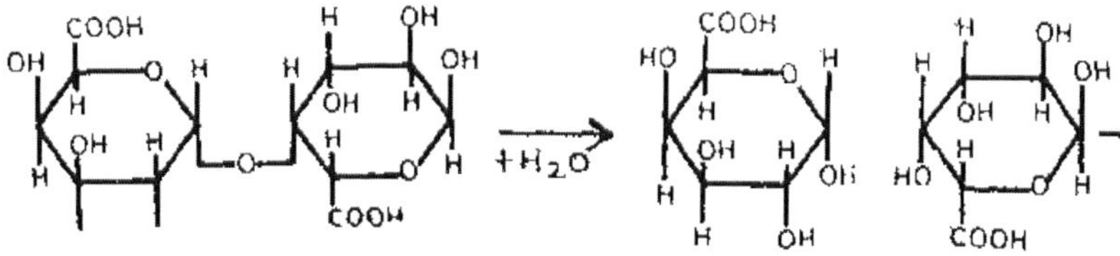

Fig. 26.7 Hydrolytic cleavage of glycosidic bond

Transeliminases or Lyases (Pectic acid lyase, PAL; Pectin lyase, PL)

Trans-eliminative cleavage of the glycosidic bond involves removal of the proton at C_5 of one residue, resulting in the formation of an unsaturated bond between C_4 and C_5 (Fig. 26.8). The unsaturated product absorbs light in the UV range, maximally at 230/235 nm, and at 547.750 nm on treatment with thiobarbituric acid. The trans-eliminases (or lyase enzymes) are also classified, like the hydrolytic enzymes, on the basis of substrate preference viz. pectic acid (pectic acid lyase) or pectin (pectin lyase), and random or terminal site of action (endo-PAL or endo-Pl and exo PAL or exo-PL). The endo-enzymes are assayed viscometrically, by measuring the loss in viscosity of the substrate, while the exo-enzymes are measured by the release of reducing groups in the reaction mixture.

Classification of Glycosidases and Lyases

(A) Hydrolytic bond cleavage

1. Random cleavage

a - Pectin preferred substrate Endo – **PMG**

b – Pectic acid preferred substrate — Endo - **PG**

2. Terminal cleavage

a – Pectin preferred substrate — Exo – **PMG**

b – Pectic acid preferred substrate — Exo – **PG**

(B) Transeliminative bond cleavage

1. Random cleavage

a – Pectin preferred substrate — Endo – **PL**

b – Pectic acid preferred substrate — Endo - **PAL**

2. Terminal cleavage

a – Pectin preferred substrate — Exo – **PL**

b – Pectic acid preferred substrate — Exo – **PAL**

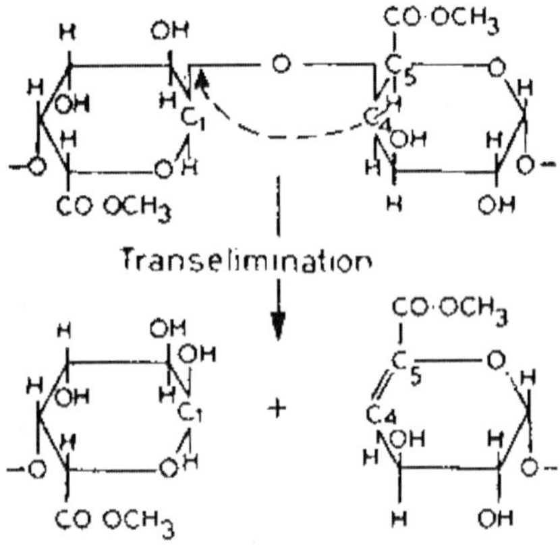

Fig. 26.8. Trans-eliminative cleavage of glycosidic bond.

Starch

Starch, the most important reserve carbohydrate of higher plants, is a good carbon source for a majority of fungi, and is often better than glucose, possibly due to the presence of some growth factors, that usually accompany native starch.

Starch is a complex polymer made of α glucose units and can also be said to be made of maltose units. It consists of two types of molecules: ***amylose*** (20 - 25%) and rest ***amylopectin***. Amylose is long, unbranched chain of about 300-1000 glucose residues, linked by α 1-4 glucosidic linkages, while amylopectin, though similar in mode of linkage, is branched and bears smaller side chains of 30 or lesser glucose units, linked by α 1-6 linkage to the main chain.

Amylose is more water-soluble than amylopectin, and is primarily responsible for the blue colour with iodine in the 'amyloid reaction'.

There are two types of extra-cellular enzymes that degrade starch, α and **β amylase** (earlier termed diastase). The α amylase attacks glycosidic linkage randomly to produce shorter chains, called **dextrins**. It is also called **endo-amylase**, and, as it does not release maltose or glucose, its action results in **liquification**. **β amylase**, on the other hand, acts terminally and removes two glucose units (maltose) at a time. It is also called **exo-amylase** and, since the products are sugars, which are sweet, the enzyme action is called **saccharification.** Maltose is then acted upon by maltase (α 1–4 glucosidase) and glucose molecules are released.

Other polysaccharides. The nutritional efficiency of chitin, glycogen, inulin, etc, is not studied much. **Chitin,** the polysaccharide of N-acetyl glucosamine units (forming exoskeleton of arthropods and cell wall of most of the fungi), is degraded by an extra-cellular **chitinase**. Glycogen, which is the reserve food of animals and fungi, is not degraded commonly by fungi. **Inulin**, the D fructose-rich polysaccharide, is degraded by some fungi by the enzyme inulase, which is active on sucrose also.

B - FATS AS CARBON SOURCES

Fatty acids or the monocarboxylic acids are produced after degradation of fats (hence called fatty acids) by the action of lipases. These are generally utilizable, but are toxic to some fungi. However, *Leptomitus lacteus* and *Apodachlya* sp which fail to grow on sugars, including glucose, utilize fatty acids avidly.

Cutin, the main component of cuticle of plants, is a polymer of C16 and C18 hydroxy fatty acids. It contains mostly ester bonds (hence called **polyester)**, and some peroxide bridges and ether linkages. There are two cutin-degrading enzymes: **cutin esterase** and **carboxy cutin esterase** (= **carboxy cutin peroxidase**). The former hydrolyses the ester bonds releasing monomers (C16 and C18 hydroxy fatty acids), while the latter hydrolyses the peroxide group.

Phospholipids are important constituents of cell membranes. Certain plant pathogenic fungi degrade cell wall membranes by phosphatidase enzymes.

C- PROTEINS AS CARBON SOURCES

Fungi degrade proteins hydrolytically by extra-cellular proteolytic enzymes, but the information about these enzymes is scanty. Proteases, which cleave peptide bonds, are of two types: **endo-peptidases** and **exo-peptidases**. The former act randomly and produce amino acid chains of shorter lengths, from which individual amino acids are released one after another, by the action of exo-peptidases attacking the terminal peptide linkages.

NITROGEN SOURCES

Like carbon, nitrogen also serves as a structural, as well as, functional element. Nitrogen is a component of amino acids, nucleic acids, vitamins, and amino sugars, like N-acetyl glucosamine, the constituent of chitin. Fungi derive their nitrogen requirements from inorganic, as well as, organic sources.

Nitrates

Numerous fungi utilize nitrates as the sole source of nitrogen. The various types of nitrates viz. NH_4NO_3, KNO_3, $NaNO_3$, Ca $(NO_3)_2$ vary in efficiency as nitrogen sources. This is due to their cations. After absorption, the nitrates are reduced to ammonia, by nitrate reductase before incorporation into amino acids.

Ammonia

Fungi that are unable to utilize nitrate, use ammonium or organic nitrogen of the same oxidation level as ammonium ion. Fungi, which utilize nitrate and ammonium both, show preference for the ammonium nitrogen. In a medium containing ammonium nitrate, the ammonium is utilized quicker and nitrate is left behind. This preferential utilization of ammonia is explained to be due to "feed-back" repression, in which the end product represses the enzymes producing it. Thus, ammonia represses production of nitrate reductase. Ammonium utilization from its salts, like NH_4 NO_3, $(NH_4)_2$ SO_4, NH_4CO_3, NH_4CI etc. results in lowering of pH, due to accumulation of anions and formation of acids. The low pH may prove inhibitory to the growth of fungi. Fungi which are unable to use ammonium e.g. *Leptomitus lacteus,* depend upon amino acids i.e. organic nitrogen.

Organic Nitrogen

Fungi, without exception, utilize organic nitrogen in the form of amino acids and amides. Usually, amides are better nitrogen sources than their corresponding amino acids. Amino acids, released from protein degradation, are utilized by fungi, as such, and are incorporated into new proteins. All amino acids are not alike in their nutritive value. Glycine, alanine, glutamic acid and aspartic acid, serve as good nitrogen sources for a large number of fungi; but cysteine and methionine, the sulfur-containing amino acids, are poor.

For a large number of fungi, the amides, e.g. asparagine and glutamine, are superior to their corresponding amino acids viz. aspartic and glutamic acids, respectively. Whether, this is due to extra nitrogen present in the amides, or is an effect of their better buffering action on the medium, is not known. Commerecial peptone and casein hydrolysate, which are mixtures of peptides of varying chain lengths, are commonly used in laboratories as nitrogen sources.

SULFUR SOURCES

Sulfur enters into composition of some amino acids (cysteine, and methionine), tripeptide glutathione, thiol group of many enzymes, vitamins (thiamine and biotin), antibiotics (penicillin, gliotoxin) and some miscell-aneous compounds.

Fungi usually derive their sulfur requirements from sulfate ions, but some take it in more reduced form, as present in organic compounds. Fischer on the basis of sulfur utilization, divided microorganisms into two groups: (1) **euthiotrophic** (those, which use sulfate ions), and (2) **parathiotrophic** (those which cannot use sulfate, and derive sulfur from amino acids or reduced inorganic or organic sulfur). Aquatic fungi are parathiotrophic.

PHOSPHORUS, POTASSIUM AND MAGNESIUM

These three macronutrients, which are taken in organic form, are needed in the concentrations ranging 0.001 to 0.004 M.

Phosphorus High concentrations of phosphorus in the medium, enhances carbohydrate utilization. Phosphorus is utilized as phosphate ions and function in this form as phosphate ester. Both inorganic and organic phosphate esters are utilized.

Potassium is not known to be a part of any organic molecule; even then, it is indispensable for growth. Its function is associated with permeability of cell membranes. Its action is known by its absence. Any alteration in permeability of cells is invariably accompanied by loss of potassium.

Magnesium is essential as a co-factor of respiratiory enzymes. In laboratory cultures, magnesium is usually supplied as magnesium sulphate.

Calcium, which is an essential macroelement for green plants, **is needed in traces or not at all** by fungi.

VITAMIN REQUIREMENTS

Vitamins are organic molecules, required in small amounts and not used as sources for energy or structural materials of the protoplasm. The term growth factor has a broader connotation than vitamins. It includes the components and derivatives of vitamins, which cannot be classified in a new category. The increased growth of some fungi on natural substrates like coconut milk, and dung etc., is due to the presence of some unidentified growth factors. Vitamins, mostly, act as coenzymes or constituents of coenzymes. Green plants synthesize all vitamins required by them; animals are completely dependent on external supply. Fungi show extreme to intermediate behaviour with respect to vitamin synthesis.

From vitamin-synthesis point of view, fungi are divided into 2 groups (1) **Auxoautotrophic** and (2) **Auxoheterotrophic.** Auxoautotrophic fungi synthesize all the required vitamins and, sometimes produce them in excess, which are then secreted into the medium. *Ashbya gossypii* secretes the excess riboflabin (vitamin B_2) into the medium and is used for its commercial production. The auxoheterotrophic fungi, on the other hand, lack the biosynthetic capacity, partially or completely. Partially-deficient species make only limited growth, in absence of vitamins, and show marked increase in growth, if the deficient vitamin is supplied. The deficiency may be genetic or due to environmental stress. For completely deficient fungi, an external supply is the only way out.

As a rule, fungi have no requirement for fat-soluble vitamins (vitamins A, D, E and K). Only water soluble vitamins e.g. vitamins of B group (vitamin B_1 or thiamine, B_2 or riboflavin and B_6 or pyridoxine) and vitamin H (biotin) are required by fungi. Vitamin C (ascorbic acid) deficiency is rare as it has no role in fungal metabolism. Rather it inhibits the growth of several fungi.

Thiamine (vitamin B_1)

Thiamine, in form of TPP (Thymine pyrophosphate) is indispensable for its role in decarboxylation of pyruvic acid into acetaldehyde and other a ketoacids. Most of the vitamin-deficient fungi show deficiency for thiamine. All species of *Phytophthora, Phycomyces blakesleeanus,* some yeasts, most basidiomycetes (e.g. *Boletus, Clitocybe, Coprinus, Exobasidium, Marasmius, Polyporus, Peniophora*, and so many) are thiamine-deficient. Thiamine consists of 2 moieties, pyrimidine and thiazaole. Biosynthesis of thiamine involves direct condensation of the two components.

Pyrimidine + Thiazole ----------→ Thiamine

Some thiamine-deficient fungi like *Phycomyces*, make-up the deficiency, if the two moieties (which they are unable to synthesize) are supplied exogenously. Others lack only one moiety and when the missing moiety is supplied, they can synthesize thiamine. In all such cases the missing moiety is pyrimidine.

Riboflavin (vitamin B_2, vitamin G)

Riboflavin forms prosthetic group of FMN and FAD (Flavin mononucleotide and Flavin adenine dinucleotide) of flavoprotein enzymes, associated with oxidation-reduction reactions. Riboflavin deficiency is rare in fungi; the lone report is for *Poria vaillantii*. It is synthesized by yeasts. *Ashbya gossypii, Ermothecium ashbyii* and *Candida* spp are used for the commercial production of riboflavin.

Pyridoxine (B_6)

Along with its two close derivatives, pyridoxal and pyridoxamine, it is referred to as vitamin B_6. *Saccharomyces cerevisiae* and several other 'yeasts' are deficient in B_6. Among the filamentous fungi, pyridoxine deficiency is reported in some ascomycetes and basidiomycetes (*Ustilago maydis*).

Nicotinic acid (= niacin)

It is an oxidation product of nicotine. It is metabolically active in form of amide, nicotinamide, which is a part of the coenzyme NAD^+ and $NADP^+$ of various dehydrogenase enzymes. Niacin deficiency is reported mostly in yeasts and several filamentous fungi. Niacin deficiency is not yet reported in basidiomycetes.

Pantothenic Acid

This growth factor was first reported in *Saccharomyces cerevisiae*. Later it was found to be very common and named as pantothenic acid (*pantos*=everywhere). It forms the major fraction (A part) of coenzyme-A (which brings about transfer of acetyl group in acetylation reaction) and in fatty acid metabolism. It is linked by an amide linkage. It is made of two components, pantonic acid and β alanine

Pantonic acid + β alanine --------→ Pantothenic acid

Pantothenic acid deficiency is more common in yeasts than in the filamentous fungi. *Polyporus texanus* is the only filamentous fungus showing absolute deficiency for pantothenic acid. As in thiamine, the deficiency is for one of the two components, usually for β-alanine, and, when this is supplied, the deficiency disappears. Thus, yeasts make up their deficiency and start producing pantothenic acid if β-alanine is supplied to them. The opposite occurs in *Polyporus texanus*, which synthesizes β-alanine and needs exogenous pantonic acid to synthesize the pantothenic acid.

Biotin (vitamin H)

It is a member of vitamin B-complex group. After thiamine, biotin is the most commonly lacking vitamin among fungi. Most of the yeasts are biotin-deficient. It is synthesized according to the following scheme:

$$\text{Pimelic acid} \longrightarrow \text{Desthiobiotin} \longrightarrow \text{Biotin}$$

The biotin deficiency in some fungi, e.g. *Ceratostomella pini* and *Sordaria fimicola,* can be overcome if the desthiobiotin is supplied exogenously.

Folic Acid

Fungi seem to be auxoautotrophic for this growth factor. It consists of three compounds–glutamic acid, *p*-aminobenzoic acid and substituted pterin. Various forms of folic acid are known, which vary in the components and mode of linkage. Whether fungi require p-aminobenzoic acid for folic acid synthesis, or it has some other role also, is not known.

Inositol

Inositol deficiency is known in yeasts. It is a carbohydrate with the empirical formula of monosaccharide i.e. $(CH_2O)_n$ Nine stereoisomers of inositol are known, but only *myo*-inositol (Fig. 26.9) shows vitamin activity.

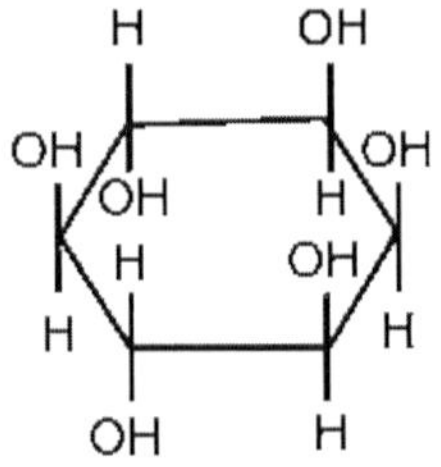

Fig. 26.9. *Myo*-inositol, a carbohydrate with vitamin action

The metabolic role of inositol as a coenzyme is not conclusively demonstrated. It is known to be a structural unit (in association with phospholipids) for the maintenance of structure of mitochondria. It is required in large amounts and, may more appropriately be classified as a metabolite, rather than vitamin.

FUNGI AS TEST ORGANISMS

Fungi are routinely used to test the presence or absence of chemicals in trace amounts. This is called **bioassay** i.e. assay or test through organisms. The organisms used are called **test organisms**. *Phycomyces blakesleeanus*, which shows absolute deficiency for thiamine, is used to test the presence of thiamine in solutions. Growth of the fungus indicates presence of thiamine. *Phytophthora* (all species) are used to test the presence of thiamine. Bioassay of pyridoxine and pantothenic acid is done by using *Saccharomyces carlsbergensis*. *S. cerevisiae* is used for the assay of biotin. Table 26.1 lists the various fungi used in vitamin bioassays.

Table 26.1. Fungi used in bioassay of vitamins

Test vitamin	Test fungi
Thiamine	*Saccharomyces cerevisiae, Phycomyces blakesleeanus, Pythium ascophallan*
Pyridoxine	*S. cerevisiae, S. oviformis, S. carlsbergensis. Neurospora sitophila, Ceratostomella pleuriannulata, C. montium, Mycoderma valida.*
Pantothenic acid	*S. carlsbergensis, S. cerevisieae.*
Inositol	*S. cerevisieae, Neurospora crassa.*
Nicotinic acid	*Zygosaccharomyces marxianus.*
P-aminobenzoic acid	*Neurospora crassa*
Biotin	*S. cerevisieae, Zygosaccharomyces marxianus, Debaryomyces matruchoti* var. *subglobosus, Ceratostomella* spp. *C. montium, C. pini, Sordaria fimicola, Neurospora crassa, N. sitophila.*

Amino acid bioassay is usually done with the help of bacteria but some amino acid-deficient *Neurospora* spp. have also been used. The presence of certain elements like phosphorus, copper, magnesium, potassium and molybdenum in amounts too small for chemical detection is smoothly detected by employing strains of *Aspergillus niger*.

Merits of Bioassay. Being least complicated, as compared to chemical analyses, bioassays are very popular in industries. Once the suitable test organism is obtained and its dose-response curve prepared, the determination and quantification is a simple exercise. Bioassays are also more sensitive and accurate. Machines make mistakes, organism never!

REFERENCES

Ridley BL *et al.,* 2001. Pectins: structure, biosynthesis and oligogalact-urunide related signaling. Phytochemistry 57: 929-967

Voragen AGJ *et al.,* 2003. Advances in pectin and pectinase research. Kluver, The Netherlands

SUMMARY

Fungi are **chemoorganotrophs** i.e. derive energy by oxidation of organic compounds. Unlike animals, which are also chemoorganotrophs, fungal nutrition is **absorptive**. Ingestion is rare and restricted only to the slime moulds. Fungi need about 17 elements. Calcium, which is an essential macroelement for green plants, is needed in traces or not at all by fungi.

Among **monosaccharides**, xylose, glucose and fructose support good growth of largest number of fungi. There are some fungi, which do not utilize glucose. **Disaccharides**, sucrose, maltose, cellobiose, and lactose are first broken by extracellular hydrolytic enzymes and the component sugars are then absorbed. Except lactose, other diasaccharides are good sources of carbon. Among **polysaccharides**, **cellulose** degradation by fungi is important for maintenance of the carbon cycle. It is a polymer of β glucose units joined by b 1-4 glycosidic linkages. There are five types of cellulolytic enzymes (= cellulalses) viz. endoglucanase, exoglucanase (CBH), cellobiase (β glucosidase), oxidative cellulases, and cellulose phosphorylase. The three main enzymes endoglucanase, exoglucanase and cellobiase by their sequential, synergistic and cooperative action bring about total breakdown of cellulose to β glucose.

Pectic substances, the most complex macromolecules in nature, are made of five polysaccharides. These are: (i) **Galacturonans** (*homogalactu-ronans*, *rhamnogalacturonan II*, *xylogalacturonan*), (ii) **Rhamnogalactu-ronan I,** (iii) **Arabinogalactan I,** (iv) **Arabinan, and** (v) **Arabinogalactan II**. They collectively form the pectic polymer structure. There is a pectic backbone comprising HG (homogalacturonans) and RG I regions. The RG I region bears several side chains viz. HG, RG II, Arabinan, AG I, AG II, XGA, and is called the hairy region. The HG is called the smooth region. The types of side chains and their arrangement on the RG I region vary with the pectin sample.

Pectolytic Enzymes degrade the homogalacturanan i.e. the "smooth region" of the pectic macromolecule. The enzymes are classified broadly into two broad categories, the chain modifying enzyme (PME) and chain splitting or depolymerisig enzymes (glycosisases and lyases). PME hydrolytically removes the methyl group. Chain Splitting Enzymes break the α 1-4 glycosidic bonds, forming smaller fragments, and ultimately releasing galacturonic acid residues. Their nomenclature is based on the preference of the substrate (pectin or pectic acid) and site of split (random or terminal).

Starch is a good carbon source for majority of fungi, and is often better than glucose, possibly due to the presence of some growth factors, that usually accompany native starch. It is a complex polymer made of α glucose units and can also be said to be made of maltose units. It consists of unbranched portion, ***amylose*** (20 - 25%) and rest of branched portion ***amylopectin***. There are two types of extra-cellular enzymes that degrade starch, **α and β amylase**. The α amylase (= endo-amylase) attacks the glycosidic linkage randomly to produce shorter chains, are called **dextrins** (causing liquification), while **β amylase** (exo-amylase) acts terminally releasing maltose (saccharification). Maltose is then acted upon by maltase releasing glucose.

Chitin, the polysaccharide of N-acetyl glucosamine units (forming exoskeleton of arthropods and cell wall of most of the fungi), is degraded by an extra-cellular **chitinase**.

Fatty acids, are generally utilizable, but are toxic to some fungi. However, *Leptomitus lacteus* and *Apodachlya* sp which fail to grow on sugars, including glucose, utilizes fatty acids avidly. **Cutin**, the main component of cuticle of plants, is a polymer of C16 and C 18 hydroxy fatty acids joined by ester bonds and some peroxide bridges and ether linkages. These are degraded by **cutin esterase** and **carboxy cutin esterase** (= **carboxy cutin peroxidase**). The former hydrolyses the ester bonds releasing monomers (C16 and C 18 hydroxy fatty acids), while the latter hydrolyses the peroxide group.

Fungi derive their **nitrogen** requirements from inorganic, as well as, organic sources. Fungi usually derive their **sulfur** requirements from sulfate ions. **Phosphorus, potassium** and **magnesium** are taken in organic form. **Calcium** is required in traces or not at all.

Vitamin requirements of fungi are only for **water-soluble vitamins** e.g. vitamins of B group (vitamin B_1 or thiamine, B_2 or riboflavin and B_6 or pyridoxine) and vitamin H (biotin). Most common deficiency (and hence requirement) is for thiamine, followed by biotin. The deficiency is more common in yeasts. Fungi rarely need external riboflavin, ascorbic acid and folic acid. Fungi are used as test organism for bioassay of vitamins. *Phycomyces blakesleeanus*, and *Phytophthora* (all species) are used bioassay of thiamine Pyridoxine and pantothenic acid assay employs *Saccharomyces carlsbergensis*. *S. cerevisiae* is used for the assay of biotin.

TEST QUESTIONS

1. Describe the structure of cellulose. Name the five cellulolytic enzymes and describe the breakdown of cellulose into β glucose.
2. Which are the five constituents of pectic substances? Describe the macromolecular structure of pectic polysaccharide.
3. Distinguish between any four: (i) α glucoe and β glucose, (ii) crystalline and amorphous cellulose, (iii) CBH I and CBH II, (iv) smooth and hairy region of pectic polysaccharide, (v) amylose and amylopectin, (vi) α amylase and β amylase, (vii) pectin, pectinic acid and pectic acid.
4. Describe the structure oh homogalacturonan and its enzymatic degradation.
5. Name the three galacturonans. Describe their backbone structure.
6. Describe the structure of composition of starch, chitin and cutin. Explain their enzymatic breakdown.
7. Distinguish between (i) liquification and saccharification of starch, (ii) chitin and cutin, (iii) glycosidases and lyases, (iv) euthiotrophic and parathiotrophic fungi in relation to sulfur utilization, (v) auxotrophic and auxoheterotrophic fungi in relation to vitamin synthesis.
8. Fill in the blanks.
   a. Fungi have ---------- mode of nutrition. Ingestion of food among fungi is found only in -----.
   b. Pectic substances are found in the -------- ------- and -------- ----- ----- of plants.
   c. Riboflavin is commercially produced from the fungus ---------- --- --------.
   d. Fungi have no requirement of --------- soluble vitamins.
   e. Vitamin deficiency is more common in -------------- than in -------- ------ fungi.
   f. Lyases cause --------------- cleavage, while glycosidases cause ---- --------- cleavage.
   g. Nitrates are reduced to --------- by enzyme ---------- ---------.
   h. Nitrates are not preferred by the fungus in presence of -------------.
   i. Euthiotrophic fungi use ---------, while parathiotrophic fungi use -- ------- as sulfur source.
   j. Deficiency of --------------- is rare in fungi.

k. Fungi are auxotrophic for -------------------.

l. *Myo*-inositol may more appropriately be classified as a ------------- than as a vitamin.

9. Answer in brief.

a. Which two structural domains of cellulases are joined by a linker?

b. Name three fungi that fail to utilize glucose.

c. Name two fungi which utilize fatty acids as preferred carbon source.

d. What is the structure of backbone of RG I, arabinogalactan I, arabinans and arabinogalactan II.

e. Which monosaccharide is a better carbon source than glucose?

f. Name the two cutinolytic enzymes.

g. Which among nitrates and ammonia are utilized preferentially as nitrogen source?

h. Among amides and their corresponding amino acids, which is a better nitrogen source?

i. What is the calcium requirement of fungi?

j. Name two organisms used for bioassay for thiamine.

k. Which organism is used for bioassay of *p*–aminobenzoic acid.

27

Secondary Metabolites
(= Special metabolites)

Fungi, like other microorganisms and higher plants, produce a variety of organic compounds with no obvious function. These are called secondary metabolites, or special metabolites, to differentiate them from primary metabolites, which are functional in the life processes. While primary metabolites of various organisms are essentially similar, the secondary metabolites include a wide variety of chemical groups. The study of these compounds has greatly contributed to organic chemistry, and some of the fungal secondary metabolites are produced in industries. Industrial mycology is largely based on these secondary metabolites of fungi.

Primary and Secondary Metabolism

Primary metabolism is the sum total of reactions, both degradative (catabolism) and synthetic (anabolism), which provide the organism with (i) energy, (ii) compounds for various syntheses, and (iii) key macromolecules, such as protein and DNA. Secondary metabolism involves mainly synthetic processes, which require energy. The end products, the 'special' or secondary' metabolites, play no role in the economy of the cell. Primary metabolism shows remarkable biochemical unity in morphological diversity. Secondary metabolites are species-, or strain-specific. Citric acid is a special case, which is produced by some fungi in excess, and is regarded also as a secondary metabolite. Secondary metabolites are derived from intermediates of primary metabolism by special routes.

Secondary Metabolites

If the secondary metabolites are of no use, why are they produced? None of the various suggestions, advanced to explain the utility of one or a few of them, holds good for all. Their ***raison d'etre*** may be explained as follows:

(i) Storage products According to this hypothesis, the secondary metabolites are reserve foodstuffs, produced when carbohydrates are in excess. Support for this hypothesis is derived from the fact that most secondary metabolites are consumed during carbohydrate shortage. This, however, fails to explain the enormous diversity of secondary metabolites. Most organisms store the excess carbon compounds in the form of fats, starch and glycogen.

(ii) Advantageous chemicals in habitat Secondary metabolites, which act as antibiotics, certainly provide advantage to the producer organism in the competitive habitat. But such cases are rare.

(iii) Products of erratic metabolism Secondary metabolites, are produced when the regulation of metabolism breaks down. But such organisms with malfunctioning regulatory processes will be swiftly eliminated by the unsparing forces of natural selection.

(iv) Exit for extra primary metabolites According to Bu' Lock (1961), *the secondary metabolites may not be important, but secondary metabolism is*. It provides an exit for excess primary metabolites, lest they become harmful. Since all secondary metabolites are produced from intermediates of primary metabolism, this is the most plausible explanation at present.

(v) Chance product According to Turner (1971), secondary metabolites are products of 'chance events'. Any function that they may show is accidental.

Whatever is their purpose or cause of production, secondary metabolites are important and some of them are now produced commercially.

Biosynthetic Pathways

Secondary metabolites originate from the intermediates of primary metabolism (Fig. 27.1), mainly from the following sources (1) Acetyl CoA or fatty acids, (2) mevalonic acid, (3) amino acids, (4) sugars, (5) shikimic acid or other aromatic acids, (6) intermediates of the TCA cycle, and (7) products of several metabolic pathways

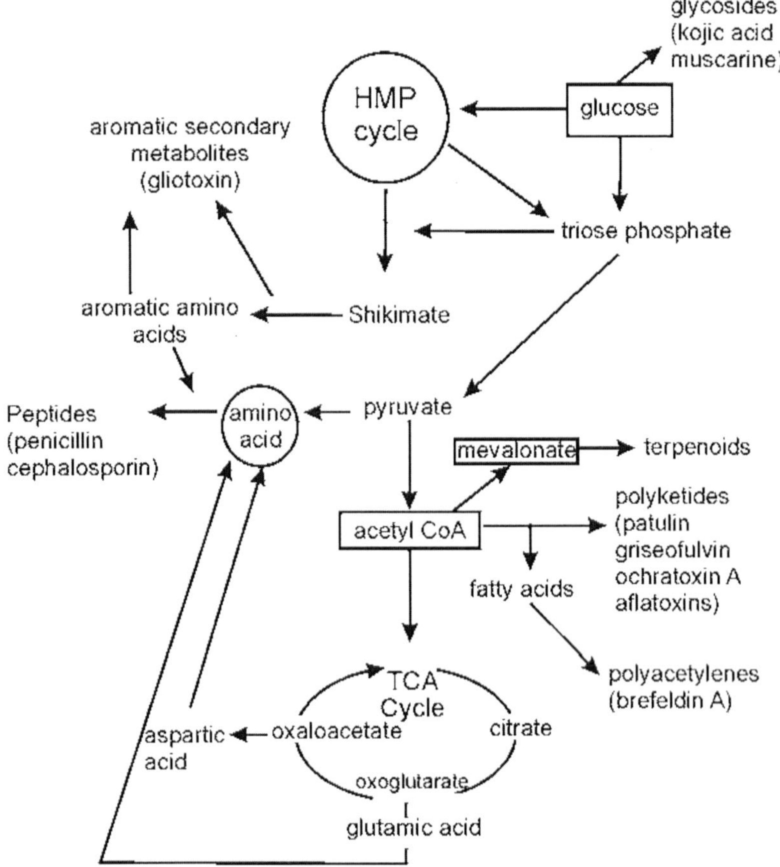

Fig. 27.1. The origin of secondary metabolites from the intermediates of the primary metabolism.

Elucidation of Pathways of Secondary Metabolism

Although techniques for studying biosynthetic pathways of secondary metabolism are same, as used for primary metabolism, their use has not been very useful.

1. **Mutation** While mutations have been successfully used in detecting biochemical blocks in primary metabolic pathways, these are not of much significance for studying secondary metabolites. Loss of ability to form a secondary metabolite goes undetected (except in case of pigment production).

2. **Use of cell free extracts** This has not found much use, because the growth of cells at the time of secondary metabolite production is in the declining phase, and cell-free extracts cannot be obtained from such cells. There is one more reason: the reason of attitude and interest.
3. **Use of labeled precursors** ^{14}C has frequently been used with success, as tracer in elucidating the secondary metabolic pathways.

The Intermediates of Secondary Metabolism

The precursors of the secondary metabolites are most difficult to pinpoint. Whether a chemical is normal, obligatory or intermediate in the biosynthesis is not easy to say. This is because of three reasons:

1. The enzyme systems are nonspecific, and accept a wide variety of substrates.
2. The suspected intermediate compound may, in fact, be in equilibrium with the true intermediate.

$$\begin{array}{l} A \rightarrow B \rightarrow C \\ \quad\;\;\downarrow \\ \quad\;\; B' \end{array} \qquad \begin{array}{l} B = \text{true intermediate} \\ B' = \text{suspected intermediate} \end{array}$$

The incorporation of B′ into C, and its presence would suggest that it is the normal intermediate for production of C.

3. The specificity of enzymes of the secondary metabolism is sometimes relative leading to a **metabolic grid** (Fig. 27.2). In this scheme, e_1 to e_4 will each catalyze the reaction of three substrates, though with varying efficiency. Consequently, there will be three pathways from A to C. The metabolic grid explains the inconsistencies in the pathways of a secondary metabolite. It also points out that the search for intermediates is a fruitless exercise in the 'zig zag' pathways, especially when it is known that several intermediates remain protein-bound.

Metabolic Pathways for Some Fungal Secondary Metabolites

Metabolic pathways, as shown in Fig. 27.1, suggest that acetyl-CoA is the most important single intermediate compound. Condensation of three molecules of acetyl-CoA produces **mevalonic acid**, the key intermediate in terpene biosynthesis. Linear condensation of acetyl-CoA with malonyl-CoA (formed by carboxylation of acetyl-CoA) leads either to **polyketides,** or to the **fatty acids**. Both give rise to several important secondary metabolites.

The intermediates of the TCA cycle provide carbon skeletons of several amino acids and of some secondary metabolites. Some important secondary metabolites and their producer organisms, given in Table 27.1, will be discussed.

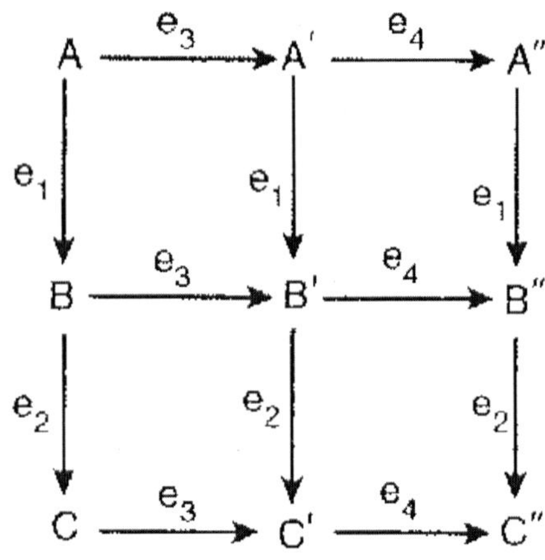

Fig. 27.2. Metabolic grid

Table 27.1 Secondary metabolites, their chemical nature and producer organisms.

Secondary metabolites	Producer organisms
A. Amino acid derivatives	
Peptides	
Penicillin	*Penicillium chrysogenum*
Cephalosporin	*Acremonium chrysogenum*
Amanita toxins (amatoxins and phallotoxins)	*Amanita phalloides, A. verna, A. virosa, Galerina* and *Lepiota* species
B. Acetate derivatives	
Polyketides	
Aflatoxins	*Aspergillus flavus, A. parasiticus, Penicillium puberulum*
Griseofullvin	*Penicillium griseofulvum, P. nigricans*
Ochratoxins	*Aspeprgillus ochraceus*
Patulin	*Penicillium expansum*
Zearalenone	*Fusarium roseum, F. moniliforme*
Terpinoids	
Gibberllic acid	*Glomerella fujikuroi*
Fusidane (Fusidic acid)	*Fusidium coccineum*

C. Sugar derivatives	
Glycosides	
Kojic acid	*Aspergillus flavus, A. oryzae* and *A. tamari*
Muscarine	*Amanita muscaria, Inocybe patouillardii, I. napipes*
D. Combination of pathways	
Ergot alkaloids	*Claviceps purpurea*

A. Amino acid derivatives: Peptides: Penicillin and Cerphalosporin (β lactam antibiotics)

Penicillin is the most important compound derived from fungi. The high penicillin-producing strains of *Penicillium chrysogenum* are now used for industrial production. In 1945, the strain of *P. chrysogenum* produced 250 units (cm^{-3}). The presently employed strains, produce as high as 1000 units (cm^{-3}). This has been possible by use of selected mutants of the organism. In addition to 'natural' penicillins (Penicillin F, G, K and X), many semi-synthetic penicillins have been prepared. Penicillin G is most commonly used. By chemical or enzymic hydrolysis of this, 6-APA (6 aminopenicillinic acid Fig. 27.3) is produced, from which semisynthetic penicillins with better therapeutic action are derived.

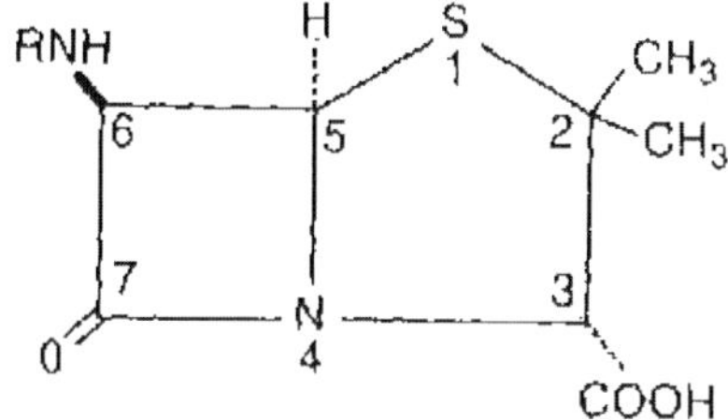

Fig. 27.3. 6-APA from which the semisynthetic penicillins are produced : (R – OH)

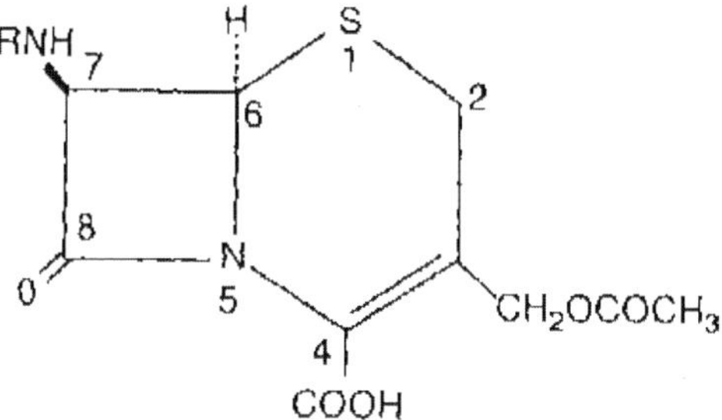

Fig. 27.4. Cephalosporin C: (7 aminccephalosporic acid, R=D – α aminodipoyl)

Cephalosporin Natural cephalosporin (Fig. 27.4) is a weak antibiotic and, like penicillins, semisynthetic cephalosporins are more effective. It is a peptide. The β-lactam ring is derived from cysteine biosynthetic pathway as lysine.

***Amanita* toxins** (amatoxins and phallotoxins) are cyclic peptides. Amatoxins are octapeptides while phallotoxins are heptapeptides.

B. Acetate-Derived Secondary Metabolites

(Polyketides and Terpenpids)

I. Polyketides. Polyketide route of secondary metabolite synthesis is most characteristic route in fungi. One molecule of acetate condenses with three or more molecules of malonate to form a polyketide (polyketomethylene) chain, which by transformation, gives rise to a host of secondary metabolites. Since malonate is derived from acetate, by carboxylation, these compounds are referred to as 'polyacetate' or 'acetate-derived' secondary metabolites. Some important polyketides are: **aflatoxins**, **griseofulvin**, **ochratoxins** and **patulin**.

$$\text{Acetate (1) + Malonate (3 or >)} \longrightarrow \text{Condensation}$$
$$\longrightarrow \text{Polyketide chain} \longrightarrow \text{Secondary metabolites}$$

Aflatoxins Primarily produced by *Aspergillus flavus* and *A. parasiticus,* aflatoxins are most prominently known as a potent liver carcinogen in humans and animals.

Griseofulvin (Fig. 27.5) is an antibiotic, active against systemic fungal infections of skin, hair and nails. Though it was first isolated from *Penicillium griseofulvum*, the antibiotic property was recognized only when it was isolated from *P. janczaewski* (=*P*, *nigricans*) by Brian *et al.* (1946). They called it "curling factor", as it induced curling in treated hyphae. Since then, griseofulvin has been isolated from about a dozen species of *Penicillium*; most important being *P. patulum*. Its several high griseofulvin-producing strains are currently used for industrial production. Griseofulvin has also been isolated from *Khuskia* (perfect state of *Nigrospora oryzae*).

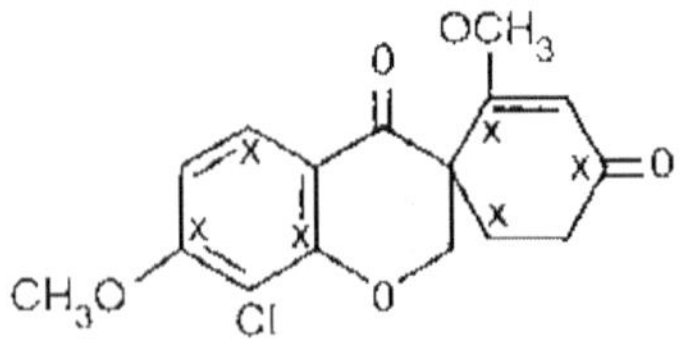

OCH_3 O O X X X X X X CH_3O Cl O

Fig. 27.5. Griseofulvin, antifungal antibiotic

Ochratoxins. These are mycotoxins, produced by *Aspergillus ochraceus*, *Penicillium viridicatum* etc. Ochratoxin A is the most important of the ochratoxins (A, B, C) and is responsible for nephrotoxicosis in sheep, cattle and pigs. It is also carcinogenic and has been found in coffee. (Details in Chapter 28).

Patulin. It is a mycotoxin, produced by *Aspergillus clavatus*, *Penicillium patulum*, *P. claviforme* and *P. expansum*. It is carcinogenic to mice and causes neurotoxicoses in cattle. (Details in Chapter 28).

Zearalenone is an aromatic compound synthesized from polyketide (or shikimic acid) pathway. It is a sex hormone of *Fusarium* which is effective on some other fungi also. It regulates perithecium formation and has estrogenic effect on mammals.

II. Terpenoids. Like polyketides, terpenoid secondary metabolites (or terpenes*) are also derived from acetate; but instead of linear condensation, the terpenes are built of branched isoperene units, formed from 3 acetate units (Fig. 27.6). Isoperene is very important unsaturated hydrocarbon, a 5-carbon molecule with a branched chain. It has two double bonds, with a single bond in between.

$$H_2C = \overset{\displaystyle CH_2 \atop \displaystyle |}{C} - CH = CH_2$$

Fig. 27.6 Isoperene molecule

The 'key' intermediate in terpene biosynthesis is mevalonate, which is most commonly used in the synthesis of terpenoid secondary metabolites. Gibberellins are diterpenes, while fusidic acid is a triterpene.

Isoperene units ⟶ mevalonate ⟶ Terpene Secondary Metabolites

Gibberellins: Gibberellic acid (GA3) and the 13 other naturally-occurring gibberellins (GA,-GA13), all have a **gibbane skeleton** (Fig. 27.7). These have been segregated in 4 groups, according to the substitutes on the basic skeleton. Gibberellic acid was discovered in rice plants, suffering from "bakane" disease, caused by *Gibberella fujikuroi*. The plants grew unusually tall and pale. The tallness was caused by this growth hormone. Gibberellic acid has found application in agriculture for improving the quality and yield of several crops, like grapes (especially the seedless varieties), oranges and lemons. It is also used in malting of barley.

---

* **Terpenes** are hydrocarbons resulting from combination of several isoperene units. **Terpenoids** (= isoprenoids) are modified terpenes in which the CH_3 groups have been moved or removed, or oxygen atoms added

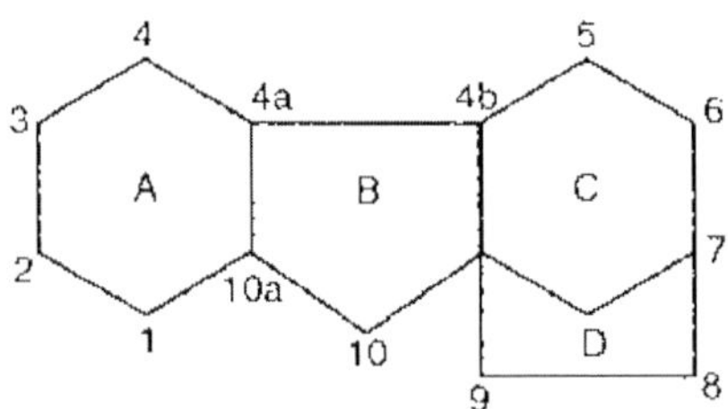

Fig. 27.7. Gibbane skeleton from which the various gibberellins are derived

1. **Fusidanes** include fusidic acid, helvolic acid and cephalosporin P1, which possess a **fusidane skeleton** (Fig. 27.8) and show antibiotic property. Only fusidic acid is clinically important; it is active against Gram+ve bacteria, especially against penicillin-resistant staphylococci. Fusidic acid is produced by *Fusidium coccineum*, *Micromucor ramanniana*, one species of *Cepholosporium*, and *Isaria kogana*.

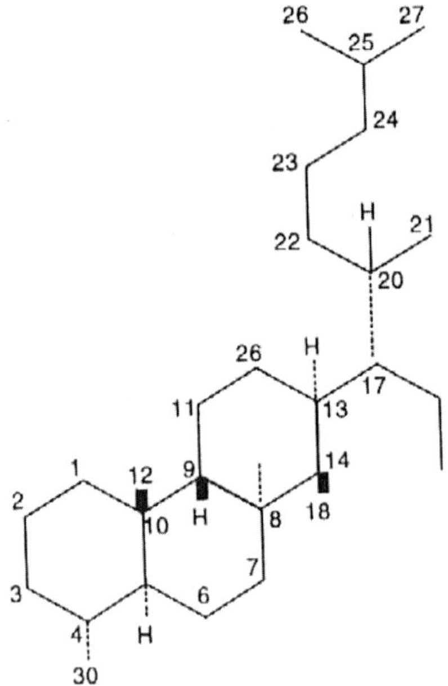

Fig. 27.8. The fusidane skeleton

C. Sugar derivatives : Glycosides: Kojic acid, Muscarine.

Kojic acid is the characteristic secondary metabolite of *Aspergillus flavus*, *A. oryzae* and *A. tamari.*

Muscarine is a group of quaternary ammonium compounds, produced by *Amanita muscaria*, *Inocybe patouillardii* and *I. napipes* (for details see Chapter 28).

D. Combination of pathways (Ergot alkaloids)

The ergot alkaloids are synthesized by several pathways. Tryptophan from the **shikimic acid pathway** is attached to an isoperinoid moiety from **mevalonate pathway** Ergot alkaloids are produced as a complex mixture of related compounds from several (branched) pathways.

Relationship between Secondary Metabolism and Growth

This aspect has been studied extensively. Borrow *et al.* (1961), while studying nitrogen-limited production of gibberellic acid by *Gibberella fujikuroi* (=*Fusarium moniliforme*), distinguished 3 phases of growth (Fig. 27.9): the **balanced phase, storage phase and maintenance phase** (the lag phase and lytic phases were discounted). During the balanced phase, the fungus grows at an exponential rate and takes up the nutrients in a constant ratio. This phase is terminated, when one of the nutrients, usually nitrogen or phosphorus, is exhausted. The fungus enters the storage phase, in which fats and carbohydrates accumulate, as evidenced by continued rise in the dry weight. After this, the fungus enters the maintenance phase, in which the dry weight becomes static. The production of secondary metabolites takes place during this phase only. The production continues until the medium gets exhausted, the endogenous reserves are depleted and autolysis or the terminal stage sets in.

Bu'Lock (1961, 1967) has used an alternative terminology for the growth phases. He divides it into 2 phases, the **trophophase** and the **idiophase**. In the trophophase (the balanced phase of Borrow *et al.*), the fungus grows exponentially and little or no secondary metabolite is produced. It terminates when nitrogen or phosphorus becomes limiting and the fungus enters the idiophase, in which the secondary metabolites are synthesized, until lysis starts. Essentially, both schemes are same.

Industrial Implications

The fact that production of secondary metabolites is inversely related to growth is made use of in the industrial production. As soon as the fungus attains some growth, the production of secondary metabolites is stimulated, by causing deficiency of carbon, nitrogen or phosphorus. In the production of 'ergot', through cultures of *Claviceps purpurea*, the balanced growth phase (or trophophase) is terminated by causing phosphate deficiency, which induces synthesis of ergot alkaloids.

In gibberellic acid production, the limiting factor is nitrogen. Penicillin production is controlled by lactose, which reduces growth and stimulates penicillin biosynthesis.

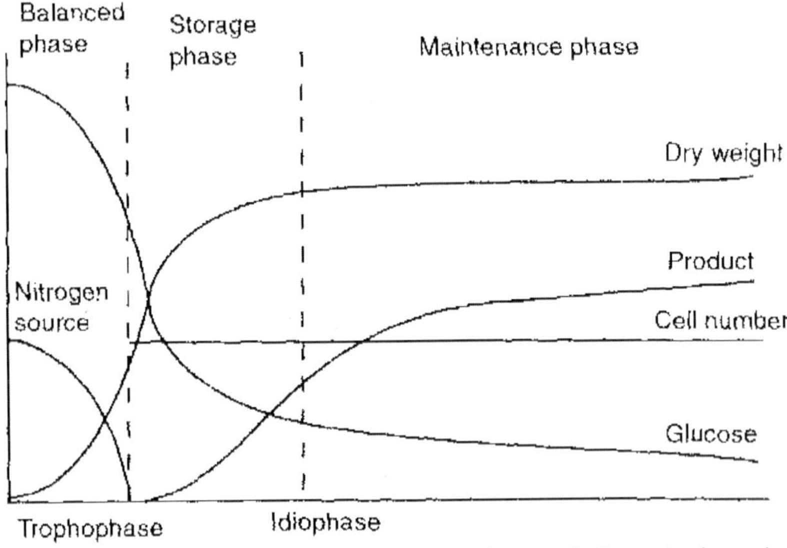

Fig. 27.9. Showing releationships between secondary metabolite production and growth phases. (The noemnclature of Borrow *et al.* is shown at the top and that of Bu Lock at the bottom of the figure)

Secondary Metabolite Production and the Morphological State of the Organism

The production of secondary metabolites is associated with the morphological state of fungus. For example, the carotenoid production occurs during the asexual phase of *Neurospora* and the sexual phase in *Blastocladiella*. Some other examples are: production of polyphenol oxidase at the time of conidia formation in *Penicillium*; tyrosinase production in the protoperithecium of *Neurospora*, production of several phenolic pigments during basidioma development in wood-rotting fungi.

The somatic differentiation is also associated with secondary metabolite production. A very good example is the association of cephalosporin production by *Cephalosporium acremonium* (= *Acromonium chrysogenum*) with the arthrospore stage. As shown in Fig. 27.10, a distinct linear relationship exists between percentage of arthrospores formed and amount of cephalosporin produced.

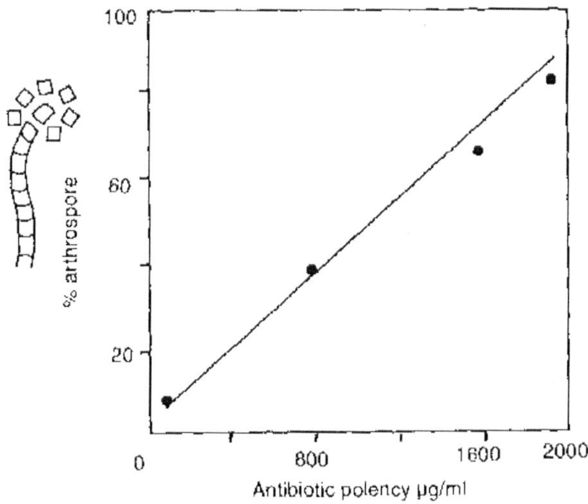

Fig. 27.10. Showing relationship between arthrospore formation and cephalosporin production in *Cephalosporium acremonium*.

REFERENCES

Borrow, A. *et al.* 1961. The metabolism of *Gibbrella fujikuroi* in stored culture. *Can J. Microbiol.,* **7:** 227.

Brian, P.W. 1960. Griseofulvin. *Trans. Brit. Mycol. Soc.*, **43:** 1 – 13.

Brian, P.W. *et al.* 1946. A substance causing abnormal development of fungal hyphae produced by *Penicillium janczaewski* zal. *Trans. Brit. Mycol. Soc.,* **29:** 173–181.

Bu'Lock, J.D. 1961. Intermediary metabolism and antibiotic synthesis. *Adv. Appl. Microbiol.,* **3:** 293 – 341.

Bu'Lock, J.D. 1967. Essays in Biosynthesis and Microbial Development. John Wiley, London.

Bu'Lock, J.D. 1974. Secondary metabolism in fungi and its relationship to growth and development. In: *The Filamentous Fungi* – 1. *Industrial Mycology.* (J.E. Smith and D.R. Berry eds.) pp. 33 – 56. Edward Arnold, London.

Nash, C.H. and Huber, F.M. 1971. Antibiotic synthesis and microbiological differentiation of *Cephalosporium acremonium. App. Microbiol.*, **22:** 6 – 10.

Turner, W.B. 1971. *Fungal Metabolites.* Academic Press, London and N.Y.

Turner, W.B. 1974. Commercially important secondary metabolites. In: *The Filamentous Fungi* Vol. I. *Industrial Mycology* (J.E. Smith and D.R. Berry, eds.) pp. 122 – 139. Edward Arnold, London.

SUMMARY

Fungi produce a variety of secondary (or special) metabolites which unlike primary metabolites have no obvious function. While primary metabolites of various organisms are essentially similar, the secondary metabolites are highly variable. Primary metabolism comprises catabolism and anabolism, while secondary metabolism involves mainly synthetic processes. Primary metabolism shows remarkable biochemical unity in morphological diversity of organisms, while secondary metabolites are species or strain-specific. Citric acid is a special case, which is a primary metabolite produced by some fungi in excess, and is regarded also as a secondary metabolite.

Secondary metabolites are regarded as (i) storage products, (reserve foodstuffs) which are consumed during carbohydrate shortage, (ii) as advantageous chemicals that provide advantage to the producer organism in the competitive habitat, (iii) product of erratic metabolism, produced when the regulation of metabolism breaks down, (iv) exit for extra primary metabolites lest they become harmful, or (v) chance products. Since all secondary metabolites are produced from intermediates of primary metabolism, secondary metabolism and secondary metabolites provide an exit. So, the exit hypothesis (iv) appears to be more appropriate explanation for secondary metabolite production.

Biosynthetic Pathways of secondary metabolites originate from the intermediates of primary metabolism, mainly from the following sources (1) Acetyl CoA or fatty acids, (2) mevalonic acid, (3) amino acids, (4) sugars, (5) shikimic acid or other aromatic acids, (6) intermediates of the TCA cycle, and (7) products of several metabolic pathways.

Elucidation of pathways of secondary metabolism by employing mutation, cell-free extracts and labeled precursors has not been possible. Similarly, the precursors of the secondary metabolites are most difficult to pinpoint. Whether a chemical is normal, obligatory or intermediate in the biosynthesis is not easy to know. The **metabolic grid** explains the inconsistencies in the pathways of a secondary metabolite. It also points out that the search for intermediates is a fruitless exercise in the 'zig zag' pathways.

Metabolic pathways suggest that **acetyl-CoA** is the most important single intermediate compound. Condensation of three molecules of acetyl-CoA produces **mevalonic acid**, the key intermediate in terpene biosynthesis. Linear condensation of acetyl-CoA with malonyl-CoA (formed by carboxylation of acetyl-CoA) leads either to **polyketides,** or to the **fatty acids**. Both give rise to several important secondary metabolites. The intermediates of the TCA cycle provide carbon skeletons of several amino acids and of some secondary metabolites. Some important secondary metabolites are classified as: (1) **Amino acid derivatives**: Peptides: penicillin and cephalosporin (β lactam antibiotics), *Amanita* toxins (cyclic peptides), (2) **Acetate-derived secondary metabolites**: (i) Polyketides (aflatoxins, griseofulvin, ochratoxins, patulin and zearalenone) (ii) Terpenoids (gibberellins and fusidic acid), (3) **Sugar derivatives**: Glycosides (kojic acid, muscarine), and (4) **Combination of pathways** (ergot alkaloids).

There is an inverse relation between growth and secondary metabolite production. When the fungus grows exponentially (in the balanced phase or trophophase), there is no secondary metabolite production. When the nutrients are exhausted, especially nitrogen or phosphorus, the growth stops and the production of secondary metabolite starts (maintenance or idiophase). The production continues until the medium is exhausted and autolysis of the fungus starts.

In the industrial production the above relation between growth and secondary metabolite production is utilized. In ergot alkaloid production the balanced growth phase (or trophophase) is terminated by causing phosphate deficiency, which induces synthesis of ergot alkaloids. Gibberellic acid production is induced by causing nitrogen deficiency. In penicillin production growth is reduced by lactose, which stimulates penicillin biosynthesis.

Secondary metabolite production is also linked with the morphological phase of the fungus: Asexual phase (carotenoid production by *Neurospora*, polyphenol oxidase by *Penicillium*, cephalosporin by *Cephalosporium*); Sexual phase (carotenoid by *Blastocladiella*, tyrosinase by *Neurospora* and phenolic pigments by basdima during wood-rotting phase

TEST QUESTIONS

1. Distinguish between primary and secondary metabolism. What are the possible reasons and purpose for the synthesis of secondary metabolites? Which explanation appears to be more true.
2. Which techniques have been used to study the pathways of secondary metabolism and with what result?
3. Why is it difficult to pinpoint the precursors of secondary metabolites? Explain the "metabolic grid".
4. Which are the four types of secondary metabolites? Give examples.
5. Answer in brief:
   a. Which is the most important compound in the biosynthesis of secondary metabolites?
   b. Which are the two acetate-derived secondary metabolites?

c. How are terpene secondary metabolites synthesized?

d. How are polyketide secondary metabolites synthesized from acetates?

e. Which is the key intermediate in terpene biosynthesis?

f. Name two secondary metabolites which are glycosides.

g. Which secondary metabolite is produced by a combination of pathways?

h. What is the relation between growth of the organism and secondary metabolite production?

i. Which two nutrients when exhausted from the medium trigger secondary metabolite synthesis?

6. Fill in the blanks:

a. Secondary metabolites are derived from the intermediates of the -------- ------------.

b. ---------------- is regarded both a primary and secondary metabolite.

c. Terpenes are made of ------------------ units.

d. Condensation of three molecules of acetyl-CoA leads to formation of -------------.

e. Linear condensation of acetyl-CoA with mevalonic acid forms either ------------ or -----------.

f. There is a linear relationship between ----------- production and ----------- synthesis by *Cephalosporium acremonium.*

g. *Amanita* toxins are -------------- derivatives.

h. Zearalenone is an aromatic compound and example of ------------ secondary metabolite.

i. The 'key' intermediate in ---------------- biosynthesis is mevalonate.

j. Gibberellins are diterpenes, while -------------------- is a triterpene.

k. Linear condensation of acetyl-CoA with malonyl-CoA leads either to polyketides, or ----------.

28

Mycotoxins and Mycotoxicoses
(Fungal Toxins Affecting Animals and Man)

That fungi produce toxic substances was suspected ever since some mushrooms were found to be poisonous. Ergot toxicity has been known since antiquity. In spite of these clues, mycotoxicoses remained "neglected diseases" until 1960, when thousands of turkey poults died in England, in one stroke, with turkey X disease (Sargeant *et al.*, 1961). The disease was caused by eating 'moldy' peanuts (groundnuts) infected with *Aspergillus flavus*. The causative agent was found to be a toxic fungal metabolite, called **aflatoxin** (toxin of *Aspergillus flavus*). Mycotoxicoses could not be neglected any more. The term **mycotoxin** was coined in 1962 in an aftermath of the mysterious turkey X disease (see Bennett and Klich, 2003). Soon the mycotoxin study was extended to include a number of previously-known fungal toxins (e.g. ergot alkaloids).

The decade following the aflatoxin discovery, was a period of intense research on mycotoxins and mycotoxicoses. The emphasis readily shifted from economic losses, due to moldy growths on grains, to problems of health hazard. Aflatoxin B_1 (AFB_1), which was toxic to many animals, was discovered to be a strong hepato-carcinogen (chemical causing liver cancer) for rats, trouts and human beings. In a world already sensitive to threatened ecology, the finding of still another menace to our food and feed supply was a serious matter indeed. Some 300-400 compounds are now recognized as mycotoxins of which about a dozen groups regularly receive attention as threats to animal and human health. Until recently, the attention was more on dietary and occupational exposure to mycotoxins, but now there is a growing interest in 'indoor molds', which are associated with respiratory disorders.

Concepts and Terminology

The diseases of animals, including man, caused by the invasion of host tissue by the fungus is called **mycoses** (sing. mycosis). **Mycotoxicoses,** on the other hand, are poisonings of the host, resulting from ingestion of toxic substances of fungal origin. These include ingestion of poisonous mushrooms, ergot sclerotia and food or feed made toxic by metabolic products of fungi. The toxic chemicals are called **mycotoxins** or fungal toxins. The toxins produced by fungi that are injurious to plants are called **phytotoxins.** Although it is logical to designate all fungal toxins as mycotoxins, and the diseases they cause as mycotoxicoses, these terms (mycotoxins and mycotoxicoses) are used only with regard to animal diseases. We shall study some mycotoxins produced in moldy food/feed-stuffs, ergot toxins, and the toxins of mushrooms. Some important mycotoxins and the organisms that produce them are listed in Table 28.1.

Table 28.1. Mycotoxins and their producer organisms.

Mycotoxins	Producer organisms
A. Toxins of food/feed stuffs.	
Aflatoxins	*Asppergillus flavus, A. parasiticus, Penicillium puberulum*
Zearalenone	*Fusarium roseum, F. verticillioides (= F. moniliforme)*
Trichothecenes	
Alimentary Toxic Aleukia	*Fusarium poae, F. sporotrichoides*
Deooxynivalenol	*Fusarium* spp.
Stachybotrys toxins	*Stachybotrys chartarum*
Fumonisins	*Fusarium verticillioides (=F. moniliforme)*
Sporodesmins	*Pithomyces chartarum*
Ochratoxins	*Aspergillus ochraceus, Penicillium viridicatum*
Yellowed-rice toxins	Species of *Penicillium* and *Aspergillus*
Tremorgens	*Aspergillus flavus, A. fumigatus, Penicillium cyclopium, P. crustosum* etc.
Penicillic acid	*Penicillium puberulum, Aspergillus ochraceus*
Patulin	*Penicillium expansum*
B. Ergot toxins	*Claviceps purpurea* and other spp.
C. Mushroom toxins	
Amanita toxins	*Amanita phalloides, A. verna, A. virosa*
Orellanine	*Cortinarius orellanus*
Gyromitrin	*Gyromitra esculenta*
Muscarine	*Inocybe, Clitocybe* and *Amaniat muscaria*
Ibatonic acid	*Amanita muscaria, A. pantherina*
Psilocybin and Psilocin	*Psilocybe mexicana, Conocybe* spp. and *Panaeolus* spp.

A. MYCOTOXINS OF FOOD/FEED STUFFS

Those mycotoxins, which appear in culture filtrates, but are not formed in moldy stuffs, will be excluded from this group. The presence of toxin in the moldy food is known either by bioassay (i.e. feeding some test animals on the suspected substrate) or by chemical means. Every moldy grain is not toxic. Also, the mere presence of a potential toxin-producing fungus does not by itself indicate toxicity. It has been reported that some toxin-producing strains of *Aspergillus flavus*, when allowed to grow on moist grain, along with other natural microflora, produced little or no toxin.

AFLATOXINS

Aflatoxins, which are the most thoroughly investigated toxins, are among the most carcinogenic substances known. It continues to be the most important mycotoxin. These are produced by *Aspergillus flavus*, *A. parasiticus* and *Penicillium puberulum*. *A. parasiticus* is the most potent producer of aflatoxins. Peanuts and their products, cotton seeds and copra are the most favourable substrates and are called "high aflatoxin risk materials".

Aflatoxins are highly oxygenated heterocyclic compounds. The eight characterized aflatoxins are: B_1, B_2, G_1, G_2 (named on the blue or green fluorescence in UV light), M_1, M_2 (first detected in milk of cows fed on groundnut meal), and B_2a and G_2a (derivatives of B_2 and G_2). Among the eight forms, B_1 B_2 G_1 and G_2 are more commonly encountered. These may occur together or independently.

Chemical Structure

The structures of aflatoxins B_1 and G_1 (the first two aflatoxins discovered and initially named as B and G, before the report of B_2 and G_2) were suggested by Asao *et al.* (1963, 1965) and confirmed by X-ray analysis (Fig. 28.1).

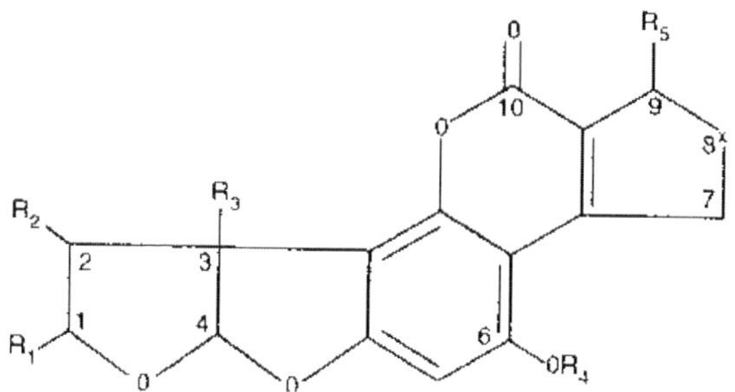

Fig. 28.1. Basic plan of affatoxin molecules.
R1 ... R5 and X vary in the various derivatives

The various derivatives vary in their toxicity. Aflatoxin B_1 is most common and also the most toxic. The furofuran ring of aflatoxin is essential for the toxic and carcinogenic activity. The double bond in the terminal furan ring determines potency. Detoxification of B_1 by rat liver involves opening of the furan ring and formation of branched aliphatic side chain with terminal aldehydic groups. The lactone portion of the molecule is also important, as illustrated by greater activity of B_1 than G_1. Reduction of the ketone of the terminal cyclopentane ring of B_1 also causes a loss of potency. Reduction of B_1 and G_1 to B_2 and G_2 also results in decreased toxicity. These on hydroxylation, give rise to B_2a and G_2a (Fig. 28.2). M_1 is produced from B_1 by hydroxylation, which on reduction forms M_2. Similarly, G_1 gives rise to GM_1, by hydroxylation, which on reduction forms GM_2.

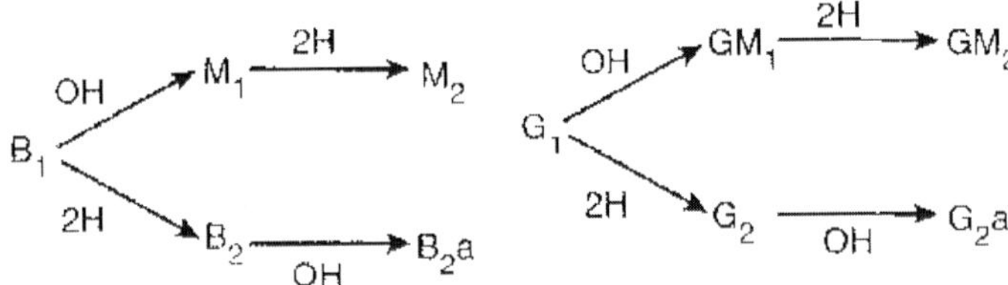

Fig. 28.2. Derivatives of aflatoxins B_1 and G_1 (M = milk aflatoxins).

Effect of Aflatoxins on Animals

The aflatoxin B_1 is one of the most important causes of liver cancer in rats, trouts and humans. AFB_1 is classified as a known human carcinogen by the International Agency for Research on Cancer. After entering the body, the aflatoxins are metabolized by the liver to a reactive epoxide intermediate or hydroxylated to become less harmful aflatoxin M1. High level exposure produces an acute hepatic necrosis, resulting later in cirrhosis, and /or carcinoma of the liver. No animal is immune to the acute toxic effects of aflatoxin, including humans. However, humans have an extraordinary high tolerance for aflatoxin and rarely succumb to aflatoxicose.

Detection in Humans

There are two main methods: In the first method **AFB_1-guanine adduct** is measured in urine. Its presence indicates exposure to AFB_1 in last 24 hours. So, this technique measures only recent exposure. This is because the adduct has a short half life. In the second method **AFB_1-albumin adduct** level is measured in blood serum. This technique measures long term exposure extending to several weeks or months.

Mechanims of Aflatoxin Action at Molecular level

The initial site of aflatoxin B_1 action is DNA of the nucleus. It binds with DNA and prevents its transcription. As a result, protein synthesis is inhibited. Degranulation of ER is seen in the electron photomicrographs of aflatoxin-treated cells.

Aflatoxin B_1 has been shown to be genotoxic *i.e.* mutagenic carcinogen. According to Groopman (see Barrett, 2000), the carcinogenic potency of AFB_1 is expressed when it is activated by cytochrome P_{450} and other oxidative enzymes. The AFB_1 is transformed into several products, but **AFB_1-exo 8,9-epoxide** is responsible for inciting the carcinogenesis in liver. It is shunted to liver cells by glutathione S-transferase. The epoxide molecules intercalate between DNA base pairs and produce **N^7 guanyl adduct**. Similar adducts have been detected in kidney or lung. Some of these adducts are resistant to DNA repair systems and, thus, bring about mutation and carcinogenesis. AFB_1-induced liver tumors show *p53* tumor suppressor gene mutation, which, thus, leads to the carcinogenesis. It has been found that AFB_1-induced liver cancer is enhanced in individuals, who are also infected by the carcinogenic hepatitis B virus. Unfortunately, there is no antidote for aflatoxin.

$$AFB_1 \xrightarrow{\text{Cyt } P_{450}} AFB_1 \text{ exo } 8{,}9 \text{ epoxide} \longrightarrow \text{Liver } N^7 \text{ guanyl}$$

$$\text{adduct} \longrightarrow \text{Carcinogenesis} \longrightarrow \text{liver tumors}$$

Detoxification of Contaminated Food Stuffs

The bacterium *Flavobacterium* removes aflatoxin completely from contaminated milk, peanut, peanut oil, butter and corn.

ZEARALENONE (ESTEROGENIC TOXIN)

Zearalenone is produced by several species of *Fusarium*, growing on corn, barley, wheat and other crops. When consumed, it causes esterogenic symptoms in animals and man. The important zearalenone-producing species are *F. roseum*, *F. graminearum*, *F. culmorum*, *F. tricinctum*, *F. oxysporum* and *F. moniliforme*; *F. roseum* being the most potent-producer.

The esterogenic symptoms in swine, most sensitive to zearalenone, involve primarily the genital organs, which get hypertrophied resulting in infertility and abortion. The toxin has feminizing effect on young males, who develop atrophy of testes and enlargement of mammary glands. These esterogenic effects are accompanied by diarrhoea, emesis, refusal to feed, loss of weight and hemorrhagia. A period of low temperature favours the disease. In addition to swine, other animals like rats, mice, guinea pigs, rabbits, turkey poults, chicks and lambs also show esterogenic syndrome when fed on 'moldy' corn infected by *Fusarium*.

Esterogenic effects on human beings are strong. Premature pubertal changes have been observed in children, who showed zearalenone or its metabolites in their blood. Zearalenone is carcinogenic and teratogenic (i.e. cause defects in embryos). The International Agency for Research on Cancer has classified zearalenone as a **2A carcinogen**, depicting the highest possible carcinogenicity. Zearalenone and its metabolic products, particularly α zearalenone and β zearalenone, bind to esterogen receptors. Dees *et al.* (*see* Barrett, 2000) have reported that zearalenone stimulates esterogen-receptor human breast cancer cells.

Zearalenone is a sex hormone of *Fusarium*, which is effective on some other fungi also that do not synthesize it (Nelson, 1971).

The biosynthesis of zearalenone is induced and activated at low temperatures (12-14°C), which is not the temperature optimum for growth. At this temperature, the organism is in a state of physiological stress. The pathway of synthesis and chemical structure of zearalenone (Fig. 28.3) have been elucidated. It is 6-β-resorcyclic acid lactone, with an empirical formula $C_{18}H_{22}O_5$. The biosynthesis can be inhibited by vapona (dimethyl-2, 2-dichlorovenyl phosphate), without affecting the vegetative growth. It is highly significant in controlling esterogenism. Zearalenone is a white crystalline solid, which exhibits blue-green fluorescence at 360 nm and more intense green fluorescence at short wavelength of UV (260nm).

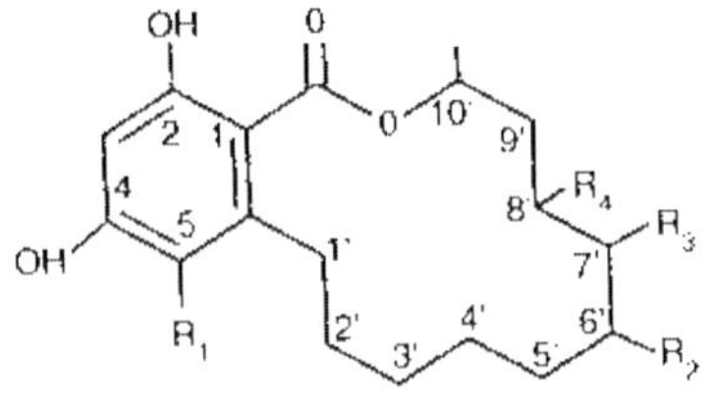

Fig. 28.3 Zearalenone (6, β resorcyclic acid lactone), responsible for esterogenic symptoms in swine.

The word toxin for zearalenone is a misnomer because zearalenone is biologically potent but hardly toxic. Rather, it sufficiently resembles **17 β estradiol** the principal hormone produced by the human ovary. So, zearalenone is better classified as a non-steroidal esterogen or **mycoesterogen**.

TRICHOTHECENE TOXINS

Trichothecenes are a group of 60 sesquiterpenoid mycotoxins, produced by several fungi, mainly species of *Fusarium*. The toxins possess a ring system named trichothecane, after the toxin trichothecin, the first reported member of this group from *Trichothecium roseum*. These are produced by species of *Fusarium*, *Cephalosporium*, *Myrothecium*, *Trichoderma*, *Trichothecium* and *Stachybotrys*. We shall examine three important trichothecene toxins viz. Alimentary Toxic Aleukia (ATA) toxin, Deoxynivalenol and *Stachybotrys* toxins.

Alimentary Toxic Aleukia (ATA) Toxin

ATA is a human disease known in Russia since 19th century. The symptoms of ATA include typical spots on the skin, inflammation of the skin, leukopenia, agranulocytosis, necrotic angina, hemorrhagic diathesis, sepsis and exhaustion of the bone marrow. It is caused by consumption of over-wintered moldy millet and other grains (wheat, rye, oats, and buck-wheat) infected with *Fusarium sporotrichiodes* (=*F. tricinctum*) and *F. poae*. Diacetoxys-cirpenol and T2 are associated with ATA. These are cytotoxic and immunosuppressive.

The T-2 toxin (the skin-irritating factor) is responsible for skin inflammation and spots. It is important to note that the exhaustion of bone marrow is an important symptom in ATA, and the T-2 and related trichothecenes are known to cause severe cellular destruction of the actively dividing cells of bone marrow.

Deoxynivalenol (vomitotoxin)

The toxin (Fig. 28.4), produced by several species of *Fusarium*, is one of the most common mycotoxins contaminating grains. It causes nausea, vomiting and diarrhoea, and therefore also called vomitotoxin or 'food-refusal factor'. Though least toxic among the trichothecenes, it has been linked to large scale poisonings, human disease and animal production problems throughout the world.

Deoxynivalenol has also been shown to have immunological effects in animal models (mice). It induces cytokines, immune system factors, causing inflammation. The experimental mice developed symptoms similar to human **Ig A--nephropathy**, a kidney disorder showing inflammation (see Barrett, 2000).

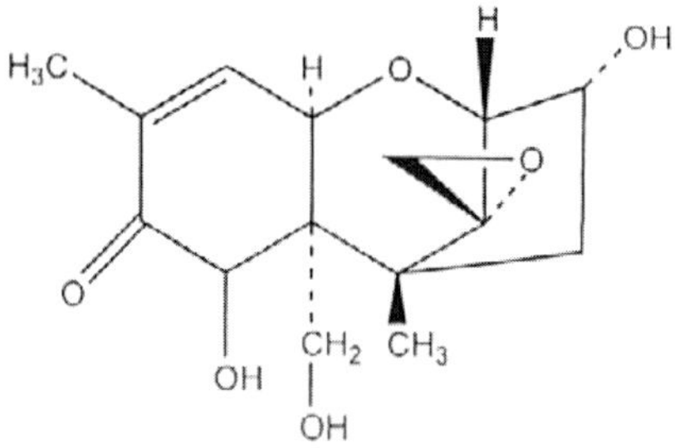

Fig. 28.4 Deoxynivalenol (vomitotoxin)

Stachybotrys Mycotoxins

Trichothecene toxins, produced by *S. atra* (formerly *S. chartarum*) are termed ***Stachybotrys* toxins**. These include satra-toxins, roridins, atranones and verrucarens. These toxins can become air-borne and cause respiratory disorders. These are responsible for the idiopathic pulmonary hemorrhage (or bleeding of lungs) in infants. This is an example of **indoor mycotoxicosis**, a new area of concern in mycotoxin research. These affect people who are atopic i.e. prone to allergy.

Though the precise mechanism of *Stachybotrys* mycotoxins are not known, they are believed to be allergens that trigger or aggravate allergy. They are potent protein synthesis inhibitors and interfere with a single binding site on the ribosome, which causes breakdown of *initiation*, *elongation* or *termination* stages of the protein synthesis.

FUMONISINS

Fumonisins were first described and characterized in 1998. These are synthesized by condensation of the amino acid alanine with an acetate-derived precursor. Fumonisins (B_1, B_2, B_3) are produced by several species of *Fusarium*, mostly by *F. verticillioides* (formerly *F. moniliforme*,= *Gibberella fujikuroi*), that grow on corn.

These cause "hole in the head syndrome' (**l**euco**e**ncephalo**m**alacia, LEM) in horses, besides brain hemorrhage and necrosis. Liver and kidney damage are pronounced in rodents (moles, rats, squirrels) sheep and rabbits. In swine, it causes porcine pulmonary edema, (a fatal condition in which fluid collects in lungs). Fumonisins are carcinogenic in rodents, although response differed with species and sex. Male rats, fed with Fumonisins B_1 developed liver and kidney cancers, while females developed only liver cancer. Fumonisins have been linked with oesophegal cancer and neural tube defects, NTB (in embryo) in humans, besides acute diarrhoea and abdominal pain.

Fumonisin B_1 is the most potent member of fumonisins. In India high levels of fumonisin B_1 have been observed in maize infected by *Fusarium moniliforme* (Chatterjee and Mukherjee, 1994). Highest fumonisin level was found in rain-affected maize samples. An outbreak of fumonisin toxicosis occurred in Andhra Pradesh in 1997, in which about 100,000 hens, in the age group of 36-64 weeks, were affected, out of which 10% died (Vasanthi and Bhat, 1998).

Fumonisin toxicity is associated with inhibition of *ceramide synthase*, a key enzyme in the biosynthesis of **sphingolipids**. The entire metabolism of sphingolipids is disrupted. The sphingolipids and their derivatives serve as messengers for many life or death decisions, including whether the cell enters apoptosis (cell death) or replication and cell division.

SPORIDESMINS (=Pithomycotoxicosis)

The sporodesmins, produced by *Pithomyces chartarum* (=*Sporodesmium bakeri*) on pastures, cause **facial eczema** of grazing cattle (Pithomycotoxicosis). The disease, which is common in North Island, New Zealand and parts of Australia and France, produces **inflammation and scabbing** of eye lids, lips, ear and face. Isolation and chemical characterization resulted in the discovery of four types of sporodesmins, viz. sporodesmin, sporodesmin B, C and D. These are identified as **e**pidi**t**hio**p**iperazine (ETP). Sporodesmin is the major constituent and is about five times more active than sporodesmin B (Fig. 28.5); sporodesmins C and D are least toxic. The toxin is cytotoxic. It disrupts cellular functions via oxidative alteration of cysteine residues of key proteins. Thus, it affects many enzyme systems.

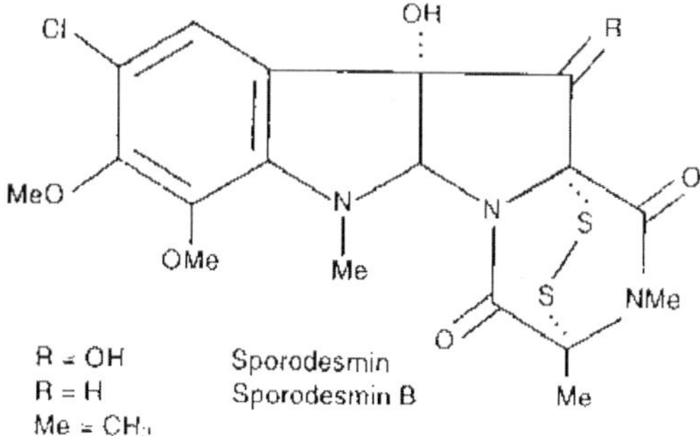

Fig. 28.5 Sporodesmins, the toxins causing facial eczema of cattle

OCHRATOXINS

Among *Aspergillus* toxins, only ochratoxin is potentially as important as aflatoxin. Ochratoxins A, B and C (OA, OB and OC, Fig. 28.6) are produced by *Aspergillus ochraceus* and *Penicillium viridicatum* on stored moldy maize, corn, peanuts, beans and mixed animal feeds. *P. viridicatum* is more potent producer of ochratoxins. Ochratoxin B is non-toxic. OA is a **carcinogen, nephrotoxin** and **hepatotroxin**. As carcinogen, it has been linked to tumors in human urinary tract. It causes tubular necrosis of the kidney, mild degeneration of liver and enteritis of the small intestine in ducklings, mice and rats. It has been implicated in endemic nephropathy in humans.

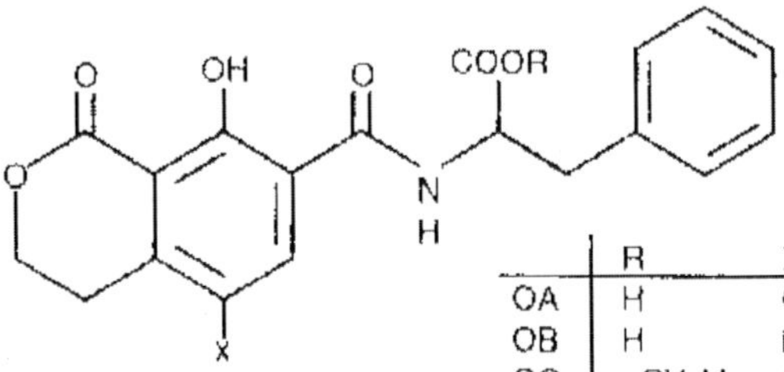

	R	X
OA	H	C1
OB	H	H
OC	$-CH_2Me$	C1

Fig. 28.6 Ochratoxins A, B and C (OA, OB, and OC) damage kidneys and liver of several animals

Ochratoxin has been detected in blood and other animal tissues and in milk including human milk. It is frequently found in pork for human consumption. The International Agency for Research on Cancer has rated ochratoxin as a possible human carcinogen.

In a study of ochratoxin contamination of coffee, undertaken by National Institute of Nutrition, Hyderabad (Vasanthi and Bhat, 2000), 60% of the test samples showed ochratoxin contamination. However, majority (53%) of the samples contained the ochratoxins in amounts, which were under the permissible limit of 5 mg/kg. Higher levels (above 10mg) were observed in 6% of coffee beans. High levels, up to 176 mg/kg were observed in visibly moldy coffee samples.

YELLOWED RICE TOXINS

Stored rice, turned yellow by growth of various species of *Penicillium*, causes mycotoxicoses (yellowed rice toxicoses) of human beings, which include such diseases as 'beriberi', hepatitis, liver cirrhosis, hepatoma, degeneration of kidneys and disorders of the circulatory system. The important toxins are: citreoviridin, citrinin and luteoskyrin (Fig. 28.7).

Citreoviridin is a polyene toxin, produced by *P. citreoviride*. It is the actual cause of cardiac beriberi in man and animals. Deficiency of vitamin B_1 and C_1 has a secondary role.

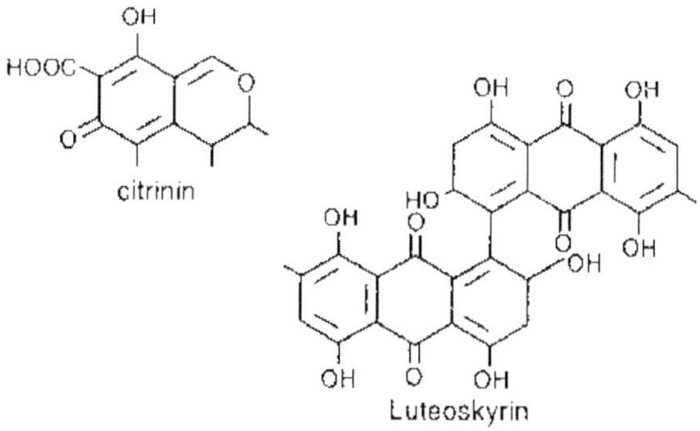

Fig. 28.7 Citrinin and luteoskyrin, the yellowed-rice toxins

Citrinin, a toxic yellow pigment is produced by over a dozen species of *Penicillium* (including *P. citrinum, P. viridicatum* and *P. camemberti*) and several species of *Aspergillus* (*A. oryzae, A. terreus* and *A. niveus*). *P. camemberti* strains are used in the preparation of cheese while *A. oryzae* is used in the preparation of soya sauce. Recently it has also been reported from species of *Monoascus*. In addition to yellowed rice, it is found in moldy-barley, wheat, rye and dried fish. *P. citrinum* causes yellowing of polished wheat, which fluoresces on UV irradiation. Citrinin acts as a **nephrotoxin** in all animals tested. It causes nephropathy and other renal disorders in animals. It acts synergistically with ochratoxin-A to inhibit RNA synthesis.

Luteoskyrin (LS), produced by *P. islandicum*, is carcinogenic, primarily affecting liver and causing hepatitis (yellowed rice disease).

TREMORGENS (TREMORGENIG TOXINS)

As is evident by the name, these toxins cause tremors in the body of animals (sheep, horses, cows) and humans, accompanied by diuresis. Finally, convulsive seizures, which are not inhibited by drugs, occur and prove fatal. The toxins, which are neurotoxins, are produced on a variety of foods and feeds. The tremorgens that have been isolated and chemically characterized are: Alfatrem, Cyclopiazonic acid, Fumitremorgin B, Janthitrems, Lolitrems, Paxilline, Paspaline, Paspalicine, Paspalinine, Paspialitrem, and Penitrem. Most of the fungal trremorgens contain a **modified indole moiety** and are produced by certain species of *Aspergillus*, *Penicillium* and *Claviceps*.

Cyclopiazonic acid (CPA) is produced by *Penicillium cyclopium* (now *P. aurantigriseum*), and many other species of *Penicillium* as well as *Aspergillus flavus*. It was originally isolated from same sample of mouldy groundnut that had caused the turkey X disease. It may have contributed to the severity of the aflatoxicoses. It is an indole tetramic acid, which is a specific inhibitor of Ca-dependent ATPase and disrupts ion transport across cell membranes.

PENICILLIC ACID

The toxin penicillic acid (PA) was isolated in 1913 by Alsberg and Black, from corn meal, overgrown by *Penicillium puberulum.* Its structure (Fig. 28.8) was established by Brikinshaw *et al.* (1936). It exists in two tautomeric forms : a substituted keto acid and a hydroxylactone.

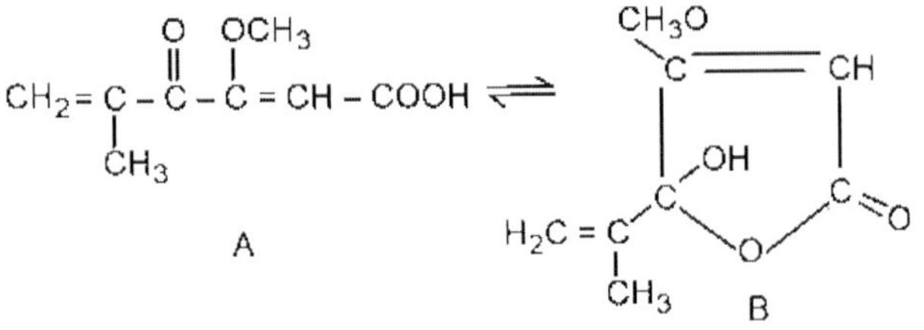

Fig. 28.8 Penicillic acid, tautomeric forms.
A, as substituted ketoacid; B, hydroxylactone

It is a carcinogen, which incites local tumors in experimental rats and mice. Several species of *Penicillium* and *Aspergillus* are capable of producing penicillic acid, the most potent being *P. puberulum*, *P. cyclopium* (now *P. aurantigriseum*) and *Aspergillus ochraceus*.

PATULIN

Patulin, which is also **carcinogenic**, has been isolated from species of *Penicillium* and *Aspergillus*, under different names e.g. clavacin, clavatin, claviformin, expansin, mycoin and penicidin. Patulin is formed in apple fruits infected by *Penicillium expansum*, a common fruit rot pathogen. Commercial apple juices have been found containing patulin. It is also found in moldy-bread and other bakery products. It is carcinogenic to mice, and causes neurotoxicosis in cattle Thus, patulin is a potential health hazard. It also shows antibacterial and antifungal properties. The structure (Fig. 28.9) was elucidated by Woodword and Singh (1949).

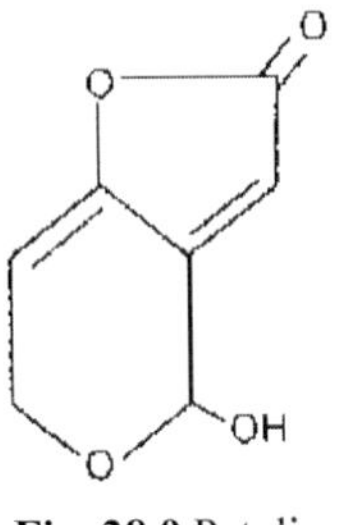

Fig. 28.9 Patulin

B. ERGOT TOXINS

Ergot, the sclerotia formed by various species of *Claviceps* on rye and different grasses, is a 'treasure house' of pharmaceutically-active compounds. *C. purpurea*, which infects 'rye' (*Secale cereale*), is one of the oldest known producers of mycotoxins. Epidemics of ergot-poisoning (egotism) are known since middle ages in the West and Middle Europe, where rye has been used for making bread. The disease was known as "holy fire" or "St Anthony's fire" because of the high fever developed by the patients, and the faith that it could be cured by a visit to St Anthony's church. Ergotism is of two types: **gangrenous** and **convulsive.** Gangrenous ergotism is accompanied by fiery pain, swelling and gangrene in the limbs. Convulsive ergotism is accompanied by convulsions and hallucinations.

Ergotism is a rare event now due to the modern milling operations, but not completely eradicated. It is still an important veterinarian problem. Ergot shows that the line between toxin and drug is thin. Ergot has changed from a feared poison to a substance highly useful in medicine. Constituents of the ergot are now chemically characterized. The most important are alkaloids, some of which have been also found outside *Claviceps* (in *Aspergillus*, *Penicillium* and some plants).

The ergot alkaloids are most fascinating fungal metabolites. These alkaloids are divided into 2 groups (1) amides of lysergic acid (Fig. 28.10) found chiefly in sclerotia of *Claviceps purpurea*, and (2) clavine alkaloids or clavines, found mainly in the sclerotia of other species of *Claviceps*, formed on wild grasses.

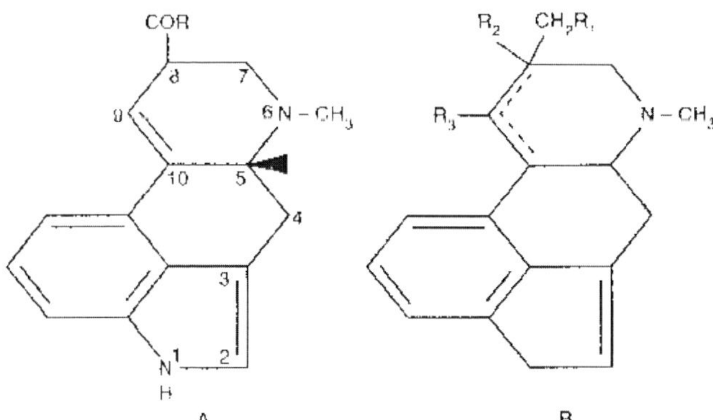

Fig. 28.10 Ergot toxins. A, lysergic acid derivatives (R = tripeptide or smaller units, OH in lysergic acid); B, clavine alkaloids (R = H, OH)

The amides of lysergic acid derivatives occur as inter-convertible pairs epimeric at C_8 position. The amine part of the amide ranges in complexity from ammonia to complicated cyclic tripeptides. These fall under 5 groups, viz. Erogotamine, Ergotoxine, Ergoxine, Ergobasine and miscellaneous.

Ergotoxine alkaloids, especially ergocornine when orally administered to female rats and mice, inhibit implantation of the ovum and prevents pregnancy despite fertilization. If this principle could be made applicable to humans, the consequences might be tremendously beneficial.

The clavines have the general formula as shown in Fig. 28.10 (B), where R_1, R_2, R_3 are H. OH, or COOH. As in the case of lysergic acid derivatives, pairs of C_8 epimers occur.

Ergot of rye was once the only source of the medicinal ergot alkaloids, but now semi-synthetic alkaloids are made from the lysergic acid, produced by *C. paspali* in culture media.

LSD (Lysergic Acid-Dimethylamide)

LSD, called "the drug of the century" by Bove (1970), is a semi synthetic compound (Fig. 28.11), which is a prototype of hallucinogens. It is widely used in the treatment of mental disorders by psychiatrists; but is more widely used for "taking a trip"!

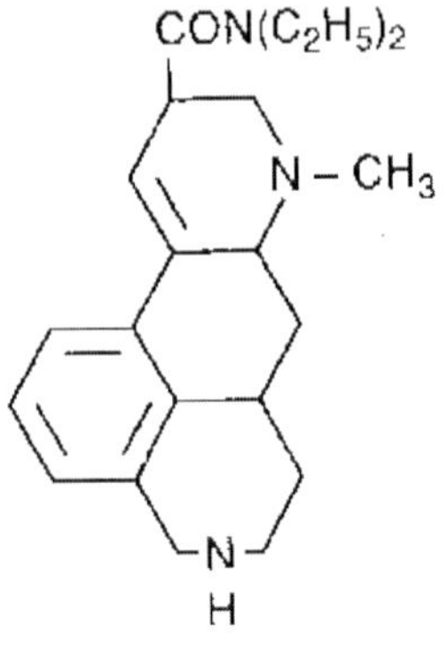

Fig. 28.11 LSD, the 'drug of the century'.

C. MUSHROOM TOXINS

The causal risk of macrophagy is often greatly disproportional to the gastronomic rewards.

Benedict, 1972

Mushroom toxicity is known ever since mycology started. The mycophagists soon discovered the poisonous nature of mushrooms and assigned them names like the "death cap" or "deadly agaric" for *Amanita phalloides*, "destroying angel" for *A. virosa*. Poisonous mushrooms are far less in number than the innocuous or edible ones, but such data cannot give relief to the victim or his relatives. About 95% of the fatal mushroom poisoning is caused by species of *Amanita*, and a wealth of knowledge is available about their toxins.

THE *AMANITA* TOXINS

Amanita toxins, can be divided into 2 main groups: (1) amatoxins (= amanita toxin), and (2) phallotoxins. Both are bicyclic peptide toxins, but while amatoxins are octapeptides with an indole-R-sulphoxide bridge, the phallotoxins are heptapeptides with an indole-thio-ether bridge. Nine amatoxins and seven phallotoxins have so far been identified. These are:

Amatoxins: α-amanitin, β-amanitin, γ-amanitin, amanin, amaninamide, amanulin, amanullinic acid and proamanulin.

Phallotoxins: phalloidin, phalloin, prophalloin, phallisin, phallacin, phallacidin and phallisacin

Compared to the amatoxins, the toxicity of phallotoxins to liver cells is quite low. It is now generally agreed that phallotoxins are not involved in the syndrome. On the other hand, most of the effects are ascribed to amatoxins. Poisoning by amatoxins is characterized by a long latent period of 6-18 hours. The symptoms appear as sudden severe abdominal pain, persistent vomiting and watery diarrhoea, extreme thirst and lack of urine production. If this early phase is survived, the patient may appear to recover for a short period. It shows up again as a rapid and severe weakness and restlessness. Death in 50-90% cases occurs due to progressive and irreversible damage to liver kidney, cardiac and skeletal muscles.

The damage to the liver is primarily on nuclei of liver cells, specifically inhibition of RNA-polymerase B. Amatoxins are known to inhibit RNA-polymerase B in all eukaryotes, except amatoxin-accumulating fungi and a few *Drosophila* species. Because of their specific inhibition of RNA-polymerase B; amatoxins have become an important tool in biochemical studies. About 5-7 mg of amatoxin would be a lethal dose, depending on the body weight of the victim. Ingestion of the equivalent of 50g of fresh weight of the fungus or a single mushroom can be fatal.

Protein conjugates of amatoxins are ten times more toxic than natural amanitins and yet, these can be used for raising amatoxin-specific antibodies. It would now facilitate use of polyclonal antibodies for amanitin immunoassays. A monoclonal antibody, then, could be a potential antidote in the immunotherapy of human poisoning by the deadly amanitas.

Amatoxins are produced by A. *phalloides*, *Amanita verna* and *A. virosa*. Amatoxins are also formed by several species of *Galerina*, *Conocybe* and *Lepiota*. These are *G. autumanalis*, *G. marginata*, *G. venenata*, *G. sulciceps*, *Conocybe filaris* and *Lepiota helveola*.

OTHER MUSHROOM TOXINS

ORELLANINE

This toxin has been isolated from *Cortinarius orellanus,* which was reported to cause fatal poisoning in Poland in 1952 and has been reported from several parts of Europe. An unusual feature is the long latent period of 3-15 days for the symptoms to appear. The toxin has a bipyridial structure and has also been synthesized. Orellanine is a nephrotoxin, causing damage to kidneys.

GYROMITRIN

This highly fatal toxin is produced by species of *Gyromitra* (an ascomycete, earlier considered as *Helvella*), most important being *G. esculenta*. Benedict (1972), however, cautions against all species. The symptoms of gyromitrin toxicoses are similar to *Amanita* poisoning. Its chemical structure (Fig. 28.12) was given by List and Luft (1967).

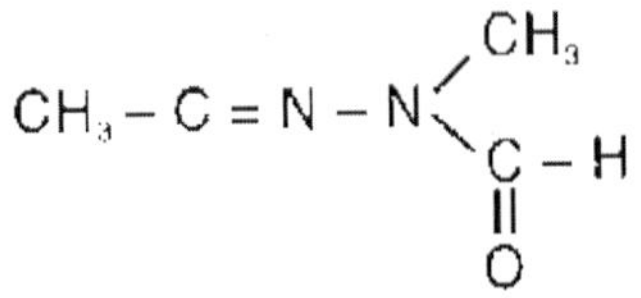

$$CH_3-C=N-N(CH_3)-C(=O)-H$$

Fig. 28.12 Gyromitrin, a fatal toxin produced by *Helvella* (= *Gyromitra*) spp

MUSCARINE

The toxic compound is named after *Amanita muscaria* (the 'fly agaric)' but, species of *Inocybe* and *Clitocybe* are more active producers of muscarine than *A. muscaria*. *I. napipes* is the most potent producer. Kogl *et al.* (1957) determined the chemical nature (Fig. 28.13), which was confirmed by its synthesis. Muscarine occurs in 8 isomeric forms of which the L(+) isomer is most active.

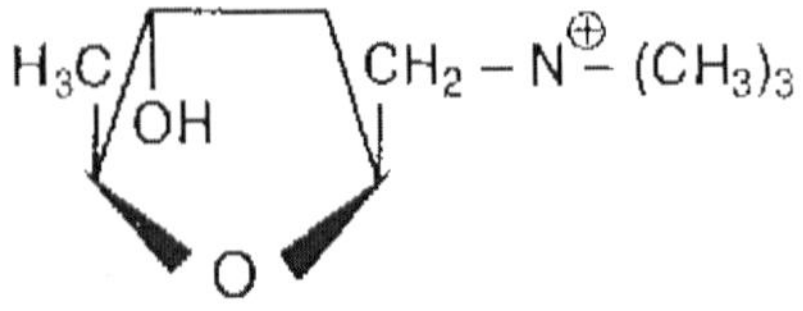

H$_3$C, OH, $CH_2-N^{\oplus}-(CH_3)_3$, O

Fig. 28.13 Muscarine, produced by *Inocybe, Clitocybe* and *A. muscaria.*

Muscimol and Ibatonic acid

In Siberia tribesmen have been eating *A. muscaria* and *A. pantherina* since long time for the hallucinogenic enjoyment after eating these mushrooms. The Siberians' experiences have been documented by travellers and historians. Wason (1968) firmly believes that the 'fly agaric' is the 'Soma' (**somras**) of the Vedas.

The intoxicant, hallucinogenic action is now known to be due to four compounds, but muscimol and ibatonic acid are more important. These are isoxazole derivatives (Fig. 28.14) and found in the genera, *Amanita* and *Panaeolus*. These are *A. muscaria* and *A. pantherina*, *A. cookeri* and *Panaeolus companulatus.* The predominant effect of these toxins is inhibition of motor functions. Muscimol is 5 to 10 times more potent than ibatonic acid. Muscimol is stereochemically γ-aminobutyric acid (GABA) and so its effect on the nervous system stems from stimulation of GABA – receptors of the cells of the central nervous system.

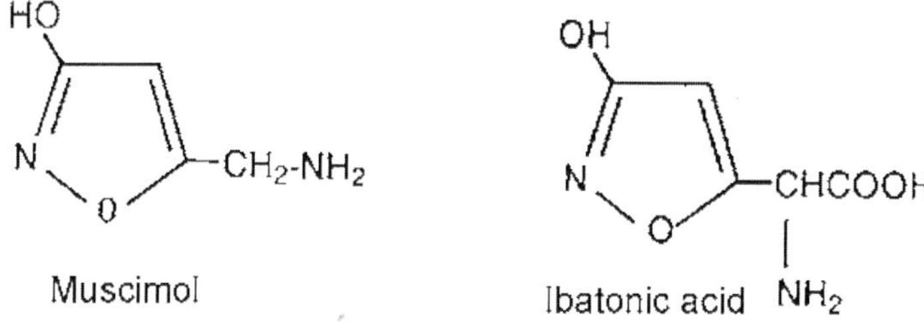

Fig. 28.14 Muscimol and Ibatonic acid produced by *A. muscaria*

A. muscaria was named 'fly agaric' because of its fly-killing property. In 1965, several groups of workers reported that the chemicals responsible for narcotic-toxicant action, had insecticidal properties also.

Psilocybin and Psilocin

The sacred hallucinogenic mushrooms, consumed in religious ceremonies in remote parts of Mexico, belong primarily to the genus *Psilocybe*, *Panaeolus* and *Conocybe*. *Psilocybe* mushrooms have been called "flesh of Gods". The hallucinogens were isolated from *Psilocybe mexicana* and named as psilocybin and psilocin. These toxins (Fig. 28.15) cause stimulation of the central nervous system.

Several of the species can be cultured and cultivated. Psilocybin is active in humans in very low doses: 4 mg, taken orally, causes mild intoxication, but 6-20 mg can elicit marked psycotropic response. Overdose can be dangerous and cause psychoses, leading to suicide mania. Primarily, *Psilocybe* hallucinogens act as **serotonin antagonists**: therefore, they are used in psychotherapy and neurological research.

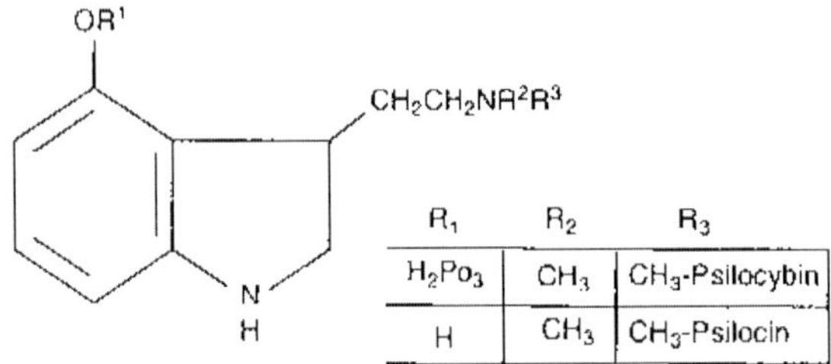

Fig. 28.15. Psilocybin and Psilocin, the hallucinogens of *Psilocybe mexicana*

REFERENCES

Barrett, J.R. 2000. Mycotoxins : of molds and maladies. *Environmental Health Perspectives*. 108 (1) A20–A23.

Benedict, R.G. 1972. Mushroom toxins other than *Amanita*, In : *Microbial Toxins* (S. Kadis *et al.,* eds) Vol VIII pp 281-320. Academic Press NY and London

Bennett JW and Klich M 2003. Mycotoxins. Clin Microbiol Rev. 16(3) 497-516

Bhat, R.V. *et al.,* 1976. The nature of alkaloids of ergoty pearl millet or bajra and its comparison with alkaloids or ergoty rye and ergoty wheat. *Toxicol.* and *appl. Pharmaeol.,* **36:** 11-17.

Bove, F.J. 1970. *The Story of Ergot*. S. Karger, Basel.

Chatterjee, D. and Mukherjee, S.K. 1994. Contamination of Indian maize with fumonisin B_1 and its effects on chicken macrophage. *Lett. Appl. Microbiol.* **18:** 251–253.

Krishnamachari, K.A.V.R. *et al.,* 1975. Hepatitis due to aflatoxicoses: An outbreak in Western India. *Lancet,* **10:** 1061–1063.

Nelson, R.R. 1971. Hormonal involvement in sexual reproduction in the fungi, with special reference to F-2, a fungal esterogen. In: *Morphological and Biochemical Events in Plant-Parasite interaction.* (S. Akai and S. Ouchi. Eds.) *Phytopathological Soc.* Japan.

Sargeant, K. *et al.,* 1961. Discussion. *Nautre,* **192:** 1096-1097.

Subramanian, C.V. 1995. Mushrooms : beauty, diversity, relevance. *Curr. Sci.* **69** (12) 986–998.

Vasanthi, S. and Bhat, R.V. 1998. Mycotoxins in foods – occurrence, health and economic significance and food control measures. *Indian J. Med. Res.* **108:** 212–224.

Vasanthi, S. and Bhat, R.V. 2000. Ochratoxin in Indian coffee and its significance in international trade. *Nutrition News*. **21**: 4, 1–6.

Wason, R.G. 1968. Soma, divine mushroom of immortality. *Harcourt Brace and World.* N.Y.

Woodword, R.B. and Singh, G. 1949. The structure of patulin. *J. Amer. Chem. Soc.,* **71:** 758.

SUMMARY

In 1960 thousands of turkey poults died in England by eating 'moldy' peanuts (groundnuts) infected with *Aspergillus flavus*. It was due to a toxic fungal metabolite, called **aflatoxin**. The term **mycotoxin** was coined in 1962. The presence of toxin in the moldy food is known either by bioassay (i.e. feeding some test animals on the suspected substrate) or by chemical means.

Aflatoxins are among the most carcinogenic substances known. The eight characterized aflatoxins are: B_1, B_2, G_1, G_2 (named on the blue or green fluorescence in UV light), M_1, M_2 (first detected in milk of cows fed on groundnut meal), and B_2a and G_2a (derivatives of B_2 and G_2). Among the eight forms, B_1 B_2 G_1 and G_2 are more commonly encountered. These may occur together or independently.

Aflatoxin B_1 is one of the most important causes of liver cancer in rats, trouts and humans. After entering the body, the afllatoxins are metabolized by the liver to a reactive epoxide intermediate or hydroxylated to become less harmful aflatoxin M1. Aflaltoxin is detected in humans by presence of AFB_1-guanine adduct in urine (only recent exposure) and AFB_1-albumin adduct level in blood serum (long term exposure). Aflatoxin at molecular level binds with DNA and prevents its transcription, and also acts as a mutagenic carcinogen.

Zearalenone is produced by several species of *Fusarium* growing on corn, barley, wheat and other crops. When consumed, it causes esterogenic symptoms (in animals and man), accompanied by diarrhoea, emesis, refusal to feed, loss of weight and hemorrhagia. The word toxin for zearalenone is a misnomer because zearalenone is hardly toxic. Rather, it resembles 17 β estradiol, the principal hormone produced by the human ovary. So, zearalenone is better classified as a non-steroidal esterogen or mycoesterogen.

Trichothecenes are sesquiterpenoid mycotoxins, produced by several fungi, mainly species of *Fusarium*. The toxins possess a ring system named trichothecane after *Trichothecium roseum*. Three important trichothecene toxins are: Alimentary Toxic Aleukia (ATA) toxin, Deoxynivalenol and *Stachybotrys* toxins.

ATA is a human disease showing typical spots on the skin, inflammation of the skin, leukopenia, agranulocytosis, necrotic angina, hemorrhagic diathesis, sepsis and exhaustion of the bone marrow. Diacetoxyscirpenol and T2 are associated with ATA. These are cytotoxic and immunosuppressive. The T-2 toxin (the skin-irritating factor) is responsible for skin inflammation and spots.

Deoxynivalenol causes nausea, vomiting and diarrhea, and therefore also called vomitotoxin or 'food-refusal factor'. Though least toxic among the trichothecenes, it has been linked to large scale poisonings throughout the world. Deoxynivalenol induces cytokines, immune system factors, causing inflammation.

Stachybotrys toxins produced by *S. atra* that are air-borne and cause respiratory disorders e.g. idiopathic pulmonary hemorrhage (or bleeding of lungs) in infants. This is a new area of concern in mycotoxin research. These are indoor molds causing mycotoxicoses. These toxins are allergens and trigger or aggravate allergy. They are potent protein-synthesis inhibitors and interfere with a single binding site on the ribosome, which causes breakdown of *initiation*, *elongation* or *termination* stages of the protein synthesis.

Fumonisins are synthesized by condensation of the amino acid alanine with an acetate-derived precursor. Fumonisins (B_1, B_2, B_3) are produced by several species of *Fusarium*, mostly by *F. verticillioides* (formerly *F. moniliforme,= Gibberella fujikuroi*), that grow on corn. These cause "hole in the head syndrome' (leuco-encephalomalacia, LEM) in horses, besides brain hemorrhage and necrosis. Liver and kidney damage are pronounced in rodents (moles, rats, squirrels) sheep and rabbits. In swine, it causes porcine pulmonary edema, (a fatal condition in which fluid collects in lungs). Fumonisin disrupts sphingolipid metabolism.

Sporodesmins, produced by *Pithomyces chartarum* (=*Sporodesmium bakeri*) on pastures, cause **facial eczema** of grazing cattle (Pithomycotoxicosis). Sporodesmins, viz. sporodesmin, sporodesmin B, C and D. These are epidithiopiperazine (ETP).

Ochratoxins are potentially as important as aflatoxin. Ochratoxins A, B and C are produced by *Aspergillus ochraceus* and *Penicillium viridicatum* on stored moldy maize, corn, peanuts, beans and mixed animal feeds. OA is a carcinogen, nephrotoxin and hepatotoxin. The International Agency for Research on Cancer has rated ochratoxin as a possible human carcinogen.

Yellowed rice toxins include citreoviridin, citrinin and luteoskyrin. *Citrinin* is produced by species of *Penicillium* and *Aspergillus*. Citrinin acts as a nephrotoxin in all animals tested. It causes nephropathy and other renal disorders in animals. It acts synergistically with Ochratoxin-a to inhibit RNA synthesis.

Patulin is formed in apple fruits infected by *Penicillium expansum*. It is found in commercial apple juices and in moldy bread. It is carcinogenic to mice, and causes neurotoxicosis in cattle.

Ergot toxins are alkaloids and are divided into 2 groups (1) amides of lysergic acid found chiefly in sclerotia of *Claviceps purpurea*, and (2) clavine alkaloids or clavines, found in the sclerotia of other species of *Claviceps*. The amides of lysergic acid derivatives fall under 5 groups, viz. Erogotamine, Ergotoxine, Ergoxine, Ergobasine and miscellaneous. Ergotoxine alkaloids, especially ergocornine, when orally administered to female rats and mice, inhibit implantation of the ovum and prevents pregnancy. If this principle could be made applicable to humans, the consequences might be tremendously beneficial. LSD (Lysergic Acid-Dimethylamide) "the drug of the century" is a semi synthetic compound which is a prototype of hallucinogens. It is widely used in the treatment of mental disorders but is more widely used for "taking a trip"!

The mushroom toxins include *Amanita* toxins, Orellanine, Gyromitrin, Muscarine, Ibatonic acid and Muscimol, Psilocybin and Psilocin.

Amanita toxins comprise: amatoxins and phallotoxins. Both are bicyclic peptide toxins, but while amatoxins are octapeptides with an indole-R-sulphoxide bridge, the phallotoxins are heptapeptides with an indole-thio-ether bridge. Phallotoxins are not involved in the syndrome. Most of the effects are ascribed to amatoxins. The damage to the liver is primarily on nuclei of liver cells, specifically inhibition of RNA-polymerase B.

Orellanine from *Cortinarius orellanus* was reported to cause fatal poisoning in Poland in 1952 and has been reported from several parts of Europe. It is a nephrotoxin, causing damage to kidneys.

Gyromitrin, a highly fatal toxin is produced by species of *Gyromitra* (an ascomycete), The symptoms of gyromitrin toxicoses are similar to *Amanita* poisoning.

Muscarine is named after *Amanita muscaria* (the 'fly agaric)' but, species of *Inocybe* and *Clitocybe* are more active producers of muscarine. *I. napipes* is the most potent producer.

Ibatonic acid and Muscimol are isoxazole derivatives and found in the genera, *Amanita* and *Panaeolus*. The predominant effect of these toxins is inhibition of motor functions. Muscimol is 5 to 10 times more potent than ibatonic acid. Muscimol is γ-aminobutyric acid (GABA), and so its effect on the nervous system stems from stimulation of GABA–receptors of the cells of the central nervous system.

Psilocybin and Psilocin were isolated from *Psilocybe mexicana* and named as psilocybin and psilocin. These toxins cause stimulation of the central nervous system. Primarily, *Psilocybe* hallucinogens act as serotonin antagonists: therefore, they are used in psychotherapy and neurological research.

TEST QUESTIONS

1. Describe the discovery of aflatoxin. How many types of aflatoxins are known? Explain the mechanism of aflatoxin as a mutagenic carcinogen leading to liver cancer.
2. What is ergotism? Which alkaloids produced by ergot? How is it useful in medicine.
3. Name the various mushroom toxins. Describe the *Amanita* toxins.
4. Which mushrooms were eaten in Siberia and Mexico for hallucinogenic pleasures? Which are the hallucinogenic toxins in these mushrooms.
5. Answer in brief:
   a. Which compounds in the human urine and blood serum indicate short term and long term aflatoxin consumption?
   b. Which bacterium can remove aflatoxin from milk?
   c. Why is the term toxin a misnomer for zearalenone?
   d. Which are the three types of trichothecenes? What is ATA?
   e. Which trichothecene causes indoor mycotoxicosis?
   f. Which are the three yellowed-rice toxins?
   g. What is penicillic acid?
   h. What are the two types of ergot alkaloids and their producers?
   i. What is LSD?
   j. Which are the two types of *Amanita* toxins?
   k. Which are the potent producers of muscarine?
   l. Which toxins are produced by *Amanita muscaria*?
   m. Why is *Amanita muscaria* called 'fly agaric'?
   n. Which agaric has been called by Wason as 'soma' (somras) of the *Vedas*?
   o. Which 'sacred' hallucinogenic mushrooms were eaten by Mexicans at religious ceremonies?
   p. Why ergotism was called "holy fire' and "St' Anthony's fire"?
   q. What are the two types of ergotism?
   r. Why ergot is no more a feared poison?

6. Fill in the blanks:
   a. The turkey X-disease was caused by moldy grains infected by ---- ----------.
   b. In liver AFB1 is converted to less harmful aflatoxin ------.
   c. Aflatoxin binds with ------------- and prevents ----------------.
   d. Carcinogenic activity of aflatoxin-B1 is activated by-------.
   e. The esterogenic symptoms caused by zearalenone appears in ------ ----- and ---------------.
   f. Zearalenone is a sex hormone of ---------------.
   g. The trichothecene toxin ------------- is also called vomitotoxin.
   h. Fumonisins disturb the ---------------- metabolism.
   i. Sporodesmins cause ------------- of grazing cattle.
   j. Cyclopiazonic acid (CPA) is a ------------- toxin produced by ------ ------------.
   k. CPA inhibits Ca-dependent -------------.
   l. Amatoxins inhibit ----------------.

29

Homothallism and Heterothallism

A.F. Blakeslee, an American geneticist, in the year 1904, observed that while some species of *Rhizopus* formed zygospores freely, others e.g. *R. nigricans* (=*R. stolonifer*), formed these only rarely. He examined zygospore formation in several *Mucorales* and discovered the phenomenon of **heterothallism** in fungi. When two isolates of *R. nigricans* or *Mucor hiemalis* were grown together in a Petri dish (Fig. 29.1), zygospores appeared at the point of contact between hyphae of the two isolates. Blakeslee called these fungi, which required another isolate for zygospore formation, as heterothallic and the phenomenon as heterothallism. He designated the two isolates, or mating types, the (+) and (–) strains, as their gametangia were morphologically similar and indistinguishable into male and female individuals. The opposite condition, in which an individual, originating from a single asexual spore, was capable of forming zygospore independently, was called **homothallism**.

In heterothallism, the two mating partners are derived from two genetically-unlike spores, designated as (+) and (–) spores. Subsequently, several common species of *Mucorales* were examined and described as homo- or heterothallic. The heterothallic species always consisted of two types of individuals, the (+) and (–) mating types.

No physiological differences could be found between the gametangia of (+) and (–) strains and, therefore, we do not designate the (+) and (–) as male and female sexes but as mating types.

Since Blakeslee's discovery, heterothallism has been reported from all major groups of fungi, though in varied forms. Several patterns of inter-mycelial reactions are recognized in heterothallic fungi, but all share one characteristic feature *viz.* the inter-mycelial contact. In some heterothallic fungi, the sexual differences are distinct and sex organs are formed on different thalli. In others e.g. in hermaphrodite heterothallic fungi, the reasons for dependence on another individual are physiological and more subtle.

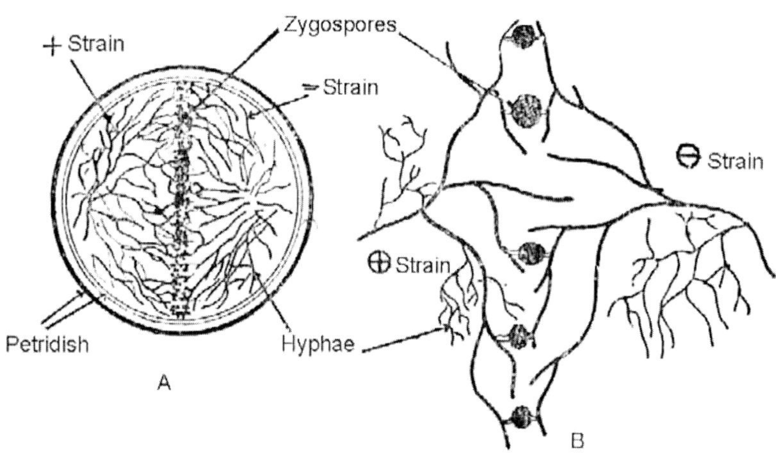

Fig. 29.1. Heterothallism in *Mucor hiemalis*. A, Petri dish culture showing zygospores along the zone of contact between the two mycelia of opposite mating types; B, magnified view of the zygote formation.

Morphological and Physiological Heterothallism

Whitehouse (1949) distinguished two major types of heterothallism: morphological and physiological. When the two interacting thalli produce morphologically distinct male and female sex organs (or gametes), the heterothallism is called morphological heterothallism. In physiological heterothallism, the interacting thalli differ in mating type or incompatibility, irrespective of the presence or absence of the sex organs or gametes. Homothallism was accepted by Whitehouse in the original sense i.e. sexual fusion between elements of the same thallus; or in unicellular organisms, between cells of the same clone. In addition, he added a new term, the **secondary homothallism**, which is explained like this. In some heterothallic fungi e.g. *Exobasidium japonicum*, it so happens that two genetically different nuclei, formed during meiosis and which ought to have gone to different spores, are regularly contained in a single spore. Such spores on germination give rise to a mycelium, which contains nuclei of the opposite strains, and the fungus behaves as if homothallic.

Genetic Determinants of Heterothallism

We have seen above that heterothallism may be due to: (1) lack of both male and female sex organs on the same thallus, or (2) incompatibility between the sex organs formed on the same thallus, with regard to karyogamy. The two conditions are genetically controlled by factors, called, respectively, the **sexual** and **incompatibility factors**. Sexual factors are responsible for the formation of sex organs and gametes. The maleness or femaleness is determined by the direction of nuclear migration during plasmogamy. The male serves as the donor of the nucleus and the female as the recipient. Incompatibility factors determine the mating capacity in addition to, or in absence of, sexual factors. In presence of sexual factors, and consequently of sex organs, the incompatibility factors determine the possibility of mating between the male and female sex organs or gametes. In absence of sex factors, when no sex organs or gametes are formed, as happens in most of the higher asco-and basidiomycetes, the incompatibility factors are the sole determinants of sexual fusions

The segregation of these factors during meiosis determines the sexual status (i.e. whether sex organs or gametes are formed) and the mating behaviour (i.e. whether homothallic or heterothallic). Three types of segregations may occur during meiosis (1) segregation of sexual factors (2) segregation of incompatibility factors, and (3) no segregation of either factor. The segregation pattern is exhibited by the thalli that develop from the spores containing the meiotic nuclei. The spores of type (1) and (2) give rise to thalli that are self-sterile i.e. heterothallic. The spores of type (3), on the other hand, give rise to homothallic thalli.

HOMOTHALLIC FUNGI

Homothallic species are more common than heterothallic in all groups of fungi, except *Basidiomycota*. The sexual or incompatibility factors do not segregate during meiosis and differentiation of compatible sexual elements is intra-mycelial. All individuals are alike, and morphologically and/or functionally hermaphrodite. Is not the sexual fusion between genetically similar nuclei, in homothallic fungi, in terms of genetic recombination, a biological absurdity? No! Nature does not encourage biological absurdities; they are swiftly eliminated. The widespread occurrence of homothallic fungi in nature is eloquent testimony to this. During the extensive somatic phase, the nuclei undergo genetic changes due to mutations, and become heterokaryotic (see Chapter 30).

HETEROTHALLIC FUNGI

Raper (1966) distinguished six types of heterothallic fungi. The segregation of the sexual and incompatibility factors of these various heterothallic fungi fall into two categories (1) two alleles at one locus (bipolar), and (2) multiple alleles at one or two loci (bipolar and tetrapolar).

In two alleles at one locus: Bipolar heterothallism

Fungi in this category have two mating types; each containing genetically different nuclei. The sexual compatibility is controlled by a pair of genetic factors, A and a (which may be sexual or incompatibility factors), located at a single locus on different chromosomes. The use of (+) and (–) designations seems preferable because A/a implies dominance of A over a, which has never been demonstrated for these alleles. This is, therefore, also called as **two-allele heterothallism**. During meiosis, the two chromosomes, containing the (+) and (–) alleles are separated in the haploid spores (viz. germ spores, ascospores or basidiospores). The spores give rise to two types of thalli, which must fuse in order to bring the two compatible (+) and (–) nuclei together. The two mating types are designated as (+) and (–) strains.

When the two alleles represent sex factors, the two types of thalli formed are unisexual, producing male or female sex organs or gametes. The heterothallic hermaphrodite (= self-sterile) individuals also come under two-allele heterothallism. The sexual factors, as in homothallic species, are not segregated, while the incompatibility factors segregate. This results in the production of self-sterile, hermaphrodite individuals. Examples of this type of heterothallism are found in majority of ascomycetes (e.g. *Neurospora*) and the rust fungi. Esser (1963) has explained the genetic basis of heterothallism of the hermaphrodite fungi as shown in Fig. 29.2.

The rectangles represent the two mating types (+) and (–); the differentiated or undifferentiated sex organs contain nuclei which are incompatible. This is indicated in the diagram by broken lines, which are blocked. Sexual fusion requires the presence of individuals of different mating types (represented by heavy arrows) and reciprocal crosses occur.

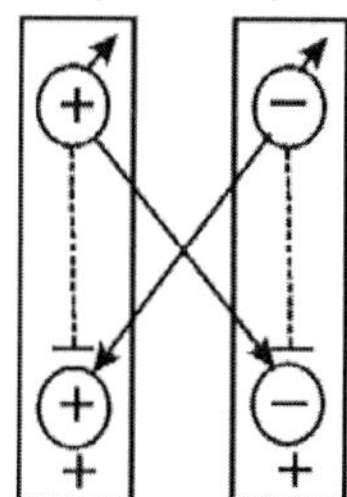

Fig. 29.2. Bipolar, 2 allele heterothallis.

Multiple Alleles at One Locus (Bipolar, Multiple allele Heterothallism)

This type of heterothallism is controlled by multiple alleles at a single locus, instead of a pair of alleles. If we name the locus as A, the multiple alleles can be designated as A1, A2 A3....An. The meiotic products give rise to thalli, which are of several mating types (equal to the number of the alleles). The thallus containing the allele A1 can mate with a thallus of any mating types, except A1 (Fig. 29.3). Like the two-allele heterothallism, here also, the genetic factors may involve sexual or incompatibility factors. Incompatibility factors are more commonly involved. It is a characteristic of the *Pezizomycotina*. The mating is reciprocal i.e. each cell of the thallus is capable of donating or accepting the nucleus of the compatible mating type.

	A_1	A_2	A_3	A_4
A_1	—	+	+	+
A_2	+	—	+	+
A_3	+	+	—	+
A_4	+	+	+	—

Fig. 29.3. Bipolar, multiple-allele heterothallism. A1–A4 represent multiple alleles at a single locus (+, compatible cross, – incompatible cross)

When the alleles involve sexual factors, several types of strains are formed, each typically self sterile but capable of mating with the other types. This type of heterothallism has been found in all the studied members of *Saprolegniales, Leptomitales* and *Peronosporales*. The sexual strains constitute a linear series in which only those strains, which lie at the two ends are strictly male or female. The rest have relative maleness or femaleness. Raper (1947) collected 10 such sexual strains of *Achlya ambisexualis*, and arranged them in a linear series on the basis of their sexual (male or female) potential (Fig. 29.4). Except the terminal pure male or female isolates, the rest of the isolates behaved as female with those on their left and as male to those on the right. To quote Raper (1959) these show "sexual ambivalence in which maleness or femaleness is determined in each mating type by common consent of the mated.

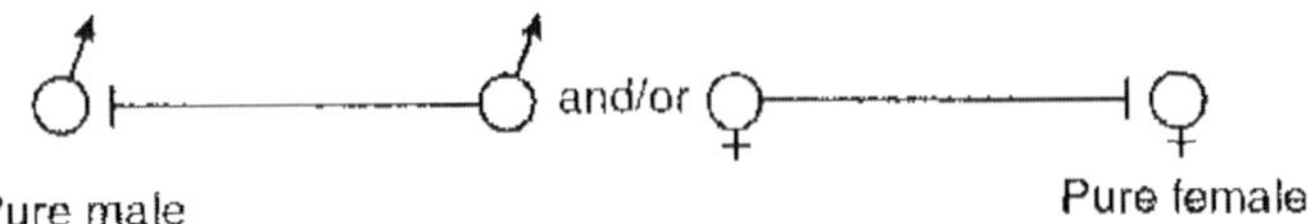

Fig. 29.4. The sexual strains of *Achlya ambisexualis* linearly arranged according to their male and female sexual potentialites. Each strain can react as male, female or both, depending on juxtaposition (♂ male; ♀, female)

Multiple Alleles at Two Loci (Tetrapolar, Multiple-allele Heterothallism)

This type of heterothallism is found in *Basidiomycota.* It involves only incompatibility factors and, therefore, sex organs are not formed. The incompatibility factors are present at two loci, say A and B, on different chromosomes. Each locus comprises of multiple alleles, as many as 100 or even more. In *Schizophyllum commune*, 122 alleles of A and 61 of B factor have been identified in laboratory. In natural populations, the number of alleles may be even more. Locus A may have as many as 350-450 alleles, while B about 65. As a rule, the multiple factors of the two loci are unequal; locus A always has more alleles than B. The A and B allelic incompatibility factors segregate independently at meiosis and give rise to thalli of numerous mating types. Mating occurs (Fig 29.5) between thallli which have nuclei carrying different alleles at both loci. The diploid nucleus comes to have the alleles A_1, B_1, A_2, B_2.

	A_1B_1	A_1B_2	A_2B_1	A_2B_2
A_1B_1	–	FL	B	+
A_1B_2	FL	–	+	B
A_2B_1	B	+	–	FL
A_2B_2	+	B	FL	–

Fig. 29.5. Mating reactions between the four possible isolates of a tetrapolar fungus. (+ compatible cross; — incompatible cross: FL, flat reaction; B, barrage reaction)

In this type of heterothallism the out-breeding is complete, (100%), against 25% in bipolar. This is due to enormous increase in the number of the possible mating type of the thalli.

Flat and Barrage Reactions

When two thalli with a common A or B alleles are mated, though the plasmogamy takes place, and heterokaryons are formed, fruiting never occurs. If the heterokaryons have common A alleles (and dissimilar B alleles), the mating is termed **flat reaction** (see Fig. 29.5). Such heterokaryons show poor growth; hyphae are curly and irregularly branched. The presence of common A alleles prevent formation of clamp connections; disrupt nuclear distribution and also cause morphological and metabolic abnormalities. The mating between thalli, containing similar B alleles (but different A alleles), show what is called a **barrage reaction**. The hyphae at the point of contact show cessation of growth, resulting in a "barrage" or 'zone of no growth', between the colonies of the two thalli. Common B factors prevent karyogamy and meiosis.

REFERENCES

Blakeslee, A.F. 1904. Sexual reproduction in the Mucorineae. *Proc. Amer. Acad. Arts. Sci.*, **40:** 205–319.

Burnett, J.H. 1956. The mating systems of fungi. I. *New Phytol.*, 55: 50–90.

Esser, K. 1963. *Genetics of Fungi.* Springer Verlag, New York *Inc.*

Raper, J.R. 1959. Sexual versatility and evolutionary process in fungi. *Mycologia*, **15:** 107–125.

Raper, J.R. 1966. Life cycles, basic patterns of sexuality, and sexual mechanisms. In: *The Fungi. An Advanced Treatise.* Vol. **II.** (G.C. Ainsworth and A.S. Sussaman, eds.). pp. 473–511. Academic Press, N.Y. and London.

Raper, J.R. *et al.* 1960. The genetic structure of the incompatibility factors of *Schizophyllum commune*: the A factor, *Natl. Acad. Sci.* U.S., **46:** 833–842.

Whitehouse, H.L.K. 1949. Heterothallism and sex in the fungi. *Biol. Rev.,* **24:** 411–447.

SUMMARY

A.F. Blakeslee (1904) discovered **heterothallism** in fungi. Fungi which required another isolate for zygospore formation were called heterothallic and the phenomenon as heterothallism. The opposite condition, in which an individual, originating from a single asexual spore, was capable of forming zygospore independently, was called **homothallism**.

Heterothallic fungi share one characteristic feature viz. the inter-mycelial contact. There are two types of heterothallism: morphological and physiological. In morphological heterothallism, the two interacting thalli produce morphologically distinct male and female sex organs (or gametes). In physiological heterothall-ism, the interacting thalli differ in mating type or incompatibility, irrespective of the presence or absence of the sex organs or gametes. In addition, there is **secondary homothallism**. In this two genetically different and compatible nuclei are contained in a single spore, which gives rise to apparently homothallic fungus.

Heterothallism may be due to: (1) lack of both male and female sex organs on the same thallus, or (2) incompatibility between the sex organs formed on the same thallus, with regard to karyogamy. The two conditions are genetically controlled by factors, called, respectively, the **sexual** and

incompatibility factors. Three types of segregations may occur during meiosis (1) segregation of sexual factors (2) segregation of incompatibility factors, and (3) no segregation of either factor. The spores of type (1) and (2) give rise to thalli that are self-sterile i.e. heterothallic. The spores of type (3), on the other hand, give rise to homothallic thalli.

Homothallic species are more common than heterothallic in all groups of fungi, except *Basidiomycota*. The sexual or incompatibility factors do not segregate during meiosis. All individuals are alike, and morphologically and/or functionally hermaphrodite.

Heterothallism is of six types which are controlled by: (1) two alleles at one locus, and (2) multiple alleles at one or two loci.

In two alleles at one locus, the fungi have two mating types; each containing genetically different nuclei. The sexual or incompatibility factors are located at a single locus on different chromosomes. When the two alleles represent sex factors the two types of thalli formed are unisexual. However, if the sex factors do not segregate and incompatibility factors segregate, then self-sterile hermaphrodite thalli are formed.

In multiple alleles at one locus several types of thalli, equal to the number of alleles, are formed. The thallus having a particular allele can mate with all except that which has same allele. Here also sexual and compatibility factors are involved, but incompatibility factors are more commonly involved. It is characteristic of basidiomycetes.

In multiple allelels at two loci only incompatibility factors are involved. This is found in baidiomycetes. The incompatibility factors are located at two different chromosomes. Each locus contains many alleles (in hundreds). The allelels at one particular locus is always more than on the other locus. The multiple alleles at the two loci segregate independently and produce numerous types of thalli. Mating occurs between thalli which have different alleles at both loci. In this out breeding is 100%

Flat and Barrage reactions occur when two mated thalli have the same allele at one or the other locus. In flat reaction, the heterokaryons show poor growth (flat, not fluffy), while in barrage reaction the mating thalli show least growth (barrage or zone of no growth) at the point of contact.

TEST QUESTIONS

1. Fill in the blanks:
   a. Heterothallism was discovered by ------- in the year-------.
   b. In --------- --------------, single spores contain genetically different nuclei.
   c. ----------- factors control formation of sex organs.

d. Homothallic fungi are common in all fungi except--------.

e. Heterothallism is determined by -------- and ----- factors.

f. There is no segregation of -------- and ----------- in homothallism.

g. Heterothallism due to multiple alleles at two loci is found in ----------------.

h. Two alleles at one locus give rise to ---------------- heterothallism.

2. What are the two factors that control heterothallism? Describe their three segregation patterns.

3. Describe heterothallism governed by multiple alleles at one locus, **or** multiple alleles at two loci

4. Write notes on: (i) Secondary homothallism, and (ii) Flat and Barrage reactions.

30

Heterokaryosis and Parasexual Cycle

Heterokaryosis

The term heterokaryosis (hetero = dissimilar, *karyons*= nuclei) has been defined as the coexistence of genetically-different nuclei in a single cell. The phenomenon, discovered by Hansen and Smith (1932) in *Botrytis cinerea*, is a near-universal feature in fungi. Heterokaryosis plays the major role in bringing about variability and sexuality in haploid fungi. Herein lies a possible explanation for the degenerate condition of sex organs or their complete absence in many fungi. It is the pre-requisite for parasexual cycle, in the same way as heterozygosis is for sexual reproduction.

Heterokaryosis has been given different dimensions by different authors. In the most limited sense, it may be applied to a single cell, but usually it is applied to the thallus. Some regard the dikaryons of a heterothallic fungus as representing heterokaryosis, while others feel that the term heterokaryosis should be restricted to conditions in which the heterokaryons lie in unrestricted ratios. According to Parmeter *et al.* (1963) such special usage of the term can lead to confusion. According to them the dikaryotic cell, containing dissimilar nuclei e.g. the spores of rusts, is clearly heterokaryotic, representing only one of the several types of heterokaryotic associations.

FORMATION OF HETEROKARYONS

The heterokaryotic condition arises by (1) mutation, (2) anastomosis, and (3) inclusion of dissimilar nuclei in spores after meiosis, in heterothallic fungi (Fig. 30.1).

Mutation

A high frequency of mutation, which is characteristic of fungi, is the main source of variability. One who has experience of handling fungal cultures, especially species *Penicillium, Aspergillus, Verticillium, Fusarium* etc., knows how difficult it is to maintain a fungus in homokaryotic condition. Mutation frequently renders it heterokaryotic. It has been clearly demonstrated that new races of *Phytophthora* in USA, in absence of sexual reproduction, have arisen by mutation. Hansen (1943) observed that the single spore of *Fusarium*, by frequent mutations, produced variants of the wild type in cultures. To this Hansen gave the name "dual phenomenon".

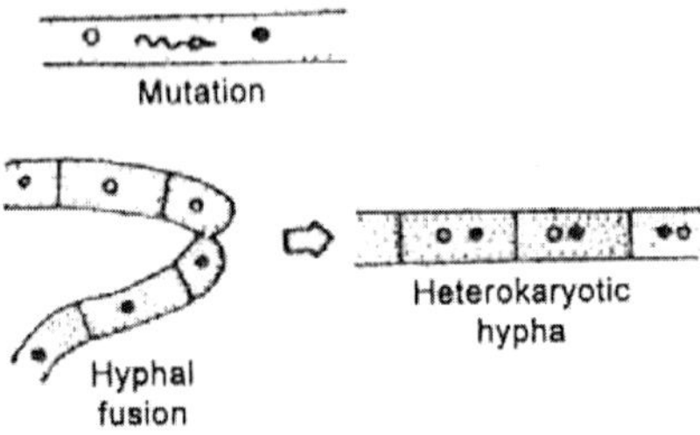

Fig. 30.1. Mutation and anastomosis render homokaryotic hyphae heterokaryotic

Anastomosis (The Fusion of Hyphae)

This is a feature of common occurrence in fungi, and is a big source of heterokaryosis. In general, fusion is mostly intra-specific, though inter-specific and inter-generic fusions have been reported in unmistakable terms. But, there are reports of cytoplasmic deterioration of fused cells, and antigen-like reactions are reported to be the cause of cytoplasmic incompatibility. Unfortunately, the physiological processes responsible for success or failure of anastomosis are not clear.

Nuclear migration from the point of fusion to the remainder of the mycelium (involving daughter nuclei of the entrant nucleus) takes place and gives rise to a heterokaryotic mycelium. Alternatively, hyphal growth may initiate at the point of fusion, containing nuclei derived from both the parent hyphae. In nature, the most familiar example of anastomosis is the development of dikaryon in *Basidiomycota*. Anastomosis is the method used for "making" heterokaryons in laboratory.

Inclusion of Dissimilar Nuclei in Some Spores

Meiosis results in the production of genetically different nuclei sharing common cytoplasm. In some fungi e.g. *Neurospora tetrasperma*, *Podospora anserina* and other secondary homothallic fungi, dissimilar nuclei are contained in the same spore, which on germination, gives rise to a heterokaryotic thallus. In the asexual phase, this occurs frequently in multinucleate spores.

SIGNIFICANCE OF HETEROKARYOSIS

1. Substitute for Heterozygocity and Variability

Heterokaryosis provides a substitute for heterozygocity rather than for sex. It is important for maintaining variability (by virtue of the presence of geneticially dissimilar nuclei), rather than creating it, which occurs during sexual reproduction by genetic recombination. The significance of heterokaryosis, as the dispenser of variability, was first emphasized by Hansen and Smith (1932) in *Botrytis cinerea*. They visualized the heterokaryons and the separation of the different nuclei during conidial formation as substitutes for syngamy and meiosis. Thus, as said above, heterokaryosis is a substitute for heterozygocity, rather than for sex. The variability gained through heterokaryosis provides plasticity to the fungi to face the environment with greater success.

2. Heterokaryosis and Pathogenicity

Heterokaryosis plays an important role in the pathogenic activities of rusts and smuts. It is essential for (rather it conditions) pathogenicity in smuts. In heteroecious rusts, heterokaryotic (dikaryotic) condition is essential for infection of one group of hosts.

3. Origin of New Races

Nelson *et al.* (1955) inoculated a resistant wheat variety with different races of rust, which individually had failed to cause infection. Lesions appeared after some time. It was made possible by anastomosis between the inoculated races, which resulted in the origin of a new pathogenic race. Similar origin of new races of rusts has been suggested in *Melampsora lini*. Fusion of urediniospore germ tubes has been seen in various rust fungi and has been taken as indication of the origin of new races.

4. An Initial Step in Parasexual Cycle

Heterokaryosis is an integral, first step of parasexual cycle. But all heterokaryotic fungi are not necessarily parasexual.

The Parasexual Cycle (=Parasexuality)

If only one outstanding contribution of microbial genetics to biological thinking had to be singled out, it would be this: the realization that transfer of genetic information from one individual or cell to another is not the monopoly of sexual reproduction. **G. Pontecorvo**

G. Pontecorvo

For a long time, sexual reproduction was thought to be the only mechanism of achieving genetic recombination. The genetic systems of the myriads of asexual microorganisms, like bacteria and "imperfect fungi", made no sense. However, when several 'alternatives' to sexual reproduction were discovered in bacteria, in quick succession, biology got a new vision. J.B.S Haldane (1952) called these new discoveries as "novel methods of genetic recombination". The newly-discovered alternatives to sex are: transformation, conjugation, transduction, lysogeny and sexduction. Thus, the transfer of genetic information from one individual to another, is not only by sexual reproduction.

One such novel alternative to sexual reproduction was discovered in fungi (*Aspergillus nidulans*) by Pontecorvo and Roper (1952). This they named as the **parasexual cycle**. In this process, the genetic recombination is achieved through "mitotic crossing over" and "haplodization". It is also called as **somatic recombination**. In *A. nidulans*, the parasexual cycle occurs in addition to the normal or standard sexual reproduction. While the events of sexual reproduction are extremely uniform, having a fine coordination between recombination, segregation and reduction, there is no such coordination in the parasexual cycle. The steps of the parasexual are independent of each other, and the frequency of each is very low. The karyogamy and haplodization are accidental events not bound by space and time. The parasexual phenomenon was next reported in *Aspergillus niger*, which, unlike *A. nidulans*, lacks a sexual cycle. Since its discovery in 1952 in *A. nidulans,* the parasexual cycle has been demonstrated to operate in atleast eight other *Aspergillus* species (Cutterback, 1996). The phenomenon is now reported in several fungi belonging to *Asco-*, and *Basidiomycota*. So far, it has not been reported in coenocytic fungi. However, indications of parasexuality do occur in *Phytophthora coctorum* and *Phycomyces blakesleeanus*.

THE STEPS OF PARASEXUAL CYCLE

The various steps of the parasexual cycle have been worked out in detail for *Aspergillus nidulans, A. niger* and *Penicillium chrysogenum*. In *A. nudulans*, which serves as the typical example, the parasexual cycle (Fig. 30.2) occurs in the following four steps:

1. Establishment of heterokaryosis.
2. Formation of heterozygous diploids,
3. Occasional mitotic crossing-over, during multiplication of the diploid nuclei, and
4. Haplodization through aneuploidy.

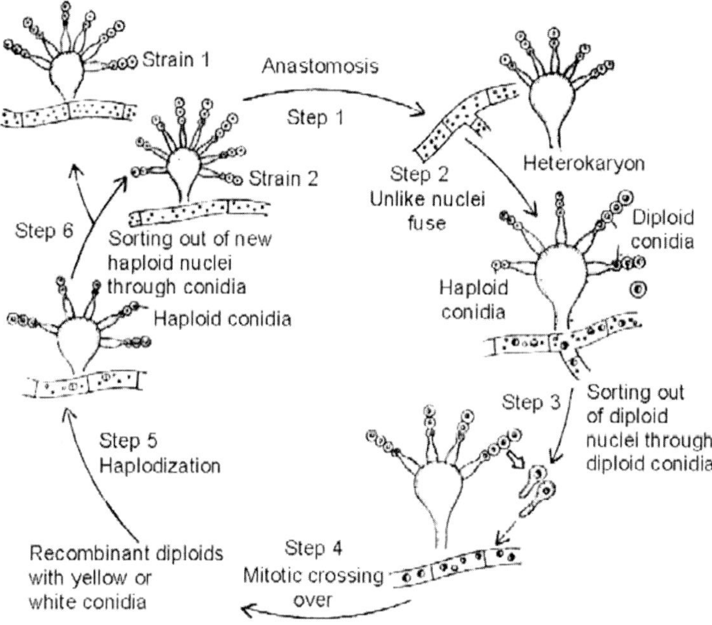

Fig. 30.2. Steps in the parasexual cycle

Establishment of Heterokaryosis

The presence of haploid nuclei of dissimilar genotypes in the same cytoplasm, is a pre-requisite for recombination. This is achieved by heterokaryosis, which is brought about by (1) mutations, and (2) anastomosis, between hyphae of different origin. This step is equal to plasmogamy of sexual reproduction.

Formation of Heterozygous Diploids

Nuclear fusion in heterokaryotic somatic cells was first noted by Roper (1952) in *Aspergillus nidulans*. The nuclear fusions between dissimilar nuclei result in the formation of heterozygous diploid nuclei or "zygotes"- a rare event, occurring at the rate of one in a million. Fusions between identical nuclei too occur. But, since the homozygous diploid nuclei, thus formed, ultimately give rise to non-recombinant haploid nuclei, such fusions have no genetic relevance.

The heterozygous diploid nuclei, which are fairly stable, by multiplication and sorting out through conidia, give rise to diploid thalli. The diploid colonies are recognized by: (1) higher DNA content of their nuclei, (2) the bigger size (1–3 times) of their conidia, and (3) certain phenotypic characteristics of the colony. This is typical of parasexual phenomenon.The prolonged diploid phase, involving repeated nuclear divisions, enhance the chances of "mitotic crossing-over".

Occasional Mitotic Crossing-over during Multiplication of Diploid Nuclei

This is the most important or 'key event' in parasexual cycle. It is during this step that genetic recombination takes place. Crossing-over during mitosis, is a rare phenomenon, first noticed in 'fruit fly' (*Drosphila melanogaster*) by Stern in 1936. In *Aspergillus nidulans*, mitotic crossing-over occurs only rarely, with a frequency of 10^{-2} per nuclear division. Interestingly enough, in some fungi, like *Penicillium chrysogenum* and *Aspergillus niger*, it is as frequent as during meiosis. Mitotic crossing-over results in development of new heterozygous, recombinant haploid nuclei. Reciprocal exchanges occur between homologous chromosomes, at the four-strand stage. It is a characteristic feature of mitotic crossing-over that the exchange, or chiasmata formation, is confined to a single chromosome pair out of the whole complement of chromosomes. In meiosis, the crossing-over occurs simultaneously in all the chromosomes. The subsequent splitting of chromosomes and segregation of strands is same as it occurs in mitosis.

Crossing-over, followed by segregation of strands, gives homozygosis for all markers distal to the point of chiasma. Markers proximal to the point of exchange and markers on other chromosomes remain heterozygous (Fig. 30.3).

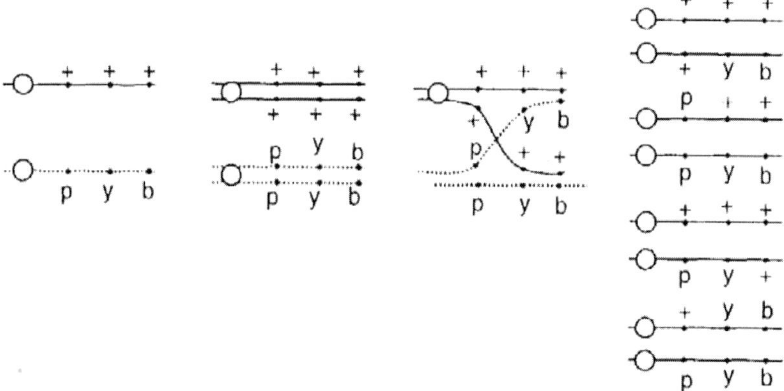

Fig, 30.3. Showing mitotic recombination

Occasional Haplodization Through Aneuploidy

The diploid nuclei give rise to hapolpid nuclei by gradual loss of chromosomes during successive mitotic divisions (Fig. 30.4). This is called **haplodization**. Meiosis is not involved. The haplodization, which occurs at a constant frequency of 10^{-3} per nuclear division, is the result of aneuploidy. During mitotic divisions, non-disjunction of the chromatids of one chromosome pair, results in aneuploid nuclei which, can be represented as (2n -1) at the beginning, and (2n - n) i.e. (n) or haploid, at the end. The aneuploids are genetically unstable, and once the loss of chromosomes has started, the selection favours the development of the fully balanced haploid nuclei.

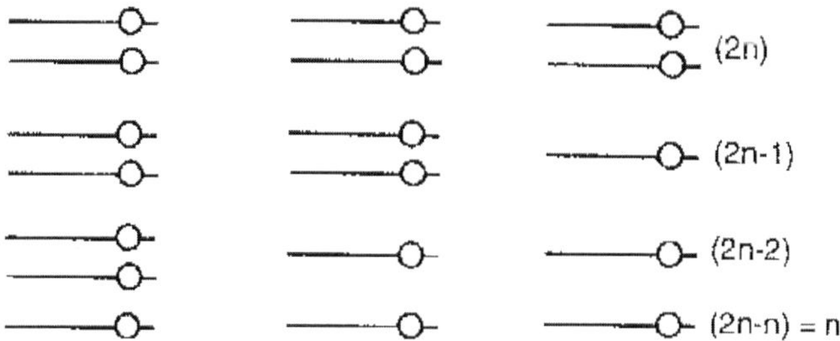

Fig. 30.4. Steps in the process of haplodization

Parasexual cycle closely resembles the events of the sexual cycle. It involves everything: plasmogamy, karyogamy and haplodization, but in a modified form, and without any fixed plan with regard to time and space. The two processes, nevertheless, end up with similar results i.e. development of recombinant haploid nuclei. A summarized comparison of the sexual and parasexual cycles has been made in Table 30.1.

Table 30.1. Comparison between Sexual and Parasexual Cycles of Fungi.

Sexual Cycle	**Parasexual Cycle**
1. Nuclear fusion in specialized structures	1. Nuclear fusion rare, occurs in somatic cells
2. Zygote usually persists one generation only	2. Zygote persists through many mitoses
3. Recombination by meiosis: crossing over in all chromosome pairs, reduction in chromosome number, random assortment of members of each pair	3. Recombination by rare "accidents" of mitosis: (a) mitotic crossing over at each event confined to one exchange in single chromosome arm, (b) haplodization independent of crossing over, random assortment of members of each chrosome pair
4. Products of meiosis readily recognized and isolated	4. Recombinants occur among somatic cells, recognized only by use of suitable markers.

Parasexual Cycle in *Aspergillus nidulans*

Let us consider a typical example of parasexual cycle. Cytological evidences for the various steps are difficult to make, because of two main reasons: (1) the fungal nuclei are extremely small (lie at the border line of resolution of light microscope), and (2) the chromosomes have low affinity for stains, and are extremely small. Notwithstanding these difficulties, the various steps, were elegantly tracked with the help of suitable "markers", like colour of the conidia, deficiency of certain vitamins, growth factors etc.

Roper (1952) 'prepared' two homokaryotic, mutant strains of *Aspergillus nidulans* (Fig. 30.5). Strain 1 had the genotype, ad^-, w (ad^-, requirement for exogenous adenine; w, white conidia, against green conidia of the wild type). Strain 2 had the genotype, $paba^-$, y, bi^- ($paba^-$, requirement for exogenous paraaminobenzoic acid; y, yellow conidia, and bi^-, requirement for exogenous biotin). The conidia of *A. nidulans* are uninucleate and the condial colour alleles w and y were autonomous. Also, w is epistatic to colour, i.e. it prevents expression of any colour gene at another locus.

Roper 'synthesized' a 'balanced' heterokaryon from these two strains, which resembled the wild type, in having no requirement for exogenous vitamins. For this, he allowed a mixture of the conidia of the two strains to grow together on a minimal medium, to support initial growth and to enable anastomosis. The heterokaryotic hyphae formed, grew on the minimal medium without any addition of vitamins. The heterokaryon formed both yellow and white conidia on same conidial head. Occasionally, green conidia appeared in the culture, which produced the wild type colony on minimal medium i.e. prototrophic fungus producing green conidia. The sudden appearance of prototrophic individuals, producing green conidia, was the result of accidental nuclear fusions in the heterokaryotic mycelium. Conidia containing heterozygous diploid nuclei gave rise to these prototrophic individuals with green conidia. The diploid nature of these individual was proved by the fact that (1) they had the phenotype expected of the heterozygous diploid, (2) the conidia were bigger than the haploid conidia (the volume being in the ratio of 2 : 1), and (3) the nuclei contained DNA which was double the amount present in the haploid nuclei. Mutation, as the cause of origin of green conidia, was ruled out; on the ground that it involved several mutations in one stroke. The second step, viz. the formation of heterozygous diploids was thus accomplished. The full genotyupe of the diploid can be written as shown below. The linkages and the positions of the centromere were based on studies of standard meiotic analyses.

$$---o---\frac{\text{Pyba}}{+}--\frac{\text{y}}{+}--\frac{\text{bi}}{+} \qquad \frac{+}{\text{w}}---o---\frac{+}{\text{ad}}---$$

The diploid individuals were stable, which multiplied through diploid conidia. Nevertheless, occasionally, sectors forming yellow and white conidia reappeared. Unmistakably, these were recombinants, and, in tests for ploidy, proved to be diploid. These were the products of accidental mitotic crossing-over during divisions of the diploid nuclei.

The diploid recombinant colonies were of various types. All white conidial diploids were prototrophic, while the yellow-conidial diploids required exogenous biotin, and some of them also had requirement for p-aminobenzoic acid.

Obviously, the yellow homozygous recombinants were products of crossing-over between y locus and the centromere. These are homozygous at bi locus. If the crossing-over occurred somewhere between pyba locus and the centromere, then they would be homozygous for the paba locus also. The markers w and ad remain heterozygous.

From the recombinant diploids were isolated aneuploid individuals, with genotypes ranging from (2n-1) to (2n–n). The haploids (with yellow and white conidia) were found to be of the following types;

1. The white haploids, all required adenine; approximately half of these had also requirement for p-aminobenzoic acid and biotin, in addition to adenine. The latter haploids were new, recombinant haploids.

2. The yellow haploids, all required p-aminobenzoic acid and biotin, but never adenine.

The various steps are shown in Fig. 30.4.

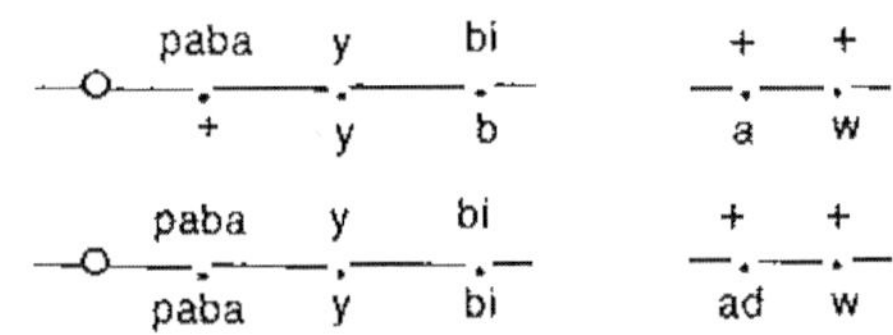

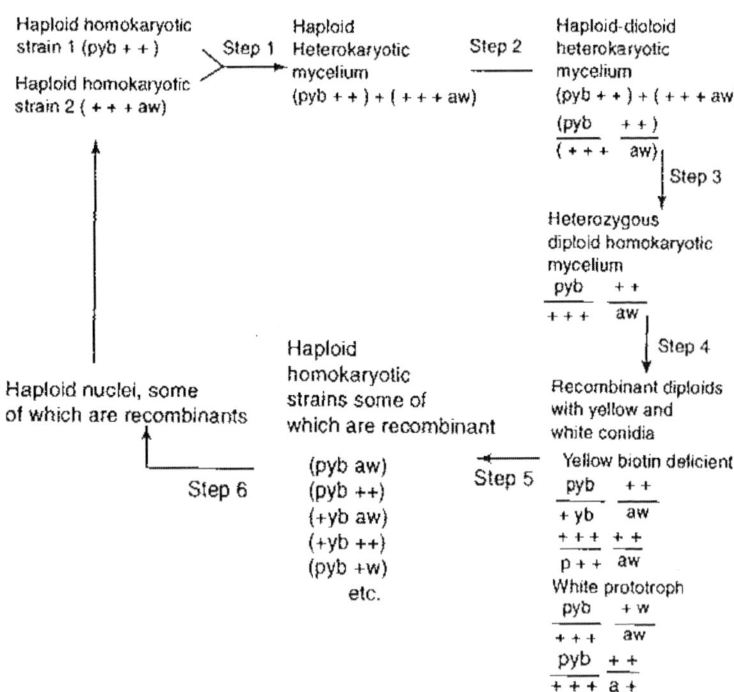

Fig. 30.5. Parasexual cycle in *Aspergillus nidulans*. Step 1, hyphal anastomosis; Step 2, 2×10^{-2} unlike nuclei fuse; Step 3, sorting out of heterozygous diploid nuclei through conidia; Step 4, mitotic crossing over; Step 5, haplodization in 10^{-3} diploid nuclei; Step 6, sorting out of the haploid nuclei.

Parasexual cycle in *Dictyostelium*

Sinha and Ashworth (1969) decribed parasexual cycle in the slime mould *Dictyostelium discoideum*. It initiates (Fig. 30.6) with the fusion of two amoebae and the formation of a binucleate cell. In some of these, karyogamy takes place resulting in the formation of a diploid cell. It is a rare occurrence. The diploid cells multiply by mitosis and establish a diploid progeny. Some of the diploid cells are not stable and give rise to aneuploid cells, with the nuclear state 2n – 1 or 2n + 1. This continues until 2n – n condition is attained and haploid cells are formed. Some of these are recombinants, showing characters different from the haploid parent cells. This is explained to be due to the occurrence of **mitotic crossing-over**, during division of the diploid nuclei of the zygote nucleus.

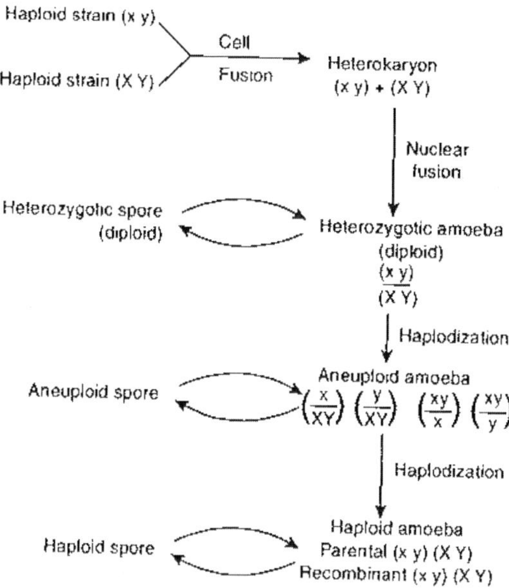

Fig. 30.6 Parasexual cycle in *D. discoideum* (After Sinha and Ashworth, 1969). x/X and y/Y represent two unlinked genes. The genotypes of various kinds of nuclei arc in parenthesis.

Occurrence and Role of Parasexual Cycle in Nature

Parasexuality was known only as a laboratory phenomenon until Ingram (1968) suggested that it may be widespread in nature. However, the current belief is that parasexual recombination is a rare event in nature (Cutterbuck, 1996). It is partly because of the prevalence of heterokaryon incompatibility in many of the wild fungi. Though it has been demonstrated only in *Asco-* and *Basidiomycota*, there are strong evidences of its occurrence in some coenocytic fungi, like *Phytophthora cactorum*. The phenomenon is linked with heterokaryotic condition, which is very frequently encountered in natural populations. But every heterokaryotic species may not necessarily be parasexual.

The more loaded question is whether the parasex has any significance in face of its being a rare, accidental phenomenon. The answer is that the role of parasexuality in the origin of new individuals is not clear

Use of Parasexual Phenomenon in Antibiotic Industry

Some of the most important antibiotic producing fungi, like *Penicillium chrysogenum* (penicillin) and *Acremonium chrysogenum* (cephalosporin), lack sexual reproduction. The discovery of parasexual phenomenon in them has provided suitable techniques to obtain higher antibiotic strains. Some antibiotic producing fungi in which parasexual genetic recombination has been established, are the following.

Species	**Antibiotic**
Aspergillus nidulans	Penicillin G
Acremonium chrysogenum	Cephalosporin C
Emericellopsis salmosynnemata	Penicillin N
Penicillium chrysogenum	Penicillin G, O, V
Penicillium patulum	Griseofulvin, Patulin

Hetetozygotic diploids, formed by haploid parent strains, were used for production of strains with increased penicillin production. But, the amounts of the penicillin formed by the diploid recombinants (produced by mitotic recombination) remained same as produced by the parent haploid strains. However, the recombinant haploids (formed by aneuploidy) gave higher yields of the antibiotic (Crueger and Crueger, 1989).

REFERENCES

Crueger, W. and Crueger, A. 1989. *Biotechnology*. 2nd ed. (Indian Edition, 2000). Panima, New Delhi.

Cutterbuck, A.J. 1996. Parasexual recombination in fungi. *J. Genet*. **75**(3): 281–286.

Haldane, J.B.S. 1955. Some alternatives to sex. *New Biology*, **19:** 7–26.

Hansen, H.N. 1943. The dual phenomenon in the Fungi Imperfecti. *Mycologia*, 442–455.

Hansen, H.N. and Smith, R.E. 1932. The mechanism of variation in the Fungi Imperfecti *Phytopathology*. **22:** 953–964.

Hansen, H.N. and Snyder, W.C. 1943. The dual phenomenon and sex in *Hypomyces solani & cucurbitae. Am. J. Botany*, **30:** 419–422.

Nelson, R.R. *et al*. 1955. Heterokaryosis as a basis for variation in *Puccinia graminis* var. *tritici, Phytopathology*, **45:** 639–643.

Parmeter, J.R. *et al.* 1963. Heterokaryosis and variability in plant-pathogenic fungi. *Annu. Rev. Phytopath.*, **1:** 51–76.

Pontecorvo, G. *et al.* 1953. Genetic recombination without sexual reproduction in *Aspergillus niger. J. Gen. Microbiol.*, **8:** 198–210.

Pontecorvo, G. *et al.* 1954. Parasexual recombination in *Penicillium chrysogenum. J. G. Gen. Microbiol.*, **11:** 94–104.

Roper, J.A. 1952. Production of heterozygous diploids in filamentous fungi *Experientia*, **8:** 14–15.

Sinha UK and Ashworth JM 1969 Evidence for the existence of elements of parasexual cycle in the cellular slime mould *Dictyostelium discoideum* in axenic culture. *Biochem J.* 119:171

Stern, C. 1936. Somatic crossing over and segregation in *Drosophila melanogaster. Genetics*, **21:** 625–730.

SUMMARY

Heterokaryosis is the coexistence of genetically-different nuclei in a single cell. It is a near-universal phenomena in fungi. The heterokaryotic condition arises by (1) mutation, (2) anastomosis, and (3) inclusion of dissimilar nuclei in spores after meiosis, in heterothallic fungi. Heterokaryosis (i) provides a substitute for heterozygocity and variability due to presence of genetically dissimilar nuclei (ii) plays an important role in the pathogenicity of rusts and smuts (iii) brings about origin of new races, and (iv) constitutes the first step in parasexual cycle.

A parasexual phenomenon was discovered in *Aspergillus nidulans* by Pontecorvo and Roper (1952). In this process, the genetic recombination is achieved through "mitotic crossing over" and "haplodization". It is also called as somatic recombination. In *A. nidulans*, the parasexual cycle occurs in addition to the normal or standard sexual reproduction. The parasexual phenomenon was next reported in *Aspergillus niger*, which, unlike *A. nidulans*, lacks a sexual cycle. The parasexual cycle is different from standard sexual cycle in several ways.

While the events of sexual reproduction are extremely uniform, having a fine coordination between recombination, segregation and reduction, there is no such coordination in the parasexual cycle. The steps of the parasexual cycle are independent of each other, and the frequency of each is very low. The karyogamy and haplodization are accidental events not bound by space and time. The phenomenon is now reported in several fungi.

The parasexual cycle occurs in four steps: (i) establishment of heterokaryosis (ii) formation of heterozygous diploids (iii) occasional mitotic crossing-over, during multiplication of the diploid nuclei, and (iv) haplodization through aneuploidy. The various steps in the parasexual cycle in *A. nidulans* and *Dictyostelium discoideum* has been described (see details).

The extent of occurrence of parasexual cycle and its role in creating new organisms in nature is yet to be known. Nevertheless it has become very important in the industrial production of antibiotics. Parasexual phenomenon has provided suitable techniques to obtain higher antibiotic-producing strains. The recombinant haploids (formed by aneuploidy) gives higher yield of the antibiotics.

TEST QUESTIONS

1. What is heteokaryosis? Describe its establishment (formation) and significance.
2. What are the steps in parasexual cycle? Describe in brief.
3. Describe the synthesis of balanced heterokaryon of *A. nidulans* by Roper. How are the diploid heterokaryotic conidia identified?
4. Describe the parasexual cycle in *Dictyostelium discoideum.*
5. Write briefly on any three: (i) Use of parasexual cycle in antibiotic production, (ii) Haplodization, (iii) Role of parasexual cycle in nature, (iv) Differences between standard and parasexual cycle.
6. Fill in the blanks;
   a. Parasexual cycle was first discovered in the fungus -------.
   b. Parasexual cycle also occurs in *Aspegillus* -------- which does not have sexual reproduction.
   c. Haplodization is the result of ------------------.
   d. The genetic recombination of parasexual cycle is also called ------ --- --------.
   e. Mitotic crossing over was known in ----------, before it was discovered in parasexual cycle.
   f. Haplodization occurs at constant frequency of ------- per nuclear division.
   g. The gradual loss of chromosomes during haplodization is called -- ----------.
   h. The 'key event' in parasexual cycle is -------- ----------.
   i. Mitotic crossing over occurs at the rate of --- per nuclear division.
   j. The homozygous diploids have ---- role in parasexual cycle.

31

Sex Hormones and Sex Morphogens in Fungi

The term **hormone**, as originally defined by Bayliss and Starling (1904), is a product of internal secretion. It is produced in one part of the body of an organism and is transported to another part, where it shows the effect. Chemicals secreted out side by an organism, affecting another organism, strictly speaking, are not hormones. Another term **pheromone** was coined for such products by Karlson and Luscher (1959). Pheromone is a substance secreted outside by one individual and received by another individual of the same species, in which it induces a specific behavioural or developmental response. The term pheromone is presently commonly used for invertebrate animals (particularly insects) while for fungi the term sex hormone is more popular. Some (Dyer *et al.,* 1992) prefer to distinguish sex hormones as (i) pheromones and (ii) sex morphogens. The **pheromones** (or sex hormones used as synonyms) are those that induce gametangia or gamete attraction, while **sex morphogens** trigger fruit body formation in asco-, and basidiomycetes. The participation of hormones in the sexual reproduction of fungi, which was proposed by de Bary (1881) for *Achlya*, is now an established fact.

DEFINITION AND TERMINOLOGY

Van den Ende (1976) has defined sex hormone as a *diffusible substance playing a specific role in the sexual reproduction of the organism that produces it.* Machlis (1972) coined the terms, **erotactins, erotropins** and **erogens,** imaginatively derived from the word *Eros* (the name of the god of love and son of aphrodite). **Erotactin** attracts motile cells (commonly called "sperm attractant"); **erotropin,** induces chemotropic growth of sex organs; and **erogen** controls the induction and differentiation of sex organs. These terms only refer to the phenomenon and not the mechanism of response, which vary with the different organisms. Van den Ende (1976) doubts the usefulness of such clear-cut segregation of the hormones. Where to put those, which show more than one type of action? For example in the fungus *Achlya*, the hormone eliciting the formation of sexual structures also influences their direction of growth.

A. SEX HORMONES (=PHEROMONES)

There are good evidences of the involvement of sex hormones in almost all groups of fungi. We shall study only those which have been chemically characterized and extensively investigated. These are: **sirenin, antheridiol, trisporic acid** and **yeast α-factor.** These do not show any marked similarity, either in chemical structure or biological activity. Sirenin (from *Allomyces*) is a sperm attractant *i.e.* (an erotactin sensu Machlis), Antheridiol and trisporic acid induce differentiation of sex organs and fall the category of erogens. Antheridiol also functions as a chemotactic agent i.e. erotropin. Table 31.1 lists the better-known pheromones, the organisms that produce them, the site of their synthesis, the morphogenetic actions they induce, and the empirical formula, where known.

Table 31.1. Fungal sex hormones (= pheromones) and their properties

Sex hormone	Produced by	Emperical formula (mol. Wt)	Site and control of synthesis	Morphogenetic action
Sirenin	*Allomyces* sp	$C_{15}H_{24}O_2$ (236)	Female gametes	Chemotaxis of male gametas
Antheridiol	*Achlya* sp	Sterol $C_{29}H_{42}O_5$ (470)	Female cells	Antheridia formed by male hyphae, attracts antheridial branches
Oogoniol	*Achyla* sp	Sterol (Ca 500)	Male cells in response to antheridiol	Oogonia formed by female cells
Trisporic acid	Several mucroales	$C_{18}H_{26}O_4$ (306)	(+) and (–) cells in collaboration	Zygophore formed by (+) and (–) hyphae
Yeast α factor	*Saccharomyces* sp	Peptide (Ca 1400)	α yeast cells	Elongation of **'a'** cells

SIRENIN

This is a sperm-attractant (Fig. 31.1), produced by the water mold *Allomyces*. *A. arbuscula,* which has been employed in the study of sex hormones, is characterized by isomorphic alternation of generations (see Chapter 6, *Blastocladiomycota*).

The gametothalli produce male and female gametangia in pairs. The male gametangia are orange-coloured and, depending on the species, lie above or below the colourless female gametangia. When mature, the motile male and female gametes are released into the water from the gametangia. The smaller, orange-coloured male gametes are attracted by the female gametes, which are twice as big as the male gametes, and sluggish in movement.

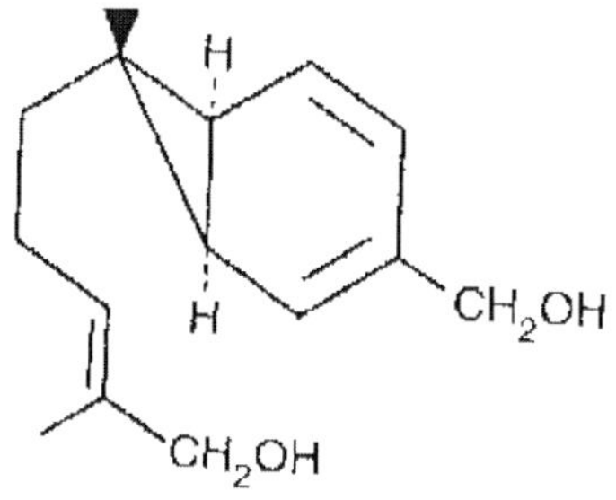

Fig. 31.1. Sirenin

The female gametes release **sirenin**, while still inside the gametangium. The horomone diffuses into the surrounding water, and male gametes start collecting around the female gametangia. The female gametes are faced with numerous male gametes as soon as released. The purpose of sirenin production is achieved. The female gametes, by anisogametic copulation, give rise to diploid motile zygotes.

The synthesis of sirenin by female gametes, and their 'sperm attractant' activity was demonstrated by Machlis (1958). The fact, that sirenin is not extractable from the male gametes after its up take, suggests that the hormone is metabolized and converted into inactive form.

Bioassay of Sirenin: the test solution is taken in a small glass tube with cellophane base and immersed in vessel containing water and male gametes. The clustering of male gametes around the cellophane indicates the presence of sirenin in the test solution. The number of male gametes collected is proportional to the concentration of sirenin.

Isolation and Chemical Characterization: For isolation, interspecific female hybrids of *A. macrogynous* and *A. arbuscula*, which produce exclusively (96-99%) female gametangia, are used. The methods of collection and purification of the pherormone are highly technical. The chemical structure of sirenin was reported by Machlis *et al.* (1968). It is an oxygenated sesqueterpene and may be species-specific. Two forms (D and L) have been isolated, but only the L-form is biologically effective. The L-sirenin has also been synthesized.

Two kinds of Chemotaxis (by sirenin and amino acids)**:** Five types of motile cells are formed in the life cycle of *Allomyces* (viz. haploid and diploid zoospores, male and female gametes, and the zygotes), but only the male gametes respond to sirenin. The two kinds of zoospores and the zygote show no response; instead, they are attracted towards amino acids. Immediately after fertilization, the male gametes cease to respond to sirenin and the resulting zygote shows positive chemotactic response towards amino acids.

Note: The female gametes, which are not attracted by sirenin are, however, attracted to a lesser-studied pheromone, **parisin** secreted by male gametes (Ashby and Dyer, 1992). The structure of parisin is not known (Lee *et al.,* 2010), but it is thought to be similar to sirenin and stable to chemical and environmental stresses.

ANTHERIDIOL AND HORMONE-B (=OOGONIOL)

J.R. Raper, in his classical studies (1939-1959), demonstrated multi-hormonal regulatory system in *Achlya ambisexuals* and *A. bisexualis*. He envisaged four hormones, named A, B, C and D but now it has been conclusively proved that only two hormones viz. antheridiol and oogonial are involved, as shown in Fig. 31.2.

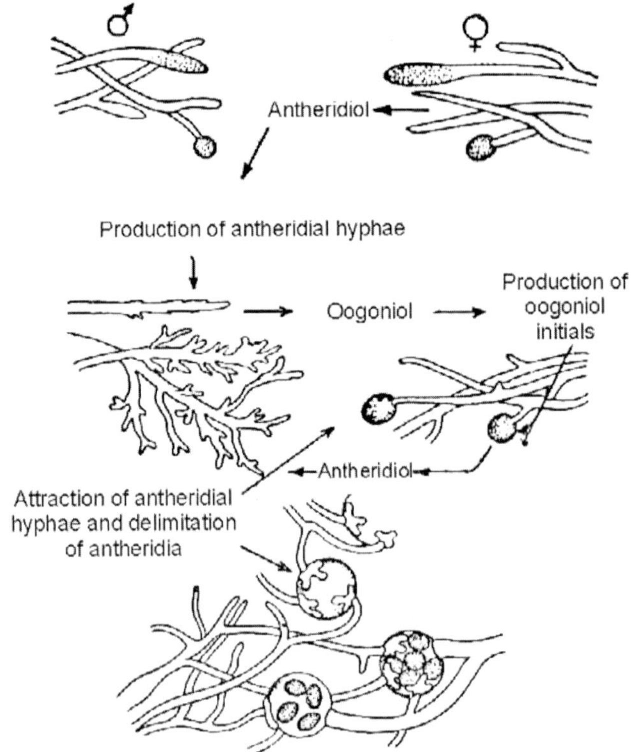

Fig. 31.2. Hormonal regulation of sexual reproduction in *A. ambisexualis* involving antheridiol and oogoniol

Antheridiol: Antheridiol, produced by female hyphae, elicits the: (i) induction of antheridial hyphae, (ii) chemotropic stimulation of male hyphae to produce oogoniol, and (iii) delimitation of antheridia. McMorris and Barksdale (1967) isolated the hormone in a crystalline from and named it antheridiol. Its structure (Fig. 31.3) was proposed by Arsenault *et al.* (1969). It is a steroid (the only steroid-hormone known outside animal kingdom). It is active at concentrations down to 10^{-10} M and elicits distinct responses.

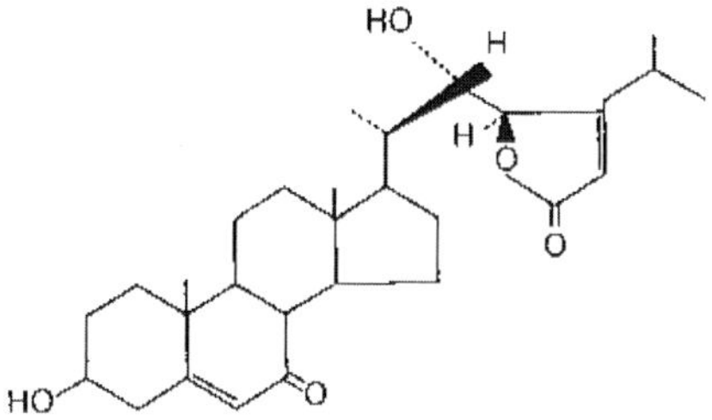

Fig. 31.3. Antheridiol

Bioassay of Antheridiol: The test sample should induce formation of antheridial hyphae in pure male *Allomyces ambisexualis*, in absence of the female isolate. The number of antheridial branches produced is directly related to the concentration of the hormone, and this forms the basis of estimating the amount of the hormone. The fate of the antheridiol in the male thallus is not known.

Oogoniol

It is produced by the male isolates only in presence of antheridiol. Barksdale *et al.* (1974) reported that oogoniol is produced by some hermaphrodite strains without the stimulus of antheridiol. McMorris and his coworkers (1975) isolated two crystalline sterols from the culture filtrates of

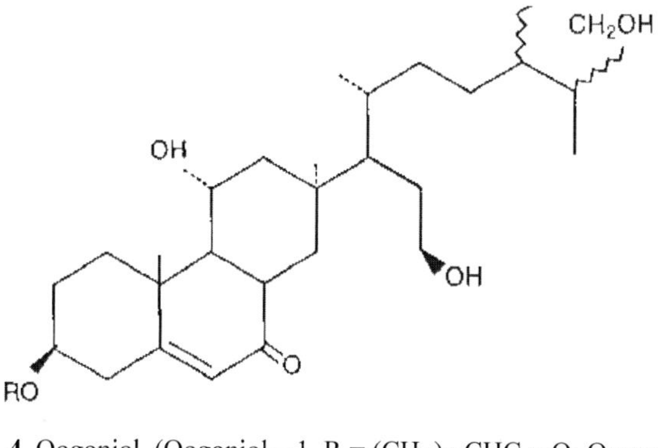

Fig. 31.4. Oogoniol. (Oogoniol – 1, R = $(CH_3)_2$ CHC = O; Oogoniol 2)

Achlya heterosexualis, which possess oogoniol ativity. The following structures have been proposed to these compounds, which have been named oogoniol-1 and oogoniol-2 (Fig. 31.4).

TRISPORIC ACID

Trisporic acid, a terpenoid (Fig. 31.5), triggers sexual reproduction of several *Mucorales* e.g. *Mucor mucedo*, *Blakeslea trispora* and *Phycomyces blakesleeanus*. Sexual interactions involve mutual stimulation for zygophore formation. The hormone is synthesized only in presence of the compatible partner; otherwise, only asexual reproduction takes place. It suppresses the formation of asexual spores and triggers sexual reproduction.

O
H
COOH
OH

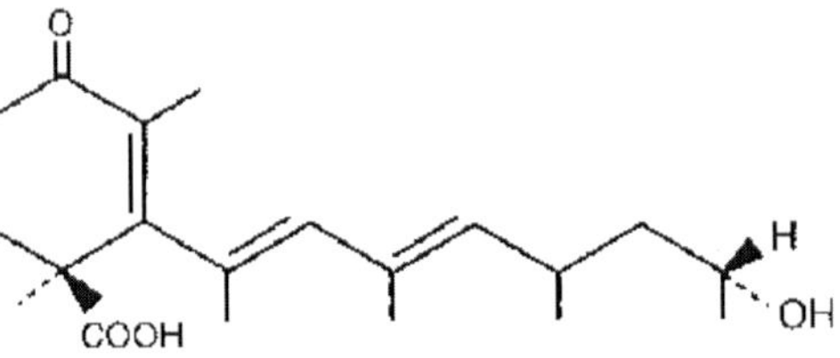

Fig. 31.5. Trisporic acid

Production of Trisporic Acid by Mucorales

When two compatible mating types, having incomplete but complementary biosynthetic pathways, grow together, they institute a "collaborative" biosynthesis of trisporic acid (Fig. 31.6). They mutually 'switch on' the biosynthesis of each other through volatile inducers. There are two routes of trisporic acid synthesis (Fig. 31.7), but neither of the route could be accomplished by either the + or –strain alone. (Details of the routes are shown in Fig. 31.7). The specific precursors, 4-dihydrotrisporic acid and its methyl ester (IV) or trisporol (V), formed by the + and –strain, respectively, are volatile and are exchanged by the two mating types. These diffuse into the nonproducer strain and initiate trisporic acid synthesis (*see* Gooday and Carlile, 1997). These same precursors are also responsible for zygotropism i.e. growth of the + and – zygophores towards each other and their hyphae. Trispoaric acid and its precursors appear to be zygophore-inducing hormones throughout the *Mucorales*. This has been demonstrated not only in another self sterile species, *Phycomyces blakesleeanus*, but also in the self fertile *Zygorhyncus*. This broad biological specifity was a key to the elucidation of trisporic biosynthesis and its hormonal activity. *Blakeslea trispora,* giving at least a thousand-fold yield of trisporic acid to that of *Mucor mucedo*, was used in the studies of biosynthesis, while *M. mucedo* was helpful as the sensitive bioassay organism.

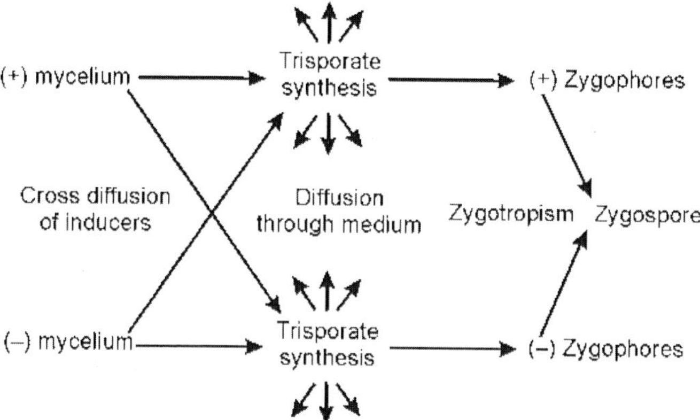

Fig. 31.6. Collaborative synthesis of trisporic acid by the compatible mating types of Mucoarales

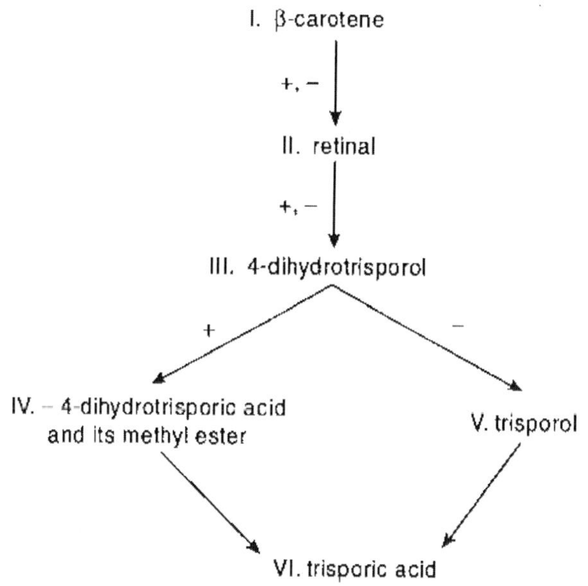

Fig. 31.7. Collaborative biosynthesis of trisporic acid by cross-feeding of intermediates between + and – mating types of *Blakeslea trispora*. β carotene (I) is produced by both + and – strains and metabolized by both *via* retinal (II) to 4-dihydrotrisporol (III). The + strain metabolizes it to 4-dihydro trisporic acid and its mythyl ester (IV), while the – strain to trisporol (V). These are metabolized to trisporic acid (VI) only by the other strain, after diffusing to the – and + strain, respectively. (Based on Gooday and Carlile, 1997)

Relationship with β Carotenes

Trisporic acid, originally identified as a stimulator of β carotene production, is itself derived from β carotene. Thus, by this action trisporic acid gives great amplification to its own synthesis, by a positive feed back mechanism (Bu'Lock *et al.*, 1976). So, the system is self-amplifying. The localized accumulation of β carotene is very evident in maturing cultures in Petri dishes and in the developing sexual hyphae. The trisporic acid and β-carotene are inseparable and go together. This is supported by (i) coinhibition of carotene and trisporic biosynthesis by diphenylamine, and (ii) impairment of sexuality in carotene-deficient mutants of *P. blakesleeanus*. β–carotene, in addition to being the precursor of trisporic acid, forms **sporopollenin,** the resistant protective material of the zygospore wall. Sporopollenin is found also in the pollen grains of angiosperms. It is an oxidatively polymerized product of β carotene.

YEAST ALFA FACTOR

There are evidences suggesting the involvement of some hormone in the sexual reproduction of *Saccharomyces cerevisiae*. The haploid cells are of two mating types (**'alfa'** and **'a'**), which conjugate and give rise to diploid cells. Levi (1956) showed that the '**alfa**' cells produce a diffusible chemical, the **alfa-factor**, which induces the formation of copulatory processes by the compatible '**a**' cells. (As yet, no **a** factor has been demonstrated). Physical contact is not necessary for the production of copulatory processes. The '**a**' cells,under the influence of the **alfa-factor**, stop growth and budding. They, instead, swell and turn into giant cells of various shapes; occasionally 10 or more times bigger than the normal cells. The giant cells show a dry weight, 30 or more times greater than that of the haploid somatic cells. The **alfa-factor** is specific; it acts only on **a** cells and has no effect on the **alfa** cells. It inhibits DNA replication in the **a** cells.

alfa cells ⟶ alfa factor ⟶ **'a'** cells ⟶ form copulatory processes

Duntze *et al.* (1970-1973) purified and partly characterized the alfa-factor. It is a peptide with a molecular weight of 1400. The peptide is complexed with copper ions, which do not dissociate even during purification.

Bioassay. The test sample should induce the formation of copulatory processes specifically in the '**a**' cells, and transforms into giant cells

B. SEX MORPHOGENS

A second group of sex hormones called **sex morphogens**, trigger development of the sporocarps (ascoma and basidioma), independent of sex.

Sex morphogens are especially interesting because they control the transition from a relatively undifferentiated form to a well developed, three-dimensional structure. They control morphogenesis, and, therefore, called 'morphogens'. This transition may explain how shape is determined during the development of various types of fruit bodies. Some of the important morphogens are described below.

1. **SF morphogen.** It is a lipid, produced by mated cultures of a plant pathogen *Pyrenopeziza brassicae,* which switches the fungus from asexual to sexual growth. Addition of the SF to a culture of one mating type of *P. brassicae* blocked the formation of asexual spore and triggered sporocarp development.
2. **Psi** (Precocious sexual inducing), **factors**. These are long chain fatty acids, produced by sexually-reproducing *Aspergillus nidulans* that can suppress the formation of asexual spores and turn on sexual development in somatic cultures.
3. **Cerebrosides.** These morphogens, which are lipid in nature, can trigger the sporocarp development in *Schizophyllum commune.*

Some other morphogens, which can initiate sporocarp development, are linolic acid and cyclic AMP

REGULATION OF PRODUCTION OF SEX HORMONES AND SEX MORPHOGENS

While we know much how the sex hormones and sex morphogens control the sexual processes, little is known about what controls their production. Some good leads are available, which are as follows.

The heterothallic fungi have **mating type genes**, which are *regulatory genes* and serve as "master switches". Sexually compatible nuclei have the complementary mating type genes, which direct the production of complementary polypeptides by *structural genes*. When the compatible hyphae anastomose and fuse, these polypoeptides, produced by the two compatible and complementary nuclei, form a novel hybrid polypeptide that binds to a specific site in the DNA of the two nuclei and block the genes that produce sex hormones or morphogens.

We can see a few examples. The first studies on functioning of the *mating type genes* were made with *Saccharomyces cerevisiae*, in which the cells are of two mating types, *alpha* (***a***) and **a**, as determined by the two mating type genes. These genes were cloned in 1981. When the two types of cells fuse, a hybrid regulatory protein is formed by the union of the two polypeptides, coded by ***a*** and **a** mating type genes. The hybrid protein switches off the production of α factor, which controls sexual reproduction in unfused haploid nuclei, allowing further sexual development.

Similar 'mating type genes' have been found functioning in the two mating types ***a*** and **A** of *Neurospora crassa*. These genes control both, the incompatibility of similar mating types and the development of ascoma. Both genes are required for production of fully fertile reproductive structures.

Lastly, how are the mating type genes themselves controlled?

There are a range of environmental stresses that activate these genes. Other phenomena, like interaction between mitochondria of two fused hyphae, too are important (Raymer and Ross, 1991; Ashby and Dyer, 1992).

REFERENCES

Arsenault, *et al.* 1968. The structure of antheridiol, a sex hormone in *Achlya bisexualis*. *J. Am. Chem. Soc.*, **90:** 5635–5636.

Ashby, A. and Dyer, P.S. 1992. The secret sex lives of fungi. New Scientist (7th March) 32–35.

Barksdale, A.W. 1963. The role of hormone–A during sexual conjugation in *Achlya ambisexualis*. *Mycologia*, **55:** 627–632.

Barksdale, A.W. *et al.* 1974. Response of *Achlya ambisexualis* E 87 to the hormone antheridiol and certain other steroids. *J. Gen. Microbiol.*, **82:** 295–299.

Bu'Lock *et al.* 1976. The apocartenoid system of sex hormones and prohormones in Mucorales. *Pure and Applied Chemistry*, **47:** 191–202.

De Bary, A. 1881. Cited by H van den Ende, 1976.

Duntze, W. *et al.* 1973. Purification and partial characterization of a factor, a mating type specific inhibitor of cell reproduction form *Saccharomyces cerevisiae*. *Europ. J. Biochem.*, **35:** 357–365.

Dyer, P.S. *et al.* 1992 The control of sexual morphogenesis in the Ascomy-cotina. Biol. Rev. 67(4): 421-458.

Edward, J.A. *et al.* 1969. The synthesis of the fungal sex hormone antheridiol. *Jr. Am. Chem. Soc.*, **91:** 1248–1249.

Gooday, G.W. and Carlile, M.J. 1997. The discovery of fungal sex hormones: III trisporic acid and its precursors. *Mycologist*. **11**(3): 126–129.

Kochert, G. 1978. Sexual pheromones in algae and fungi. *Annu. Rev. Plant. Physiol.* **29:** 461–486.

Lee, S.C. *et al.* 2010 The evolution of sex: a perspective from the fungal kingdom. Microbiol Mol Biol Rev **74**(2) 298-340

Levi, J.D. 1956. Mating reaction in yeast. *Nature*. **117:** 753–754.

Machlis, L. 1958. Evidence for a sexual hormone in *Allomyces*. *Physiolo. Plant.*, **11:** 181–192.

Machlis, L. 1972. The coming of age of sex hormones in plants. Mycologia, **64**: 235–247.

Machlis, L. *et al.* 1968. The structure of sirenin. *Jr. Am. Chem. Soc.*, **90:** 1674–1676.

McMorris, T.C. and Barksdale, A.W. 1967. Isolation of a sex hormone from the water mould *Achlya bisexualis*. *Nature*. **215:** 320–321.

McMorris, T.C. *et al.* 1975. Structures of oogoniol –1, –2. and –3, steroid sex hormones of the water mould *Achlya*. *Jr. Amer. Chem. Soc.*, **97:** 2544–2545.

Raper, J.R. 1970. Chemical ecology among lower plants. In: *Chemical Ecology* (I. Sondheimer and J.B. Simeone eds.). pp. 21–42. Academic Press, N.Y.

Rayner, A. and Ross, I. 1991. *New Scientist.* 30 March.

SUMMARY

Hormone is defined as a product of internal secretion. Chemicals secreted out side by an organism, affecting another organism are called **pheromone.** The term pheromone is commonly used for hormones of insects and, therefore, the term 'sex hormone' is used for pheromone-like substances in fungi. The participation of hormones in the sexual reproduction of fungi (*Achlya*) is known since the times of de Bary. Van den Ende (1976) has defined sex hormone as a *diffusible substance playing a specific role in the sexual reproduction of the organism that produces it.* Machlis (1972) coined the terms, **erotactins, erotropins** and **erogens. Erotactin** attracts motile cells (commonly called "sperm attractant)**; erotropin,** induces chemotropic growth of sex organs; and **erogen** controls the induction and differentiation of sex organs. Van den Ende (1976) doubts the usefulness of such clear-cut segregation of the hormones.

The hoormones which have been chemically characterized and investigated are: **sirenin** (in *Allomyces*), **antheridiol and oogoniol** (in *Achlya*), **trisporic acid** (in mucors) and **yeast α-factor** (in *Sacharomyces cerevisiae*). These hormones do not show any marked similarity, either in chemical structure or biological activity. Sirenin (from *Allomyces*) is a sperm attractant *i.e.* (an erotactin), Antheridiol and trisporic acid induce differentiation of sex organs and are erogens. Antheridiol also functions as a chemotactic agent i.e. erotropin. Sirenin, an oxygenated sesqueterpene, attracts only male gametes and has no effect on any of the other four motile cells. **Paricin** produced by male gametes of *Achlya* attracts female gametes but not much is known about this hormone. Antheridiol and oogoniol (produced by *Achlya*) are sterols, and are the only sterol hormones known outside the animal kingdom. Trisporic acid, a terpenoid hormone, is synthesized only when stimulated by the other partner. The two partners mutually switch on the trisporic acid synthesis in each other by volatile inducers. Trisporic acid is derived from β carotene and also stimulates its synthesis. Both are inseparable and go together. β carotene also leads to sporopollenin biosynthesis (an oxidatively-polymerized product of β carotene) present in zygospore wall. The **yeast alfa factor** is a diffusible peptide produced by **alfa cells**, which induce copulatory processes in **'a'** cells.

A second group of 'sex hormones' are the **Sex morphogens**, which trigger the development of ascoma and basidioma. Some of the important morphogens are: (1) SF morphogen, a lipid produced by mated cultures of *Pyrenopeziza brassicae,* which switches the fungus from asexual to sexual growth. (2) Psi (Precocious sexual inducing), factors, long chain fatty acids that suppress the formation of asexual spores in sexually-reproducing *Aspergillus nidulans.* (3) Cerebrosides, lipid in nature, trigger sporocarp development in *Schizophyllum commune.* Some other morphogens which can initiate sporocarp development are linolic acid and cyclic AMP.

The control of production of sex hormones and sex morphogens is through "mating type genes", which serve as **regulatory genes**. Sexually compatible nuclei have the complementary mating type genes, which direct the production of complementary polypeptides (by structural genes). When the compatible hyphae anastomose and fuse, these polypoeptides, produced by the two compatible and complementary nuclei, form a novel **hybrid polypeptide** that binds to a specific site in the DNA of the two nuclei and block the genes that produce sex hormones or morphogens.

For example, *Saccharomyces cerevisiae* has cells of two mating types, ***a*** and **a**, as determined by the two mating type genes. When the two types of cells fuse, a hybrid regulatory protein is formed by the union of the two polypeptides, coded by ***a*** and **a** mating type genes. The hybrid protein switches off the production of ***a*** factor, which controls sexual reproduction in unfused haploid nuclei. Similar *mating type genes* have been found functioning in the two mating types ***a*** and **A** of *Neurospora crassa.* The mating type genes themselves are activated by environmental factors and interaction among the mitochondria of the two mated hyphae or cells.

TEST QUESTIONS

1. Distinguish between: (i) hormone and pheromone, (ii) fungal sex hormones (pheromones) and sex morphogens, (iii) erotactin, erotropin and erogen
2. Name four sex hormones (= pheromones) and their producer organisms. Describe hormonal regulation of sexual reproduction in any two of them.
3. Name some morphogens. Describe the control of sex hormone production in yeast cells by complementary mating type genes.

4. Describe the collaborative biosynthesis of trisporic acid in *Mucorales.*
5. Answer in brief:
   a. What is the relation between trisporic acid and β carotene?
   b. What is sporopollenin?
   c. What are the three functions of antheridiol?
   d. Which two species of *Achlya* produce the sex hormones?
   e. At what stage is sirenin released by the female gametes?
   f. How will you assay sirenin?
6. Fill in the blanks:
   a. Sirenin attracts only ----------, while ------- ------ attract other motile gametes in *Allomyces.*
   b. The male gametes of *Allomyces* produce --------- which attract ---- gametes.
   c. The yeast alfa factor induces formation of ------------ ------- by 'a' cells.
   d. Sirenin is an oxygenated -----------------.
   e. The hyphae of mucors release --------- compounds that initiate ---- ----- ---- synthesis in the complementary hyphae.
   f. Antheridiol is secreted by the -------- hyphae of ------------ -------.
   g. ------------- is the only steroid hormone known outside the animal kingdom.
   h. Oogoniol is secreted by ---------- only in presence of -----------.
   i. In mucors both compatible hyphae are needed for ---------- ------- in either of the partners.

Author Index

A

Ahmadjian V 474, 483
Aime MC 372,375,392
Ainsworth GC 21,22,27,52,54,146,148, 248,250,277,292,293,294,348,366
Ajello LD 102,117
Albersheim P 11
Alexander C 462,467
Alexopoulos CJ 21,22
Aliouat-Denis CM 156,162
Alsberg CL 534
Alvarez E 101,118
Arsenault GP 573,578
Asao J 525
Ashby A 572,578
Ashworth JM 441, 564,567

B

Banno I 391,392
Barkley FA 16,27
Barksdale AW 573,579
Barr DJS 66,74
Barrett JR 527,528,540
Barron GL 251
Batsch 4
Bauer R 350,366
Bayliss WM 569
Beadle GW 10,227
Begerow D 350,366
Benedict RJ 536,538,540
Bennett JE 523,540
Berbee ML 177,281,294
Beschel RE 477,482,483
Bessey EA 301,344
Bhat RV 531,540
Bisby GR 52,54
Bjorkman E 467
Black OF 534
Blakeslee AF 10,546,552
Booth C 263,277
Borrow A 517,518,519
Bove FJ 536,540
Brasier CM 230,242
Brefeld O 6, 24
Brian PW 9, 514
Brinkshaw JH 534
Broadbent D 9, 12
Brodo IM 474, 483
Brown W 260, 277
Buchner E 7
Buller AHR 2, 12, 256, 287, 294, 310, 322, 344
Bu'Lock JD 509,517,518,519,576,578
Burns TD 19,27,397,400
Butler EJ 84,88,423,426

C

Cagniard de la Tour 7
Carlile MJ 574,575,578
Carmichael JE 252

Catanzariti AM 373,392
Cavalier-Smith T 69,74, 396,397
Chatterjee D 531,540
Christensen CM 201,242
Clusius CE 2
Copeland HF 16,27
Corda ACG 5
Corliss 397
Corner EJH 288,294
Couch JN 386,392
Crueger A 566
Crueger W 566
Cutterback AJ 558,565,566

D

De Bary A 5,6,7,8,,11,12,259,430,455, 467,569,578
Dick MW 18,397,407,426
Dioocorides 1
Dodge BO 10
Doolittle WF 27
Drechsler C 113,118,257
Duddington CL 113,118
Duntze W 576,578
Dyer PS 569,572,578

E

Edgerton CW 242,
Ellis MB 251
Ericksson J 224
Eriksson OE 178
Esser K 549,552

F

Faegri 482
Fleming A 8,9
Flor HH 382,392
Frank AB 456,467
Fries EM 4,5,337
Fuckel L 5

G

Garrett SD 8,10,11,50
Gaumann E 463,467
Gooday GW 574,575,578
Grab D 209
Grove JF 271

H

Haeckel EH 16,27
Haldane JBS 558,566
Hall BD 177,242
Hansen HN 263,278,555,556,557,566
Hawksworth DL 23,26,27,52,146, 148, 252, 292,293,294,348,366,430
Hesseltine CW 101,118
Heywood 57
Hibbett DS 25,27,77,79,283,294
Hooke R 3
Hughes SJ 251,277

I

Ingold CT 265,277,291,294,295,311, 328,329,344,430
Ingram R 565

J

James TY 28,65,74,80,88,106,118,154
Jirovec Otto 156
Jursieu 2

K

Karelyn Cuz Martinez 154
Karlson P 569
Kendrick WB 251,252,278
Kirk PM 23,24,26,28,52,59,74,78, 90, 92,125,131,147,148,153,168,174,178,179,185,218,224,252,292,293,301, 349,350,366,371,374,393,397,400, 407,430,441,445,446,473,483
Kirk TK 11
Klich M 423,540
Koch R 7
Kogl F 538
Koning 12
Kurtzman CP 157,162
Kutzing FT 7

L
Landvik S 155,162
Languebeck B 255,278
Lee SC 572,578
Leeuwenhoek A van 2,3
Letscher 65,74
Levi JD 576,578
Liebig J von 7
Link 4
LindegrenYJ 10
Linnaeus C von 3,4,16,337
List PH 538
Litschauer V 386
Liu 177
Lowe B 1,12
Luft P 538
Luscher M 569
Luttrell ES 251
Lutzoni Ff 20,28,88,215,242
M
Machlis L 569,570,571,578,579
Malik VS 362
Malpighi M 2
Mandels M 11
Manns MM 8,10,12
Margulis L 147,148
Martin GW 52
Mason EW 251
Mathney PB 300,301,344
Mathre DE 362
Mauenchhausen O'van 4
Mc Dougal WB 456,467
Mc Morris TC 573,579
Melin E 461,467
Meyer 337
Micheli PA 3
Millardet PME 8
Mitra M 362
Moore RT 38,41,59
Morten JW 456,467
Mukherjee SK 531,540
Mundkur BB 362
N
Nelson RR 263,278,528,540,557,566
Nicander 2
Nilson H 251
Nishida H 147,148,153,163
O
Oedummans 10
Orpin CJ 77,79
O'Donnell K 183,242
P
Parmeter 555,567
Pasteur L 7,8
Patel JD 423,427
Patouillard NT 386
Pegler DN 344
Perry BA 184,242
Persoon CH 4
Pliny 2
Poggeler S 191,242
Pontecorvo G 11,41,54,200,242,557, 558, 567
Porta G della 2
Prez-Martin J 358,366
R
Raghukumar S 402,403,405
Raper JR 39,54,548,550,552,572,579
Raulin 8
Raymer A 578,579
Reese ET 11
Reinking OA 8,10,12,263,278
Ridley BL 493,504
Roper JA 11,41,54,200,242,558, 567, 560,562
Rosling A 153,154,163,
Ross I 578,579
S
Saccardo PA 6,12,21,248,249,252,278
Saksena R 102

Saksena SB 102,118
Sargeant K 523,540
Schoch CL 177,178,179,212,242,
Scholander PE 472, 483
Schuessler AD 90,118,125,127,130,131
Schwarzott A 126,127,131
Schweinitz 4
Schwendener S 471,484
Sedin M 215,242,473,484
Singh G 534,541
Singh R 107,118
Sinha UK 441,564,567
Smith M 301,324,344,555,557,566
Snyder WC 263,278
Sparrow FK 407,427
Spatafora JW 177,179,191,234,242
Starling 569
Steinberg G 358,366
Stern C 560,567
Stoll M 350,366
Strinrger JR 156,163
Subramanian CV 251,263,278
Sugiyama J 147,148,153,163
Sung 234
Swann EC 292,293,295

T

Tanaabe Y 90,106,109,118
Tatum EL 11,227
Taylor JW 177,281,292,293,294,295
Thind KS 318,337,344
Thom C 9
Tournefort 2,3
Toussoun TA 263,278
Trappe 206
Tubaki K 251
Turner WB 509
Tuslane LR & C 5

V

van Sterbeeck 2
van den Ende H 569
Van Tiegham 8
Vasanthi S 531, 540
Venkata Ram 364, 366
Von Hohnel 386
Voragen AGJ 493, 504
Vuillemin P 250

W

Walker C 127,131
Waksman SA 8,9,10,50,54
Ward HM 259,278
Wasson RG 1,12,539,541
Wasson VP 1,12
Wehmer S 8
Whisler HC 87,88
Whitehouse HLK 547,552
Whittaker RH 16,17
Wilhelm S 255,267
Wilson PA 102,118
Woese CR 16
Wollenweber HW 263,278
Woodward RB 534,541
Woronin MS 6
Wostemeyer 191,242

Y

Yartlett N 77,79
Yarwood CE 239,242

Organism Index

Absidia 93,95
A.glauca, A. spinosa 95, 96
Acaulopage 113, 114
Acaulospora 125, 127, 129
Achlya 11, 406, 410, 570,
A. ambisexualis 550, 572, 573
A. bisexualis 572
Acrasis (*A. rosea*) 416,417
Acremonium chrysogenum 512, 518, 519, 566
Actinomucor 92
Aessospora salmonicolor 249
Agaricus (=*Psalliota*) 304—309
A. bisporus 309
A. campestris 306--309
Ajellomyces (*A. capsulatum, A. dermatitidis*) 208
Albugo (=*Cystopus*) 45,424-426
A. candida 425
Allomyces 39,47,49,81,84,570
A.arbuscula 85,86,570,571
A. heterosexualis 474
A. macrogynous 86,571
Alternaria 249,258
A. solani and other spp. 258
Althronia 403
Amanita 316-318,460
A. citrina 318
A. cookeri 538
A. muscaria 316,318,513,517,524, 538, 539
A. pantherina 316,524,539
A. phalloides 1,316,317,320,512,537
A. verna 316,512,524,537
A. virosa 512,524
Ambispora 125
Amoebophilus 114
Amorphomyces 213
Ampelomyces quisqualis 51
Aplectosoma 114
Ancylistes 104
Aphanomyces 410
Aplanes 410
Aplanochytrium 403
Aplectosoma 114
Apodachlya 406,408,409,489,497
Archaeospora 125,126,129,130
Armillaria 323
A. gallica 300,324
A. mellea 323-325,463
Arthrobotrys 180
Arthroderma 209
Aschochyta 249,274
Ascobolus 181
A. crenulatus 181,
A. stercorarius *182*

Ascoidea 167

Asellaria 112,113

Aspergillus 41,199-202,556

A. clavatus 200, *A. fischerianus* 200, *A. flavus* 200,512-514, 516, 524, 525, 533, *A. fumigatus* 200,524, *A. glaucus* 199,201, *A. halophilicus* 201, *A. nidulans* 11,200,558-560, 562, 564-567, 577, *A. niger* 8, 200, 559, *A. niveus* 533, *A. ochraceous* 512,524,532, *A. oryzae* 200,513, 516, 533, *A. parasiticus* 512, 514, 524, 525, *A. restrictus* 200, *A stellatus* 198, *A. tamari* 513,516, *A. terreus* 200,533

Auricularia 330

A. auricular-judae 330,331

B

Basidiobolus 104,107

B. ranarum 107,108

Basidiophora 421

Batrachochytrium dendrobatidis 60

Bdellospora 114

Bipolaris 264,

B. sorokiniana 249

Blakeslea 92,93,97

B. trispora 97,574,575

Blastocladia 81,83

B. pringsheimii 83,84

B. ramosa 83,84

Blastocladiella 81,518

B. emersonii 81-83

B. simplex 81

B. stubenii 81

Blastocladiopsis 81

Blastomyces 208

B. zygonema (*B. dermatitidis*) 208

Blumeria 119

Boletus 309,325,326,460,466

Bombardia 144

Botryodiplodia 274

Botryosphaeria 249

Botryotinia 249

Botrytis 249,259

B. cinerea 259,260,557

Bremia 421

Byssochlamys 197

C

Calonectria 249

Calvatia (*C. gigantia*) 310

Candida 253,255

C. albicans 249,254,255

Cantharellus 332,460

C. cebarius 332,333

Cephalosporium 516

C. acremonium 518,519

Ceratiomyxa 432,433

Ceratobasidium 333

C. cornigerum 334,457,462

Ceratocystis 237

Ceratostomella montium 503

C. pini and *C. pleuriannulata* 503

Cercospora 249,260

C. arachidicola, *C. musae*, *C. nicotinae* and *C. personata* 260

Chaetomium 112,225

C. globosum 225

C. thermophile 225

Choenophora 93,97

C. cucurbitarum 97,98

Chromelosporium 187

Chrysonilia 227

Chrysosporium 209

Cladosporium 251

Clavaria 318

C. vermicularius 318

Claviceps 232

C. microcephala 232
C. paspali 536
C. purpuria 232,233,323,513,517,535
Clitocybe 524,538
Cochliobolus 249,265
C. sativus 249,265
Cochlonema 114,115
Coelomyces 86
C. anophelescia 86
C. indiana 86,88
C. psorophorae 86,87
Coemansia 109,110
C. pectinata 110
Colletotrichum 249,272
C. carchorum, *C. gloeosporioides* & *C. graminicola* (=*C. falcatum*) 272
Comoclathris 249
Completoria 104
Coniodiobolus 104
Conocybe ; *C. filaris* 538
Conotrema 473
Coprinus 287,304,311
C. comatus 311,312
Cordyceps 234
C. militaris 234,235
Corticium 463
Cortinarius 460
C. orellanus 524
Craterellus 333
C. cornucopioides 333
Cryptococcus 255
C. neoformans 249,256,342
Ctenomyces 209,210
Cunninghamella 93,98
C. ehinulata 98
Curvularia 249,261
Cyathus 312
C. striatus 312,313
Cystopage 113
Cystopus (see *Albugo*)
Cyttaria 216,217

D

Dactylella bembicoides 180
Debariomyces matruchoti 503
Diaporthe 249, 276
D. citri 276
Dictyostelium (*D. discoideum*) 437-441, 564, 565
Dictyuchus 410,411
Didymella 249,275
Diplodia 249,274,275
D. macrospora 275
Dipodascus 47,167
D. albidus 167,168
Diversispora 124-128
Drechslera 249,264
Dimargares 112
Dispira 112

E

Elsinoe 249,273
Emericella 198,199
Emericellopsis salmosynnemata 566
Endarluca caricis 51
Endocochlus 114-116
Endocronartium harknessii 374
Endogone 103,457,460
Entoloma 460
Entomophthora 104,105
E. maimaiga 105
E. muscae 105,106
Entrophospora 125,127,465
Eremascus 47
Erysiphe 218,219
Eupenicillium 204,207
Eurotium 198,199,202,203
Euryancale 114

Exerohilum 264
Exobasidium 364
E. japonicum 547
E. vaccinii 363
E. vexans 364

F

Filobasidiella 342
F. neoformans 249,342
Fomes 336,463
F. fomentarius 336,337
Fusidium coccimeum 512,516
Fusarium 45,249,262,515,528,529,556
F. culmorum 527
F. graminearum 527
F. moniloforme 512,527,531
F. oxysporum (f.spp.) 27,263,527
F. poae 524,529
F. roseum 512,524,527
F. sporotrichoides (=*F. tricinctum*) 524, 529
F. tricinctum 527

G

Galactomyces 167
Galerina (4 spp) 537,538
Gamphidius 460
Ganoderma 335
G. applanatum 336
G. lucidum 335
Geastrum (*G. triplex*) 327,328
Genea 185
Genestillospora 111
Geoglossum 212
Geolegnia 410, 411
Geosiphon pyriformis 123, 125, 126, 129, 130
Gibberella 249
G. fujikuroi 515,517,530
Gigaspora 125,128,464
Gloeosporium 272
Glomerella 241,249
G. cingulata 241
G. fujikuroi 512
Glomus 126,127,129,130,131
Golvinomyces 218,219
Gonapodya 69
Graphium 230
Guignardia 249,275
Gymnoascus 210
Gymnosporangium 379--381
G. juniperi-virginianae 373,374,379-381
Gyromitra esculenta 524,528

H

Haemonectria 262
Harpella 111
Harpochytrium 69
Hasegawea 157
Hebeloma 460
Helminthosporium 264
H. vetulinum 264
Helvella 182
H. crispa and others 183
Hemileia 384
H. vastratrix 384, 385
Histoplasma (*H. capsulatum*) 208
Hydnum 334
H. repandum 334, 335
Hyphochytrium 398, 399
H. catenoides 398
H. hydrodictyii 398, 399
Hypocrea 235
H. rufa 234, 235

I

Inocybe 524,560

I. nappies 513,517,538
I. patouillardii 513,517
Isaria kogna 516
Isoachlya 410

J

Japanochytrium 403

K

Khuskia 249,266,514
Kickxella 109,110

L

Labyrinthula (*L. macrocystis*) 403
Lactarius 460
Lasidiplodia (*L. theobrome*) 274
Lemonniera 265
Lepiota 309
L. helveola 538
Leptomitus 406,408
L. lacteus 408,409,489,497,498
Levillula 119
Lewia 249
Ligniera 451
Lycopeerdon 314-316
L. perulatum 315

M

Magnaporthe 249,267
M. grisea 267
Malassezia 365,366
M. furfur and other spp. 365
Malbranchea 211
Marasmius 463
Masseeella 385,386
Massospora 104
Melampsora 382
M. lini 373,382,383,557
Meliola 239,240
Microallomyces 81
Microbotryorum 389,390
M. violaceum 389,390
Micromucor ramanniana 516
Microsphaera 218
Microsporum 209
Monoblepharella 69,72
M. taylori 72,74
Monoblepharis 69,70
M. polymorpha 70-72
Monilia 222,249,266
M. fructigena 266
Monilinia 222,223,249
Monoascus 533
Morchella 183,184
M. esculenta and other spp. 184
Mortierella 103
Mucor 46,47,92-94
M. hiemalis 546,547
M. mucedo 21,574
M. ramosissimus 93
Mycoderma valida 503
Mycospharella 192,193,249
M. arachidicola, *M. berkelyii*, *M. musae* and *M. nicotinae* 260

N

Nectria 236,249
Neocallimastix 78,79
Neolecta 153,155
Neoartorya 198,199
Neovossia 362
N. indica (=*Tilletia indica*) 362,363
Neurospora 10,48, 227-229,518
N. crassa 227,503,578
N. sitophila 27,228,503
N. terasperma 227,556
Nigrospora 249,266
N. oryzae 267,514

O

Octomyxa 450
Oedocephallum 188
Olpidium 67 (*O. brassicae*) 67-69
Onygena 211,212
Ophiostoma 230
O. novoulmi 230-232
Orbilia bembicoides 180

P

Pacispora 124-128
Panaeolus spp. 524
P. companulatus 539
Paraglomus 126,130,131
Paxillus 460
Pellicularia 334
Penicillium 9,41,204-206,556
P. aurantigriseum (-*P. cyclopium*) 533
P. camamberti 204,533
P. chrysogenum 512,513,559,560
P. citrinum 533
P . citrioviridae 532
P. crustosum 524
P. cyclopium 524,533
P. digitatum 204
P. expansum 512,515,524
P. griseofulvum 512,514
P. islandicum 533
P. italicum 204
P. janczaewski (=P. nigricans) 514
P. nigricans 514
P. notatum 9
P. palitans 524
P. patulum 512,514,515
P. puberulum 524,525,534
P. roquefortii 204
P. viridicatum 514,524,532,533
Peronospora 417
Pestalosphaeria 249,273
Pestalotiopsis 249,272,273
P. mangiferae, P. theae 273
Peziza 185,188
P. ammophila, P. aurantia 186
Phanerochaete chrysosporium 11
Phoma 249
Phomopsis 249,276
P. citri 276
Phragmidium 383
P. mucronatum 3
Phycomyces 93,98
P. blakesleeanus 92,98,99,503, 559, 574, 576
Phyllachora 241
Phyllactinia 219
Phyllosticta 249,275,276
Physarum 434
P. polycephalum 434--437
Phytophthora 417,421-423,556
P. cactorum 559,565
P. infestans 419,420
Pilobolus 93,99
P. kleinii 100,101
Piptocephalis 116,117
Pisolithus 460
P. tinctorius 457
Pithomyces chartarum 524,531
Plasmodiophora 449,450-453
P. brassicae 451,453
Plasmopara 421,422
P. viticola 422--423
Pleurotus 319
P. ostreatus and other spp. 319
Pneumocystis 153,156
P. cairnii 156
P. jirovecii 156,157
Podosphaera 218,219
Podospora 119

P. anserina 556
Poitrasia 97
Polymyxa 451
Polyphagus 62
P. euglenae 62,63
P. parasiticus 62
Polyporus 337-339,
P. squamosus 338
Protomyces 153,160
P. macrosporus 338
Psalliota(see *Agaricus*)
Psilocybe 322,323,540
P. mexicana 524,539,540
P. semilenciata 322,323
Puccinia 48,285,287,376-379
P. graminis tritici and other spp 6, 373, 374, 376-379
Pyrenopeziza brassicae 577
Pyrenophora 265
Pyricularia 249,267
P. oryzae (= *P. griseae*) 267
Pythiopsis 410
Pythium 415-417
P. aphanidermatum 415
P . ascophalan 503
P. debaryanum, P. graminicolum, P. Myriotilum and *P. thalassium* 415

R

Ravenelia (*R. emblicae*) 383
Rhizidiomyces 399
R. apophysatus 399,400
Rhizoctonia 249,268,462
R. solani 268
Rhizomucor 93
Rhizophydium 47,65
Rhizopogan 460
Rhizopus 92-95
R. niger 95
R nigricans (*R. stolonifer*) 92,95,545
Rhodosporidium 391
R. sphaerocarpon 92
Rhodotorula 136,391
Russula 460

S

Saccharomyces 169,570
S. boulardii lyo 170
S. carlsbergensis 503
S. cerevisiae 41,169-172,503,576,577
S. oviformis 503
Saccharomycodes 173
S. ludwigii 173,174
Saitoella 153
Saksenaea 101
S. vasiformis 101-102
Saprolegnia 399,405,411-414
S. ferax, S. parasitica 411
Schizochytrium 403
Schizophyllum 321
S. commune 320-322,551,557
Schizosaccharomyces 41, 136, 153, 157, 158
S. octosporus, S. pombe 158
Sclerocystis 126,465
Scleroderma 326,460
S. citrinum 327
Sclerogone 103
Sclerophthora 424
Sclerospora 421,423
S. graminicola 423,424
Sclerotinia 222
S. sclerotiorum 222
Sclerotium 222
Scutellospora 125,128,129,464,465
Sebacina 463
Septobasidium 386-389
S. burtii 388,389

Septoria 249,277
S. chrysanthemella 277
S. lycopersici and other spp 277
Septosphaeria 264
Sirobasidium 341
S. sanguineum 341,342
Smittium 111
Sordaria 226
S. fumicola 226,489,502
Sorodiscus 450
Sorosphaera 450
Sphaceloma 249,273
S. roseum273
Sphacelotheca (*S. reliana*) 390
Sphaerobolus 328
S. stellatus 328 ,329
Sphaerotheca 144,218
Sphaerulina 249
Spilocaea pomi 194
Spongospora 451
Sporobolomyces 256
S. salmoniocolor 249,256,257
Sporopachydermia 167
Sporothrix 230,231
Stachybotrys 529,530
S. chartarum 524
Stachylina 111
Stereum 340
S. hirsutum 340.343
Stictis 473
Stigmatomyces 213,214
Stylopage 113,114
S. rhicnarca, S. haploe 114
Syncephalis 116,117
Synchytrium 64
S. endobioticum 60,64
S. taraxaci 64
Syzygites 93

T

Talaromyces 204,205,207
Taphrina 153,161
T. deformans 161,162
T. epiphyla 161
T. pruni 161
Tetramyxa 450
Thaminidium 93
T. elagans 97
Thanetophorus 249
T. cucumeris 268
Thraustrochytrium 403-405
Thraustotheca 410
Tieghemiomyces 112
Tilletia 360
T. caries (=*T. tritici*) 360-362
T . levis (=*T. foetida*) 360-362
Tolyposporium 355
T. penicillariae 355,356
Torula 251
Trametes versicolor 339
Tremella 342
T. aurantia 343
T. mesenterica 343
Trichoderma 11,529
Trichophyton 209
Trichothecium 269
T. roseum 269,529
Tuber 188-191
T. aestivum and other spp 188,189
Tulasnella 463

U

Uncinula 218,219
U. necator 219
Urocystis 354
U. agropyri 354
Uromyces 381

U. ciceris 381
U. dianthi and other spp.373
Ustilago 356
U. maydis (*U. zeae*) 351,358,359
U. tritici (*U. segetum* var. *indica*) 356
Utharomyces 101

V

Venturia 193
V. inaequalis and other spp. 194
V. pyrina 194
Verticillium 269-271,556
V. albo-atrum 269,270
V. dahliae 269,270
Volvariella 320
V. volvacea 309,320,321

W, X, Y, Z

Woronina 451
Xerotinus 463
Xylaria 238,239
X. hypoxylon 4
Yarrowia 167
Youngiomyces 103
Zografia 86
Zoopage 113,114
Zoophagus 113
Zygorhyncus 93,95,96
Zygopleurage 144

Subject Index

Acellular (true) slime moulds 430-441
Acervular fungi 252,271
Acrasid 445-447
Acrasin 439,440
Aecium 378,379
Aflatoxins 512,514,523-527
AFTOL project 24,147
Agglutination 169,171
Air spora 51
Aleuriospores 198
Alpha (α) conidia 276
Alimentary toxic aleukia (ATA) 529
Amanita toxins (Amatoxins) , 512, 514, 537,538
Amoeboid cell 433
Amylase (α, β) 497
Amyloidity 289
Amylose 496
Amylopectin 496
Anamorph/ teleomorph 249
Anamorphic fungi 25,248-277
Anastomosis 555, 556
Aneuploidy 561
Anisokont flagella 406
Antabuse 311
Anther smut 390
Antheridiol 570,572,573
Anthracnose (fungi) 272
Antibiotic 9
Aphanoplasmodium 36
Apple scab 194
Aplanospore 42
Appressorium 419
Aquatic fungi 50
Aquatic hyphomycetes 255,256
Arbuscular mycorrhizae (AM) 462
Arbuscules (= pelotons) 124,464
Arbutoid mycorrhizae 465
Arthrospores 252
Ascal proliferation 167
Ascogenous cells/hyphae 137
Ascolocular ascomycetes 144
Ascohymenial ascomycetes 144
Ascoma 143
Ascoma ontogeny/development 143
Ascomycetous yeasts 166
Ascostroma 143
Asexual reproduction 41
Aspergillus genome 200
Athelete's foot 204
Auxoautotrophic 500
Auxoheterotrophic 500
Azole antibiotics 38

B

Bakane disease of rice 515

Bacterivory 404
Baker's and Brewer's yeast 169
Balanced heterokaryon 562
Ballistospore 291
Barrage reaction 551
Basal groups 147,154,179
Basidiobolomycosis 107
Basidiole (= paraphyses) 290
Basidioma(= basidiocarp) 283,288
Basididospore discharge 291
Bell morel 183
Beta (β) carotene 576
Beta (β) conidia 276
Beta (β) 1-3 glucan 157,166
Beta (β) lactam antibiotics 513
Beta (β) orsellinic acid 479
Beta (β) tubulin gene sequences 25,155
Binding hyphae 288
Bioassay 503
Biodegradation by fungi 11
Biodiversity (fungal) 52
Biological clock (= circadian rhythm) 101
Biological species 27
Bioluminescence 325
Bipolar heterothallism 549
Biosynthetic pathways 510
Bird's nest fungus 312
Black(= sooty) mildews 239
Black wart of potato 64
Black yeasts 195, 196
Blast disease of rice 267
Blastomycosis 208
Blastospore 136,161,252
Blepharoplast 42
Blister blight of tea 364
Boletes 325
Bothrasomes 402-404
Bread mold 95
Budding (holo-, enterogenous) 136, 171, 391
Buller phenomenon 287
Bunts of wheat 360

C

Calvacin 310
Candidiasis 255
Capillitium (capillitial hairs) 211,315
Carcinogenic 523,526,528,534
Carotenoid production 575,576
Cellular slime moulds 445,446
Cellulin plugs 408
Cellulolysis 102
Cellulolytic enzymes 10,491,492
Cellulose 490
Cellulosomes 78,492
Cell wall 37,406
Cell wall scales 402
Centrum 144
Cephalosporin 512-514,519,566
Cerebroside morphogens 577
Cheese (mold-ripened) 9,204
Chemoorganotrophs 488
Chemotaxis 570
Chitin 37,409,497
Chitinase 497
Chromistan fungi 20,396
Chronometer 19
Chytrids 57
Cigar end of banana 269
Circadian rhythm (biological clock) 101
Citreoveridin 532
Citrinin 533
Clade 19,348
Clamp connection 281,283
Classification 20-24
Clavines 536
Cleistothecium 143

Clone 433
Club/coral fungi 318
Club root of crucifers 451
Coffee rust 384
Commensalism 455
Competitive saprobic ability 10
Complementation 352
Conidia (types) 44
Conidial ontogeny (origin) 250
Conidioma136
Conjugate division 137
Contronema-Sticta connection 215
Coprine 311
Coprogen 101
Coremium 136
Crook cell 138
Crossing over 560, 562
Cruciform cell division (promitosis) 431, 448
Cryptococosis 255,256
Cutin/ cutinolytic enzymes 497
Cutis; epi-, and subcutis 322
Cyanophily 291
Cyclic AMP 439,440,577
Cyclosporin 234
Cyphelle 475
Cyst (= resting spore) 435,478
Cystosorus 449
Cystidia 290
Cytoskeleton proteins 19

D

Damping off 415
Deadly agaric 536
Death cap 536
Death chair 316
"Deep Hypha" project 147
Dermatomycosis 209
Deoxynivalenol (vomitotoxin) 529
Depsides 479
Depsidones 479
Destroying angel 537
Dextrinoidity 291
Dictionary of Fungi 24
Dikarya (Dikaryomycotina) 281
Dikaryon 46
Dikaryophase 46
Di-Mon mating 287
Dimorphism 36,161
Dimorphism of zoospores 409
Diplobiontic life cycle 173
Distribution of fungi 52
Dolipore-parenthesome septum 281,284
Domain 16
Downy growth 423
Downy mildew 421
Dry spores 252
Dual phenomenon 556
Dual nomenclature 248
Dutch elm disease 230,231

E

Early blight of potato 259
Earth balls 326
Earth star 327
Earth tongues 155,211
Ecological species 27
Eel grass wasting disease 403
Egg fungi (*Oomycota*) 406
Elicitors 273
Elongation factor 19
Endocyanosis 129
Endosymbiosis 398
Enterogenous budding 391
Environmental DNA 154
Epibasidium 372
Epicutis 322
Epispore/perispore 282

Ergosterol 406
Ergot 232
Ergot alkaloids 513,517
Ergotism (convulsive, gangrenous) 232, 535
Ergot toxins 535
Ericoid mycorrhizae 462
Esterogen 528
Estradiol -17 β 528
Exogenous/enterogenous budding 171
Extraradical (external) hyphae 465
Euthecia / Pseudothecia 143
Eye spot 402

F

Facial eczema 531
Fairy ring 305
False morels 182
Fermentation 7
Fibrolytic activity 77
Fibroplasmodium 37
Fission 41
Fission yeast 158
Fissitunicate ascus 141
Flagellum, (9+2) structure 43,44
Flag smut of wheat 354
Flat reaction 551
Flax rust 382
Floc (of yeast) 172
Floculence (autoaggregation) 169,172
Fly agaric 538
Fly cholera 105
Food spoilage 9
Forma specialis (f.sp.) 27
Fossil fungi 123
Free cell formation 138
Friesian approach 4
Fucosterol 406
Fumonisins 524,530,531
Fungi: definition 20
Fungal artillery 328,329
Fungal bridge 466,467
Fungal dinosaurs 155
Fungal garden 386
Fungal hyperparasites 51
Fungal sheath 456,457
Fungal tissues 36
Fungus flowers 288
Funiculus 311
Fusidic acid (fusidane) 512,516

G

Gamma (γ) particles 80
Gangliospore 252
Generative cells 111
Generative hypha 288
Germinal wall 124
Giant puff ball 310
Gene for gene hypothesis 382
Genome sequence of *Aspergillus* 200
Gibbane skeleton 515,516
Gibberellic acid 512
Gibberellins 515
Gills (=lamellae) 303
Gill fungi 304
Gleba 315
Glomerospore 124
Glycosides 513,516
Gram rust 381
Gray blight of tea 273
Green ear (downy mildew) of bajra 424
Grex 437
Griseofulvin 512,514,566
Gyromitrin 538

H

Habitat 50
Hairy fungi 421

Hamathecium (interascal tissue) 140,142
Haplodization 561
Hartig net 456,459
Hat thrower (*Pilobolus*) 99
Haustorially expressed secreted proteins (HESP) 373
Hedgehog fungus 324
Heteroecious 374
Heteroecism 6
Heterokaryons 555
Heterokaryosis 39,556,557
Heterokont flagella 406
Heterothallism 546-552
Histoplasmosis 208
Holobasidium 281
Hole-in-Head syndrome 530
Holobiont 455
Hologenous budding 136
Homoiomerous/Heteromerous 304,474
Homoplasy 288
Homothallism 546
Honey fungus 323
Hormone 569
Horn of plenty 333
Host shifts 381
Hughesian classification 250
Hulle cells 198
Hydrogenosomes 78
Hymenium 289
Hymenophore 289
Hyperparasites 51
Hyphal analysis 288
Hyphal bodies 104
Hyphal pegs 466
Hyphal system (=mitic system) 288,289
Hyphal types 288
Hyphidium 290
Hypobasidium 372

I

Ibatonic acid 539
Ig-A nephropathy 529
Imperfect fungi 25,248
Imperfect yeasts 252,253
Incompatibility factors 548
Incompatibility system 285
Indoor mycotoxicoses 530
Inoculum potential 10,324
Infection threads 357
Infection types (smuts) 352
Ink caps 311
Interaction zones (in smuts) 348
Interascal tissue (hamathecium) 140,142
Inulin 497
Isidia 478
Isoperene 515

J

Jelly fungi 299
Jew's ear 330

K

Karnal bunt 360,362
Karyochorisis 38
Keratinophilic fungi 211
KIID cladc 108
Kojic acid 513,516

L

Late blight of potato 418
Leptospores 385
Leuco-encephalo-malacia (ECM) 530
Liberty cap 322,323
Lichen acids 479
Lichenin 481
Lichenicolous fungi 179
Lichenizing (lichen-forming) fungi 179

Lichenometry 482
Life cycles 39,53,374
Living fossil (=*Neolecta*) 155
Lomasomes 35
Loose smut of wheat 356
LSD 536
LSU rDNA sequences 155
Luteoskyrin 533
Lysergic acid 535

M

Macrocyst 441
Magic mushroom 335
Maize smut 358
Maize smut genome 358
Mating type genes 577,578
Medula/rind 135
Merosporangia 91
Metabolic grid 511,512
Mevalonic acid 511,517
Microsclerotia 270,271
Microsporidia 25
Mirror picture yeast 256
Mitic system 289
Mitochondrial cristae (shapes) 430,431
Mitospores 42
Mitosporic fungi 25,248
Mitotic crossing over 560,562
Molecular clock 123
Monophyletic/ polyphyletic 18
Monotropoid mycorrhiza 466
Morels 183
Morphological species 27
Mucoromycosis (phycomycosis) 91
Muscarine 513,517,538
Muscimol 539
Mushroom cultivation 309
Mushroom toxins 536-539
Mutualism 455
Mycelial strand 135
Mycelia sterilia 250
Mycelium (primary, secondary, tertiary) 282
Mycobiont 471,473
Mycolaminnarin 396,406,
Myconet 147
Mycorrhiza 455
Mycorrhizal types 456
Mycoses 523
Mycotoxicose 423
Mycotoxin 423
Myxamoebae 433,434

N

Natural /artificial classification 20
Nematophagus fungi 133,180
Nephrotoxin 532
Nomenclatural types 26
Nuclear cap (ribosomal) 80
Nuclear division 38
Nuclear status 38
Nutrrition 34,477
Nutritional yeast 169

O

Ochratoxins 512,514,531,532
Operculum 58
Oogoniol (Hormone-B) 570,573,574
Optional lichenization 215,473
Orcein 481
Orchid mycorrhizae 462
Orellanine 538
Oyster mushroom 319,320

P

Paddy straw mushroom 320
Paraphyses (=basidiole) 142
Parasexual cycle 555

Parasexuality 41,557-566
Parenthesome types 299,300
Parisin 572
Pathovar 27
Patulin 512,515,534,566
Pectolytic enzymes 494
Pectic polysaccharides 492
Peloton (=arbuscules) 124,462
Penicillic acid 534
Penicillin 9,512,513,566
Peridiole 313
Periplasm (of oogonium) 416,417
Perispore/epispore 282
Perithecium 143
Peroxisomes 396
Phallotoxins 512
Phaneroplasmodium 37,436
Phase separation symbiosis 463
Pheromone 569
Phialide 252
Phialospore 252
Photobiont 471,472
Phragmobasidium 281
Phycomycosis 91
Phylogenetic relationships 18,294
Phylogenetic species 27
Pietra fungaria 337
Pinnotes (effuse sporodochia) 262
Pityriasis versicolor (tinea versicolor) 365
Plasmodiophorids 431,448
Plasmodoium (= acellular thallus) 36
Plasmodium types 434,447,453
Plectenchyma 36
Pleurotoid 319
Pneumo-Cystis Pneumonia (PCP) 156
Polyene antibiotics 38
Polyketides 411,514
Polymorphism 5
Polyphyletic/ monophyletic 18
Polythetic species 27
Pore mushroom 325
Porospore 252
Powdery mildew fungi 217
Predacious fungi 113
Primary, secondary, tertiary mycelium 282
Primary sugar fungi 411
Probasidium 372
Probiotic 169,170
Progrnote 17
Promitosis 448
Prosenchyma 36
Protoplasmodium, aphanoplasodium 36
Protozoan fungi 17,20,430
Pseudofungi (*Oomycota*) 405
Pseudomycelium 37
Pseudoparenchyma 36
Pseudoplasmodium 445
Pseudothecium 143
Psi-factor (morphogen) 577
Psilocin 323,539
Psilocybin 323,539
Pucciniomycetous smuts 389
Puff ball 310,314
Puff ball mushrooms 310
Pycnidial fungi 252,273
Pycnium 378

R

Red bread mold 227
Residual asexual fungi 248
Resting mycelium 270
Rhizines 475
Rhizoids 57
Rhizomycelium 57
Rhizomorph 324
Rind/medula 135

Ringworm 209
Root-inhabiting fungi 10
rRNA 18S, gene sequences 23,77,124, 126,133,374
RPB 1/2 protein gene sequences 108, 123
Rubigalia 373
Rumposome 69

S

Sacred fungi 322
Saddle fungi 322
Scale insects 385
Sclerotium 135, 434, 436
Secondary(= special) metabolites 508
Secondary homothallism 547
Secondary mycelium 282
Septa (types of) 35
Septal pores 350
Septal pore organelles 135
Serotonin antagonists 540
Setae (haplo-, and macrosetae) 290
Sex hormones 569-576
Sex morphogens 569, 576-578
Sexual ambivalence 550
Sexual cycle 562
Sexual factors 548
Sexuality 41
Sexual reproduction 46
Sewage fungus 408
SF-morphogen 577
Shaggy ink cap 311
Shelf (bracket) mushroom 335
Shikimic acid 510, 517
Shoe string fungus 323
Shot gun fungus 99
Siderophores 101
Sirenin 570-572
Sister groups 20,177,191,281,294, 300, 348
Skeletal hypha 288
Slime moulds 430-441
Slimy spores 252
Slug 437
Soil clone group 1 (SCG 1) 154
Soil inhabitants/invaders 10
Soma ('som-ras' of Vedas) 539
Somatic recombination 558
Sooty molds 195, 196
Soralium 477
Soredium 477
Sorocarp 437, 445
Sorophore 437
Species concept 26
Spermatia 48
Spermatization 48
Sphaerocysts 304
Spherules/Spherulation 437
Sphingolipids 531
Spindle pole bodies 294
Split gills 320, 321
Sporangiole 91
Spore balls 351
Spore repetition 294
Spore wall structure 124
Sporocarp 49, 437, 445
Sporodesmins (-Pithomycotoxins) 531
Sporodochium 136
Sporophore 433
Sporopollenin 576
Stab culture 254
Stachybotrys-toxins 530
St'Anthony's fire 535
Starch 496
Statismospores 291
Sterigma 291,292
Sterol 406

Straminopila 17, 18
Stroma 135
Sufu (Chinese cheese) 92
Sugar fungi 408, 411
Swarm cells 434
Symbiosis (=mutualism) 51, 455
Sympodium 258
Sympodula 258
Synapolymorphy 109
Synascus 159
Synnema 136

T

T2 toxin 529
Teleomorph/anamorph 249
Teliospore 282
Telial hairs/horns 385, 386
Tempeh 92
Terpenes 515
Terpenoids (=Isoprenoids) 515
Tertiary mycelium 283
Teterapolar heterothallism 551
Thallospores 44
Thallus 34
Three-tier (tripartite) symbiosis 467
Thrush 255
Tinea 209
Toadstool 302
Top / bottom yeast 173
Torula condition 93
Trama 304
Transfer cells 467
Tree of Life Project 17
Tremorgenic toxins 533
Trichospores 111
Trichothecene toxins 524, 529
Trisporic acid 570, 574-576
Trophocyst 100
True fungi 25
True slime moulds 430-441
Trufflle /truffle-like 185, 188
Turkey X-disease 423

U

Uni-perforated septa 109
Urediniospore 282
Usnic acid 481

V

Vesicles 124,465
Vital theory 7
Vitamin requirements 500
Vitamin bioassay 503
Vomitotoxin (deoxynivalenol) 529

W

Wart of potato 64
Water molds 407
Wax tapers 318
Woronin bodies 135

X

Xanthocroic reaction 291

Y

Yeast 166
Yeast alfa (α) factor 570, 576
Yeast genome project 10, 170
Yellowed rice toxins 532

Z

Zearalenone 512, 515, 524, 527
Zoosporangial proliferation 409, 413
Zoosspore dimorphism 409
Zoospores (primary, secondary) 407
Zymase 7